AF671957

ALLE ZEIT WACH
1842

Dimitrios P. Tassios

Applied Chemical Engineering Thermodynamics

Springer-Verlag
Berlin Heidelberg GmbH

Prof. Dimitrios P. Tassios

National Technical University of Athens
Zographou Campus
15780 Zographos
Athens, Greece

Formerly:
New Jersey Institute of Technology
323 Dr. Martin Luther King Jr. Blvd.
Newark, N.J., 07102, USA

Additional material to this book can be downloaded from http://extras.springer.com

ISBN 978-3-662-01647-3 ISBN 978-3-662-01645-9 (eBook)
DOI 10.1007/978-3-662-01645-9

Originally published by Springer-Verlag Berlin Heidelberg New York in 1993

Typesetting: Camera-ready by author;

61/3020 5 4 3 2 1 0 Printed on acid-free paper

To my students,

Past, Present, and Future

The Hazards of Literary Life

"The hardships endured by authors are typified by a letter bound inside the flyleaf of a volume in exhibition of annotated rare books on view at the Smithsonian Institution. .....

The book in question is "History and Progress of the Steam Engine", published in London in 1837. It was written by civil engineer Elijah Galloway, who addressed the letter inside to Charles Manby, a prominent English engineer. It reads:

'Dear Sir:

I am just on the eve of completing a new work on "The Mechanic Powers and Their Application to Machinery". But literary work is only paid for when the Manuscript is completed and what with ill health and the all absorbing nature of this kind of labor I am in the awful predicament of ***wanting a bit of bread!***

I have by me an only copy of my well known "History of the Steam Engine", which Mrs Galloway will herewith hand you. Will you in charity purchase it of her? This is the only legitimate source by which I can avert the horror of hunger and cold for we have neither food, fire, nor a penny in the world!

I am, Dear Sir, Your ever Obliged E. Galloway.'"

(Chem. & Eng. News, Nov. 10, 1986, p.40.)

This is by way of expressing my appreciation to NJIT for granting me a Sabbatical leave for the 1985-1986 academic year, during which most of this book was written.

Preface

There are many excellent books on Chemical Engineering Thermodynamics, but their use in teaching the subject in the classroom leads to two serious difficulties:

First, most books tend to concentrate either on undergraduate or graduate level material only. This creates problems in the review of undergraduate material, which I have found essential for an effective graduate course.

Second, their presentation of the subject is too extensive for the number of credits allocated to it (about six semester hours in the undergraduate level and three in the graduate one). As a result, Chapters - or Sections of them - must be skipped, making the subject - that is by its very nature complex - still more difficult for the students to comprehend. Thus, not only learning suffers, but a fear of Thermodynamics is embedded into the student's mind, which hinders his or her future learning and use of the subject. (This fear is apparent to anyone teaching the graduate level course; and felt by many practicing engineers and chemists, leading a well known thermodynamicist to state a few years ago: "The most under-utilized subject in the chemical and petroleum industries is Thermodynamics.")

These problems are, hopefully, avoided with this book for the material included in it **covers both levels and can be presented in the aforementioned semester hours**.

Obviously, less material is thus presented, but the book does provide the undergraduate and graduate student with the material needed in industrial practice and for future study; in addition, through the extensive list of pertinent references, it provides for:

* a more in depth study of the different subjects;
* practical applications;
* the development of awareness of data sources, and
* making the student appreciate the importance of using the literature in problem solving.

The emphasis is, of course, placed on covering the undergraduate level material, while that for the graduate one is presented in Sections of Chapters 3,5,7, 8, and 12, with the Problems numbered over 50 assigned to it; and in Chapters 10, 16 and 17. The presentation is made in such a fashion, however, that minimizes any loss of continuity in the undergraduate material, while the latter's review in the graduate course is facilitated with the assignment of some of the aforementioned Problems (51, 52...).

The short length of the book should also make it useful to the practicing engineer and chemist that needs a relatively quick review of the subject, as well as references for a more detailed study of items of his/her interest.

The development of the subject is based on the Laws of Thermodynamics, i.e. it follows the *historical* approach, which seems easier for most students to understand than the *postulatory* one. (The latter is based on a small number of postulates, that cannot be proved but only disproved by showing that consequences resulting from them are in disagreement with experimental evidence, and from which the laws of thermodynamics can be derived.)

We start the presentation of Chemical Engineering Thermodynamics by taking a Glimpse at the Subject and proceed with a discussion of its Laws, followed by a brief Historical Account of its development. (A Nation cannot exist without a sense of its history, why should a field of Science and Technology?)

We continue with a discussion of Efficient Energy Utilisation - a subject of immense importance that became, unfortunately, apparent only after the energy crisis of the early 1970's - followed by a brief presentation of the Broader Implications of the Second Law. (We are educating the technological leaders of tomorrow, not just technologists.)

We proceed with a limited discussion of Intermolecular Forces, which will help in understanding the Physical and Thermodynamic Properties of Pure Fluids which are considered next and, of course, the rest of the subject matter.

Cubic Equations of State are then considered for they have become a major tool in describing quantitatively the properties of Pure Fluids, as well as those of Mixtures, which are discussed next.

We continue with a discussion of Equilibrium and Stability, that prepares the ground for the ensuing presentation of the two topics of paramount importance to chemical engineers: Phase and Chemical Reaction Equilibria.

We close with a brief discussion of Statistical Mechanics that should prepare the student for further study in this area, which is becoming progressively more and more important in applied thermodynamics.

The inclusion of computer Programs in the enclosed diskette should help the students in developing familiarity with the determination of fluid properties and phase equilibria calculations.

The presentation of the material has been influenced by the excellent books of Professors K. Denbigh (*The Principles of Chemical Equilibrium*, 4th Ed., Cambridge University Press, 1981), J.M. Prausnitz (with R.N.

Lichtenthaler and E.G. de Azevedo, *Molecular Thermodynamics of Fluid Phase Equilibria*, 2nd Ed., Prentice Hall, 1986) and H.C. Van Ness (with J.M. Smith, *Introduction to Chemical Engineering Thermodynamics*, 4th Ed., McGraw-Hill, 1987), that are often mentioned in the book. The philosophy of presentation, however, has been shaped by many students, who always want to know the why and the relevance of what they are taught, and to whom this book is dedicated; and from my own conviction, that we should strive to make the topic more attractive to students by exposing them to the broader world of Thermodynamics. If they like this book, the effort will have been more than worthwhile.

In closing I would like to gratefully acknowledge: Professor D. Cardwell for permitting the use of material from his book: *From Watt to Clausius*; Professor H. Van Ness for providing the program that generates the Steam Tables; and Academic Press for permitting the use of material from W. Kenney's book *Energy Conservation in the Process Industries*; and, of course, to express my appreciation to NJIT for granting me a Sabbatical leave for the 1985-1986 academic year, during which most of this book was written.

Dimitrios P. Tassios
Athens, 1992

Contents

3 The Second Law of Thermodynamics 61

Appendices

1
A Glimpse at Thermodynamics

The modern steam engine was designed and first used to facilitate the mining of coal. As mines had to be sunk deeper into the ground to extract available supplies, it became more difficult to ventilate the mines and to face still another problem. At a certain depth the water table was reached and drainage became a priority. All of these problems required a technological solution. The steam engine was the answer. The first steam pump was patented in 1698 by Thomas Savery.

... The answer to the transportation (of coal) crisis was the invention of the steam locomotive and rail road tracks. Like the steam pump, the steam locomotive was a direct technological response to the needs created by the new coal environment. ***Together, the steam pump and the steam locomotive laid the technological base for the industrial era that followed.*** *(Emphasis added,* Rifkin, 1980, p.74.)

1.1 Introduction

Imagine yourself living in the first half of the 19th century. The society you live in is changing rapidly as a result of the Industrial Revolution made possible by the Steam Engine, which replaced the limited human, animal, and waterfall power, by the - what it seemed then - unlimited heat power.

Yet, even though the steam engine has been around for over a century - it was first introduced by Thomas Savery in 1698 - the scientific principles underlying its operation are not established yet.

It was not until the year 1850 that Clausius - on the basis of the pioneering work of several individuals, especially Carnot and Joule - expressed the foundations of the conversion of heat into work, i.e. the First and Second Laws of Thermodynamics.

A new, thus, field of Science and Technology was established: *Thermodynamics* (from the Greek: therme = heat; dynamis = force, for in those days the terms work and force were used interchangeably).

Finally, in the latter part of the 19th century, J. Willard Gibbs extended the application of thermodynamics to Chemical Reaction and Phase Equilibria, laying the foundations of - what we call today - *Chemical Engineering Thermodynamics*.

As the profession of Chemical Engineering grew into a separate discipline in the early part of the 20th century, thermodynamics played an increasingly important role in its development. Today, it represents - together with Chemistry, Mathematics, Physics, Transport Phenomena, Kinetics, and Economics - the cornerstones of our Profession.

On the Laws of Thermodynamics

> *The Laws of Thermodynamics control, in the last resort, the rise and fall of political systems, the freedom or bondage of Nations, the movements of commerce and industry, the origins of wealth and property, and the general welfare of the race.* (Frederick Saddy, Nobel-Prize winning chemist.)

1.2 Objective and Approach

Our main objective in this Chapter is to take a quick look - a glimpse - at thermodynamics: its *Vocabulary, Foundations*, and *Purpose*.

We start with a very brief description of the majority of the terms used in this field, the Vocabulary of the field, so that we can begin to learn and understand them and, eventually, communicate through them.

We proceed with a discussion - again brief - of the Foundations of thermodynamics: the Laws and the Thermophysical data, so that we become aware of their importance from the very beginning of this endeavor.

We continue with a brief outline of the Purpose of thermodynamics, i.e. the role that thermodynamics plays in the Chemical and Petroleum Industries, and close with a series of Examples that give a brief preview of the type of problems to be addressed in future Chapters.

1.3 The Vocabulary of Thermodynamics

There is a variety of terms that are often used in thermodynamics and

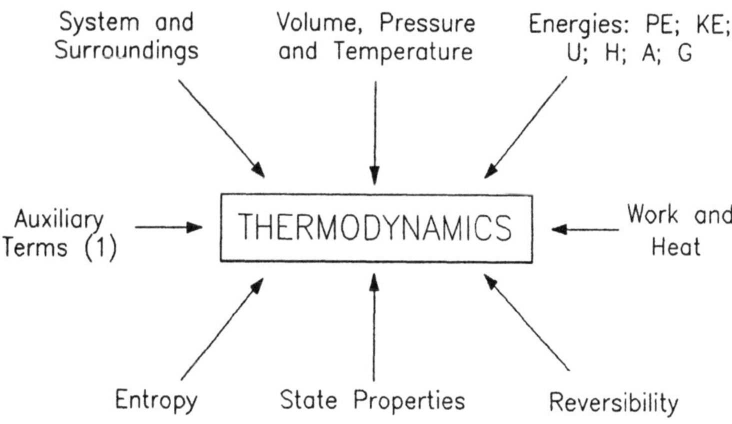

(1) Heat capacities, Ideal gas, Residual properties, Ideal solution, Excess properties, Activity coefficients, Chemical Potential and Fugacities

Figure 1.1 The Vocabulary of Thermodynamics.

constitute its *vocabulary*. Figure 1.1 presents the most common among them in a pictorial fashion, while a brief description of them is presented in Sections 1.4 through 1.13.

Our objective here is not, of course, to develop familiarity with - and understanding of - these terms. Rather, it is to expose you to them and, hopefully, to refresh your memory, since you have already used several of them in Chemistry and Introductory Chemical Engineering courses.

As we proceed through the ensuing Chapters, familiarity and understanding will be developed. After all one learns a language by using it, not by just memorizing words.

There is another similarity between learning a language and learning thermodynamics. The same way we can visualize certain objects in a language - for example, chair, or table - we can visualize volume, or temperature in thermodynamics.

But when it comes to concepts - such as, love, hate, sorrow, etc. in a language; or, entropy, energy, fugacity, etc., in thermodynamics - we only develop a 'feeling' or a 'sense' about them through repeated exposure to and use of them. We cannot visualize them. (It should be kept in mind, however, that the thermodynamic concepts are well defined mathematically.)

1.4 Pressure, Temperature and Volume

Our understanding of Pressure comes from Physics and it is defined as the ratio of the perpendicular force F exerted on a given area A, divided by this area:

$$P = \frac{F}{A} \tag{1.4.1}$$

Consider, next, Temperature. A close examination of this term suggests that, even though we grow up with it - from the "warmth of our mother's womb" to medicinal applications and weather forecasting - what we really become familiar with is a 'physiological' sense of 'hotness' and 'coldness' measured with a mercury thermometer in degrees Celsius (°C) or Fahrenheit (°F), not with a thermodynamic quantity.

The latter is a more complex term that will be developed in the next two Chapters, leading to the concept of *absolute thermodynamic* or just *absolute temperature*, measured in Kelvin (K) or Rankine (R).

The relationship among these scales is:

$$t(^\circ\mathrm{C}) = T(\mathrm{K}) - 273.15$$
$$t(^\circ\mathrm{F}) = T(\mathrm{R}) - 459.67$$
$$T(\mathrm{R}) = 1.8T(\mathrm{K})$$

The units for the quantities: F, P, T and V (volume) are given in Table 1.1 for the two main systems: the English System (ES) and the International System (Système Internationale, SI).

For conversion from the one set of units to the other see Appendix A.

Von Guericke, the Power of the Atmosphere and the Feeble Horses.

The credit for making the pressure exerted by the atmosphere entirely explicit belongs to Otto von Guericke who, in 1672, published the famous book in which he described his air pump and the experiments that he had made with it from the mid-1650s onwards. A dramatic picture in this book, which was subsequently widely reproduced, showed two teams, each of eight horses, straining ineffectually to pull apart two small hemispheres just over a foot in diameter. These had been evacuated of air by means of a pump and were held together in the form of a sphere solely by the weight of the atmosphere. In this way von Guericke graphically illustrated the power that could be generated by the weight of the atmosphere. It should be added that either

Table 1.1 *F, P, V, T* Units in ES and SI

Quantity	ES	SI
Force	lb_f	Newton: N
Pressure	psia: $lb_f\ in^{-2}$	Pascal: $N\ m^{-2}$
Temperature	°F, R	K
Volume	ft^3	m^3

the sixteen horses were remarkably feeble or von Guericke was exaggerating. The latter is the more probable conclusion. (Cardwell, 1971, p.12)

1.5 Work

1.5.1 Definition

Our earliest, probably, experience with work is associated with the lifting of a certain object. In general, we define work (W) through:

$$dW = Fds \tag{1.5.1}$$

where F is the force, and s the displacement, both in the same direction. [For limitations of this definition, and a more rigorous one, see Hatsopoulos and Keenan (1965).] In the ES system work is expressed in $ftlb_f$; in the SI, in J (Joule).

In thermodynamics, work involves - in the typical case - a change of the system (Section 1.8.1) volume due to difference in its pressure and that of the surroundings, as demonstrated in the following two Examples.

1.5.2 Example 1.1

A certain amount of air is placed inside a metallic cylinder-and-piston arrangement. The piston, assumed to be frictionless, is held in its position with latches so that the air volume is $V_1 = 5$ lit. The system is then placed into a constant temperature (25°C) bath and when thermal equilibrium is reached, the pressure of the air becomes 5 atm. The latches are now removed and the piston is allowed to move against some external pressure P_0 until the air volume becomes V_2.

Develop an expression for the work produced in terms of the air volume.

In this case: $$F = P_0 A \quad \text{(A)}$$

where A is the cross sectional area of the piston. To express the displacement ds in terms of the corresponding volume change of the air dV, we note that:

$$ds = dV/A \quad \text{(B)}$$

Introduction of Eqs (A) and (B) into Eq.1.5.1 yields:

$$dW = P_0 dV$$

and:

$$W = \int_{V_1}^{V_2} P_0 dV = P_0(V_2 - V_1)$$

1.5.3 Example 1.2

Let us now evaluate the work produced in Example 1.1 for the following three cases:

a. there is a vacuum outside the piston;

b. the pressure outside the piston is 1 atm;

c. the pressure outside the piston is controlled in such a way, that it is always barely (infinitesimally) smaller than that of the air inside it.

In all three cases the final volume is twice the initial one.

a. $P_0 = 0$, and no work is produced.

b. $P_0 = 1$ atm, which remains constant throughout the movement of the piston. Hence:

$$W = P_0(V_2 - V_1) = (1)(10 - 5) \text{ atm lit} = 506.6 \text{ J}$$

c. In this case, P_0 is practically the same as the pressure of the air P throughout the movement of the piston, and, consequently:

$$W = \int_{V_1}^{V_2} P dV \quad \text{(A)}$$

Since P varies with the expansion of the air, we need a relationship between P and V so that the integration can be carried out. Such a relationship is provided by the ideal gas law:

$$PV = NRT \quad \text{(B)}$$

The number of moles of air in the cylinder (N) can be calculated from the initial condition: $P = 5$ atm; $V = 5$ lit; $T = 298.15$ K, giving a value of $N = 1.022$ moles.

Before we introduce Eq.(B) into Eq.(A), we must consider the following question: Does the air temperature T remain constant during the expansion process?

Since the opposing pressures are so close to each other, the process is very slow and it is reasonable to expect that the air temperature will remain constant and equal to that of the bath, i.e. that the system is continuously in a state of internal equilibrium (see Section 1.8.5).

Eq.(A) yields, therefore:

$$W = NRT\ln\left(\frac{V_2}{V_1}\right) = NRT\ln 2 = 1756\ \mathrm{J}$$

Comments:

1. We will see in Section 1.12 that cases (a) and (b), where the driving force - i.e. the difference between external and internal pressures - is large, are referred to as *irreversible*; and case (c), where the driving force is very small, as *reversible*.

2. Notice that as we move from the most irreversible case (a) to the reversible case (c), the amount of work produced is increasing.

3. Actually, case (c) produces the maximum possible work. Any attempts to produce more work, by increasing further the external pressure, would lead to compression, i.e. the consumption of work.

4. Only in the reversible case (c), can we replace the external pressure P_0 with the air pressure P.

1.6 Energies

In its broadest sense, *energy* reflects the ability of a body, or system, to produce work, for example, to raise an other body to some elevation h above its original position. Notice that we use the term 'sense' for energy, because it is a concept.

Several energy terms are encountered in thermodynamics, from the more comprehensible ones, such as potential, kinetic, internal and enthalpy to the more abstract ones, such as the Helmholtz and the Gibbs free energies.

1.6.1 Potential Energy, *PE*

This energy reflects the force of gravity exerted on a body of mass m and it is defined in terms of the elevation of this body h, relative to some reference plane:

$$PE = mgh \tag{1.6.1}$$

where g is the local acceleration of gravity.

1.6.2 Kinetic Energy, *KE*

This is the energy that a body of mass m possesses as a result of its velocity u:

$$KE = \frac{1}{2} mu^2 \tag{1.6.2}$$

1.6.3 Internal Energy, *U*

This energy reflects the total energy contained *within* a body, i.e. independently of its potential and kinetic energies, and is essential in the formulation of the first law of thermodynamics. (We shall return to this subject in Section 1.9.)

1.6.4 Enthalpy, *H*

It is defined through the internal energy:

$$H = U + PV \tag{1.6.3}$$

and its use, among others, simplifies the expression of the first law when applied to a flow process.

1.6.5 The Helmholtz, *A*, and the Gibbs, *G*, Free Energies

They are defined through the internal energy U and entropy S (Section 1.10):

$$A = U - TS \tag{1.6.4}$$

$$G = U + PV - TS = H - TS \tag{1.6.5}$$

The Helmholtz free energy is of more theoretical importance, used often in statistical mechanics.

The Gibbs free energy, on the other hand, is very important from the practical point of view, for it represents the basic quantity in chemical and/or phase equilibrium calculations.

1.7 Heat

Here is an other term that we are exposed to from very early in our lives, yet its clear meaning is by no means apparent. (It took, for example, hundreds of years for people to even differentiate between heat and temperature, as we will see in Chapter 4.)

Consider a hot body A that comes into contact with a cold body B. We know from experience that, with time, the temperature of the hot body decreases and that of the cold body increases until they become the same.

What is responsible for this change? Something, obviously, flows from the one body to the other that causes their temperatures to become the same.

We will see in Section 1.9 that this 'something' is just the energy of the molecules of the hot body that is transferred into the molecules of the cold one. It is referred to as *heat*. Notice, however, that heat exists only as a result of energy flow. It did not exist when the two bodies were not in contact. (For a more rigorous definition, that does not involve the use of temperature, see Hatsopoulos and Keenan.)

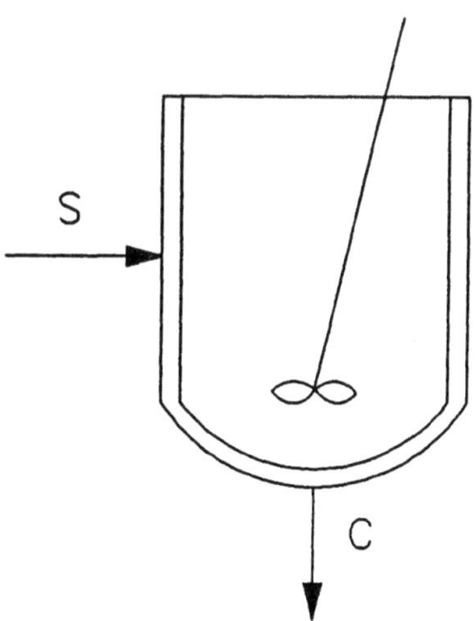

Figure 1.2 The heating of benzene with steam: benzene is the system, everything else represents the surroundings.
S: Steam; C: Condensate

1.8 System, Surroundings, and Related Terms

1.8.1 System and Surroundings

Consider the jacketed vessel of Figure 1.2, where benzene is heated from 20°C to 60°C. The stirrer is used to facilitate the heating process through mixing. We refer to benzene as the *system*. The steam, vessel, and the atmosphere are the *surroundings*.

In general, the part of the universe that we focus our attention to represents the system. Everything else, the surroundings. Clear identification of these two parts is essential before we proceed with the solution of a problem.

1.8.2 Modes of Interaction Between System and Surroundings

If a system can exchange thermal energy with its surroundings, it is considered to be in *thermal contact* with them and the wall enclosing it, as *diathermal*. If not, as *adiabatic*.

If a system can exchange mechanical energy with its surroundings, it is considered to be in *mechanical contact* with them.

If a system can exchange mass with its surroundings, it is referred to as an *open system*; if not, as a *closed system*.

Thus, the aforementioned benzene system is a closed one, in:

* thermal - through the vessel walls, and
* mechanical - through the stirrer,

contact with the surroundings.

Finally, a system that cannot exchange with its surroundings thermal energy, mechanical energy, and mass is referred to as *isolated*.

1.8.3 State

It is known from experience that if the temperature T and pressure P of a gas with given composition are known, then its molar volume V is also known.

Actually for a given amount of a homogeneous system with specified composition, knowledge of any two of these three properties (P, T, V) suffices to determine the condition - or *state* - of the system. (For a more comprehensive definition of the state of a system see Section 1.11 and, especially, Chapters 9, 12, and 15.)

1.8.4 Process

A sequence of states that brings a system from its initial state to its final one is referred to as *process.*

1.8.5 Equilibrium: Internal and External

When we refer to a property of a system - say pressure or temperature - we assume that the system is uniform and, therefore, this property reflects the whole system.

A system within which there are no non-uniformities, that act as driving forces for change, is said to be in a state of *internal equilibrium.* This is the type of systems that we will be dealing with in this book (see Example 1.2.c).

And, if there are no driving forces that can bring about a change between a system and its surroundings, the system is said to be in a state of *external equilibrium.*

1.9 Internal Energy and Molecular Energy

When we say that the internal energy of one mole of steam is U Joules, where is this energy stored? Or when Q Joules of energy flow from system A to system B, where was this energy stored in system A? Where will it be stored in system B?

The energy is stored within the molecules of the steam or of the systems in two forms:

1. kinetic; and
2. potential.

The *kinetic* energy is contained within the individual molecules and, for each molecule, it is independent of that of the others. For a given species, or mixture of them, it depends on temperature only, and it involves two types of energy:

1. translational, and

2.internal (rotational, vibrational, nuclear, and electronic).
(We will see in Chapter 17, however, that the rotational and vibrational energies depend on density too but to a much smaller degree than on temperature).

The *potential* energy is contained by all the molecules collectively as a result of the forces exerted among them - referred to as *intermolecular forces* - and depends on:

* the temperature, and
* the distance among them, i.e. the system density.

The intermolecular forces are of two main types:

1.attractive; and
2.repulsive,

and are discussed in Chapter 7.

1.10 Entropy

It is the most difficult, probably, concept in thermodynamics, introduced in conjunction with the second law. We will use entropy to determine if a process can take place or not; and as the 'yardstick' for measuring how efficiently energy is utilized in a given process. Entropy is also used in Information Theory, Economics, even to describe the development of our civilization (Chapter 6).

1.11 State Properties

Assume that we want to determine the value of the volume V of one kg of steam from the Steam Tables presented in Appendix B. To this purpose, we need only specify the state of the steam by specifying (Section 1.8.3), the temperature T and pressure P. The way that the steam was prepared, i.e. the path followed to arrive at this kg of steam, has no bearing on its volume value. Thus, at 400 kPa and 500°C, V = 889.2 m^3/kg, independent of the path followed in making it. And the same applies for the internal energy U (3129.2 kJ/kg), entropy S, etc.

In general, for a homogeneous system of given mass and composition, specification of any two of its properties (P, T, V, U, S, etc.) suffices for the determination of the others, i.e. for the determination of the state of the system, independently of the path followed to arrive at this state. They are referred to, appropriately, as *state properties* or *functions*. (For an extensive discussion of the specification of the state of a system see Chapters 9, 12, and 15.)

Let us now contrast state properties to two very important quantities in thermodynamics: work (W) and heat (Q).

To this purpose, consider the following question: how much work and/or heat are required to make the aforementioned kg of steam? i.e. what are the values of W and Q associated with this steam?

To answer this question, we need not only specify the starting point, but also - as we will see in the next Chapter - the path followed to arrive at this steam. We will conclude, for example, that if we start with steam at 1 bar and 100°C, and follow two paths:

1. constant pressure heating to 500°C, followed by constant temperature compression to 400 kPa; or
2. constant temperature compression to 400 kPa, followed by constant pressure heating to 500°C;

we arrive at two different sets of values for W and Q; they are, in other words, *path depending* quantities.

The characteristics of these two types of thermodynamic quantities, state properties, and work and heat, are contrasted in Table 1.2, which explains why, for example, we use ΔU but not ΔQ; or why, rigorously speaking, we should use δQ and δW, and not dQ and dW. (In the interest of simplicity, dQ and dW will be used in this book, but it should be kept in mind that they refer to infinitesimal amounts, not changes.)

1.12 Reversibility

Processes that proceed so that displacements from equilibrium, i.e. the driving forces, are infinitesimally small, are referred to as *reversible* processes; if not, as *irreversible*. We examine, next, the practical implications of reversibility.

Consider body A, with a weight of $F_A = 20$ N and at an elevation of 2 meters from the ground, which will be used through a pulley to raise another body B of weight F_B from the ground to the same elevation. We assume, for simplicity, that the pulley is frictionless, and proceed to determine the efficiency of the operation:

Table 1.2 Characteristics of State Properties and of Work and Heat

State Properties (S,T,U, etc.)	Work and Heat
Path independent	Path dependent
Instantaneous properties	Energy in transit
Differential: infinitesimal change	Infinitesimal amount
Integral: change in property	Amount

$$n = \frac{\textit{Actual Work Produced}}{\textit{Maximum Producable Work}} \tag{1.12.1}$$

as a function of the difference $(F_A - F_B)$, i.e. as a function of the driving force of the process. The results are presented in Table 1.3.

Comments:

1.As the driving force decreases, the efficiency of the process increases, i.e. more work is obtained.

2.At the limit $F_B \longrightarrow F_A$, where the process becomes reversible, the efficiency assumes the maximum value equal to 1.0.

3.We conclude, therefore, that a reversible process provides the maximum possible amount of work, just as we did in Example 1.2. Therefore Eq.1.12.1 becomes:

$$n = \frac{\textit{Actual Work Produced}}{\textit{Reversible Work}} \tag{1.12.2}$$

4.If work is consumed, it is easily shown that a reversible operation requires the minimum one and the efficiency is defined by:

$$n = \frac{\textit{Reversible Work}}{\textit{Actual Work Consumed}} \tag{1.12.3}$$

5.For the reversible operation, a small increase in F_B would result in a reversal of the process, thus, the name reversible.

6.Actual - real life - processes are of course irreversible. The presence of friction, for example, would prevent the movement of the pulley, or of the piston, if the driving force is infinitesimally small; and a heat exchanger operating under such a small temperature difference would require an infinite area for heat transfer.

7.Reversibility as a concept, however, is of paramount importance in thermodynamic applications, such as in the area of efficient energy utilization, as we will see in later Chapters.

Table 1.3 Amount of Work Produced by the Falling Body A with Weight F_A = 20 N as a Function of the Driving Force.

F_B(N)	$F_A - F_B$	Work (J)	Efficiency
0	20	0	0
5	15	10	0.25
10	10	20	0.50
12.5	7.5	25	0.625
15	5	30	0.75
17	3	34	0.85
19	1	38	0.95
19.5	0.5	39	0.975
20	0	40	1.00

1.13 Auxiliary Terms

1.13.1 Heat Capacity

The amount of heat required to raise the temperature of one mole of a substance by one degree under conditions of:

1. constant volume, or
2. constant pressure,

is referred to as *heat capacity*, C_V and C_P, respectively and is defined by:

$$C_V = \left(\frac{\partial U}{\partial T}\right)_V$$
$$C_P = \left(\frac{\partial H}{\partial T}\right)_P \tag{1.13.1}$$

where U and H are the molar internal energy and enthalpy respectively.

1.13.2 Ideal Gas

This is a hypothetical gas where:

1.there are no intermolecular forces; and
2.the molecular volume is negligible as compared to the volume of the gas.

In theory, such a behavior is approached by real gases when their pressure, and thus the density, tends to zero. In practice, however, gases can be assumed to behave ideally at pressures up to one bar (Example 1.5). There are some exceptions, however, mainly the organic acids.

1.13.3 Real Fluid versus Ideal Gas

The main difficulty in evaluating thermodynamic properties is the presence of intermolecular forces. (Nature would be very obliging if it could do without them, but could it?)

An ideal gas - where such forces are absent - represents, therefore, a useful reference fluid for isolating their effect on the properties of real fluids. Two terms are used for this purpose:

* Residual properties:

Real fluid (T, P, x) *property - Ideal gas* (T, P, x) *property.*

* Departure functions:

Real fluid (T, P, x) *property - Ideal gas* (T, P°, x) *property.*

In the first case, we compare the properties of the real fluid - at the given temperature T, pressure P, and composition x - to those of a hypothetical ideal gas, at the same conditions; in the second, the pressure of the ideal gas is set equal to a specified value P°, dictated by convenience (Chapter 9).

1.13.4 Ideal Solution

An ideal solution is formed when:

1.the forces between like and unlike molecules are the same; and
2.the size and shape of the molecules are the same.

As in the case of the ideal gas, an ideal solution is also a hypothetical fluid. Here, however, intermolecular forces are present, but we assume that those between unlike molecules are the same with those between like molecules.

If an ideal liquid solution is in equilibrium with the vapor above it that behaves as an ideal gas, the relationship between the compositions of the two phases is described analytically by *Raoult's* law:

$$\bar{P}_i = y_i P = x_i P_i^s \tag{1.13.2}$$

where,

$\bar{P}_i$: partial pressure of component i, i.e. the part of the total pressure due to component i;

x_i,y_i: mole fraction of component i in the liquid and vapor phases respectively;

P : total pressure over the solution;

P_i^s : vapor pressure of pure liquid i, i.e. the pressure that would exist over the liquid if the latter was pure i (at the temperature of the solution).

In practice, Raoult's law is valid, approximately, for mixtures of similar compounds only, such as, hexane-heptane or ethanol-propanol.

1.13.5 Real versus Ideal Solution

In Section 1.13.3, we used the difference between the properties of a real fluid and an ideal gas to describe the effect of intermolecular forces on the thermodynamic properties of the fluid. Here we use the difference between real and ideal solution behavior of a fluid, at the same T, P and composition x, to describe the effect of the differences in:

* intermolecular forces; and
* size and shape;

among like and unlike molecules on the properties of the fluid.

The comparison is done at the same temperature, pressure, and composition, and the resulting difference is called the *excess function* (Chapter 11):

$$\textit{Excess Function} = \textit{Real solution property}\ (T,\ P,\ x) - \textit{Ideal solution property}\ (T,\ P,\ x)$$

1.13.6 Activity Coefficient

The difference in the behavior of the component of a real solution, as compared to that determined by Raoult's law (Eq.1.13.2), is described by its *activity coefficient* (γ_i):

$$y_i P = \gamma_i x_i P_i^s \tag{1.13.3}$$

The activity coefficient is of paramount importance in chemical engineering thermodynamics applications, for it represents a key quantity in separation calculations (Example 1.8).

1.13.7 Chemical Potential and Fugacity

The presentation of these two very important quantities - in chemical and phase equilibrium calculations - is not possible here. They will be discussed in Chapters 9 and 11.

1.13.8 Intensive and Extensive Properties

Properties that are independent of the amount of material present, such as pressure and temperature, and properties that are defined per unit of mass, such as heat capacities, are referred to as *intensive* properties.

On the other hand, properties that depend on the total amount of material present, such as volume, are referred to as *extensive* properties.

1.14 The Foundations of Thermodynamics

Having finished with the Vocabulary of thermodynamics, we turn next to the second objective of this Chapter: a brief discussion of the *Foundations* of thermodynamics, depicted pictorially in Figure 1.3:

1. the *laws* of thermodynamics; and
2. the *thermophysical* data.

The Laws provide the *framework of the necessary equations* for the solution of thermodynamic problems; the Data, *the information* required to arrive at numerical answers through these equations.

1.14.1 The Laws of Thermodynamics

1.The Zeroth Law: *When bodies A and B are in thermal equilibrium with body C, they are in thermal equilibrium with each other.* It is used to establish temperature as a state property.

2.The First Law: *Energy is conserved.* The basis for energy balances, it relates the state functions: internal energy and enthalpy, to the path dependent ones: work and heat.

3.The Second Law: *Any process whose only effect is the transfer of heat from one temperature level to a higher one, is impossible.* It introduces

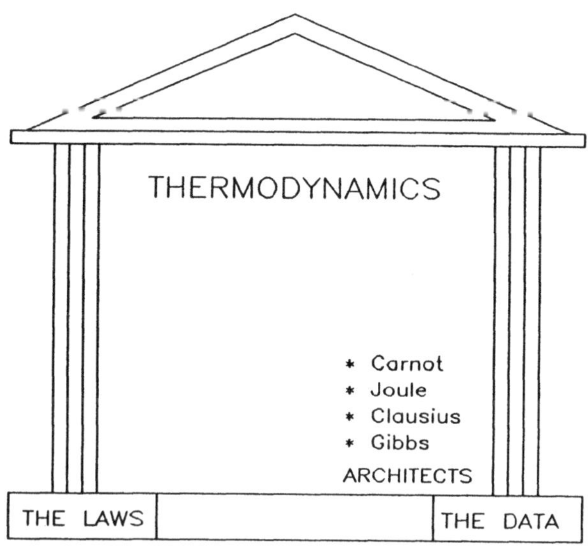

Figure 1.3 The Foundations of Thermodynamics.

the concepts of thermodynamic temperature and of entropy; it also establishes the feasibility of processes and, in combination with the first law, the degree of efficient energy utilization in them.

4.The Third Law: *The absolute entropy of a perfect crystal at zero absolute temperature, is zero.* It is used to determine absolute values of entropy, that are essential in calculations involving chemical reactions.

Note:The development of thermodynamics based on these laws follows the 'historical' approach. For other approaches, see the Preface.

On the Laws of Thermodynamics

It must be admitted, I think, that the laws of thermodynamics have a different feel from most of the other laws of the physicist. There is something more palpably verbal about them, they smell more of their human origin. The guiding motif is strange to most of physics: namely a capitalizing of the universal failure of human beings to construct perpetual motion machines of either the first or second kind. Why should we expect nature to be interested either positively or negatively in the purposes of human beings, particularly purposes of such an unblushingly economic tinge? (P. W. Bridgman, *The Nature of Thermodynamics*, Harvard University Press, 1941.)

1.14.2 The Thermophysical Data

Properties, such as: density, enthalpy, entropy, etc., for given species or mixture of them, are referred to as *thermophysical properties* and their values, as *thermophysical data*. A typical example is the Steam Tables. Such data involve two types of quantities:

1. Measurable; and
2. Derived.

The first type refers to the properties that can be measured in the laboratory:

* Pressure-Volume-Temperature (*PVT*);
* Heat capacities;
* Vapor pressures (considered part of the *PVT* measurements); and,
* Enthalpies of vaporization.

We will refer to them as *physical properties*.

The second type involves all other properties such as, enthalpy, entropy, etc., that cannot be determined experimentally. There is no such a thing, for example, as an 'entropymeter'.

Their values are derived from the measurable quantities through relationships based on the laws of thermodynamics. They are, therefore, 'conceptual' quantities and this is why they are more difficult to comprehend. We will refer to them as *thermodynamic* properties and their values as *thermodynamic* data.

The availability of thermophysical data is essential to the application of thermodynamics in Chemical Engineering calculations. The first law, for example, provides the necessary relationship for the evaluation of the amount of heat involved in changing the temperature of a given fluid. Unless, however, the required heat capacity data is available, no numerical answer can be obtained.

The situation can be simply summarized as follows: "You cannot get much use out of a beautiful *959-Porsche* without gasoline; You cannot get much information for a given process from the laws of thermodynamics, without the appropriate thermophysical data".

1.15 The Purpose of Chemical Engineering Thermodynamics

We turn now to the third objective of this Chapter; a brief outline of the

Purpose of chemical engineering thermodynamics: the study of the three broad areas of the field, outlined in Figure 1.4.

1.15.1 Feasibility of Processes and Efficient Energy Utilization

The feasibility of processes and the efficient energy utilization in carrying them out represent the original objectives of thermodynamics and the reason for its name. Actually, the second objective has been traditionally of interest to power engineers, mostly mechanical, while chemical engineers paid only lip service to it, when energy cost was very low.

The drastic increase in energy cost following the Oil Embargo of 1973, however, and its impact on the product cost of the chemical and petroleum industries, made the subject of efficient energy utilization a major concern of chemical engineers as well.

A combination of the first and second laws will be used to establish the criteria for:

1. the feasibility of, and
2. the degree of efficient energy utilization in,

a process under consideration.

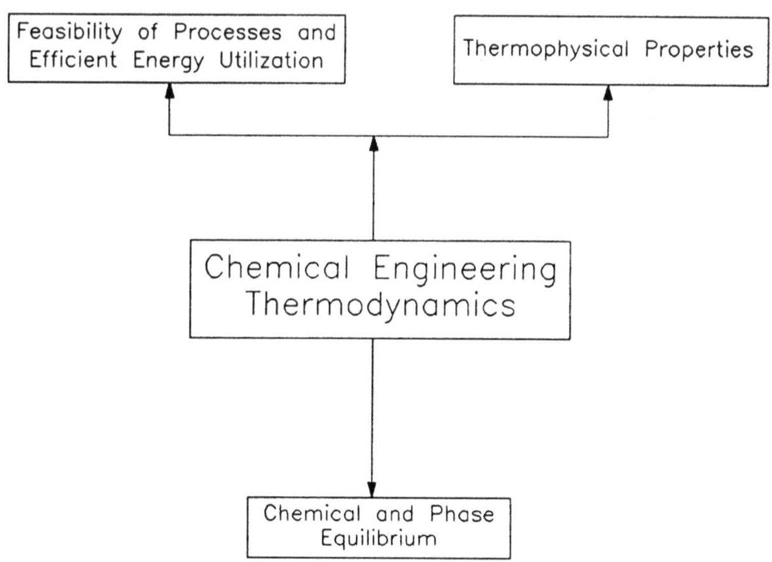

Figure 1.4 The Three Broad Areas of Chemical Engineering Thermodynamics.

1.15.2 Thermophysical Properties

Knowledge of the values of these properties is essential in obtaining numerical results in the other two areas. Our objective here will be two-fold:

1. evaluation of the thermodynamic - i.e. derived - properties: internal energy, enthalpy, entropy, etc., from data on the measurable ones: pressure, volume, temperature, and heat capacities: and
2. estimation of thermophysical properties.

1.Evaluation of Thermodynamic Properties

Starting with a combination of the first and second laws we will develop the necessary framework of equations that can be used, in combination with physical property data, in the determination of thermodynamic properties. Because the available in the typical case data cover a limited range of pressure and temperature, the obtained values are not the absolute ones, but relative to some arbitrarily chosen reference state.

The entropy and enthalpy values in the Steam Tables, for example, have been developed using these equations, combined with extensive experimental pressure-volume-temperature (*PVT*) and ideal gas heat capacity data. The given values, however, are relative to zero values for saturated water at 273.16 K.

2.Estimation of Thermophysical Properties

On the other hand, considering:

* the plethora of compounds involved in the chemical and petroleum processes;
* the practically infinite number of mixtures resulting from them; and
* the large cost, in time and money, required for the experimental determination of the necessary data for these compounds and mixtures;

it becomes apparent that, in the typical case, we must resort to some estimation, i.e. approximate evaluation, of some - or even all - properties, physical and thermodynamic.

The importance of thermophysical properties in Chemical Engineering is demonstrated in Table 1.4, using information from a major chemical company.

An excellent reference for such estimation techniques is the book of R. Reid, J. Prausnitz, and B. Poling, *Properties of Gases and Liquids*, 4th Ed., McGraw-Hill, New York, 1987.

Table 1.4 Physical Property Errors Are Costly

Property	If % Error in Property is	Resulting % Error in Equip. Size	Equip. Cost
Activity Coefficient	10		
Separation:			
* Very easy		3	2
* Easy		20	13
* Difficult		50	31
* Very Difficult		100	100
Heat Capacity	20	6	6
Heat of Vaporization	15	15	15
Density	20	16	16
Viscosity	50	10	10
Thermal Conductivity	20	13	13
Diffusivity	20	6	4
	100	40	23

Data compilations are also available through the *Design Institute for Physical Property Data (DIPPR)*, operating under the auspices of the American Institute of Chemical Engineers (AIChE) and supported by several chemical and petroleum industries; and by several Computerized Data Banks.

1.15.3 Chemical and Phase Equilibrium

The importance of these areas becomes apparent the moment one visits a chemical plant or a petroleum refinery: chemical reactors and separation units are the essential blocks of the processes involved in both places, as the typical example shown in Figure 1.5 demonstrates for the manufacture of methanol. [Carbon monoxide and hydrogen react at moderate pressures (50-100 atm) and temperatures (250-270°C) to form methanol. Following condensation, the unreacted gas is recycled, while methanol is recovered from the liquid mixture by distillation.]

Starting with a combination of the first and second laws we will arrive at a framework of equations which, in combination with the appropriate

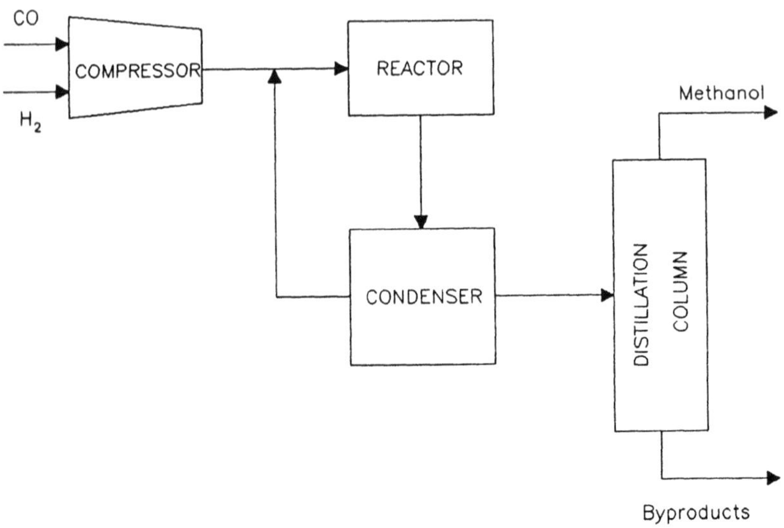

Figure 1.5 The manufacture of methanol.

thermophysical data, provide the necessary information for designing these units.

Note. The separation of Chemical and Phase Equilibrium from the general subject of Thermophysical Properties is intended to emphasize the importance of the former.

1.16 Examples 1.3 through 1.9

We close this Chapter with a series of Examples that represent a brief preview of the type of problems that chemical engineering thermodynamics deals with:

* Examples 1.3 and 1.4 will help develop some familiarity with the SI and ES systems of units. The first is the accepted system in Europe and is becoming slowly, but steadily, the same in the United States, where the second still dominates.

* Example 1.5 discusses one method of estimating thermophysical pro-

perties - gas densities in this case - the ideal gas assumption, and demonstrates the approximate character of, and the errors involved with, this approach.

* Examples 1.6 and 1.9 deal with the problem of energy use, the second with efficient energy utilization.

* Finally, Examples 1.7 and 1.8 deal with a subject of major importance to chemical engineering thermodynamics applications: distillation.

1.16.1 Example 1.3

Two systems of units are used today in the U.S.: the traditional English (ES) and the SI (Système International d'Unites), towards which the country is moving, albeit slowly. (The SI system was introduced in 1960 by the General Conference of Weights and Measures, an international body that meets every six years in Paris.)

a. Why is the adoption of the SI system so slow?
b. What are the *base* units in the SI system?
c. What are *derived* units?

a. The slow adoption of the SI system in the U.S. must be attributed to two main reasons:

1. The force of habit. The English system has been used in the U.S. for a long time, and we all know that "habit is a second nature".

2. Economic. It has been estimated that the cost of retooling the U.S. industry from the English to the SI system will be well over $10 billions (middle 80's).

b. See Table 1.E.3a.

Table 1.E.3a Base Units in the SI System

Quantity	Unit	SI Symbol
length	meter	m
mass	kilogram	kg
time	second	s
electric current	ampere	A
thermodynamic temperature	kelvin	K
amount of substance	mole	mol
luminous intensity	candela	cd

c. These are units obtained by combination of the base units. For examples, see Table 1.E.3b.

Table 1.E.3b Examples of Derived Units

Quantity	Unit	SI Symbol	Formula
density	kilogram per cubic meter	-	kg/m^3
force	newton	N	$kg\ m/s^2$
energy	joule	J	N m
entropy	joule per kelvin	-	J/K
pressure	pascal	Pa	N/m^2

1.16.2 Example 1.4

What is the weight of 1 kg of water in a place where the local gravity acceleration is $g = 9.70\ m\ s^{-2}$ in the English and the SI systems of units.

a. SI Units
From Newton's law: $F = mg = (1\ kg)(9.70\ m\ s^{-2}) = 9.70\ N$

b. English Units
Using the conversion factors of Appendix A:
$F = (9.70\ N)(0.2248\ lb_f/N) = 2.18\ lb_f$

1.16.3 Example 1.5

Gas densities are important in chemical engineering calculations such as in the sizing of vessels, pipes for gas transport, etc. As a first approximation, such densities can be calculated using the ideal gas law.

a. What factors effect the accuracy of the obtained results?

b. To test your answer in part (a), calculate the density of steam at:

* 3500 kPa, as a function of temperature; and
* 300°C, as a function of pressure;

and compare the obtained values to those given in the Steam Tables.

a. The assumptions behind the ideal gas behavior become more valid as the volume per unit mass increases, i.e. as the density decreases (Section 1.13.2). We should expect, therefore, increasing accuracy:

* at a given pressure, as the temperature increases; and
* at a given temperature, as the pressure decreases.

b. According to the ideal gas law the density of steam is given by:

$$d = 18.015/V = 18.015\ P/(RT)$$

where $R = 8.314\ m^3\ kPa(kmol\ K)^{-1}$, is the ideal gas constant. The results, pre-

Table 1.E.5 Steam Densities (kg/m^3) Calculated with the Ideal Gas Law

P = 3500 kPa				t = 300°C			
	Density				Density		
t, °C	Calc.	Exp.	% Error	P, kPa	Calc.	Exp.	% Error
250	14.50	17.04	-14.9	100	0.38	0.38	- 0.2
275	13.83	15.67	-11.7	200	0.76	0.76	- 0.5
300	13.23	14.61	- 9.5	250	0.95	0.95	- 0.6
325	12.68	13.75	- 7.8	350	1.32	1.33	- 0.8
350	12.17	13.02	- 6.5	500	1.89	1.91	- 1.2
375	11.70	12.39	- 5.6	1000	3.78	3.88	- 2.5
400	11.27	11.84	- 4.8	2000	7.56	7.97	- 5.1
450	10.49	10.88	- 3.6	3000	11.34	12.32	- 8.0
500	9.81	10.09	- 2.8	4000	15.12	17.00	-11.0
600	8.69	8.84	- 1.7	5000	18.90	22.07	-14.4

sented in Table 1.E.5 and graphically in Figure 1.E.5, support our findings in part (a).

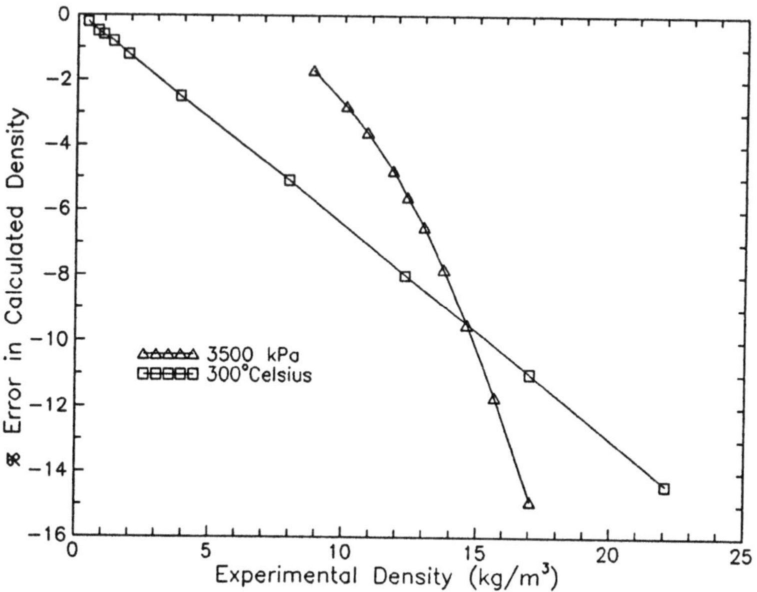

Figure 1.E.5 The effect of density on the ideal gas assumption for steam.

Comments:

1. As the density approaches zero so does the error, which is consistent with the definition of an ideal gas.

2. Notice the pronounced effect of temperature as compared to pressure on the accuracy of the obtained results.

* An increase in temperature by a factor of 1.7, from 523 K (250°C) to 873 K (600°C), leads to the same reduction in error as a 10-fold decrease in pressure from 5000 kPa to 500 kPa.

* At densities below about 15 kg m^{-3} the errors for $t = 300°C$ are larger (in absolute value) than those for 3500 kPa, because the temperatures involved in the latter case are larger than 300°C.

* The trend is reversed at densities above 15 kg m^{-3}, because here the temperatures for $P = 3500$ kPa are lower than 300°C.

3. The explanation for this phenomenon will become apparent when we discuss intermolecular forces. It suffices to say, here, that it has to do with the strong effect that temperature has on the dipole-dipole interactions.

1.16.4 Example 1.6

Water at a pressure of 1 atm and a temperature of 100°C enters a boiler where it is vaporized at a rate of 100 kg/min.

a. How much heat, in kJ/min and Btu/min, is needed?

b. If the heating is done by condensing saturated steam of 1.5 bar, how much steam is needed in kg/min?

c. What role does thermodynamics play in solving this problem?

a. From the Steam Tables, the heat of vaporization for water at 1 atm and 100°C is 2256.9 kJ/kg. Hence, heat required:

$$Q = 100\text{x}2256.9 = 2.26\cdot10^5\,\text{kJ/min} = 2.14\cdot10^5\,\text{Btu/min}$$

b. The heat of condensation of saturated steam at 1.5 bar is 2226.3 kJ/kg. Hence:

$$\text{steam} = [2.26\cdot10^5\,\text{kJ/min}]/[2226.3\,\text{kJ/kg}] = 101.5\,\text{kg/min}$$

c. The heats of vaporization and condensation are thermophysical data for steam. In addition, the calculation of the steam requirement is based on the equality:

heat absorbed by the water = heat lost by the steam

which follows from the first law.

1.16.5 Example 1.7

One of the most important applications of thermodynamics in our profession is in the design of distillation columns. Such columns are used to effect the separation of a mixture into its components with desirable purity levels.

Consider, now, the separation of a binary mixture of components 1 and 2 into products of, say, 99.9% purity.

a. What are the major parts of a distillation unit?
b. What are the commonly used types of distillation columns?
c. Describe briefly the operation of the unit.

a. A simple schematic diagram of such a unit is shown in Figure 1.E.7 and it consists of three major parts:

1. the column and packing or trays
2. the boiler and
3. the condenser.

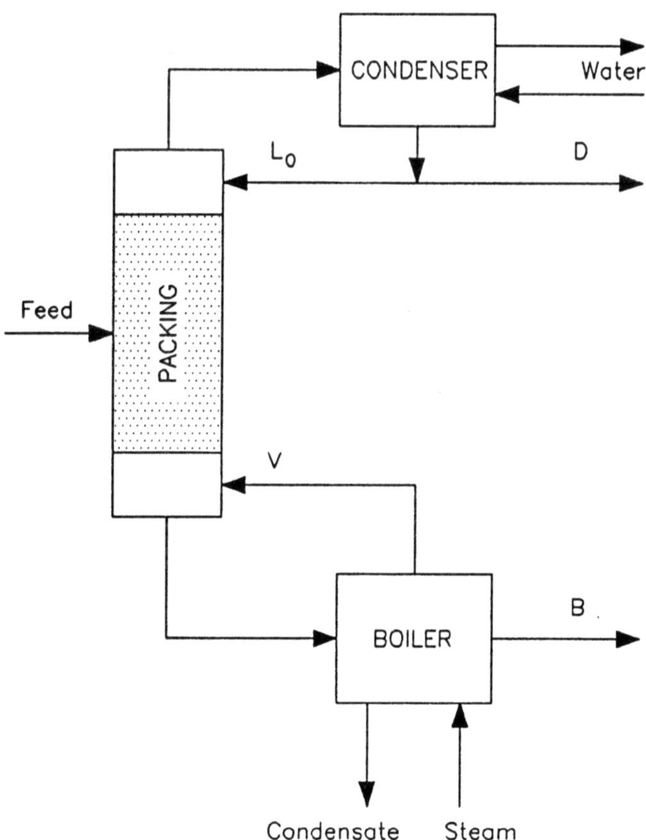

Figure 1.E.7 Schematic diagram of a packed distillation column.

b. Distillation columns come into two main types:
1. packed; and
2. tray (mostly sieve).

c. The operation of the column is as follows: Vapor (V), generated in the boiler, rises through the column coming into contact with the liquid descending from the condenser (L_0, called the 'Reflux'). During this contact, the vapor loses some of its less volatile component, which goes into the liquid phase, and the reverse occurs with the liquid. It is apparent that this exchange of constituents between the two streams requires a large contact area, which represents the purpose of the packing or trays.

Thus, as the vapor reaches the top of the column, it has lost most of its heavy (less volatile) constituent; and, as the liquid reaches the bottom, it has lost most of its light constituent.

If components 1 and 2 are the light and heavy components, then:
* stream B - known as the 'Bottoms' - can be practically pure 2; and,
* stream D - known as the 'Distillate' - can be practically pure 1;

provided that the height of the column and the reflux are sufficient.

The separation, thus, of the feed stream into its components with the specified purity has been accomplished. (Are there any exceptions to this?)

1.16.6 Example 1.8

The essential quantity involved in determining the required height and reflux of a distillation column, such as the one used for the separation of the mixture of components 1 and 2 in the previous Example, is the *relative volatility* of 1 with respect to 2, α_{12}. It is defined as follows:

$$\alpha_{12} = \frac{y_1/x_1}{y_2/x_2} \tag{A}$$

where y and x represent the vapor and liquid phase compositions at some point in the column.

Assuming that the two phases are at equilibrium with each other, what information is needed to evaluate α_{12}?

From Eq.1.13.3:

$$\frac{y_i}{x_i} = \frac{\gamma_i P_i^s}{P} \tag{B}$$

Introduction of Eq.(B) into Eq.(A), yields:

$$\alpha_{12} = \frac{\gamma_1 P_1^s}{\gamma_2 P_2^s} \tag{C}$$

Evaluation, therefore, of the relative volatility requires knowledge of:

1. vapor pressures, and
2. activity coefficients,

for the two components.

We will see in later Chapters how these quantities are determined.

1.16.7 Example 1.9

A new chemical plant is expected to have three distillation columns. In the one the boiler operates around 90°C and in the other two around 110°C.

The following discussion reflects the thinking of three members of the design team.

Monica: Let us buy a boiler that generates 3 bar steam. Its temperature is somewhat over 130°C, good enough for all three boilers. Besides, being a low pressure one, it should be cheap.

George: I agree, but I think it would be better to get a high-pressure boiler with a superheater. Expand the steam to 3 bar, so we can generate some electricity, and use the exhaust steam for the boilers. It will obviously be more expensive to buy, but the electricity will come very cheap.

Dave: You sound like the Government. You take a simple problem and make it complicated. Monica's boiler will require fuel; George's scheme will just require extra fuel to produce the electricity. Remember, however, that "a Joule, is a Joule, is a Joule", and there is no free lunch. Let us just use electricity.

Can you offer anything to this debate?

Not yet, but you should, after Chapter 3.

The Social Implications of the Industrial Revolution

In the van of the industrial revolution were the twin English towns of Manchester and Salford. Rapid industrial growth meant equally rapid population increase. In 1801 the combined population of the two towns was about 90,000; by 1831 it had grown to 230,000, and by 1851 it stood at 367,000. This enormous expansion was accompanied by severe social stress. In August 1819 the 'Peterloo Massacre' took place, when the yeomanry - roughly equivalent to the National Guard - killed eleven people and injured many more while dispersing a large crowd of demonstrators. (Cardwell, 1983, p.45)

References

Cardwell, D.S.L., 1971. *From Watt to Clausius, the Rise of Thermodynamics in the Early Industrial Age*, Cornell University Press, Ithaca, N.Y.

Cardwell, D.S.L., in *Springs of Scientific Creativity,* Aris, R., Davis, H.T., Stuewer, R.H., Eds, University of Minnesota Press, Minneapolis, Min., 1983.

Hatsopoulos, G.H., Keenan, J.H., 1965. *Principles of General Thermodynamics*, Wiley, New York.

Rifkin, J., 1980. *Entropy,* Viking Press, New York.

Problems

1.1 With reference to system and surroundings, present an example of your own where you:

a. identify system and surroundings, and

b. demonstrate if the system is in thermal and mechanical contact with the surroundings, and if it is open or closed.

1.2 Express your weight in SI and English units.

1.3 Assuming that inside von Guericke's hemispheres (insert before Section 1.5) there was an absolute vacuum:

a. calculate the total force exerted by the atmosphere;

b. which of Cardwell's characterization do you agree with, exaggeration or feeble horses?

1.4 In what areas of thermodynamics are the following properties: H; U; G; and S; important?

1.5 Discuss what you mean when using the terms:

a. temperature; and

b. heat.

1.6 What are some of the practical applications of Entropy?

1.7 Discuss the differences between: state properties, and work and heat.

1.8 Is the heat capacity:

a. a path dependent, or path independent, quantity?

b. an extensive, or intensive, property?

1.9 With reference to gas heat capacities:

a. is C_P larger or smaller than C_V?

b. can you explain, qualitatively or quantitatively, your answer?

1.10 A certain fluid undergoes the following change:

$$(P_1,V_1,T_1)\rightarrow(P_2,V_2,T_2)\rightarrow(P_3,V_3,T_3).$$

Is knowledge of the path required for the evaluation of:

a. the entropy change of the system? and

b. the amount of heat involved?

1.11 What is the difference between:

a. the kinetic and potential energy of a body, and

b. its internal kinetic and potential energies.

1.12.a. What are residual properties?

b. What is their importance?

1.13 What is - in terms of intermolecular forces - the difference between an ideal gas and an ideal solution?

1.14.a. What are excess properties?

b. What is their importance?

1.15 In discussing the concept of reversibility we concluded that a reversible operation yields the maximum possible work. Using an example of your own demonstrate that, when work is consumed, a reversible operation requires the minimum one.

1.16 What is meant by the terms: saturated steam and vapor pressure?

1.17 One kg of saturated water at 100°C is placed into a cylinder-and-piston arrangement, which is in a constant temperature bath at 100.1°C. The pressure outside the piston is equal to 1 atm. The water is vaporized slowly, absorbing heat from the bath, pushing the piston until it becomes saturated vapor.

a. How much work is produced?

b. Could more work be produced, for the same final conditions?

c. If yes, how much?

1.18 For what properties is information given at the back of the book *Properties of Gases and Liquids*? Is this experimental or estimated information?

1.19 Consider the separation of a mixture of n-hexane and n-heptane, using distillation, into products of 99.99% purity. If the pressure at the top of the column is 800 mmHg, and in the bottom 840 mmHg, estimate the temperature at the top and bottom of the column.

1.20 With reference to Example 1.5:

a. why are the calculated with the ideal gas assumption density values smaller than the experimental ones?

b. how could you get a better estimate of the steam density?

c. what is a dipole-dipole interaction?

1.21 In Example 1.8 it is stated that a key quantity in determining the height of the column is the relative volatility. What else, in your opinion, is also very important to this purpose?

1.22.a. What is a reversible operation?

b. Are real life processes reversible? Explain.

1.23.a. What is the origin of the atmospheric pressure?

b. What is the weight of the atmosphere that surrounds the earth?

1.24.a. Using the Steam Tables estimate the ideal gas heat capacity of steam at constant pressure as a function of temperature

b. Give your estimate of the accuracy of the obtained values.

1.25 Why do we define C_V as $(\partial U/\partial T)_V$ and not as $(\partial Q/\partial T)_V$?

1.26 What are the foundations of chemical engineering thermodynamics?

1.27 What is the purpose of chemical engineering thermodynamics?

1.28 For the design of the methanol plant described in Figure 1.5, what is the contribution of:

a. the laws of thermodynamics, and

b. the thermophysical data.

1.29 What are the perpetual motion machines of the first or second type? (See insert before Section 1.14.2.)

1.30 The heat of vaporization of methanol at 100°C is needed. Could its value be determined from available *PVT* data including vapor pressure as a function of temperature?

2
The Zeroth and First Laws of Thermodynamics

Law: The body of rules governing the affairs of man within a community or among states; common law: the law of nations. Often plural: Principles of conduct perceived to be of natural origin.
(Am. Heritage Dict. of English Language)

2.1 Introduction

Just as Society is built upon the foundation of its laws, so is a field of Science. There is, however, a difference: Society's laws express man's collective wisdom; Science's laws express Nature's 'wisdom'. Thus the Laws of Thermodynamics express Nature's wisdom, as of course perceived by humans, and were developed in two stages:

First, observations of a variety of physical phenomena led men to conclude that there is a 'common reason' behind all of them, dictated by Nature and not by the way they are carried out. They arrived, thus, at a verbal statement of this common reason such as, "Energy is Conserved".

Second, this verbal statement was expressed analytically so that quantitative results can be obtained. Thus, "Energy is Conserved" becomes: $Q = \Delta U + W$.

These analytical statements of the laws led eventually to the development of a 'framework' of equations that are used to solve all thermodynamic problems.

The whole process - from the collection and interpretation of the observations to the development of the framework of equations - was a very lengthy one, as we will see in Chapter 4, where we consider the historical development of the laws of thermodynamics. And this should be kept in mind as we proceed: it is not easy to comprehend and assimilate in a limited amount of time, what took hundreds of years to develop.

We will discuss the Zeroth and First Laws in this Chapter, and the Second Law in the following one. The Third Law is of more limited application; it will be stated in Chapter 4 and used in Chapter 15.

2.2 Objective and Approach

There are two main objectives in this Chapter:

First, to introduce, through the zeroth law, temperature as a *state property* rather than a physiological sense of 'hotness' and 'coldness'.

Second, to develop familiarity with the first law, its formulation and its applications.

We will start with the statement of the zeroth law, use it to establish temperature as a state property, and proceed to develop the ideal gas temperature scale.

We move then to the first law and discuss its development and analytical expression, especially the form that is convenient for steady-state flow processes that are typical in Chemical Engineering practice.

Since applications of the first law involve calculations of work and heat through evaluation of internal energy and enthalpy changes, we proceed to review briefly the fluid for which such changes are the easiest to determine: the ideal gas.

We close with some Examples and a few Concluding Remarks.

2.3 The Zeroth Law and the Ideal Gas Temperature

2.3.1 The Statement

We know from experience that a piece of wood and one of copper sitting in our back yard for a long time have the same 'hotness' as that of the atmosphere around them. Or, more generally: *If bodies A and B are in thermal equilibrium with a body C, then they are in thermal equilibrium with each other.*

Thermal equilibrium results when two bodies are brought into thermal contact through a diathermal wall, i.e. a wall that allows the flow of energy between the two bodies. The important point to notice here is that bodies A and B are in thermal equilibrium without having been brought into thermal contact with each other.

This statement, though intuitively obvious, represents a fact of Nature and is referred to as the *Zeroth Law* of thermodynamics, introduced by Fowler and Guggenheim in 1939. Its practical value is to introduce the concept of temperature as a state property rather than a sense of 'hotness' or 'coldness'.

2.3.2 Is it Really a Law?

You may find this statement 'too obvious' to be raised to the level of a Law. Indeed it has been debated in the past, as suggested by the following statement from one of the giants of thermodynamics, James Clerk Maxwell (as quoted by Hatsopoulos and Keenan, 1965):

"This law is not a truism, but expresses the fact that if a piece of iron when plunged into a vessel of water is in thermal equilibrium with the water, and if the same piece of iron, without altering its temperature, is transferred to a vessel of oil, and is found to be also in thermal equilibrium with the oil, then if the oil and water were put into the same vessel they would themselves be in thermal equilibrium, and the same would be true of any other three substances. This law, therefore, expresses much more than Euclid's axiom that 'Things which are equal to the same thing are equal to one another'".

Redlich (1968), on the other hand, expresses strong objections to this law, while Hatsopoulos and Keenan discuss exceptions to it (Problem 2.51).

2.3.3 Temperature: The Common Property of Systems in Thermal Equilibrium

Consider a large bath - large enough that its 'hotness' is not effected by the transfer to, or from it, of a small amount of energy - in which we place a cylinder-and-piston arrangement containing one mole of nitrogen. After thermal equilibrium is reached, we record the pressure (P_1) and volume (V_1) of the gas inside the cylinder.

We now move the piston so that the volume changes to V_2, and record the corresponding pressure P_2 when thermal equilibrium is reached. We repeat this process with volume values: V_3, V_4, ... and record the corresponding pressure values: P_3, P_4, ...

We repeat next the experiment replacing nitrogen with helium, and record the volume values: V_1', V_2', ... and the corresponding pressure values: P_1', P_2', ...

Now, according to the zeroth law, all the states of N_2: (P_1,V_1), (P_2,V_2), ... are in thermal equilibrium with:

* each other; and
* with those of He: (P_1',V_1'), (P_2',V_2'), ...

All these states of N_2 and He, therefore, share a common property which we call *temperature*. Notice that this property is not some abstract quantity; rather, it has the effect of determining in this case the value of pressure, once the value of the volume is specified.

We have, thus, moved - through the zeroth law - from a sense of 'hotness' to the establishment of a *state property*, the *temperature*.

2.3.4 The Ideal Gas Temperature Scale

The next problem is the assignment of numerical values to this property. Evaluation of the products: P_1V_1, P_2V_2, ... and: $P_1'V_1'$, $P_2'V_2'$, ... indicates that, at low pressures, the values are very close independently of the gases involved. More important, if the products are plotted against the corresponding pressures and then extrapolated to zero pressure, i.e. to the ideal gas state, the lines for the two gases converge to a common value,

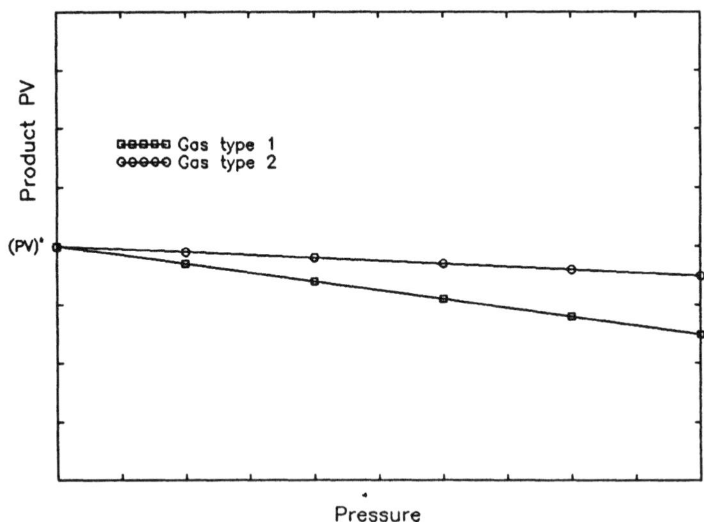

Figure 2.1 Extrapolation of the product (PV) to $P = 0$.

which we refer to as $(PV)^*$, as shown in Figure 2.1.

If the procedure is repeated using a bath of different hotness, the lines again converge to a common, but different from the previous case, value of $(PV)^*$.

We will demonstrate next how these ideal gas values, obtained by extrapolating the product (PV) to $P = 0$, are used to establish a temperature scale that, in contrast to liquid-phase thermometers (mercury, ethanol, etc.), is *independent* of the fluid used.

The procedure, which leads to the *ideal gas temperature scale* (T), involves the following two arbitrary steps:

1. The product $(PV)^*$ is a linear function of T:

$$(PV)^* = RT \tag{2.3.1}$$

2. The value of T at the triple point of water is set equal to 273.16 K.

Experimental determination of the product $(PV)^*$ for several gases at the triple point of water has resulted in an accepted value of 22711.6 bar cm^3 mol^{-1} and, consequently:

$$R = 22711.6/273.16 = 83.14 \text{ bar cm}^3 \text{ mol}^{-1} \text{ K}^{-1}.$$

To establish the value of the ideal gas temperature at some other conditions, the value of the product $(PV)^*$ is evaluated and the corresponding value of T is calculated from Eq.2.3.1 (Example 2.1).

The ideal gas temperature scale is thus established; and it is the same, as we will see in the next Chapter, with the *absolute thermodynamic* temperature scale, arrived at through the second law of thermodynamics.

Notice, finally, that the ideal gas constant R is a derived quantity from the definition of the ideal gas temperature. Values of R in different units are given in Table 2.1.

Table 2.1 Value of the Ideal Gas Constant in Different Units

R =	1.987 cal mol^{-1} K^{-1} = 1.987 Btu lb-mol^{-1} R^{-1}
=	8.314 J mol^{-1} K^{-1}
=	8314 cm^3 kPa mol^{-1} K^{-1}
=	83.14 cm^3 bar mol^{-1} K^{-1}
=	82.06 atm cm^3 mol^{-1} K^{-1}
=	10.73 psia ft^3 lb-mol^{-1} R^{-1}
=	0.7302 atm ft^3 lb-mol^{-1} R^{-1}
=	1.314 atm ft^3 lb-mol^{-1} K^{-1}

2.3.5 Example 2.1

The *P-V* data of Table 2.E.1 were obtained using steam in a piston-and-cylinder arrangement, placed in a large bath of constant 'hotness'. What is the corresponding ideal gas temperature of the bath?

Figure 2.E.1 presents a plot of (PV) versus P. By extrapolation to $P = 0$, $(PV)^* = 4764.7\ \text{cm}^3\ \text{MPa mol}^{-1}$. Hence, from Eq.2.3.1:

$$T = 4764.7\ \text{cm}^3\ \text{MPa mol}^{-1}/8.314\ \text{cm}^3\ \text{MPa mol}^{-1}\ \text{K}^{-1} = 573.1\ \text{K} = 300°\text{C}$$

Table 2.E.1 Pressure-Volume Data for Example 2.1

P (MPa)	V (cm^3/g)	(PV) (cm^3 MPa mol^{-1})
0.001	264500	4765.0
0.010	26440	4763.2
0.030	8810.8	4761.8
0.040	6606.5	4760.6
0.050	5283.9	4759.5
0.075	3520.5	4756.6
0.100	2638.7	4753.6

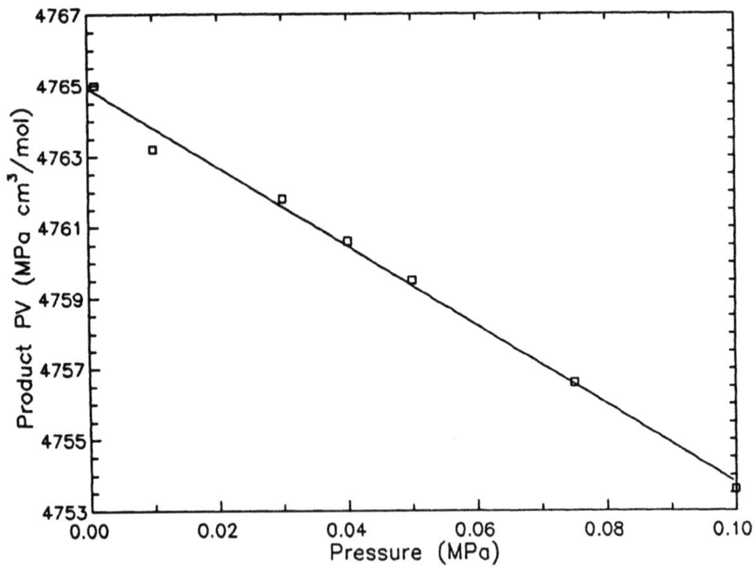

Figure 2.E.1 Extrapolation of the product (PV) for steam at a constant 'hotness'.

2.4 First Law

2.4.1 Background and Statement

The first law expresses our empirical evidence that energy is conserved and its origin lies with the experiments that Joule carried out between 1843 and 1850.

Joule used several methods, whereby mechanical work was converted into heat absorbed by water placed into a well insulated container. He determined the amount of heat produced by measuring the corresponding increase in water temperature, and concluded that - in all cases: the ratio of heat produced to work consumed was the same.

His fascinating work, extremely difficult even with today's capabilities (see Example 2.2), included (Denbigh, 1981, p.15):

* paddle-and-wheel experiments, where the work produced by a falling weight was converted into heat through the friction between paddles and water, as shown schematically in Figure 2.2;

* experiments where mechanical work was expended on an electric machine, the resultant electric current being passed through a coil immersed in the water;

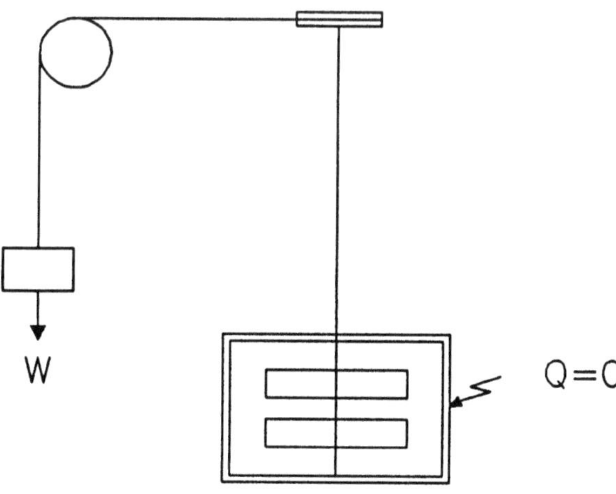

Figure 2.2 Joule's paddle-and-wheel experiment.

* experiments in which mechanical work was expended in the compression of a gas in a cylinder, immersed in the water;
* experiments in which the mechanical work was expended on two pieces of iron, which rubbed against each other beneath the surface of the water.

Joule, the Brewery Business, Religion and Deflation (!)

James, the second surviving child, had an elder and a younger brother and two younger sisters, one of whom died in childhood. He was too delicate for school, so he had private tutors instead. When he was fourteen he and his elder brother, Benjamin, were sent to learn mathematics and science under John Dalton, the doyen of Manchester science. The Joule boys studied for three years under John Dalton at the Manchester Literary and Philosophical Society, after which they assisted their father in the brewery. I doubt if the work was onerous, for young Joule was able to carry out his early scientific work there before his father built him a laboratory at home.

This looks like a rather indulged childhood and adolescence. But considering Joule's social status I doubt if it was untypical. As the family increased in wealth and security, so the sons gave up the dissenting church of their father - Dr Roby's Independent Chapel - and joined the Church of England, the established church; a transition that was fairly common among those rising in the world. By the time he was thirty-six Joule had given up work at the brewery, which in any case the family were soon to sell.

The brothers became gentlemen of leisure. Benjamin had estates in Ireland and spent his time training church choirs, playing the organ, and composing hymns and chants. John, the youngest, retired to the vacation Isle of Man, where he died comparatively young. James's career was outwardly similar to those of many other nineteenth-century British scientists who, having made or inherited a comfortable sum of money, went on to devote themselves to science. Such was the common career pattern of, for example, W. H. Perkin, David Edward Hughes, Henry Wilde and James Nasmyth. You made your hundred thousand pounds or more, invested it in sound public utilities, railroad stock, or government securities to yield, say, three per cent, and retired to a comfortable life of science in the secure knowledge that things were getting better and cheaper all the time. (Cardwell, 1983, pp.45-47.)

As a result of Joule's work, the first law was developed and can be stated as follows: *Even though energy can be converted from one form to an other, the total amount is conserved.*

We will discuss Joule's contributions in detail in Chapter 4. It should be mentioned here, however, that by demonstrating the convertibility of work to heat, he tore down the long held axiom that heat, or 'caloric' as it was also called then, is conserved.

According to this axiom, caloric consisted of very small particles and was, therefore, neither created nor annihilated. Thus, the steam engine operated through the 'flow' of caloric, not by conversion of caloric to work. This is why there was such an opposition to the conversion of heat to work as demonstrated by Mayer's tragedy.

J. R. Mayer, the Establishment, and a Tragedy in Thermodynamics.

Actually, just before Joule, J. R. Mayer suggested - but did not clearly demonstrate - the convertibility of heat to work. But so strong was the conviction about the conservation of heat in that period, that Mayer - who also had the misfortune of not being part of the then scientific establishment - was ignored, even scorned, leading to his mental collapse. He finally achieved the recognition he deserved, due to Clausius, recovered from his breakdown and was voted into the Royal Society in 1862. (See Chapter 4.)

2.4.2 The Analytical Expression

To obtain an analytical expression for the first law, we will accept the following convention:

1. Heat added to a system is positive;
2. Work done on a system is negative.

Let us now consider a closed system that exchanges an amount of heat Q with the surroundings, but no work. The resulting change in the internal energy of the system U, as it goes from the initial state 1 to the final state 2, is:

$$\Delta U = U_2 - U_1$$

Notice that if Q is a positive number - i.e. heat is added to a system - we would expect that its internal energy will increase; and if negative, that it will decrease. According, therefore, to the first law statement and the accepted convention:

$$\Delta U = Q \tag{2.4.1}$$

If now an amount of work W is exchanged between system and surroundings, but no heat exchange is involved, then following again the first law statement and the convention:

$$\Delta U = -W \tag{2.4.2}$$

In this case, a positive value for W indicates that the system does work on the surroundings consuming its internal energy which, consequently, decreases; and a negative one, the reverse, both in agreement with Eq. 2.4.2.

A combination of Eqs 2.4.1 and 2.4.2 gives the analytical statement of the first law:

$$Q = \Delta U + W \tag{2.4.3}$$

2.4.3 Comments

1.When small - infinitesimal - amounts of heat and work are involved, Eq.2.4.3 becomes:

$$dQ = dU + dW \tag{2.4.4}$$

Remember, however, that dQ and dW refer to small amounts, not small changes, as is the case with the internal energy (Section 1.11).

2.The term W contains all types of work: pressure-volume (P-V), i.e. work of compression or expansion; and kinetic KE and potential PE energy changes.

3.In thermodynamics, however, the term W is reserved for P-V work only. Hence Eq.2.4.3 becomes:

$$Q = \Delta U + W + \Delta(KE) + \Delta(PE) \tag{2.4.5}$$

where W refers to P-V work only.

4.In case of nuclear reactions, where conversion of mass to energy takes place according to Einstein's expression, then the sum: mass plus energy is conserved.

Two simple applications of the first law are presented in Examples 2.2 and 2.3.

2.4.4 Example 2.2

To get some idea about the experimental difficulties, and required accuracy, in the temperature measurements of Joule's work, let us assume that in the paddle-and-wheel experiment (Figure 2.2):

* the fall of a body with 1 kg mass for 10 meters was used, and
* 1 kg of water was stirred by the paddle.

Assuming: $C_V = 4.184$ kJ/kg K, calculate the corresponding temperature increase. Use $g = 9.8067 \text{ m sec}^{-2}$.

From the first law: $\Delta U = -\Delta(PE)$. The potential energy change is:

$$\Delta(PE) = (1 \text{ kg})(-10 \text{ m})(9.8067 \text{ m sec}^{-2}) = -98.067 \text{ J}$$

and from the definition of the heat capacity (Eq.1.13.1): $\Delta U = C_V \Delta t$. Thus: $\Delta t = 0.023°C$, a very small temperature change. (See also Problem 2.33.)

Indeed, Joule had an excellent instrument maker for whom Cardwell (1971) writes: "John Benjamin Dancer, of Cross Street in Manchester, the inventor of micro-photography. I can confirm from personal experience, the superb quality of Dancer's instruments".

2.4.5 Example 2.3

Steam at 125°C and 1 atm, state 1, is placed into a well insulated cylinder-and-piston arrangement and heated to 200°C, state 2, with electricity. How much heat is absorbed per kg of steam, if the pressure outside the piston is that of the atmosphere and equal to 1 atm?

From the first law:

$$Q = \Delta U + \int_{V_1}^{V_2} P\,dV = (U_2 - U_1) + (V_2 - V_1)(P_1) \tag{A}$$

i.e. the absorbed heat not only increases the internal energy of the steam, but also does work against the atmosphere.

We turn now to the Steam Tables to obtain values for U_1, U_2, V_1, and V_2:

$U_1 = 2544.7$,
$U_2 = 2658.1$, both in kJ/kg
$V_1 = 1792.7$,
$V_2 = 2143.8$, both in cm³/gr

Thus:

$U_2 - U_1 = 2658.1 - 2544.7 = 113.4$ kJ/kg
$(V_2 - V_1)(P) = (2143.8 - 1792.7)(1) =$
$= 351.1 \text{ atm cm}^3 \text{ gr}^{-1} \times 8.314 \text{ J} / 82.06 \text{ atm cm}^3 = 35.6 \text{ J/gr}$

and: $Q = 113.4 + 35.6 = 149.0$ kJ.

The calculation can be much simplified by noticing that, since the pressure remains constant, we can write Eq.(A) as:

$$Q = U_2 - U_1 + P_2V_2 - P_1V_1 = H_2 - H_1$$

Q can be thus calculated by using the enthalpy values only:

$$Q = H_2 - H_1 = 2875.3 - 2726.4 = 148.9 \text{ kJ}.$$

2.5 The First Law Expression for a Steady-State Flow Process

2.5.1 The Problem

Application of the first law to non-flow, constant mass, systems can be carried out directly by using Eq.2.4.5, as demonstrated in Examples 2.2 and 2.3. Most chemical engineering processes, however, are of the steady-state flow type. Eq.2.4.5 applies to them of course but, as we will see next, it can be recast into a more convenient for this purpose form.

2.5.2 Development of the Expression

Consider the general steady-state flow process shown in Figure 2.3 and apply to it the first law, Eq.2.4.5, on the basis of one mole of fluid entering the system in point 1. In evaluating the *P-V* work term we notice that it involves three parts:

1. the work done by the surroundings in pushing the fluid into the system, W_1.
2. the work of expansion or compression, W_s, which is called *shaft* work, because it is communicated through a shaft.
3. the work done by the system in pushing the fluid out of the system into the surroundings, W_2.

From the definition of work:

$$W_1 = \int P dV = P_1 \Delta V$$

since the pressure at this point is constant and equal to P_1.

But what is ΔV? Simply, the molar volume of the fluid at point 1, since the pressure P_1 must push one mole of the fluid into the system. Hence:

$$W_1 = -P_1 V_1 \text{ (Work done on the system)}$$

Similarly:

$$W_2 = P_2 V_2 \text{ (Work done by the system)}$$

Thus, the total work W is:

$$W = W_s - P_1 V_1 + P_2 V_2 \tag{2.5.1}$$

Introduction of Eq.2.5.1 into Eq.2.4.5 yields:

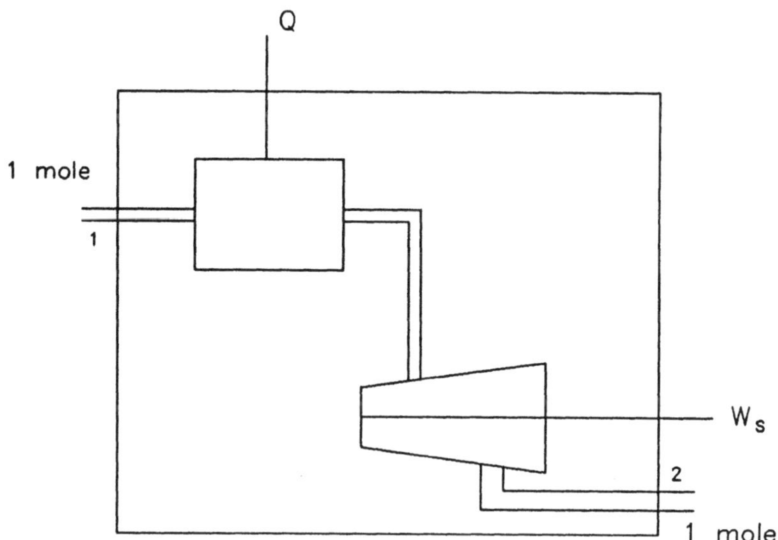

Figure 2.3 Schematic diagram for the development of the first law expression for a steady-state flow process.
System: inside the box; Surroundings: outside the box.

$$Q = (U_2 - U_1) + (W_s - P_1V_1 + P_2V_2) + \Delta(KE) + \Delta(PE)$$

or since: $(U_2 + P_2V_2) - (U_1 + P_1V_1) = H_2 - H_1 = \Delta H$

$$Q = \Delta H + W_s + \Delta(KE) + \Delta(PE) \qquad (2.5.2)$$

The convenience of using Eq.2.5.2, as against Eq.2.4.5, for a flow process is apparent. And since, as mentioned, most processes are of the flow type, enthalpy has become a very commonly used term in thermodynamics.

2.5.3 Comments

1.We will refer to *steady-state flow* processes as just *flow* processes.
2.The potential and kinetic energy changes are usually negligible as compared to the other terms.
3.The importance of Eqs 2.4.5 and 2.5.2 lies with the observation that the path depending quantities, Q and W, are related to changes in the state functions U and H.
4.Determination of internal energy and enthalpy changes - from experimental PVT and heat capacity data - is discussed in Chapters 9 and 11.

5. We turn therefore to the simplest case for the calculation of such changes, that of an ideal gas, which is considered next, and then proceed to apply the first law in Examples 2.4 through 2.7.

2.6 Thermodynamic Relationships of the Ideal Gas

2.6.1 Importance of Ideal Gases

As mentioned in Section 1.13.2 an ideal gas is a hypothetical fluid since it represents the behavior of a real gas at zero pressure. We did use this hypothetical fluid, however, to establish the ideal gas temperature through the process of extrapolation to $P = 0$ of the behavior of real gases.

The ideal gas concept is not just useful for the aforementioned purpose. It is a very valuable tool in thermodynamics, because:

1. With very few exceptions ideal gas behavior can be assumed at pressures up to 1-2 bar and can be thus employed for quick estimates with reasonable accuracy (Example 1.5). (For accurate results, of course, experimental data should be used. If not available, more reliable - than ideal gas behavior - estimation techniques that will be discussed later, should be considered.)
2. Ideal gas properties are easier to determine than those of real fluids. By focusing, thus, attention to the difference in the properties of these two types of fluids (Section 1.13.3), we can develop a better understanding of the factors leading to real fluid behavior; and, estimation techniques for the latter (Chapters 8, 9, and 11).

In this Chapter, we will take advantage of this easier determination of ideal gas properties by using them in applications of the first law. We can thus enhance our understanding of this law without being concerned with real fluid behavior.

We review, therefore, next some relationships of ideal gases.

2.6.2 Equation of State

The term equation of state means the functional relationship among pressure, volume, and temperature, which for an ideal gas has the form:

$$PV = RT \qquad (2.6.1)$$

V is the molar volume, and R is the ideal gas constant, whose value depends on the units used for the other quantities in the equation (Table 2.1).

2.6.3 Internal Energy and Enthalpy Changes Due to Heating or Cooling

From the first law for a constant volume process involving one mole:

$$dQ = dU$$

and from the definition of heat capacity at constant volume (Eq. 1.13.1):

$$dU = C_V^* \, dT \tag{2.6.2}$$

where the superscript (*) indicates ideal gas behavior.

Again from the first law, but this time for a constant pressure reversible process involving one mole:

$$dQ = dU + PdV = dU + d(PV) = dH$$

and from the definition of heat capacity at constant pressure (Eq. 1.13.1):

$$dH = C_P^* \, dT \tag{2.6.3}$$

2.6.4 The Difference in Heat Capacities

From the definition of enthalpy applied to one mole of an ideal gas:

$$dH = dU + d(PV) = dU + RdT$$

Introduction of Eqs 2.6.3 and 2.6.2 into this expression yields:

$$C_P^* - C_V^* = R \tag{2.6.4}$$

2.6.5 The *P-V-T* Relationship for an Adiabatic and Reversible Change

Consider one mole of an ideal gas that undergoes an adiabatic and reversible change from the initial condition: P_1, V_1, and T_1, to a final pressure P_2. What are the corresponding values of T_2 and V_2?

From the first law for an adiabatic process: $dW = -dU$. Since the process is reversible, $dW = PdV$, which for an ideal gas becomes:

$$dW = PdV = RdT - VdP = RdT - \frac{RT}{P}dP$$

From Eq.2.6.2: $$dU = C_V^* \, dT$$

Hence: $$-C_V^* dT = RdT - \frac{RT}{P}dP$$

or: $$C_P^* \, d\ln T = R \, d\ln P$$

Integration from T_1 and P_1, to T_2 and P_2, assuming C_P^* constant, yields:

$$\ln\frac{T_2}{T_1} = \frac{R}{C_P^*}\ln\frac{P_2}{P_1}$$

After rearrangement:

$$T_2 = T_1\left(\frac{P_2}{P_1}\right)^k \tag{2.6.5}$$

where: $k = \dfrac{\gamma - 1}{\gamma}$, $\gamma = \dfrac{C_P^*}{C_V^*}$

For the final volume V_2, it can be easily seen that:

$$P_2 V_2^\gamma = P_1 V_1^\gamma$$

which also suggests that the product (PV^γ) remains constant during the process:

$$PV^\gamma = C \tag{2.6.6}$$

where, $C = P_1 V_1^\gamma$ = constant.

2.6.6 Comments

1. The ideal gas heat capacity is a function of temperature only, which becomes more pronounced with increasing number of atoms in the molecule as shown in Figure 2.4.

2. The internal energy and enthalpy of an ideal gas are, therefore - according to Eqs 2.6.2 and 2.6.3 respectively - also functions of temperature only.

3. Information on C_P^*, expressed as:

$$C_P^* = a + bT + cT^2 + dT^3 \tag{2.6.7}$$

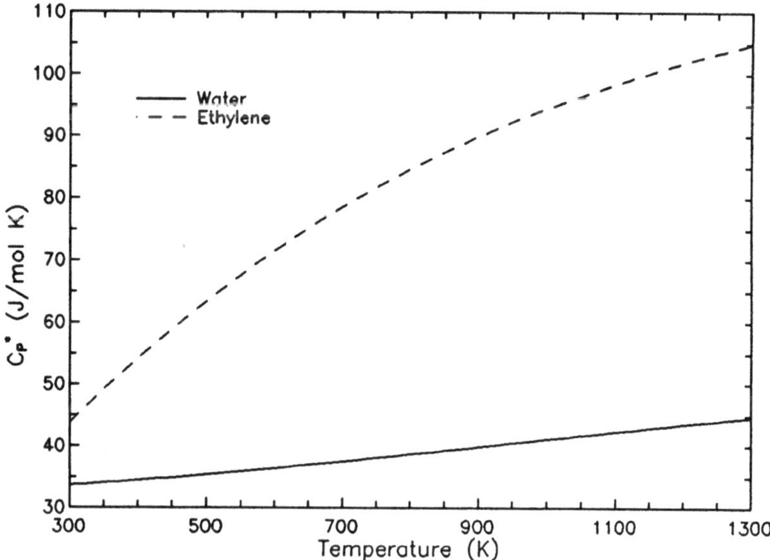

Figure 2.4 Ideal gas heat capacities.

is given, among others, by Reid et al (1987) for a variety of gases, and parameter values for some selected ones are presented in Appendix C. C_P^* information is also given by Daubert and Danner (1985; 1986).

4.Eqs 2.6.5 and 2.6.6 apply to ideal gases of constant heat capacity undergoing reversible and adiabatic changes *only*.

2.7 Examples 2.4 through 2.7

Some simple applications of the first law are presented in the next Examples, which consider:

* Example 2.4, the uncertainties involved in assuming ideal gas behavior;
* Example 2.5, an irreversible case;
* Example 2.6, the feasibility of a process; and
* Example 2.7, the efficiency of water-wheels.

2.7.1 Example 2.4

Steam is heated from $t_1 = 200°C$ to $t_2 = 550°C$ in a constant pressure flow process. Assuming ideal gas behavior, calculate the amount of heat required and compare it to the actual amount as a function of pressure from 100 to 1000 kPa.

From the first law for a flow process, neglecting changes in *PE* and *KE*, and using $W_s = 0$, since no shaft work is involved:

$$Q = H_2 - H_1 = \int_{T_1}^{T_2} C_P^* \, dT$$

where, from Appendix C:

$$C_p^* = a + bT + cT^2 + dT^3, \quad T\text{: K; } C_p^*\text{: J mol}^{-1}\text{ K}^{-1}$$

$$a = 32.24;\ b = 1.908 \cdot 10^{-3};\ c = 1.057 \cdot 10^{-5};\ d = -3.602 \cdot 10^{-9}$$

Thus:

$$Q = a\,(T_2 - T_1) + (b/2)(T_2^2 - T_1^2) + (c/3)(T_2^3 - T_1^3) + (d/4)(T_2^4 - T_1^4)$$

$$= 12940 \text{ J mol}^{-1} = 718 \text{ kJ/kg}$$

We calculate, next, the actual Q values as a function of the operating pressure obtaining the H_2 and H_1 values from the Steam Tables. The results are presented in Table 2.E.4 and indicate that reasonable accuracy is obtained with the ideal gas assumption up to about 4 bar.

2.7.2 Example 2.5

One mole of oxygen is compressed in a flow process from state 1 (1 atm and 25°C) to state 2 (6 atm). The compression is adiabatic but not reversible, with an efficiency of 0.75. Assuming oxygen to behave as an ideal gas under these

Table 2.E.4 Actual Q Values as a Function of Pressure, Example 2.4

P (kPa)	H_1 (kJ/kg)	H_2 (kJ/kg)	Q (kJ/kg)	% error
100	2875.4	3595.6	720.2	-0.2
200	2870.5	3594.7	724.2	-0.8
300	2865.5	3593.7	728.2	-1.3
400	2860.4	3592.8	732.4	-1.9
600	2849.7	3590.9	741.2	-3.1
800	2838.6	3589.0	750.4	-4.3
1000	2826.8	3587.1	760.3	-5.5

conditions, calculate the:

a. work requirement;

b. discharge temperature, T_d; and

c. amount of heat that must be removed to cool the gas back to 25°C at the pressure of 6 atm.

Comment on the accuracy of the obtained results.

Given: $C_p^* = 26.625 + 0.0049\ T$ (J mol^{-1} K^{-1}); T, in K, and: $\gamma = 1.41$ (assumed constant).

a. We calculate first the reversible and adiabatic work $W(RA)$, and then calculate the real work, $W(real)$. From the first law:

$$-W(RA) = \Delta H = (1) \int C_p^* dT = 26.625(T_2 - T_1) + (0.0049/2)(T_2^2 - T_1^2)$$

To evaluate T_2, notice that for a reversible and adiabatic process (Eq.2.6.5):

$$T_2 = T_1\ (P_2/P_1)^k = 502.0\ \text{K}$$

Hence: $W(RA) = -5827.2$ J and, since work is consumed, from Eq.1.12.3:

$$W(real) = W(RA)/0.75 = -7769.6\ \text{J}$$

b. From the first law:

$\Delta H(real) = -W(real)$. Thus:

$$26.625\ (T_d - T_1) + (0.0049/2)(T_d^2 - T_1^2) = 7769.6\ \text{J}$$

Solution of this quadratic equation gives $T_d = 568.4$ K.

Notice that the extra work, over the $W(RA)$ one, is converted into enthalpy of the gas. This leads to the larger value for the discharge temperature T_d, as compared to 502.0 K for the reversible case.

c. From the first law:

$$Q = \Delta H = H(P_2, T_d) - H(P_2, T_1)$$

For an ideal gas, however: $H(P_2, T_1) = H(P_1, T_1)$ and consequently all the energy imparted into the gas through the compression must be removed:

$$Q = W(real) = -7769.6\ \text{J}$$

<u>Comment:</u>

Two assumptions are made here:

* ideal gas; and
* γ is temperature independent.

To evaluate the impact of these two assumptions exactly, we need actual experimental data for oxygen. In their absence, the following observations can be made:

1. Considering the temperature and pressure range involved, the error due to the ideal gas assumption should not be more than 5%.
2. Evaluation of γ at 298 K and 568 K gives values of 1.421 and 1.394 respectively, with an average value of 1.408, close to the one used. The answer, hence, should be within 5% of the correct one.

2.7.3 Example 2.6

Consider the following device proposed by an inventor: Air enters his device at atmospheric conditions, say: $T_0 = 300$ K and $P_0 = 1$ atm, and two streams, 1 and 2, of equal amounts are emerging: stream 1 at $T_1 = 280$ K, and stream 2 at $T_2 = 320$ K, both at 1 atm. No exchange of heat or work with the surroundings is involved.

Is this device possible? (Assume that due to the small temperature range involved, the air heat capacity is temperature independent.)

To this purpose we examine if the first law, applied to this device, is satisfied. Using as a basis the flow of one mole of air entering the device, and since $Q = W_s = 0$:

$$\Delta H = (0.5\, H_2 + 0.5\, H_1) - (1)H_0 = 0 \ (?)$$

or:

$$0.5(H_1 - H_0) + 0.5(H_2 - H_0) = 0 \ (?)$$

or:

$$0.5\, C_P\, [T_1 - T_0] + 0.5\, C_P\, [T_2 - T_0] = 0 \ (?)$$

which is apparently valid and, consequently, the device is possible according to the first law. (We will see in the next Chapter if it also satisfies the second law.)

2.7.4 Example 2.7. The Efficiency of Water-Wheels

We have considered, so far, systems where potential and kinetic energy changes have been negligible. A prime example where these are important are the water-wheels. These devices were used extensively as the motive power for the textile industry, one of the early accomplishments of the Industrial Revolution, and were later supplemented by steam engines. To this purpose consider a stream of water that:

* is available at an elevation h_1 with a velocity u_1;
* enters the water-wheel at an elevation h_1' with a velocity u_1';
* leaves at an elevation h_2' with a velocity u_2'; and
* hits the ground at an elevation h_2 with velocity u_2;

as shown in Figure 2.E.7.

Identify the conditions under which the maximum work is obtained.

From the first law applied to the flow of one mole of water:

$$Q = W_s + \Delta H + \Delta(PE) + \Delta(KE)$$

Assuming no exchange of heat with the atmosphere, $Q = 0$:

$$\begin{aligned} W_s &= -\Delta H - \Delta(PE) - \Delta(KE) \\ &= C\,(T_1 - T_2) + g(h_1' - h_2') + 0.5(u_1'^2 - u_2'^2) \end{aligned}$$

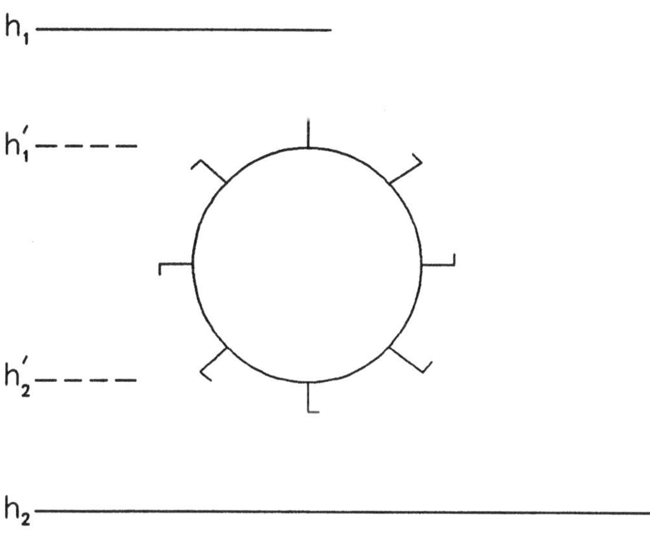

Figure 2.E.7 The water-wheel for Example 2.7.

where C is the water heat capacity, and T_1 and T_2 the entrance and exit temperatures respectively. To obtain, therefore, the maximum work:

1.$(T_2 - T_1) = 0$, i.e. no heat must be generated through water turbulence and, consequently, $u_1' = u_1$; $u_2' = u_2$.

2.$(h_1' - h_2') = (h_1 - h_2)$, i.e. the whole height of fall should be utilized.

Notice that, as expected, these requirements for obtaining the maximum work are those for a reversible operation:

$$u_1 - u_1' = 0; \qquad h_1 - h_1' = 0; \qquad h_2 - h_2' = 0; \qquad u_2 - u_2' = 0.$$

Comments:

1.To produce the maximum possible amount of work, the exit velocity u_2 must be set equal to zero, requiring a wheel moving with an infinitesimally small velocity. This, in turn, requires that $u_1 = 0$ (why?), i.e. that all kinetic energy is converted to potential, before the water enters the wheel.

2.The conditions for obtaining the maximum work from water-wheels were identified by the 17th century (see Cardwell, 1971). It was, thus, possible to determine exactly their efficiency.

3.This, however, was not the case with the efficiency of the steam engine. More than hundred years after its introduction, and even though it had become the dominant source of power, the efficiency of the steam engine was still measured in terms of "work produced per bushel of good coal(!)". They did not know what the maximum obtainable work was.

4.We will see in the next two Chapters that Carnot, whose father was a hydraulic engineer, not only was aware of the requirement that for maximum efficiency all the fall of water in a water-wheel must be utilized, but that he used it in establishing the requirement for obtaining the maximum work from a steam engine.

5.We will also see how Clapeyron, using Carnot's brilliant approach, arrived at an analytical expression for the determination of the efficiency of a steam engine; and, that this eventually led to the formulation of the second law of thermodynamics.

2.8 Concluding Remarks

1.The zeroth law establishes temperature as a state property rather than a physiological sense of 'hotness' or 'coldness'.

2.Use of the ideal gas behavior exhibited by real gases in the limit of zero pressure leads to the establishment of the ideal gas temperature scale.

3.Application of the first law to a given process provides for the calculation of the important - from the chemical engineering point of view - heat and work requirements through changes in the state variables of the system involved: internal energy or enthalpy.

4.If the process involves gases, use of the ideal gas assumption for such calculations allows for reasonable approximations up to moderate pressures (2-5 bar) in most cases.

References

Cardwell, D.S.L., 1971. *From Watt to Clausius, the Rise of Thermodynamics in the Early Industrial Age*, Cornell University Press, Ithaca, N.Y.

Ibid., 1983. In *Springs of Scientific Creativity*, Aris, R., Davis, H.T., Stuewer, R.H., Eds, University of Minnesota Press, Minneapolis, MN.

Daubert, T.E., Danner, R.P, 1985; 1986. *Data Compilation Tables of Properties of Pure Compounds*, AIChE, New York.

Denbigh, K., 1981. *The Principles of Chemical Equilibrium*, Cambridge University Press, Cambridge.

Hatsopoulos, G.N., Keenan, J.H., 1965. *Principles of General Thermodynamics*, Wiley, New York.

Redlich, O., 1968. *Reviews Modern Phys.*, **40**, 556.

Reid, R.C., Prausnitz, J.M., Polling, B., 1987. *The Properties of Gases and Liquids*, 4th Ed., McGraw-Hill, New York.

Problems

2.1 Propane at 2 atm is heated from 300 K to 400 K:
a. in a constant pressure reversible process;
b. at constant volume; and
c. in a constant pressure flow process.
How much heat is required per mole in each case?
2.2 Why do we write the analytical statement of the first law as: $Q = W + \Delta U$, and not as $Q = W - \Delta U$?
2.3 What is the difference between W and W_s?
2.4 Does the expression: $Q = \Delta U + W + \Delta(KE) + \Delta(PE)$ apply to a flow process? Explain your answer.
2.5 Starting with Eq.2.4.5, develop the expression of the first law for a flow process.
2.6.a. What is the enthalpy of an ideal gas a function of? Justify your answer.
b. Why is enthalpy important in chemical engineering calculations?
2.7 It has been suggested that the kitchen in your house could be cooled in the summer by closing it from the rest of the house and opening the door of the electric refrigerator. Show if this is feasible. Make sure you specify system and surroundings, and use data of your choice to demonstrate your conclusion.
2.8 An ideal gas flows through a horizontal tube which has nonconducting walls. No heat is added, or removed, nor is any shaft work involved. The cross-sectional area of the tube changes with length and this causes the velocity to change. Develop an expression relating the temperature and velocity of the gas at any point of the tube to their entrance values.
2.9 A gas at a pressure of 5 atm and a temperature of 300 K flows through a well insulated expansion valve with an exit pressure of 1 atm (expansion through a valve is referred to as 'throttling').
a. Calculate the exit temperature, if the behavior of the gas can be approximated by that of an ideal one.
b. Will the same answer hold for a real gas? Explain.
Neglect kinetic energy effects.
2.10.a. What are the assumptions behind ideal gas behavior?
b. Why are these assumptions failing as the density increases?
2.11 What makes ideal gases important in chemical engineering calculations?
2.12 A certain gas is heated from 100°C to 200°C at a constant pressure of 1 atm in a flow process involving no shaft work. How much heat is involved per mole of gas? In this temperature range and for this gas:
$U = U_0 + 28(T - 373)$ J/mol; T,K.
2.13 Methane goes from 273 K and 10 bar to 350 K and 1 bar.
a. Devise a non-flow reversible process - any number of steps - to accomplish this change;
b. Calculate the enthalpy change per mole of methane.

2.14 A gas at a pressure of 3 atm is contained in a vessel of 5 l. The vessel is connected through a stop cock with the atmosphere. The stop cock is opened and the gas leaks slowly so that the temperature of the vessel remains constant and equal to that of the atmosphere:

a. how much work is produced?

b. how much heat is exchanged with the atmosphere?

2.15 Ethylene is compressed from $V_1 = 20$ l, $P_1 = 1$ bar and $T_1 = 273$ K to $V_2 = 5$ l and $P_2 = 6$ bar. Calculate any of the following quantities that you have sufficient information for:

a. heat requirement;

b. work requirement;

c. internal energy change;

d. enthalpy change.

2.16 Carbon dioxide at 40 psia and 100°F and at a rate of 400 ft^3/min is compressed to 200 psia and 300°F. The compressor cooling jacket removes heat at a rate of 15 Btu/lbm. What is the compressor's horse power?

2.17 In a steady flow process water at 20°C and saturated steam at 2 bar are mixed to produce 50 kg/min of saturated water at 1.5 bar. In the mixing operation heat is lost at the rate 20 kJ/kg of saturated water produced. What is the flow rate of water entering the mixer?

2.18 One mole of air at 40°C and 10 bar is in a vertical cylinder confined by a piston held with latches. The weight of the piston and the pressure of the atmosphere are equivalent to a downward pressure of 2 bar. The latches are suddenly removed and the gas expands to twice its original volume, whereupon the piston hits a stop. The system consisting of the cylinder, piston and gas, is allowed to return to its initial temperature of 40°C. How much heat is transferred to, or from, the surroundings?

2.19 A steel tank whose volume is 3 m^3 contains 23 kg of steam and the rest is water at 200°C. The contents of the tank are heated to 250°C. How much heat is added to this purpose?

2.20 One mole of nitrogen undergoes the following series of reversible changes in a cylinder: Isothermal expansion from 10 bar and 90°C to 4 bar, constant volume cooling, and then adiabatic compression to its initial state. Find Q and W for each step and for the entire cycle.

2.21 Consider two well insulated tanks A and B with volumes: $V_A = 20$ l and $V_B = 50$ l. Tank A contains N_2 at 10 bar and 60°C and tank B methane at 3 bar and 20°C, and are connected through a valve. The valve is opened quickly, allowing the two pressures to become equal, and then closed. Determine:

a. the common pressure; and

b. the temperature and number of moles in each tank.

2.22 One kg of steam at 180°C and having a volume of 0.2 m^3 is heated at constant volume to 250°C:

a. what is the final pressure of the steam?

b. how much heat has been absorbed?

2.23 A centrifugal pump delivers 600 gal/min of water at 80°F from a pressure of 1 atm to a pressure of 3 atm. The diameter of suction inlet to the pump is 6 in and of the outlet 4 in. If the suction and discharge are at the same level, and the pump has an efficiency of 0.7:

a. what is the required work?

b. if there is no change in the internal energy of the water, how much heat is exchanged with the surroundings?

Hint. Show first that for a reversible pump: $W_s = \int_{P_1}^{P_2} V\,dP$.

2.24 A very large reservoir, containing nitrogen at 12 bar and an unknown temperature T, is connected through a valve with a small tank, whose volume is 0.2 m^3 and contains initially 0.2 kg of nitrogen at 27°C. The valve opens and the gas flows into the tank until the pressure in it becomes 3 bar and the temperature 140°C. Calculate the value of T assuming, for simplicity, ideal gas behavior with $C_P^* = 29.3$ J/mol K.

2.25 A well insulated vessel is divided into two parts. The left contains 1 kg of saturated steam at 1 bar. The right, 2 kg of steam at 6 bar and 300°C. What will be the temperature and enthalpy of the steam if the divider is removed?

2.26 One mole of an ideal gas goes from 1 bar and 20°C to 10 bar and 100 °C. Develop two paths to accomplish this change, and:

a. calculate, for each path, the amounts of heat Q and work W involved, and the changes in internal energy and enthalpy, ΔU and ΔH;

b. compare Q, W, ΔU, and ΔH for the two paths, and comment on their values.

Assume $C_P^* = 7$ cal/mol.

2.27 A gas is heated from T_1 to T_2 at a constant pressure P. If we assume ideal gas behavior to calculate the amount of heat required, what factors will influence the accuracy of the obtained results?

2.28 Starting with the first law, develop Eq.2.6.5.

2.29 One mole of propane at 300 K is compressed from 0.1 MPa to 1 MPa in a batch process. If the compression is adiabatic but irreversible with an efficiency of 0.8:

a. how much work is required?

b. what is the temperature of the compressed gas?

c. how much heat must be removed to cool the gas to its initial temperature?

State any assumptions you make and estimate their impact on the accuracy of the obtained results.

2.30 Repeat the previous problem, assuming a heat loss of 3 J/mol during the compression.

2.31 Air enters an apparatus at 1 atm and 30°C and two streams, both at 1 bar emerge: one containing 40% of the inlet air at 0°C and the other at 45°C. If the apparatus does not exchange any energy with the surroundings, do you see any uncertainty in the data?

2.32 Assume that in Example 2.5 the compression is done with a two-stage compressor with intercooling: the gas leaves the first stage at an intermediate pressure P_i, is cooled to the entrance temperature of 25°C and then is compressed to the final pressure of 6 atm. Using the same assumptions as in Example 2.5:

a. determine the value of P_i so that the total work consumed is the minimum;

b. explain the difference between this work and that consumed in Example 2.5;

c. determine the heat removed in the intercooler.

2.33 In Example 2.2 we assumed for simplicity that the heating of the water takes place at constant volume, while in reality it takes place at constant pressure. What is the actual temperature change assuming that the water is initially at 25°C?

2.34 The following data for propane were measured in a bath of constant 'hotness':

Density (kg/m^3)	5	10	15	20	25	30	35
Pressure (MPa)	0.2896	0.5571	0.8027	1.0268	1.2295	1.4112	1.5721

What is the corresponding ideal gas temperature?

2.35.a. What are in your opinion the important items you have learned in this Chapter?

b. Give a practical application in each case.

2.51 Discuss exceptions to the zeroth law.

2.52.a. Demonstrate that the introduction of temperature through the zeroth law can be made without the prior knowledge of heat.

b. Proceed then to introduce heat through the first law.

(See Denbigh.)

3
The Second Law of Thermodynamics

A good many times I have been present at gatherings of people who by standards of the traditional culture are thought highly educated and who have with considerable gusto been expressing their incredulity at the illiteracy of scientists. Once or twice I have been provoked and have asked the company how many of them could describe the Second Law of Thermodynamics. The respond was cold: it was also negative. Yet I was asking something which is the scientific equivalent of: Have you read a work of Shakespeare's? (C.P.Snow: 'The Two Cultures and the Scientific Revolution'.)

3.1 Introduction

The origin of the Second Law lies with Sadi Carnot's work: *Reflexions sur la Puissance Motrice du Feu*, published in 1824, dealing with the establishment of the maximum amount of work that can be obtained from a steam engine. Thus, its efficiency could be rigorously expressed, instead of the method used then: "work ($ftlb_f$) per bushel of good coal(!)".

Carnot used in his development the erroneous - but strongly held then - axiom of the *conservation of caloric* (he did have, however, some doubts about it later, but never articulated them). He died young, of cholera, and his work laid dormant until it was revived a few years later by Clapeyron.

In the meantime, the advent of Joule's work in the 1840's - showing that caloric was not conserved - led Kelvin, an advocate of the Caloric theory, to declare in 1849 that Carnot and Joule could not be both right.

Clausius and Kelvin, who later abandoned the caloric theory, eventually reconciled the work of these two giants of thermodynamics and established the first and second laws in the 1850's and 1860's. (We will discuss the fascinating historical development of these two laws in the next Chapter.)

3.2 Rationale for a Second Law

Let us now return to Carnot's work and consider the following question: Is the maximum work produced by the steam engine sufficient reason for a chemical engineer to get involved with the second law and, especially, with the associated concept of entropy that seems to haunt anyone dealing with thermodynamics?

While the development of the second law started with Carnot's work on the efficiency of the steam engine, the law covers a broad spectrum of phenomena. Let us consider some of them.

3.2.1 The Direction of Natural Phenomena

We all know from experience that:

* if a pipe carrying steam is not insulated, heat flows from the steam to the environment, and not from the latter to the former;
* if the valve separating two gases - one at low and the other at high pressure - is opened, the flow occurs from the high pressure to the low one, and not in the reverse direction;
* once the wall that separates two pure gases in a vessel is removed, the gases will be soon completely mixed; but if we start with a mixture of these gases, they will not separate by themselves.

Nature, thus, seems to dictate a certain direction to processes: heat flow from a hot body to a cold one; gas flow from high to low pressure; mixing rather than unmixing. *Yet, the first law does not object to the reversing of any of these processes.*

3.2.2 The Feasibility of Processes

But these observations, and other similar ones, could still hardly suggest the need for a second law. One could argue that experience suffices to predict these phenomena.

Let us, however, consider the following device proposed by an inventor (and discussed in Example 2.6): Air enters his device at atmospheric pressure and temperature, say 300 K, and leaves in two streams that have the same flow rate: one at 320 K and the other at 280 K, both being at atmospheric pressure. The device requires no external energy, thermal or mechanical, and it would obviously be of immense practical use in providing cooling during the summer and heating during the winter. Is this device possible?

We concluded, in Example 2.6, that the device does not violate the first law: the enthalpy lost by the cold stream can be found in the hot one. Notice, also, that the device does not involve flow of heat from a cold to a hot body; or gas flow from a low pressure to a high one; or any unmixing of gases.

It should, therefore, be feasible. Yet, our intuition suggests that this may not be the case.

3.2.3 The Big Question: Where is the Energy Gone?

Let us next raise the following very important question: If energy is conserved, according to the first law, how come the world is running out of it? Why can't we use all the energy surrounding us in the air and sea? Is it that we have not developed the proper devices yet, or is it really impossible?

3.2.4 Summary

All these observations and questions - and of course a variety of others - can be summed up as follows:

1. The first law is a *necessary but not sufficient condition* for the occurrence of a process. It makes no provision with respect to the direction in which the process will proceed.
2. Natural (spontaneous) processes, such as the flow of heat or the mixing of gases, etc., cannot be reversed on their own.
3. The direction of a process seems to be determined by the initial and final states of the substances involved: their pressure, temperature, and composition. It is possible, therefore, that some state function is involved, whose characteristics determine the direction of the process.
4. Experience and intuition may not always suffice for the establishment of the feasibility of a given process. We need a more definite and quantitative criterion.

3.3 Objective and Approach

The overall objective of this Chapter is to present the second law and to introduce, through it, the concept of *entropy* as the quantitative criterion - the 'yardstick' - for:

1. the establishment of the feasibility of a given process, and
2. the efficient utilization of thermal energy.

More specifically, to understand:

1. The concept of a *reversible heat engine*, as it is described by the Carnot cycle, and that it establishes the upper limit for the amount of mechanical energy that can be produced from a given amount of heat flowing from a given temperature level to a lower one.
2. The concept of *absolute thermodynamic temperature* and its relationship to that obtained through an ideal gas fluid.
3. The approach and methods used in the calculation of entropy changes.
4. The application of the second law in achieving the overall objective of this Chapter, i.e. in:
   * establishing the feasibility of a given process, and
   * obtaining the maximum possible amount of mechanical energy from a given amount of thermal one.

The latter is, by far, the most important one, for - in the final analysis - thermal energy is the motive power of our civilization.

We start with the development of the second law, which will be carried out in the following five steps:

1. Presentation of two - but equivalent - statements of the law that represent a verbal description of our experiences with respect to the way certain processes occur (or, do not).
2. Discussion of the conversion of heat to work with a reversible heat engine - the Carnot cycle - essential for the development of an analytical statement of the second law; (and, to bring this 'ideal' engine closer to reality, the very 'real' power plant cycle - for the production of electricity from fuels - is described briefly).
3. Presentation of four Propositions, that will be proved through a combination of the Carnot cycle and the verbal statements of the second law.
4. Use of the first three Propositions to introduce the concept of thermodynamic temperature and the concept of entropy.
5. Use of the latter in the fourth Proposition to develop the analytical statement of the second law, i.e. the desirable quantitative criterion for the establishment of the feasibility of a given process and for the efficient utilization of thermal energy.

Application of this criterion requires the determination of entropy changes, which are considered next, followed by a series of Examples that demonstrate how these entropy changes are used in applications of the second law, developing thus our ability to use it. (To provide for some understanding of the concept of entropy, however, a molecular interpretation of it is presented before the Examples).

We return, then, to the discussion of steam power plants, a major application of the second law, to enhance our familiarity with using this law, and close with some Concluding Remarks.

3.4 Second Law: The Verbal Statement

3.4.1 Statements

We will use the following two - equivalent - statements of the second law:

Statement A: *It is impossible to make a transfer of heat from a heat bath, at a uniform temperature, and obtain an equivalent amount of work, without causing a change in the thermodynamic state of an other body* (Denbigh, 1981, p.26).

Statement B: *Any process whose only effect is the transfer of heat from one temperature level to a higher one, is impossible.*

3.4.2 Example 3.1

Consider the following two observations:

a. An ideal gas expands isothermally in a constant temperature bath absorbing an amount of heat Q and producing an amount of work W. From the first law, noting that the isothermal change of the internal energy of an ideal gas is equal to zero, we conclude that:

$$W = Q$$

which indicates that all heat absorbed is converted into work.

b. Air-condition units transfer heat from the colder room to the hotter atmosphere.

Do you see any disagreement between these observations and the second law statements?

a. No, because one of the thermodynamic properties of the ideal gas, its pressure, has changed (Statement A).

b. Again, no, for this is not the only effect in the process. Electricity is used (Statement B).

3.5 The Carnot Cycle

> *In the annals of science Carnot's work is almost unique, for it had no discernible predecessors and was built up from the assemblage of unordered opinions and problems, concepts, theories and measurements that were available at the time. Carnot, in short, was not standing on the shoulders of giants; he saw further than his contemporaries because he had a much clearer vision* (Cardwell, 1971, p.193).

> *Indeed we cannot (explain Carnot's thought process), for he was one of the most truly original thinkers in the whole history of science. It might not be wholly inappropriate to say that in the independence of his spirit he was English, and in the rigor and clarity of his mind he was French; but in the depth and scope of his insight he transcends all such classifications* (Ibid, p.211).

We will start with a description of the cycle, develop an expression for its thermal efficiency, and then demonstrate that the cycle provides the maximum possible efficiency.

3.5.1 Description

Assume that we have:
* a heat source, large enough that removal of an amount of heat Q_1 does not change its temperature T_1; it is referred to as a heat reservoir. And,
* a heat receiver at a temperature T_2, lower than T_1, large enough that addition of an amount of heat Q_2 does not change its temperature; it is referred to as a heat sink.

We consider now a hypothetical cycle whose net effect is to:
* absorb an amount of heat Q_1 from the heat reservoir;
* produce an amount of work W; and
* reject an amount of heat Q_2 to the heat sink.

To this purpose, the cycle uses one mole of an ideal gas, which undergoes the following four changes shown in Figure 3.1:

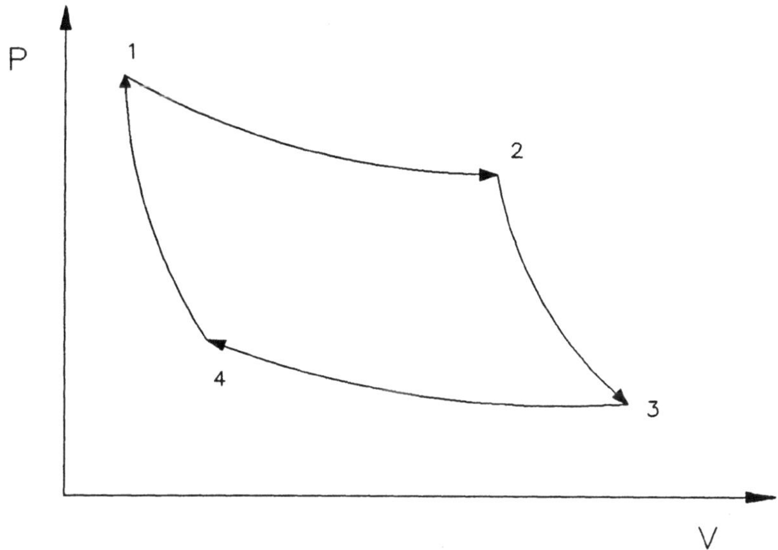

Figure 3.1 The *P-V* behavior of the gas in the Carnot Cycle.

1.It expands isothermally ($1 \rightarrow 2$), at $T_1' = T_1 - \Delta T_1$, absorbing an amount of heat Q_1 from the heat reservoir and producing an amount of work W_{12}.
2.It expands adiabatically ($2 \rightarrow 3$) to temperature $T_2' = T_2 + \Delta T_2$, producing an amount of work W_{23}.
3.It contracts isothermally ($3 \rightarrow 4$) at T_2' rejecting an amount of heat Q_2 to the heat sink and consuming an amount of work W_{34}.
4.It is compressed adiabatically to its original state ($4 \rightarrow 1$), consuming an amount of work W_{41}.

All *P-V* changes, expansions and compressions, are reversible.

3.5.2 The Expression for the Thermal Efficiency

The thermal efficiency of the cycle is defined by:

$$n = \frac{\textit{net work produced}}{\textit{heat absorbed}} = \frac{W}{Q_1} \tag{3.5.1}$$

According to the first law applied to the whole cycle, and considering the fact that the gas returns to its original state, i.e. $\Delta U = 0$: $Q = W$, where:

$$Q = Q_1 + Q_2$$

$$W = W_{12} + W_{23} + W_{34} + W_{41}$$

Hence:

$$n = \frac{Q_1+Q_2}{Q_1} = 1+\frac{Q_2}{Q_1} \tag{3.5.2}$$

To express Eq.3.5.2 in terms of temperatures we notice that for the two isothermal steps:

$$Q_1 = \int_{V_1}^{V_2} PdV = RT_1'\ln\frac{V_2}{V_1} = -RT_1'\ln\frac{P_2}{P_1}$$

and:
$$Q_2 = -RT_2'\ln\frac{P_4}{P_3}$$

while from the two adiabatic steps (Eq.2.6.5):

$$\left(\frac{P_3}{P_2}\right)^k = \frac{T_2'}{T_1'} \quad \text{and} \quad \left(\frac{P_1}{P_4}\right)^k = \frac{T_1'}{T_2'}$$

where: $k = \dfrac{\gamma-1}{\gamma}$; $\gamma = \dfrac{C_P^*}{C_V^*}$

Thus:
$$\frac{P_1}{P_2} = \frac{P_4}{P_3}$$

and Eq.3.5.2 becomes:

$$n = 1+\frac{T_2'\ln(P_4/P_3)}{T_1'\ln(P_2/P_1)} = 1-\frac{T_2'}{T_1'} = 1-\frac{T_2+\Delta T_2}{T_1-\Delta T_1}$$

It is apparent that in order for the cycle to produce the maximum work ΔT_1 and ΔT_2 must approach zero, i.e. the heat transfer must be also reversible. Thus:

$$n = 1-\frac{T_2}{T_1} \tag{3.5.3}$$

and the cycle efficiency is determined by the temperature of the heat reservoir and that of the heat sink only.

This reversible cycle was introduced by Carnot in 1824, and was described - in terms of the volumetric behavior of the gas (Figure 3.1) - by Clapeyron several years later.

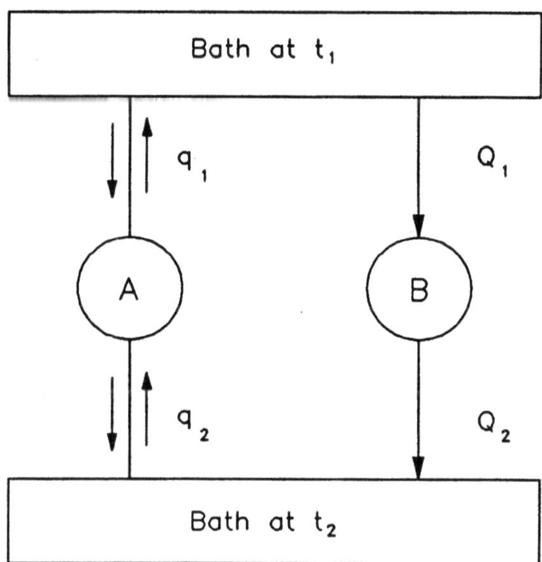

Figure 3.2 The two engines of Section 3.5.3.

3.5.3 The Maximum Thermal Efficiency

We conclude, thus, that as the cycle approaches reversible operation, its thermal efficiency approaches its maximum value. We will demonstrate next that, of all heat engines operating between two specified temperatures, the reversible one has the maximum thermal efficiency.

Consider the two heat engines shown in Figure 3.2. They operate between the two temperatures t_1 and t_2, absorbing heat at t_1 and rejecting heat at t_2 ($t_1 > t_2$), and produce the same amount of work. Engine A is a reversible one; engine B is irreversible. We will demonstrate next that if B were more efficient than A, the second law would be violated.

If B is more efficient, and both engines produce the same amount of work, then from the definition of thermal efficiency (Eq.3.5.1):

$$|q_1| > |Q_1|$$

where the absolute values are used to avoid concern with signs. Also:

$$|q_2| = |q_1| - W, \; |Q_2| = |Q_1| - W, \text{ hence } |q_2| > |Q_2|.$$

Let us now reverse engine A and drive it using the work produced by B. Because A is reversible it will now absorb an amount of heat $|q_2|$ and reject an amount of heat $|q_1|$.

We have, therefore, a process - the combination of the two cycles - whereby the only effect is the transfer of an amount of heat $q = |q_2| - |Q_2|$, from a lower to a higher temperature. This, of course, violates statement B of the second law and, consequently, the irreversible engine cannot be more efficient than the reversible one.

3.5.4 Comments

1. It is apparent that:

a. The efficiency of the Carnot Cycle is independent of the nature of the working fluid (the ideal gas was used for convenience; the same result is obtained using a real gas, Problem 8.56).

b. For maximum efficiency, there should be no temperature difference between the working fluid and the:

* heat source, and
* heat sink.

2. Notice the similarity between the cycle's efficiency and that of a water-wheel, discussed in Example 2.7. Indeed, Carnot was aware of this - his father was a military hydraulic engineer - and it influenced his thinking in working with the efficiency of the steam engine. Carnot himself states:

"In accordance with the principles we have now established, we can reasonably compare the motive power of heat with that of a head of water: for both of them there is a maximum which cannot be exceeded, whatever the type of hydraulic machine and whatever the type of heat engine employed. The motive power of a head of water depends upon its height and the quantity of water; the motive power of heat depends also on the quantity of caloric and on what may be called - on what we shall call - the height of its fall, that is on the temperature difference of the bodies between which the caloric flows." (Cardwell, p.196)

3. Notice that Carnot does not talk about conversion of caloric to work, for it was indestructible according to the caloric theory, but about flow of it.

4. Carnot did, however, recognize the difference between the two efficiencies: for the water-wheel it does not depend on the absolute values of the two heights, only their difference; for the thermal efficiency, it does depend on the absolute values of the two temperatures. The recognition of this difference reflects the genius of Carnot.

5. The cycle is, of course, an imaginary one: it represents the limit of a real cycle as the driving forces become infinitesimally small. This is

why we call it an 'ideal heat engine' but, as with the case of the ideal gas, this makes it no less useful.

6.Eq.3.5.3 leads to two very practical conclusions with respect to the amount of work produced from a given amount of thermal energy in a cyclic process:

a.it is always less than that of the thermal energy; and

b.it depends on the temperature:

* at which the thermal energy is available (the higher the better); and
* of the heat sink (the lower the better).

7.Eq.3.5.3 provides also the answer to the 'big' question posed in Section 3.2, as to 'where the energy is gone': use of the thermal energy contained by the atmosphere, or the oceans, would require a large heat sink at a lower temperature, which is not available.

8.The importance of the relationship between thermal energy and the temperature it is available at is demonstrated in Examples 3.2 and 3.3; and it is considered in detail in Chapter 5.

9.The maximum efficiency of the reversible heat engine is consistent with our observations in Chapter 1, where it was demonstrated that a reversible operation produces the maximum amount of work, such as in the expansion of a gas, when *P*(gas) is just larger than *P*(opposing); or, in raising weight *B* with weight *A*, when W_A is just larger than W_B.

10.Some authors refer to Eq.3.5.3 as defining the efficiency of the Carnot cycle without, in other words, including the word 'thermal'. This, however, can be misleading for:

* since efficiency is defined, Eq.1.12.1, as:

$$\frac{\textit{net work produced}}{\textit{maximum producible work}}$$

* and the Carnot cycle provides the maximum producible work,

its efficiency is always equal to one. Its thermal efficiency, on the other hand, is always less than one.

When, hence, we refer to the efficiency of a heat engine, we imply its thermal one.

11.Comparison of Eqs 3.5.2 and 3.5.3 yields:

$$\frac{Q_1}{Q_2} = -\frac{T_1}{T_2} \tag{3.5.4}$$

i.e. the ratio of heat absorbed to heat rejected depends on the ideal gas temperatures of the two baths only, and is equal to their ratio.

We will use Eq.3.5.4 in Section 3.9 to demonstrate that the ideal gas temperature (T) represents also the thermodynamic temperature.

To relate the somewhat 'abstract' Carnot cycle to the real world, we consider - after the Examples 3.2 and 3.3 - the very 'real' power plant cycle, used for the production of electricity from thermal energy.

3.5.5 Example 3.2

It has been suggested that the temperature difference between the surface and deep water of the oceans can be used for the production of electricity. Assuming that the two temperatures are 25°C and 10°C respectively:

a. calculate the maximum amount of electricity that can be produced per Joule of absorbed heat;

b. explain why the real amount will be less.

a. From Eq.3.5.3:

$$W = Q_1\left(1-\frac{T_2}{T_1}\right) = 0.050 \text{ J}$$

In other words, the maximum amount of produced work is only 5% of the absorbed heat.

b. In obtaining this answer, we assumed:

* complete reversibility - through the Carnot cycle - in the compression and expansion of the working fluid; and in the heat transfer.

* complete convertibility of mechanical power to electricity, which is theoretically feasible but - in reality - the efficiency is about 90%.

In practice, thus, the work produced will be still less. In spite of this low thermal efficiencies, the available amounts of energy in this source are so large, that it was estimated in the middle 1970's that it could provide from 1 to 5% of the U.S. energy needs by the year 2000 (Seltzer, 1976). This, of course, was before the low oil prices of the 1980's severely curtailed work on alternative energy sources.

3.5.6 Example 3.3

One kg of water at 80°C is available. Calculate the maximum work that can be obtained from it. The atmosphere at 20°C will be the heat sink. Assume a constant heat capacity for water of C = 1 kcal/kg K.

If the maximum work is to be produced, a Carnot cycle operating between the water and the atmosphere must be used. Notice, however, that in this case the temperature of the water does not remain constant but it varies from the initial value, T_1 = 353.15 K, to the final one, T_2 = 293.15 K. Let us assume that at a given moment:

* the temperature of the water is T; and
* a small amount of heat dq is absorbed from it.
The amount of work produced from this amount of heat will be:

$$dW = dq(1-\frac{T_2}{T}) = -(1)(CdT)(1-\frac{T_2}{T})$$

and:

$$W = -(1)(C)\left((T_2-T_1)-T_2\ln(\frac{T_2}{T_1})\right) = 5.41 \text{ kcal}$$

Since the amount of heat absorbed in this case is 60 kcal, the overall thermal efficiency of the cycle is 5.41/60 = 0.09.

Note. The small amount of work produced per unit of heat absorbed in both Examples supports the characterization of thermal energy as a 'second class' form of energy.

3.6 A Real Heat Engine: A Simple Power Plant

Most, by far, electric energy is produced by the combustion of fossil fuels. The resulting thermal energy is converted in part into mechanical energy - which is then converted into electrical one - while the remaining is discarded or, on occasion, used for process heating purposes. The process is carried out in what is referred to as a *power plant cycle*.

3.6.1 Description

For small units, a 'simple' cycle is often used, consisting of the following steps (Figure 3.3):

1. water at high pressure is vaporized in a boiler and the steam is then superheated ($1 \rightarrow 2$);
2. the superheated steam expands through a turbine, where part of its energy is converted into mechanical ($2 \rightarrow 3$), which in turn is converted into electric;
3. the exiting steam enters a condenser where it is condensed ($3 \rightarrow 4$), using cooling water; and
4. the condensate is pumped back into the boiler ($4 \rightarrow 1$).

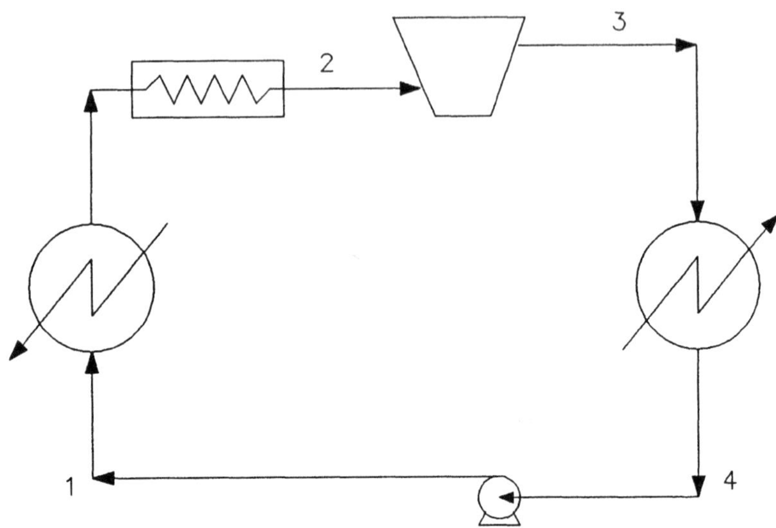

Figure 3.3 A simple power plant cycle.

3.6.2 The Expression for the Thermal Efficiency

From the definition, Eq.3.5.1, the efficiency of the cycle is given by:

$$n = \frac{W_t + W_p}{Q_1} \qquad (3.6.1)$$

where, W_t: turbine work
W_p: pump work
Q_1: heat absorbed in the boiler/superheater.

3.6.3 Comments

1.The turbine is usually adiabatic. Hence from the first law for a steady-state flow process: $W_t = -(H_3 - H_2)$.

2.In the same fashion: $W_p = -(h_1 - h_4)$, where the symbol h is used for the liquid phase and H for the vapor.

3.Again from the first law: $Q_1 = H_2 - h_1$.

4.We are not ready yet to carry out calculations involving power plants. We will do it after we discuss entropy change calculations.

We proceed now with the development of the analytical statement of the second law.

3.7 Second Law: The Analytical Statement

Once we understand the concept of a reversible heat engine, we are ready to develop the analytical statement of the second law. This is carried out with the use of the ensuing four 'Propositions' discussed briefly here (for an in depth presentation see Denbigh p.26):

I. If a heat engine performs a reversible cycle by:

* absorbing an amount of heat Q_1 at a temperature t_1, and
* rejecting an amount of heat Q_2 at a temperature t_2,

then the ratio (Q_1/Q_2) is a function of t_1 and t_2 only, i.e.:

$$\frac{|Q_1|}{|Q_2|} = f(t_1, t_2)$$

where the absolute values are used to eliminate concern with signs.

II. This ratio is given by:

$$\frac{|Q_1|}{|Q_2|} = \frac{f(t_1)}{f(t_2)}$$

where $f(t)$ is the *thermodynamic* temperature, which can be set equal to the 'ideal gas' temperature, i.e.:

$$\frac{|Q_1|}{|Q_2|} = \frac{T_1}{T_2}$$

III. The entropy S, defined by:

$$dS = \frac{dQ_r}{T}$$

where the subscript r stands for reversible, is a state function.

IV. For any natural (spontaneous) process, the total entropy change (ΔS_{tot}) equal to:

* the entropy change of the system (ΔS_{sys}), plus
* that of the surroundings (ΔS_{sur}),

is larger than zero. It is equal to zero, only when the process is reversible, and it can never be negative:

$$\Delta S_{tot} = \Delta S_{sys} + \Delta S_{sur} \geq 0$$

This expression represents the 'analytical' statement of the second law.

We proceed, next, to discuss these four Propositions.
[Undergraduate students, however, can move to Section 10 (Proposition III), and accept that: the ideal gas temperature represents also the thermodynamic temperature, called such because its establishment is based on the second law of thermodynamics.]

3.8 Proposition I

Consider two reversible heat engines A and B operating between the same two baths with temperatures t_1 and t_2 ($t_1 > t_2$). Cycle A absorbs an amount of heat q_1 at t_1 and rejects an amount of heat q_2 at t_2. Cycle B operates in the opposite direction: it absorbs an amount of heat Q_2 at t_2 and rejects an amount of heat Q_1 at t_1.

We operate engine A n times and engine B N times so that $n|q_2| = N|Q_2|$, i.e. the total amount of heat rejected to the cold bath by engine A equals the amount of heat absorbed from the same bath by engine B. We will demonstrate next that, since engines A and B have returned to their initial states (n and N are integers), then: $n|q_1| = N|Q_1|$.

Assume first that $n|q_1| > N|Q_1|$. The net result of the operation of the two cycles will be that an amount of heat equal to: $(n|q_1| - N|Q_1|)$ is absorbed from the bath at t_1. Now, according to the first law applied to the whole process, this heat has been converted completely into work, which is impossible according to statement A of the second law.

If now $n|q_1| < N|Q_1|$, we can reverse the two cycles - remember they are reversible - and end up with the first case.

We conclude, thus, that $n|q_1| = N|Q_1|$, which combined with the starting equation, $n|q_2| = N|Q_2|$, yields:

$$\frac{|q_1|}{|q_2|} = \frac{|Q_1|}{|Q_2|}$$

The ratio is, thus, the same for both cycles, independent of the operating pressures and volumes of the gas involved in each cycle. It must, therefore, be a function of the temperatures of the two baths, t_1 and t_2, only:

$$\frac{|q_1|}{|q_2|} = \frac{|Q_1|}{|Q_2|} = f(t_1, t_2) \qquad (3.8.1)$$

3.9 Proposition II

3.9.1 Approach

We will show first that the function $f(t_1,t_2)$ of Proposition I is given by:

$$f(t_1,t_2) = \frac{f(t_1)}{f(t_2)}$$

and proceed to introduce the concept of *thermodynamic* temperature, as contrasted to the ideal gas one. We will then demonstrate that the two temperatures can be set equal to each other.

3.9.2 $f(t_1,t_2) = f(t_1)/f(t_2)$

Consider three reversible heat engines that operate among three different temperatures t_1, t_2, t_3 ($t_1 > t_2 > t_3$), as shown in Figure 3.4:
* the first between t_1 and t_2;
* the second between t_2 and t_3; and

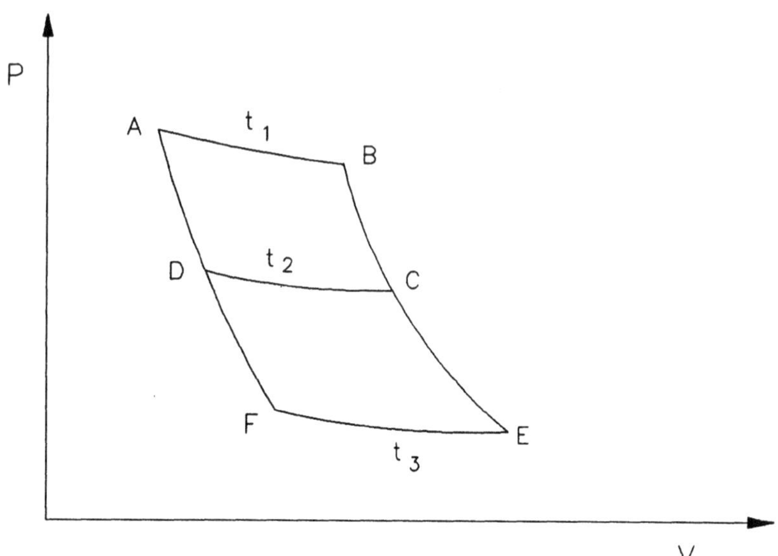

Figure 3.4 The three cycles used in the proof of Proposition II.

* the third between t_1 and t_3.

As demonstrated in Proposition I, Eq.3.8.1:

$$\text{Cycle } ABCD\text{:} \quad \frac{|Q_1|}{|Q_2|} = f(t_1,t_2)$$

$$\text{Cycle } DCEF\text{:} \quad \frac{|Q_2|}{|Q_3|} = f(t_2,t_3)$$

$$\text{Cycle } ABEF\text{:} \quad \frac{|Q_1|}{|Q_3|} = f(t_1,t_3)$$

Elimination of $|Q_3|$ between the last two equations leads to:

$$\frac{|Q_1|}{|Q_2|} = \frac{f(t_1,t_3)}{f(t_2,t_3)} \tag{3.9.1}$$

and by comparison with the first equation:

$$f(t_1,t_2) = \frac{f(t_1,t_3)}{f(t_2,t_3)} \tag{3.9.2}$$

This relationship should be valid for all values of t_3 and, consequently, t_3 should cancel out between the numerator and denominator. This would require that:

$$f(t_1,t_3) = \frac{f(t_1)}{f(t_3)}$$

$$f(t_2,t_3) = \frac{f(t_2)}{f(t_3)}$$

$$f(t_1,t_2) = \frac{f(t_1)}{f(t_2)}$$

and consequently:

$$\frac{|Q_1|}{|Q_2|} = \frac{f(t_1)}{f(t_2)}$$

$$\frac{|Q_2|}{|Q_3|} = \frac{f(t_2)}{f(t_3)}$$

$$\frac{|Q_1|}{|Q_3|} = \frac{f(t_1)}{f(t_3)}$$

3.9.3 The Thermodynamic Temperature

We have arrived, thus, at a new temperature, established by the second law and not arbitrarily through, for example, the Celsius scale or even through an ideal gas. Rather, it is defined in terms of the ratio of the heats absorbed and rejected in a reversible heat engine:

$$\frac{|Q_1|}{|Q_2|} = \frac{f(t_1)}{f(t_2)} \tag{3.9.3}$$

We have also seen, Eq.3.5.4, that:

$$\frac{|Q_1|}{|Q_2|} = \frac{T_1}{T_2}$$

where T is the ideal gas temperature. It follows then that this temperature is proportional to T:

$$f(t) = aT$$

To avoid the existence of two scales, we set $a = 1$, and we call T the *absolute thermodynamic* temperature, or just the *absolute* temperature.

Finally, since Q_1 and Q_2 have opposite signs, we recover from Eq. 3.9.3, Eq.3.5.4:

$$\frac{Q_1}{T_1} = -\frac{Q_2}{T_2} \tag{3.5.4}$$

3.10 Proposition III

Let us now take a closer look at the ratio (Q/T). We notice that no matter what the values of the heat absorbed Q_i and rejected Q_j are, when divided by the corresponding temperatures T_i and T_j, they must satisfy Eq.3.5.4:

$$\frac{Q_i}{T_i} = -\frac{Q_j}{T_j}$$

There may be, thus, more to this ratio (Q/T) than meets the eye. We define, therefore, the quantity S such that:

$$dS = \frac{dQ_r}{T} \tag{3.10.1}$$

where the subscript r depicts the reversible character of the operation, and call it *Entropy*. (As we will see in Chapter 4 this term was introduced by Clausius in 1865.)

We will demonstrate next that entropy is a *state function*, i.e. that there is indeed more to the ratio (Q/T) than meets the eye.

To this purpose consider a gas that undergoes a Carnot cycle by absorbing an amount of heat Q_1 at T_1 and rejecting an amount of heat Q_2 at T_2. To calculate the entropy changes of the gas going through this cycle, we notice that they occur only in the heat absorption and heat rejection steps since - according to Eq.3.10.1 - there is no entropy change in the two adiabatic ones:

$$\Delta S_1 = \int \frac{dQ_1}{T_1} = \frac{Q_1}{T_1}$$

$$\Delta S_2 = \int \frac{dQ_2}{T_2} = \frac{Q_2}{T_2}$$

and:

$$\Delta S = \Delta S_1 + \Delta S_2 = \frac{Q_1}{T_1} + \frac{Q_2}{T_2} \tag{3.10.2}$$

Thus - according to Eq.3.5.4 - the total entropy change of the gas going through the whole cycle is zero:

$$\oint dS = \Delta S = 0 \tag{3.10.3}$$

Eq.3.10.3 applies to any reversible cycle, i.e. other than those involving two isothermal and two adiabatic steps (for proof see, among others, Denbigh), and demonstrates that entropy *is* a state function.

Consider, for example, a body going from state A to state B following path 'a', or path 'b', reversibly in both cases. Application of Eq.3.10.3 to the cycle: A to B through path 'a' and then back to A through path 'b' gives:

$$\int_A^B dS(a) + \int_B^A dS(b) = 0; \quad or,$$

$$\int_A^B dS(a) = -\int_B^A dS(b) = \int_A^B dS(b)$$

The entropy change of the body is thus independent of the path followed - provided it is a reversible one - and, consequently, entropy is a state function.

3.11 Proposition IV

3.11.1 Objective

In discussing the concept of reversibility in Chapter 1 we noted that all real life - spontaneous - processes are irreversible. We will demonstrate next that when a system undergoes such a spontaneous (irreversible) process, the total entropy change - that of the system plus that of the surroundings - is positive. This represents the formal (analytical) statement of the second law.

3.11.2 Entropy Change of a System Undergoing a Spontaneous Process

Consider a system which undergoes a spontaneous (irreversible) change from state A to state B within an adiabatic enclosure. Since the process is irreversible (spontaneous), the entropy change (S_B - S_A) cannot be evaluated through Eq.3.10.1.

We bring now the system back from state B to state A using a reversible engine. This, of course, cannot be accomplished adiabatically, for then the process A to B would be also reversible. So, let Q_{BA} be the amount of heat involved in the process B to A, and W_{AB} and W_{BA}, the amounts of work associated with the forward and backward directions of the process. From the first law applied to the whole cycle:

$$Q_{BA} = W_{AB} + W_{BA}$$

Clearly, Q_{BA} must be negative, for:

* If it were zero, the whole process would be reversible; and
* If it were positive - i.e. heat was absorbed by the system - it would correspond to the conversion of this heat completely into work in a cyclic process - since the system returns to its initial state - and it would violate statement A of the second law.

We conclude, therefore, that returning the system reversibly from B to A involves the rejection of an amount of heat Q_{BA}.

Let, now, T be the system temperature at which the heat was rejected. Then:

$$S_A - S_B = \int_B^A \frac{dQ_r}{T} = \frac{Q_{BA}}{T} < 0$$

In other words, S_B is larger than S_A, which indicates that the original spontaneous process leads to an increase in the system entropy.

3.11.3 The Total Entropy Change

If we are now to remove this entropy from the system - i.e. if we want S_B to equal S_A - then during the spontaneous process an appropriate amount of heat must be transferred to the surroundings. This, of course, will result in an increase of the entropy of the surroundings. On the other hand, it is also apparent from our discussion that when the process A to B is reversible, there will be no entropy change for the system or the surroundings.

We conclude, therefore, that the total entropy change ΔS_{tot} - that of the system (ΔS_{sys}) plus that of the surroundings (ΔS_{sur}) - must be positive for any natural (spontaneous) process:

$$\Delta S_{tot} = \Delta S_{sys} + \Delta S_{sur} \geq 0 \qquad (3.11.1)$$

Eq.3.11.1 represents the formal statement of the second law, the equal sign referring to the case of a reversible process, and provides the desired criterion that establishes:

* the direction and feasibility of a given process, and
* the efficiency in the use of energy in this process.

3.12 Entropy Change Calculations

Application of Eq.3.11.1 to a given process requires evaluation of the entropy changes involved, which is considered next.

3.12.1 The Problem and Approach

Consider a system undergoing a spontaneous (irreversible) change from state 1 to state 2. How will the entropy change of the system be evaluated?

Since entropy is a state function, the change: $\Delta S = S_2 - S_1$ is independent of the path followed by the system in going from state 1 to state 2. *We devise, therefore, a hypothetical reversible path from state 1 to state*

2 and use it to evaluate the entropy change with Eq.3.10.1. We consider next some applications.

3.12.2 Isothermal Removal or Addition of Heat

This occurs in the case of a bath large enough that the addition or removal of an amount of heat Q does not effect its temperature T, such as a bath consisting of a system undergoing an isothermal phase change. Then from Eq.3.10.1:

$$\Delta S = \int \frac{dQ_r}{T} = \frac{Q}{T} \tag{3.12.1}$$

It takes, for example, 2256.9 kJ to vaporize one kg of saturated water at 373.15 K. Hence, according to Eq.3.12.1, the entropy of vaporization is 6.0485 kJ/kg K.

3.12.3 Heating or Cooling of a Body from T_1 to T_2

To this purpose, we combine Eq.3.10.1 with the definition of heat capacity, Eq.1.13.1, which yields on a molar basis:

a. Constant volume (isochoric) process: $dQ_r = dU = C_V\,dT$,

$$\Delta S = \int_{T_1}^{T_2} \frac{dQ_r}{T} = \int_{T_1}^{T_2} \frac{C_V}{T} dT \tag{3.12.2}$$

b. Constant pressure (isobaric) process: $dQ_r = dU + PdV = dH = C_P\,dT$,

$$\Delta S = \int_{T_1}^{T_2} \frac{dQ_r}{T} = \int_{T_1}^{T_2} \frac{C_P}{T} dT \tag{3.12.3}$$

3.12.4 Entropy Changes of an Ideal Gas

Consider one mole of an ideal gas that goes from state 1 (T_1, V_1, P_1) to state 2 (T_2, V_2, P_2). Since our hypothetical path is reversible, $dW = PdV$, and Eq.3.10.1 yields:

$$dS = \frac{dQ_r}{T} = \frac{dU + PdV}{T} = \frac{C_V^*}{T} dT + \frac{R}{V} dV$$

and:

$$\Delta S = \int_{T_1}^{T_2} \frac{C_V^*}{T} dT + R\ln\frac{V_2}{V_1} \tag{3.12.4}$$

To obtain the final answer in terms of P_1 and P_2 we introduce:

$$\frac{V_2}{V_1} = \frac{T_2 P_1}{T_1 P_2} \quad \text{and} \quad C_V^* = C_P^* - R$$

leading, after some algebra, to:

$$\Delta S = \int_{T_1}^{T_2} \frac{C_P^*}{T} dT + R\ln\frac{P_1}{P_2} \tag{3.12.5}$$

Notice that while the enthalpy and internal energy changes of an ideal gas are functions of temperature only (Section 2.6.3), the entropy change depends on both, temperature and pressure (or volume), changes.

3.12.5 Entropy of Mixing

Two pure ideal gases 1 and 2, both at the same temperature T and pressure P, are mixed isothermally with final composition y_1 and y_2. What is the entropy of mixing per mole of mixture formed?

We start by noticing that as a result of the mixing process the pressure of component 1 goes from P to its partial pressure $\bar{P}_1$, i.e. its contribution to the mixture pressure; and the same applies to component 2. Thus, according to Eq.3.12.5:

$$\Delta S = y_1 \Delta S_1 + y_2 \Delta S_2 = -\left[y_1 R\ln\left(\frac{\bar{P}_1}{P}\right) + y_2 R\ln\left(\frac{\bar{P}_2}{P}\right)\right] \tag{3.12.6}$$

But since, by definition, $\bar{P}_1 = y_1 P$ and $\bar{P}_2 = y_2 P$, Eq.3.12.6 becomes:

$$\Delta S = -R(y_1 \ln y_1 + y_2 \ln y_2) \tag{3.12.7}$$

Similarly, for N components:

$$\Delta S = -R\sum_1^N y_i \ln y_i \tag{3.12.8}$$

The same equation applies in the isothermal formation of an ideal solution, while real gases and real solutions will be considered in Chapter 11. Returning to Eq.3.12.8 we notice that, since all y's are less than one, the mixing process leads to an increase in entropy. This, of course, should be expected. Mixing is a natural (spontaneous) process, and as such it leads to an entropy increase.

We are ready now to carry out entropy change calculations and use them in applications of the second law, such as the establishment of the feasibility of processes and the efficient utilization of energy in them.

But first, to develop some sense about the physical meaning of entropy, we consider a molecular interpretation of it.

3.13 A Molecular Interpretation of Entropy

The laws of thermodynamics do not concern themselves with the molecular structure of matter. But the concept and meaning of entropy can become somewhat clearer when we consider this structure of matter, as suggested by the ensuing discussion (Denbigh).

According to Denbigh (pp.48-52), once we recognize that matter consists of molecules in a continuous motion, then we can view - in a broad sense - every large scale process as essentially a mixing one.

We can actually consider two types of mixing processes occurring separately or simultaneously: *spacial* and *thermal*. For example, when the wall separating two gases A and B is removed, the continuous motion of the molecules at the interface will lead to complete mixing (spacial). If the two gases are at different energy levels (i.e. temperatures), then a mixing of their energies will also occur (thermal).

To arrive at some qualitative relationship between the concept of entropy and the consequences of this mixing process, let us consider two crystals composed of atoms A and B respectively.

The crystals are in contact with each other and are enclosed within an adiabatic wall. For simplicity, we will assume that the two crystals are sufficiently alike in their lattice structure, so that the atoms can be interchanged without any change in the energy states of the crystals. Under these conditions we are only concerned with mixing over positions in space (spacial), but not with mixing over energy states.

We will demonstrate next that this mixing of the two crystals leads to a loss of information about them. For ease of the calculations we will assume that each crystal contains only four atoms.

At the beginning (t=0), all four atoms of A are to the left of plane P, and all four atoms of B to the right (Figure 3.5). Considering the fact that the A atoms are indistinguishable amongst themselves, and the same for the B ones, there is just one arrangement for the initial state of the system:

$$N(4,0) = 1$$

where $N(4,0)$ is the number of arrangements where there are four A atoms to the left of the plane and none to the right.

Consider now that with time, and as a result of the motion of the atoms, one A atom moves to the right of the plane and one B to the left. How many such arrangements are possible?

Each A atom can move to any of the four spaces to the right and the same applies to each B one moving to the left. Hence we have 4x4=16 arrangements, or:

$$N(3,1) = 16$$

The number of arrangements for the case when two A atoms move to the right of the plane and two B ones to the left is less obvious. One such arrangement, occurring after time $t=t_1$ has elapsed, is shown in Figure 3.6.

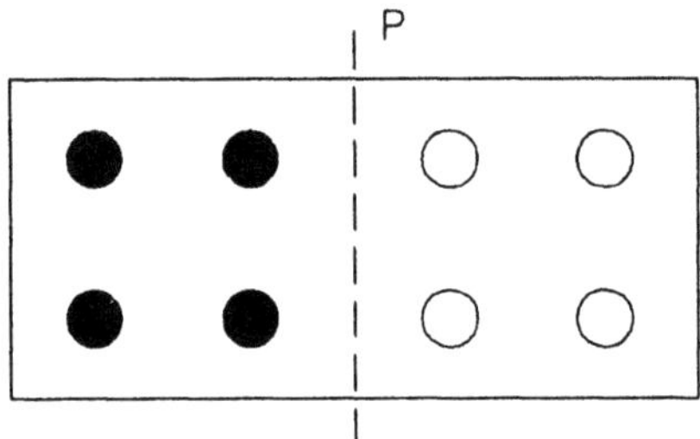

Figure 3.5 The mixing of two perfect crystals: $t = 0$.

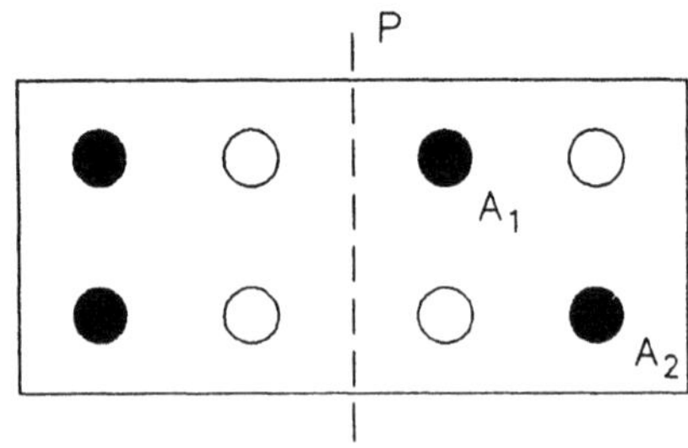

Figure 3.6 The mixing of two perfect crystals: $t = t_1$.

Notice that the second A atom that moves to the right, can occupy any of the three remaining positions, which would lead to 4(from the first atom)x3=12 arrangements. Since, however, the atoms are indistinguishable, interchange between molecules on the same side - say A_1 and A_2, Figure 3.6 - does not lead to a new arrangement. Hence, there are (4x3)/2! arrangements of A atoms to the right of the plane and, when combined with (4x3)/2! ones of B atoms to the left, they give:

$$N(2,2) = 36$$

physically distinct arrangements.

It can be easily seen that:

$$N(1,3) = 16$$
$$N(0,4) = 1$$

The total number therefore of possible arrangements, or complexions, is:

$$N = N(4,0)+N(3,1)+N(2,2)+N(1,3)+N(0,4) = 70$$

The following question arises at this point. After a long time has elapsed, which of these 70 arrangements (complexions) will the system be in?

In the absence of any reason to the contrary we assume that, after the crystals have been in contact for a long time, the system can be found with equal probability in any one of the 70 arrangements. Hence, the probability of four A to the left of the plane and four B to the right is 1/70; for three A to the left and one A to the right, 16x(1/70); and for two A on each side, 36x(1/70); etc.

The (2,2) arrangement would be called the most completely mixed state, and it is seen to be the most probable end-state of this mixing process, simply because it is made up of a much larger number of arrangements than any other.

The fact remains, however, that we do not know - at a given moment - what the state of the system is. Only the probabilities of the different states are known. Notice, for example, that while the (2,2) state has the highest probability, there is a distinct probability that the system can also be found in the (4,0) state, where the crystals are 'unmixed'.

The latter probability, however, becomes smaller as the number of atoms increases. For example, with 6 atoms each of A and B there are 924 complexions. Four hundred of the (3,3) state, probability 400/924, but only one of the (6,0) state, probability 1/924.

We conclude, therefore, that the mixing process led from one possible arrangement (4,0), to 70 ones; or, to 924 ones in the second case; and we do not know in which one of these arrangements the system is to be found at a given moment. There is a loss of knowledge about the system (crystals), in other words, as a result of this mixing process.

On the other hand, according to classical thermodynamics the adiabatic mixing of the two crystals would result in an increase of entropy suggesting, thus, a relationship between entropy increase in a system and loss of information about it.

This example should, hopefully, enhance our understanding of, and sense about, entropy:

* from the second law point of view, the mixing of these two crystals leads to an increase in entropy due to the irreversible character of the process, while
* from the molecular point of view, this entropy increase is interpreted in terms of loss of information resulting from the mixing process.

We should keep in mind, however, that entropy is a concept introduced by thermodynamics, just as internal energy and enthalpy, and not an observable quantity. Consequently, we can only improve our understanding of it, never grasp it completely.

But this should not make us any less comfortable and confident in using entropy than we are in using internal energy or enthalpy.

The second law applications in the ensuing Examples should increase our confidence in dealing with this property.

3.14 Examples 3.4 through 3.12

Examples 3.4 through 3.12 involve some typical applications of the second law:

* Example 3.4 uses the fact that the total entropy change of a reversible cycle is equal to zero to calculate the work produced by it.
* Example 3.5 demonstrates how entropy changes of the two fluids in a heat exchanger are calculated.
* Examples 3.6, 3.7, and 3.8, use the mathematical statement of the second law to establish if a process, such as the one suggested in Section 2 of this Chapter, is feasible including an interesting application.
* Examples 3.9 and 3.10 prepare the ground for the study of steam power plants. The work produced by a turbine and the properties of the exhaust steam are evaluated for three types of turbine operation:
  a.reversible and adiabatic;
  b.irreversible and adiabatic;
  c.irreversible and non-adiabatic.
* The preparation is completed with Example 3.11 which demonstrates how the pump duty is calculated.

* Example 3.12 demonstrates how calculations for a simple power plant are carried out.

We proceed then with a brief discussion of Steam Power Plants, which should help us understand some of the issues associated with the conversion of thermal energy to mechanical one. These will be discussed in more detail in Chapter 5.

3.14.1 Example 3.4

Calculate the work produced in Example 3.3, using this time the entropy changes involved.

Using the subscripts 'w' for water, and 'a' for atmosphere and for the temperature those of Example 3.3:

* Amount of heat removed from the water:

$$Q_w = (1)(C)(T_2 - T_1) = -60 \text{ kcal}$$

* Entropy change of water (Eq.3.12.3):

$$\Delta S_w = (1)(C)\ln(T_2/T_1) = -0.1862 \text{ kcal/K}$$

Since the cycle is reversible: $\Delta S_a = -\Delta S_w$ and, consequently, the amount of heat transferred to the atmosphere is:

$$Q_a = T_2 \Delta S_a = (293.15)(0.1862) = 54.59 \text{ kcal}$$

Let Q_1 and Q_2 the amounts of heat absorbed and rejected by the cycle. Then:

$$W = Q_1 + Q_2 = 60 + (-54.59) = 5.41 \text{ kcal}$$

Question: why is $Q_1 = -Q_w$, and $Q_2 = -Q_a$?

3.14.2 Example 3.5

Oxygen at 1 atm pressure and at a rate of 1 mol/s is heated isobarically from 15°C to 50°C using saturated steam of 1 atm pressure. Calculate the entropy change of:

a. the oxygen stream, ΔS_0, and

b. the heating medium, ΔS_S.

Given: $C_P^* = 26.625 + 0.0049T$; J/mol K ; T, K

a. For isobaric heating, from Eq.3.12.3:

$$\Delta S_0 = \int \frac{C_P^*}{T} dT = 26.625 \ln\frac{T_2}{T_1} + 0.0049\,(T_2 - T_1) = 3.224 \text{ J/Ks}$$

b. Calculate first the heat absorbed by the oxygen stream:

$$Q_0 = \int C_P^* dT = 26.625\,(T_2 - T_1) + \frac{0.0049}{2}\left(T_2^2 - T_1^2\right) = 984.3 \text{ J/s}$$

This amount has been lost by the condensing steam. Hence:

$$\Delta S_S = -Q_0/T_S = -2.638\ \mathrm{J/Ks}$$

Note. We have assumed here that the flow rate of steam is high enough that no cooling of the condensate occurs. Will the entropy change of the heating medium be the same, if such cooling did occur?

3.14.3 Example 3.6

We are now ready to consider the device mentioned in Section 3.2 (Rationale for a Second Law), which - according to its inventor - takes in a stream of air (A) at 1 atm and 300 K, and delivers two equal streams, also at 1 atm: one (B) at 280 K and the other (C) at 320 K, as shown schematically in Figure 3.E.6. Remember that, as demonstrated in Example 2.6, the device satisfies the first law. Does the device also satisfy the second law?

To this purpose we examine if the device satisfies Eq.3.11.1 and, since there is no heat exchanged with the environment, we consider the entropy change of the system only. Using as basis one mole of air:

$$\Delta S_{sys} = 0.5(S_B+S_C)-S_A = 0.5(S_B-S_A)+0.5(S_C-S_A) =$$

$$= 0.5\,C_P^*\left(\ln\frac{T_B}{T_A}+\ln\frac{T_C}{T_A}\right) = 0.5\,C_P^*(-0.0690+0.0645) = -0.002\,C_P^*$$

Hence the machine violates the second law. Our intuition was correct.

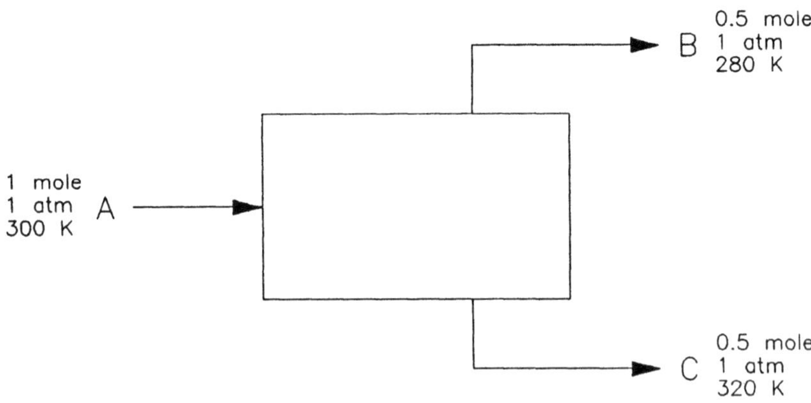

Figure 3.E.6 Schematic diagram for Example 3.6.

3.14.4 Example 3.7. The Hilsch-Ranque Vortex Tube

Before we abandon the device of Example 3.6, let us take a second look at it. How can we make the total entropy change positive? We have seen, in Section 3.12.4, that the entropy change of an ideal gas between two states depends not only on its temperature change, but also on that of its pressure. What will happen, thus, if the pressure of the entering air is - say - 5 atm, and everything else remains the same? Assume, for simplicity, ideal gas behavior with C_P^* temperature independent and equal to 28 J/mol K.

In this case from Eq.3.12.5:

$$\Delta S_{sys} = 0.5\ C_P^* \left(\ln\frac{T_B}{T_A} + \ln\frac{T_C}{T_A} \right) - 0.5R \left(\ln\frac{P_B}{P_A} + \ln\frac{P_C}{P_A} \right) = 13.32\ \mathrm{J/K}$$

and this device is possible. In fact such a device does exist. Van Ness (1969, pp. 72-74) describes it as follows:

"It was first invented in the 1930s by a Frenchman named Ranque, but no one paid any attention. It was not until the Allies entered Germany after World War II that these devices became known. During the war a German by the name of Hilsch had built a number of them and distributed them in German laboratories. Each consisted of a straight piece of tubing or pipe which was divided into two parts by an orifice as shown in Figure 3.E.7a.

Compressed air is fed through a connection adjacent to the orifice, but it is introduced in a special way, that is, tangentially to the tube. This is shown by the cross-sectional view of the tube in Figure 3.E.7b. This entering air, expanding into the tube, reaches fairly high velocities and imparts a rotary motion to the air. This is the reason it is called a vortex tube. In expanding into the tube from a high pressure to a low pressure, the air has gained kinetic energy at the expense of its internal energy.

Thus the temperature of the rotating air mass is lower than the initial air temperature, and it is more or less uniform over the cross section of the tube. However, the linear velocity, and therefore the kinetic energy of the air, varies from place to place in the air vortex, which rotates more or less as a unit. Thus air at the centre of the tube has a very low velocity and kinetic energy, and these increase as we move outward from the centre toward the wall. Very near the wall this analysis fails, but this does not destroy the general line of reasoning. The air near the centre of the tube passes through the orifice and leaves the tube as the cold air stream. The air in the outer parts of the tube is also at a low temperature, but this air has a high kinetic energy. As it swirls down the tube away from the orifice, it loses its kinetic energy as a result of turbulence, internal friction, and mixing, and this kinetic energy reverts to internal energy. Thus the gas temperature rises, and the gas comes out of the tube hotter than the initial temperature because nearly all of the kinetic energy generated by expansion reap-

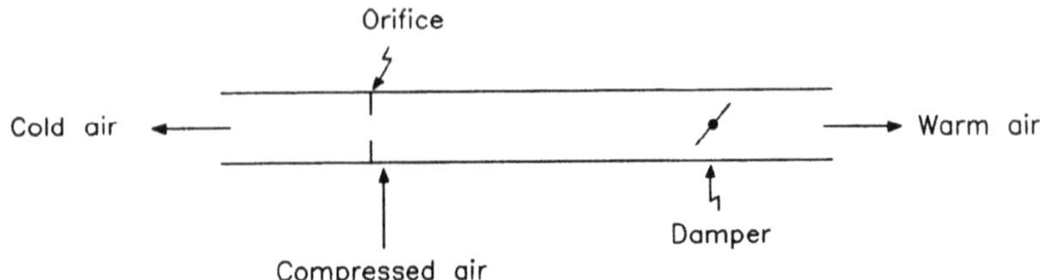

Figure 3.E.7a The Hilsch-Ranque tube.

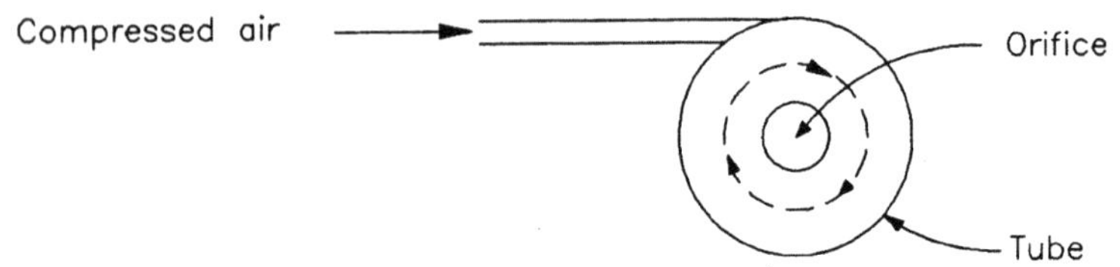

Figure 3.E.7b Cross-section of the tube.

pears as internal energy in only part of the initial gas stream. The highly irreversible nature of this process is clear, and this is the reason that data taken on actual tubes always show large increases in entropy. (See the next Example.)

The sole attractive feature of this device from a practical point of view is the absence of any moving parts. Thus the devices are cheap to build and require no maintenance, and they become practical devices only where these factors are of overriding importance. Thus we find them in use to cool drinking water on some railroad locomotives. Union work rules require that cool water be available for the crew, and unless it is, the locomotive is out of service. Downtime on locomotives is so expensive that it is a big advantage to have a foolproof water cooler. All that is needed is compressed air, and no train can operate anyway without that, for if the air fails the brakes lock."

3.14.5 Example 3.8

A test of the Vortex Tube gave the following data (Figure 3.E.6):
stream A: 5 atm, 20°C
stream B: 1 atm, -21°C
stream C: 1 atm, 26°C
moles C/moles B: $n_C/n_B = 5.4$
Determine the total entropy change using the C_P^* value for air given in Example 3.7.

Basis: 6.4 moles of A. Then, n_B = 1.0 mole and n_C = 5.4 moles.

From an energy balance:

$$(H_B - H_A) + 5.4(H_C - H_A) =$$

$$C_P^*(T_B - T_A) + 5.4C_P^*(T_C - T_A) = -8.6C_P^* = -240.8 \text{ J}$$

i.e. the tube does not operate adiabatically. Rather, 240.8 J per the 6.4 moles of our basis, are transferred into the surroundings. Using as a temperature of the surroundings 20°C, we have:

$$\Delta S_{sur} = 240.8/(273.15 + 20) = 0.82 \text{ J/K}$$

The calculation of ΔS_{sys} is done as before:

$$\Delta S_{sys} = \left(C_P^* \ln\frac{T_B}{T_A} - R\ln\frac{P_B}{P_A}\right) + 5.4\left(C_P^* \ln\frac{T_C}{T_A} - R\ln\frac{P_C}{P_A}\right) = 84.43 \text{ J/K}$$

Hence: ΔS_{tot} = 0.82+84.48 = 85.30 J/K.

3.14.6 Example 3.9

Production of electricity in steam power plants is accomplished by expanding the steam leaving the boiler/superheater (Figure 3.3) through a turbine, or a series of turbines, to some specified exhaust pressure. The produced shaft work is then converted into electricity through a generator coupled with the turbine. We will discuss next how the produced shaft work is evaluated. To this purpose, we will consider:

* intake conditions: P_1 = 35 bar; t_1 = 450°C

* outlet pressure: P_2 = 1 bar

and three cases, determined by the type of turbine involved:

1. the turbine operates reversibly and adiabatically (RA);
2. the turbine operates adiabatically, but irreversibly with an efficiency of 0.90;
3. the turbine is neither adiabatic nor reversible but, due to poor insulation, there is loss of 20 kJ per kg of steam going through. Its efficiency, as compared to the reversible and adiabatic operation, is 0.90.

In all cases we will neglect the effect of potential and kinetic energy changes.

Case 1

From the fist law for a flow process: $Q = \Delta H + W_s$, and since $Q = 0$:

$$W_s = -\Delta H = -(H_2 - H_1)$$

H_1 can be determined from the given conditions (35 bar, 450°C) using the Steam Tables: H_1=3338.0 kJ/kg. But how are we to evaluate the enthalpy of the steam at the exhaust of the turbine, where we only know the pressure P_2=1 bar?

To this purpose, we consider the entropy change of the steam flowing through the turbine:

$$\Delta S = S_2 - S_1 = \frac{dQ_r}{T}$$

We know that $Q = 0$, but is $Q_r = 0$?

The answer is yes, because the turbine is not only adiabatic but also reversible. Consequently, $S_2 = S_1 = 7.0074$ kJ/kg K, as read from the Steam Tables at (P_1, T_1). Thus, the coordinates of point 2 are: $P_2 = 1$ bar, $S_2 = 7.0074$ kJ/kg K. To evaluate H_2 we note that at 1 bar:

s(sat. liquid) = 1.3027; S(sat. vapor) = 7.3598, both in kJ/kg K;
h(sat. liquid) = 417.51; H(sat. vapor) = 2675.4, both in kJ/kg.

By linear interpolation:

$$H_2 = 417.51 + (2675.4 - 417.51)\frac{7.0074 - 1.3027}{7.3598 - 1.3027} = 2544.0 \text{ kJ/kg}$$

and $W_s = -(2544.0 - 3338.0) = 794.0$ kJ/kg.

There is, however, a simpler approach in determining H_1 and H_2, that also avoids this interpolation calculation. It utilizes the Mollier chart and involves the following steps:

1. Locate point 1 on the chart and read: $H_1 = 3338$ kJ/kg.
2. Follow a constant entropy line, by moving straight down, until you meet the pressure of 1 bar.
3. The intersection locates point 2 on the chart, where you read the corresponding enthalpy value: $H_2 = 2545$ kJ/kg.

Thus, $W_s = 793$ kJ/kg, which is close enough - within the accuracy of reading the chart - to the value calculated before.

Case 1 represents the reversible and adiabatic work, which is referred to as W_s(RA). It represents the basis for evaluating the real work through the given turbine efficiency, as we will see in the following Cases 2 and 3.

Case 2

If we follow the same as in Case 1 procedure, we notice that:

* P_1, T_1, H_1; and
* P_2

are the same as in that Case. But is $S_2 = S_1$? The turbine is still adiabatic, so $Q = 0$, but is $Q_r = 0$?

The answer is negative. The lack of complete reversibility inside the turbine leads to turbulence and friction, which results in an increase of the steam entropy through the conversion of some of its energy into heat. And this is what the given efficiency of 0.9 reflects. Thus:

$$W_s(\text{real}) = 0.9 W_s(\text{RA}) = 713.7 \text{ kJ/kg}$$

and the remaining: 793 - 713.7 = 79.3 kJ/kg, have been converted into heat. It is found in the enthalpy of the exhaust steam:

$$W_s(\text{real}) = -[H_2(\text{real}) - H_1] = 713.7; \text{ and:}$$
$$H_2(\text{real}) = 3338 - 713.7 = 2624.3 \text{ kJ/kg}$$

which is larger by: 2624.3 - 2545 = 79.3 kJ/kg than the value corresponding to the reversible operation.

Notice that once irreversibility enters the picture, thermodynamics alone does not suffice for the determination of the shaft work. The turbine efficiency must be known, usually from the manufacturer.

Case 3
Here the turbine is neither adiabatic nor reversible, but it delivers, as in Case 2, 90% of the W_s(RA). Hence, W_s(real) = 713.7 kJ/kg. The heat loss will be reflected in the properties of the exhaust stream as we will see in the next Example.

3.14.7 Example 3.10

For the three Cases of Example 3.9, determine the properties of the exhaust steam. More specifically the:

* enthalpy,
* entropy,
* quality, and
* temperature.

(Quality is defined as: 100 minus the percent of moisture, i.e. condensed water, present in the steam).

Case 1
We have already located point 2. Hence, from the Mollier chart:

$H_2(1)$ = 2545 kJ/kg
$S_2(1)$ = 7.0074 kJ/kg K
$Qual(1)$ = 94.0 %

In determining the temperature $t_2(1)$ we notice that, since condensation or vaporization of a pure fluid at a given pressure takes place of constant temperature, $t_2(1) = t$(sat. vap. at 1 bar) = 99.6°C.

Case 2
Here we have not established point 2 yet. But we know that P_2 = 1 bar; and we calculated $H_2(2)$ = 2624.3 kJ/kg. We can, hence, locate point 2 on the chart and read:

$S_2(2)$ = 7.21 kJ/kg K
$t_2(2)$ = 99.6°C (for the same reason as before)
$Qual(2)$ = 97.7%

Case 3
Here we only know P_2. We can calculate $H_2(3)$, however, from the first law:

$$Q = \Delta H + W_s$$
$$-20 = [H_2(3) - H_1] + 713.7$$

and: $H_2(3)$ = -713.7 - 20 + 3338 = 2604.3 kJ/kg. We can now locate point 2 and read:

$S_2(3)$ = 7.16 kJ/kg K
$t_2(3)$ = 99.6°C
$Qual(3)$ = 96.8%

3.14.8 Example 3.11

Show that the adiabatic and reversible shaft work is given by: $W_s = -\int VdP$. The limits of integration are the inlet and outlet pressures.

Starting with the first law for a flow process and for an adiabatic operation:

$$dW_s = -dH = -(dU + PdV + VdP)$$

Also from the first law:

$$dQ = dU + PdV = 0$$

and consequently:

$$W_s = -\int VdP$$

3.14.9 Example 3.12

We are now ready to calculate the thermal efficiency, and some other pertinent quantities, of a power plant cycle. To this purpose let us assume that the simple power plant of Section 3.6 operates as follows:

* the boiler/superheater operates at 35 bar producing steam with a temperature of 450°C;
* the turbine is adiabatic but irreversible with an efficiency of 0.9 and an exhaust pressure of 1 bar;
* the condenser operates at 1 bar and the water leaving is saturated.

Calculate the thermal efficiency and the condenser duty Q_c.

With reference to Figure 3.3, and using as a basis one kg of steam:

$$n = \frac{W_t + W_p}{Q_1}$$

From the first law:

$$W_t = -(H_3 - H_2);\; W_p = -(h_1 - h_4);\; Q_1 = (H_2 - h_1);\; Q_c = (h_4 - H_3)$$

For the enthalpy values, we note that from Example 3.9: $H_2 = 3338$; $H_3 = 2624.3$ (Case 2); and from the Steam Tables: $h_4 = 417.51$, all in kJ/kg.

The evaluation of h_1 is not so straight forward. The Chart does not cover the liquid range so that we can follow a straight line - assuming a reversible and adiabatic pump - until we hit the pressure at 35 bar. Instead we calculate the pump duty using the expression developed in Example 3.11:

$$W_p = -\int_{P_4}^{P_1} vdP$$

Here v stands for the volume of liquid water and, since liquids are incompressible for practical purposes, we use v = 1043 cm^3/kg, i.e. the volume of saturated water at 1 bar. Hence:

W_p = -1043 cm^3/kg x(35 - 1) bar = -3.55 kJ/kg, and

h_1 = 417.51 + 3.55 = 421.06 kJ/kg

We proceed now with the calculations:

W_t = -(2624.3 - 3338) = 713.7 kJ/kg

Q_1 = (3338 - 421.06) = 2916.94 kJ/kg

n = (713.7 - 3.55)/2916.94 = 0.243

Q_c = 417.51 - 2624.3 = -2206.79 kJ/kg

Comments.

1. Heat absorption in the boiler/superheater involves three steps.

a. In the boiler, the water is first heated to the saturation temperature at 35 bar, 242.54°C, absorbing an amount of heat q_1' = 1049.8 - 421.06 = 628.74 kJ/kg.

b. It is then vaporized at the constant temperature of 242.54°C absorbing an amount of heat: q_1'' = 2802.0 - 1049.8 = 1752.2 kJ/kg.

c. It enters then the superheater and absorbs an amount of heat q_1''' = 3338.0 - 2802.0 = 536.0 kJ/kg.

The significance of these amounts, and corresponding temperatures, will be discussed in the ensuing Section 3.15.1.

2. The calculated thermal efficiency represents also the power plant efficiency.

3. As we can see, the pump duty is very small as compared to the work produced by the turbine. It is thus often neglected in calculating the thermal efficiency.

4. The requirement of the second law that all heat absorbed cannot be completely converted into work in a cyclic process, i.e. that part of it must be rejected, is very painful here. About 3/4 of the heat absorbed in the boiler/superheater is rejected in the condenser. And not only do we lose it, but it is a main source of thermal pollution.

5. We will see in Example 3.13, and in more detail in Chapter 5, however, that there is an approach, called Cogeneration, whereby the enthalpy of the exhaust from the turbine steam is utilized.

3.15 Steam Power Plants

3.15.1 Operating Conditions

In Section 3.6 we described briefly a simple steam power plant and in Example 3.12 we calculated the thermal (or power plant) efficiency of such a plant. For this purpose we used an arbitrary set of operating conditions:

* boiler pressure and degree of superheat
* condenser pressure.

In practice, the choice of operating conditions is determined by:

1. achievement of the maximum efficiency; and
2. materials of construction.

For maximum efficiency, according to the Carnot Cycle:

* heat absorption must occur at the highest possible temperature, as close as possible to that of the flue gases; and
* heat rejection must occur at the lowest possible temperature.

Heat absorption, as discussed in Example 3.12, involves three steps:

1. heating the water to the saturation temperature at the boiler pressure;
2. vaporization of the saturated water at constant temperature;
3. superheating of the saturated vapor.

A close examination of the amounts of heats involved in these steps, 629, 1752 and 536 (all in kJ/kg) respectively for the power plant cycle of Example 3.12, suggests that the most important one, with respect to efficiency, is the temperature at which water is vaporized which, in turn, is determined by the boiler pressure.

The effect of the boiler pressure on the power plant efficiency is presented graphically in Figure 3.7. The results are based on a steam temperature of 475°C and a condenser pressure of 1 bar. Notice that, as

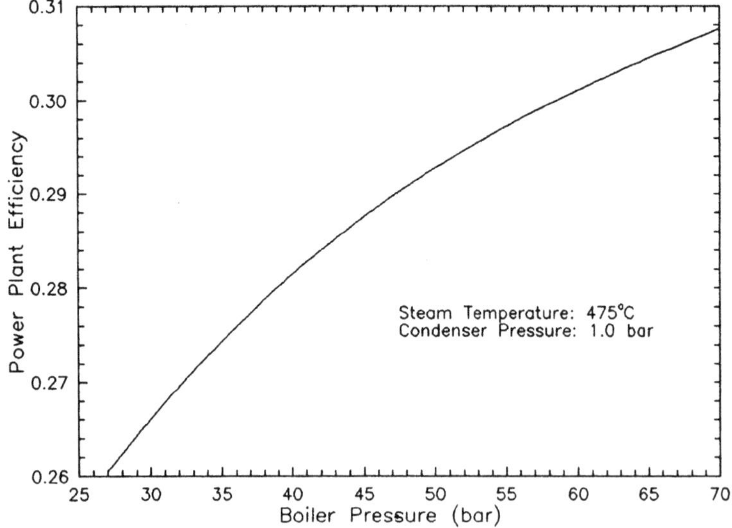

Figure 3.7 Effect of boiler pressure on power plant efficiency.

predicted by the Carnot cycle, higher boiler operating pressures, and thus vaporization temperatures, lead to increased efficiencies.

Since the degree of superheat effects also the average temperature at which heat is absorbed, in the results presented in Figure 3.8 we keep the boiler pressure constant at 35 bar, and vary the temperature of the superheated steam. The condenser operates as before at 1 bar. Again, as predicted by the Carnot cycle, higher operating temperatures lead to increased efficiencies.

Turning to the heat rejection step, we notice that the temperature at which the condenser operates is determined, as with case of the boiler, by the operating pressure. The effect on the cycle efficiency is presented in Figure 3.9. Here we keep the boiler pressure and steam temperature constant (35 bar and 475°C). Again as predicted by the Carnot cycle, low operating pressures and, thus, temperatures, lead to higher efficiencies.

Note: Calculations of efficiencies are based on a reversible and adiabatic turbine.

3.15.2 Comments

1.We conclude that, as predicted by the Carnot cycle, for increased thermal efficiency:

a.high boiler pressure and degree of superheat, and

b.low condenser pressure,

are desirable.

2.Modern power plants, thus, operate at high pressures and temperatures, within the limitations imposed by the materials of construction, but rarely much above 100 bar and 600°C (Smith and Van Ness, 1987).

3.They also operate at low condenser pressures, determined by the temperature of the cooling water, with typical values of around 0.1 bar.

4.More complex cycles are also used involving multistage turbines with part of the exhaust from each stage steam used to increase the feed water temperature (why?). See, for example, Problem 3.57.

5.Under such conditions, overall thermal efficiencies of about 0.45 are realized.

6.A new approach, actually an old one that is revived, provides improved utilization of the fuel heat by combining the generation of mechanical power with the production of process steam. This is referred to as Cogeneration and is discussed in detail in Chapter 5, where the efficient utilization of thermal energy is discussed in detail. A simple case, is demonstrated in the next Example.

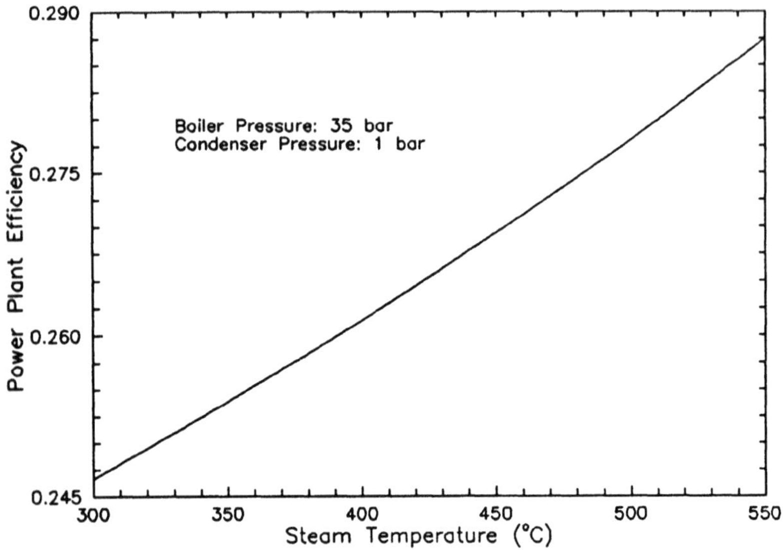

Figure 3.8 Effect of steam temperature on power plant efficiency.

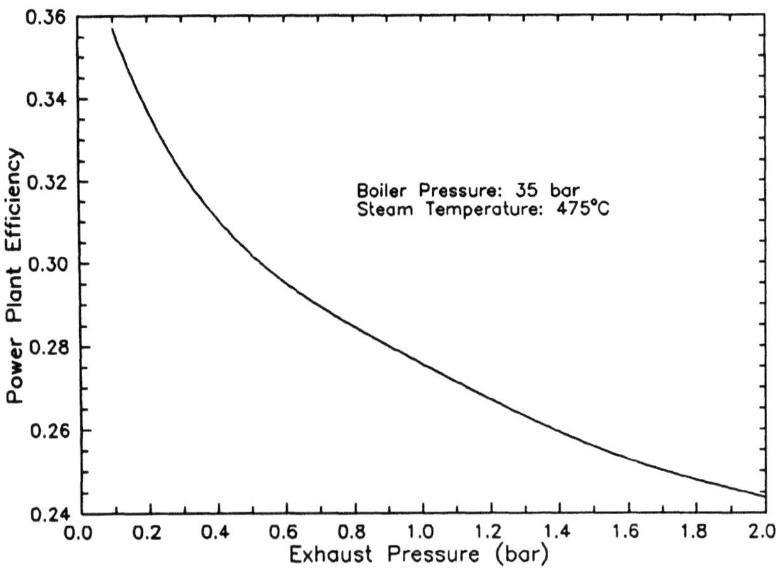

Figure 3.9 Effect of condenser pressure on power plant efficiency.

3.15.3 Example 3.13. A Look at Example 1.9

Let us consider again the debate among George, Monica, and David and see if we can enter into it.

We start with David's argument: "A Joule is a Joule". Even modern plants will give no more that 0.45 Joule of electricity per Joule of fuel heat . Here "A Joule is *not* always a Joule".

Monica's position is more valid. Her boiler should produce - in theory at least - one Joule of steam enthalpy per Joule of fuel heat. About 0.9 in practice.

George's position is much better, however. Using the same ratio of 0.9 Joule of steam enthalpy per Joule of fuel heat, as in Monica's case, he will be getting about 0.9 Joule of electricity per extra - over Monica's - Joule of fuel heat. This is about twice what modern power plants get. But George read about this approach somewhere. It is called *Cogeneration.*

To better understand George's argument, and the concept of Cogeneration, let us assume that the cycle he proposes operates as follows:

1. Steam at 100 bar and 500°C leaves the boiler/superheater;
2. The turbine is adiabatic, with an efficiency of 0.9 and an exhaust pressure of 3 bar, i.e. the desired pressure for the process steam; and
3. One GJ/hr of process heat is required.

Following Figure 3.3, with the condenser replaced by the distillation boilers where the exhaust steam will be used, we arrive at the following operating conditions:

* Point 2: P_2 = 100 bar; t_2 = 500°C; H_2 = 3374.6 kJ/kg
* Point 3: P_3 = 3 bar; H_3(RA) = 2570 kJ/kg. Hence:
  H_3 = 3374.6 - 0.9(3374.6 - 2570) = 2650.5 kJ/kg; t_3 = 133.5°C
* Point 4: Assume, for simplicity, that only the heat of condensation of the process steam is used. Hence:
  P_4 = 3 bar; t_4 = 133.5°C; h_4 = 561.4 kJ/kg
* Point 1: $W_p = -1.073\cdot 10^{-3}(P_1 - P_4)$ = -10.4 kJ/kg; Hence:
  h_1 = 571.8 kJ/kg

Summarizing the cycle:

Q_1 = 3374.6 - 571.8 = 2802.8 kJ/kg
Q_2 = 2650.5 - 561.4 = 2089.1 kJ/kg (process heat)
W_t = 3374.6 - 2650.5 = 724.1 kJ/kg
n = (724.1-10.4)/2802.8 = 0.255

To produce, therefore, the required 1 GJ/hr of process steam, the amount of fuel heat - using a boiler/superheater efficiency of 0.9 - is:

$$Q = \frac{1}{0.9}\frac{2802.8}{2089.1} = 1.49 \quad \text{GJ/hr}$$

The amount of mechanical energy produced is:

$$W = 1.49\text{x}0.9\text{x}0.255 = 0.342 \text{ GJ/hr}$$

Hence, George's approach uses (1.49-1/0.9) = 0.38 GJ/hr more fuel heat than Monica's approach. But 90% of it is converted into mechanical energy. This is the advantage of Cogeneration.

Note. The water used in power plant cycles must be very pure to avoid corrosion of the turbine blades. In practice, therefore, the exhaust steam is used to generate the process steam through a heat exchanger, not directly (see Chapter 5).

3.16 Concluding Remarks

1.The feasibility of a process should be tested against both, the first and the second law. To this purpose the latter introduces the state function *entropy*, defined by Eq.3.10.1, and requires that Eq.3.11.1 be satisfied. Note that the entropy change of both, system and surroundings, should be considered.

2.The second law establishes also the criterion for the efficient utilization of energy in a given process: the less total entropy - system plus surroundings - it generates, the more efficient it is. (This subject is considered in more detail in Chapter 5.)

3.Conversion of thermal energy - our main source of energy - into mechanical takes place through cyclic processes. The thermal efficiency (Eq.3.5.1) in these processes depends on the temperature:

* at which the thermal energy is available, and
* of the heat sink, where the thermal energy not converted into mechanical is rejected to.

4.The maximum thermal efficiency is established by the Carnot cycle, Eq.3.5.3, which indicates that complete conversion of thermal energy into mechanical is not possible.

5.Conversion of thermal energy into mechanical is carried out in power plants with maximum efficiencies of about 45%.

6.Improved efficiency in the utilization of thermal energy for the production of mechanical one can be accomplished by using Cogeneration.

References

Cardwell, D.S.L., 1971. *From Watt to Clausius, the Rise of Thermodynamics in the Early Industrial Age*, Cornell University Press, Ithaca, N.Y.

Denbigh, K., 1981. *The Principles of Chemical Equilibrium*, Cambridge University Press, Cambridge.

Seltzer, R.J., Feb. 9,1976. *Chemical & Engineering News*, 19.
Smith, J.M., Van Ness, H.C., 1987. *Introduction to Chemical Engineering Thermodynamics*, McGraw-Hill, New York.
Van Ness, H.C., 1969. *Understanding Thermodynamics*, McGraw-Hill, Paperbacks, New York.

Problems

3.1 With reference to Example 3.1:

a. how much work is produced per mole, if the expansion takes place at 350 K, from 3 atm to 1 atm?

b. how much work is required to bring the gas back to 1 atm at 350 K?

c. would $Q = W$ be valid, if the gas were not ideal? Explain.

Assume reversible operation.

3.2 Discuss the similarities and dissimilarities between the efficiencies of the Carnot cycle and a water-wheel (see Example 2.7).

3.3 Develop the expression for the efficiency of a Carnot cycle.

3.4 A Carnot cycle operates between the constant temperatures of 100°C and 50°C. For each Joule of work produced how much heat is rejected?

3.5 With reference to Example 3.3, what is the minimum amount of work required to heat the 1 kg of water from 20°C back to 80°C, using the atmosphere as a heat source?

3.6 What is the maximum amount of work that can be produced from 1 kg of saturated steam at 1 atm? The atmospheric temperature is 25°C.

3.7 Discuss the practical usefulness of the Carnot cycle.

3.8 Assume that in Section 3.13 we used two crystals containing six atoms each.

a. How many arrangements are possible in each state?

b. Which one of them will the system be in?

3.9.a. What conclusion do you reach about entropy when considering the molecular nature of matter?

b. Compare it to the classical thermodynamics view of entropy.

3.10 In Example 3.5, assume that the flow rate of steam is such that the condensate leaves at 60°C.

a. Will the entropy change be the same as it was with the case of saturated liquid? Explain your answer.

b. If the answer is no, what is the entropy change?

3.11 One mole of gas A is heated from 50°C to 100°C using gas B, whose temperature changes from 150°C to 120°C.

a. Calculate the entropy changes: gas A, gas B, and total, for co-current operation.

b. Would the values be the same, if the operation were counter-current?
Assume ideal gas behavior with: $C_P^* = 7$ cal/mol K for both gases.

3.12 One hundred kg of water are heated from 5°C to 90°C at $P = 1$ atm using the condensation of saturated steam at 1 bar.

a. Calculate the entropy change of the water and of the steam.

b. Does the process satisfy the second law?

c. If 15 kg of the steam are used for the same purpose, what will be the entropy change of the heating medium?

d. Which of the two heating processes is less irreversible?

3.13 One kg of saturated steam at 1 bar, and one at 1 bar and 200°C, are mixed adiabatically.

a. What is the entropy change of each?

b. What is the entropy of the mixture?

c. Is the process reversible?

3.14 One mole of an ideal gas is compressed from $[T_1, V_1]$ to $[T_2, V_2]$.

a. Starting with Eq.3.10.1, develop an expression for the entropy change of the gas.

b. What is the total entropy change?

3.15 One mole of methane at 300 K and 1 bar is mixed with two moles of ethane at 400 K and 2 bar. What is the entropy change?

3.16 Superheated steam at 20 bar and 550°C enters a turbine and leaves at 0.15 bar.

a. If the turbine is reversible and adiabatic, what is the temperature of the exiting steam?

b. If there is a heat loss of 7 kJ/kg of steam, and the turbine has an efficiency of 90% (as compared to the reversible and adiabatic operation), what is the enthalpy of the exiting steam?

3.17 One mole of propane is compressed in a flow process from 1 bar and 60°C to 10 bar. The compressor is adiabatic and reversible.

a. What is the entropy change of the gas?

b. What is the temperature of the exiting gas?

c. How much work is required?

Assume that in this range propane behaves approximately as an ideal gas.

3.18 With reference to the previous Problem, what would the temperature of the compressed gas be, if there were a loss of 0.2 J/mol and the compressor efficiency is 0.8 (based on the reversible and adiabatic operation)?

3.19 Starting with: $\Delta S_{tot} > 0$, show that it is impossible to develop a process whose only effect is the transfer of one Joule of heat from 273 K to 373 K.

3.20 Ethane expands from 10 bar and 100°C to 1 bar and 20°C.

a. What is the entropy change of the gas?

b. Can you calculate the work produced?

3.21 Show whether the following process is possible: Air at 7 bar and 40°C enters an apparatus that can exchange heat with the surroundings but no mechanical work. Half of the air leaves at 70°C and the other half at -20°C, and both streams have the same composition with the inlet stream. Assume ideal gas behavior with $C_P^* = 7$ cal/mol K.

3.22 Assuming steam to be an ideal gas find the entropy change per kg when the pressure and temperature change from 1 bar and 125°C to 10 bar and 450°C. What is the error resulting from the ideal gas assumption?

3.23 Steam in steady flow is expanded reversibly and adiabatically as it flows through a tube of variable cross-section. The steam enters the tube at 18 bar and 375°C with a negligible velocity, and leaves at 5 bar. What is the velocity of the steam leaving the tube?

3.24 An inventor claims that his apparatus, which exchanges no energy with the surroundings, can separate an equimolar mixture of nitrogen and oxygen at 2 bar and 300 K into practically pure products at the same temperature and a pressure of 1 bar. Is his claim valid?

3.25 Ten lb_m of ice (C_P=0.5 Btu/lb_m R) at 10°F are placed in an insulated tank containing 100 lb_m of water at 90°F. Calculate the resulting entropy change when the system reaches equilibrium. Heat of fusion of ice: 144 Btu/lb_m.

3.26 Which operating pressure in a power plant condenser will lead to higher cycle efficiency:

a. 0.1 bar;

b. 1 bar.

The boiler/superheater operation is the same in both cases. Justify your answer.

3.27 A simple power plant operates as follows:

1. steam at 600°C and 50 bar leaves the superheater;
2. the turbine is adiabatic but irreversible with an efficiency of 0.90;
3. saturated water leaves the condenser which operates at 0.1 bar.

Calculate:

a. the cycle efficiency, and

b. the condenser duty.

3.28.a. Describe the operation of the power plant cycle of Problem 3.27 on a *T-S* diagram by plotting the temperature of the water as a function of its entropy as it goes through the cycle.

b. Show on this diagram the effect of irreversibility in the turbine operation.

3.29 Process heat at a rate of 500 MJ/hr, as saturated steam at 5 bar, is needed.

a. How much mechanical energy can be produced, if a cogeneration scheme is used?

b. What is the thermal efficiency of the cycle, based on the heat used for the production of the mechanical energy only?

c. If a conventional cycle were used to produce this mechanical energy, with a turbine exhaust pressure of 0.4 bar, what would its efficiency be?

Use any of the information given in Example 3.13 that you find appropriate.

3.30 Show that entropy is a state function.

3.31 Discuss the importance of the second law in the Chemical Industry.

3.32 Explain why in the Hilsch-Ranque tube the air enters tangentially and close to the orifice of the tube.

3.33 What is the major source of irreversibility in a steam power plant using a fossil fuel as source of thermal energy?
3.34 An inventor claims that he can avoid the condenser in a power plant cycle - and the resulting loss of energy there - by compressing the exhaust from the turbine vapor directly into the boiler. Identify the problems you see with this approach.
3.35 With reference to this Chapter:

a. What are the most important items that you have learned?

b. Give an example in each case.

3.51 Show, using the verbal statement of the second law, that if two reversible heat engines operate between the same temperatures, they have equal efficiencies.
3.52.a. What is the difference between the thermodynamic and the ideal gas temperature?

b. Do they have to be the same?

3.53 Two moles of a given liquid are available. One is at 400 K and the other at 300 K. The heat capacity is 20 cal/mol K and is temperature independent.

a. What is the final common temperature of the system that yields the maximum amount of work that can be obtained from it?

b. What is this work?

c. What is the common temperature that leads to the minimum work produced?

3.54 Using the mathematical statement of the second law, prove the verbal ones.
3.55 Discuss the purpose of the four Propositions.
3.56 Show that for a real process the total entropy change is positive.
3.57 Steam at 30 bar and 600°C is expanded in the first stage of a turbine to 3 bar. At this point a portion of the steam is extracted for feed-water heating and the rest is expanded in the second stage of the turbine to 0.1 bar, the condenser pressure. The saturated liquid condensate from the condenser is pumped into the feed-water heater where it mixes with the extracted steam, the mixture leaving as saturated water at 3 bar.

a. Find the amount of steam extracted per kg of steam generated.

b. What advantage do you see in using the extracted steam in the feed-water heater?

Assume reversible and adiabatic expansion in the turbine and adiabatic operation of the feed-water heater.
3.58 With reference to the previous problem, calculate the thermal efficiency of the cycle for:

a. the proposed operation; and

b. one where all the steam expands through the second turbine.

Explain the difference.

4
Thermodynamics: A Historical Perspective

4.1 Introduction

In the last two Chapters we concluded that the first and second laws represent the basis for the development of the framework of equations necessary for the:

a. calculation of energy balances;

b. determination of the feasibility of processes;

c. establishment of guidelines for the efficient utilization of energy.

Furthermore, as we will see in later Chapters, these two laws provide also the framework of equations used for the solution of two other important chemical engineering problems:

a. the evaluation of thermodynamic properties: enthalpy, entropy, etc., from the measurable properties: pressure, volume, temperature, and heat capacities; and

b. the calculation of chemical and/or phase equilibrium.

Even though these two laws were developed over a somewhat short period of time, about 50 years, their evolution scans a period of several centuries.

The real impetus for the birth of thermodynamics, and especially the first and second laws, was the *Steam Engine*. Actually, it was exactly the conversion of thermal energy into work that gave the name to this new field of science and technology. (From the Greek: therme = heat; dynamis = force = work - for in that era, force and work were not differentiated.)

This unlimited - as it seemed then - source of energy replaced the limited human and animal ones, and provided - literally - the motive power for the establishment of the *Industrial Revolution*.

The development of the steam engine, however, while accelerated by the urgent needs of the Industrial Revolution for power, was a natural consequence of early - very crude by today's standards, of course - experiments that demonstrated the power of heat.

From the theoretical point of view, the so-called *Science of Heat* was in such an embryonic state that contributed rather little to the development of the steam engine. (Actually, some important contributions in this field took place during - and to some degree, as a result of - the work on the steam engine.) It does represent, however, the theoretical precursor of thermodynamics.

4.2 Objective and Approach

Our objective, as we take this trip in the past, from the very far to the more recent, is to:

1. Trace the development of the two laws enhancing, in the process, our understanding of them and of the quantities associated with them: heat, work, internal energy, and entropy; and of the concept of reversibility.
2. Appreciate the broader importance of thermodynamics, by recognizing the role it played in the establishment of the Industrial Age, in which we still live today - the suggestion that we are already in the 'Post-Industrial' Age not withstanding.

Our underlying objective, however, is a more fundamental one: *To provide a sense of the history of thermodynamics, for a field of Science and Technology - just like a Society - cannot function and grow without it.*

We start with a brief discussion of the early recognition of the power of heat, followed by a historical account of the development of the steam engine and of the evolution of the science of heat through the 17th century.

We proceed, then, with the establishment of the first and second laws through the contributions of Carnot, Clapeyron, Joule, Kelvin, and Clausius, relying on the excellent book of D.S.L. Cardwell (1971) *From Watt to Clausius, the Rise of Thermodynamics in the Early Industrial Age*.

We close with a brief account of the developments associated with the evaluation of the thermodynamic properties of matter, a field that was founded with the pioneering work of J.W. Gibbs.

4.3 The Power of Heat: Early Recognition

Heat is, in fact, the grand moving-agent of nature. Inertia, together with central forces, account for the local motion of planets and satellites; on the earth heat is the source of motion; without it the accidental factors of air resistance and friction would soon destroy all motion. There would be no rain, no winds, no variations in climate, no crops and indeed no life at all. (Cardwell, p.187)

Fire has been of paramount importance to humans from the beginning of their existence, its power so overwhelming that the early people worshipped it. Harnessing this power has been the aspiration of philosophers, dreamers, scientists, and engineers, through the years. Even today, the use of the 'nuclear fire' of the sun represents the biggest, perhaps, dream of the human race.

Table 4.1 presents a brief list of events that demonstrate the recognition of the fire - and the resulting heat - power, going back to the pre-Christ period. For more details on the fascinating devices associated with these events the interested reader should see: *The Great Engineers* by J.B. Hart (1928).

The First James Bond was a Thermodynamicist !!!

The Marquis of Worcester was a man of many accomplishments. In addition to a bent for scientific investigation, he was also a man of affairs. He had a somewhat chequered career. Becoming involved in the Civil War between Charles I and the Parliament, he took up the cause of his King with a vigour which, unfortunately for himself, lost him his fortune and brought him to imprisonment in Ireland. He succeeded, however, in escaping to France, and there joined the Court of the exiled Charles. Returning to England in the role of secret agent to the Royal Family, he was detected and thrown into the Tower of London. Here he had the time and leisure to exercise his bent for scientific reflection. The story is told of him, ... that whilst in the Tower his attention was attracted to the continual lifting of the lid of a saucepan in which his dinner was cooking. His reflections led him along the path of invention with which his name has since become associated. (Hart)

Table 4.1 Major Early Events in the Recognition of the Power of Heat

1st Century B.C.	Heron designs (builds?) the 'Aeolopile', a toy operating with steam.
11th Century	Gunpowder is probably used in China and India in cannons.
15th Century	Leonardo da Vinci proposes the 'Architronito', a steam using cannon.
16th Century	Blanco de Garay uses steam power to move a ship out of port(?).
1601	Giovanni Battista della Porta uses steam power to transfer water out of a vessel.
1629	Giovanni Branca describes a steam-powered machine that can be considered the forerunner of the steam turbine.
1628-1647	Marquis of Worcester erects (?) two machines that use steam power to raise water.
1678-1680	Jean Hautefeuille and Christiaan Huygens design cylinder-and-piston arrangements operating with gunpowder, which can be considered the forerunner of the internal combustion engine.

4.4 The Steam Engine

On Steam Power and Good Horses(!)

Water being evaporated by fire, the vapours require a greater space (about two thousand times) than that occupied by the water, and rather than submit to imprisonment, it will burst a piece of ordnance. But, being controlled according to the laws of Statics and, by science, reduced to the measure of weight and balance, it bears its burden peaceably (like good horses), and thus may be of great use to mankind, especially for the raising of water. { Sir Samuel Morland, first Master Mechanic of the Royal Household, 1682 (Hart, p.66)}.

Following the events outlined in Table 4.1 we arrive to the last part of the 17th century. In England , it is the dawn of the Industrial Revolution that was destined to transform the world.

Industry, however, needed power. The flooding of mines, for example, becomes a serious problem as the available horses are not sufficient to remove the water. Attention to the whole problem of steam and its practical use is becoming more rigorous, and King Charles II creates the post of 'Master Mechanic' to the Royal Household. It is in this atmosphere that the development of the steam engine, outlined in Table 4.2, took place.

The Savery engine represents the first, probably, commercial steam engine to come to existence using high pressure steam to remove the water from the mines, thus its name. It had a rather limited success however. "Exactly how many of these engines were made is unknown; whether any of them worked satisfactorily is uncertain." (Cardwell, p.13)

A major improvement came soon by Thomas Newcomen's steam engine. He introduced in 1712 a piston-and-cylinder arrangement operating by the weight of the atmosphere, thus avoiding the problems with high pressure steam. "The Newcomen engine became, therefore, the most important technological innovation of the time. Such engines were set up all over England and Scotland, as well as abroad: Sweden, Schemnitz (Slovakia) - which was at the center of one of the leading mining areas of Europe - France, Spain, Russia and the Americas... And the systematic work of John Smeaton in England led to doubling efficiency of the Newcomen engine from 4.7 to about 9.0 million $\mathrm{ftlb_f}$ per bushel of coal." (Cardwell, p.32) *Notice the way efficiency was measured at that time!*

Table 4.2 Major Events in the Development of the Steam Engine

1698	Savery presents a model of his steam engine (the Miner's Friend) to the Royal Society of England.
1712	Newcomen presents a piston-and-cylinder steam engine.
1769	Watt patents a new type of steam engine that employs a separate condenser.
1774	The firm of 'Boulton and Watt' presents the first really successful single-acting steam engine.
1781	Watt receives a patent for the double-acting steam engine.
Early 1800's	Trevithick introduces the high pressure steam engine.

4.4.1 James Watt

The consequences of James Watt's double-acting engine were far reaching. Reasonably efficient in construction, commercially successful in practice, it created a new industrial situation. By linking up the crankshaft of the engine with suitable shafting in the workshop, the large-scale use of machinery in industry became possible. The whole outlook of industrial life was changed, and parallel inventions and improvements in the machinery of the textile and other industries were rapidly effected. Thus was brought about what is known as the Industrial Revolution. A new era for industry was at hand. (Hart)

The next major development comes from James Watt (1736-1819) who was working at Glasgow University, a place that was becoming a center for the study of heat and the technology of power.

Watt was an instrument maker and a scientist, not an engineer. His first experience with a Newcomen engine was in the winter of 1763 while repairing a small replica of it, used in a natural philosophy class. He soon became aware of the tremendous loss of steam involved in heating the cylinder after the condensation step - which was more pronounced in the small engine due to 'scaling' effect. This was one of the major problems of the Savery engine which, of course, remained with the Newcomen one as well. Watt spent the following two years trying to resolve this problem. (For a detailed account of this effort - during which Watt stumbled into the phenomenon of latent heat and also determined the vapor pressure of water with remarkable accuracy - the reader is referred to Cardwell.)

His elaborate analysis of the problem led him to the following dilemma: "For maximum economy the cylinder must be kept hot all the time, but for maximum power it must be cooled down once every cycle. These desiderata are quite incompatible, for a cylinder cannot be both hot and cold at one and the same time. Or rather they can be compatible if we have two cylinders in place of one; the first to be kept hot all the time, the second cold." (Cardwell)

The introduction of the separate condenser was a major breakthrough in the development of the steam engine.

Watt obtained his patent in 1769 and turned to the construction of the first commercial engine, not an easy task. Hart (p.84) describes the commercialization of the Watt engine :

"In order to produce an engine model on these lines, Watt hired a room in an old pottery, and here, through the good offices of Dr. Black, he became acquainted with Dr.Roebuck, who had just started the famous

Carron Iron Works. ... The expenses to which Watt was put as a result of his experiments considerably impoverished him, but now Dr. Roebuck agreed to finance him, and the two went into partnership.

The first engine made by Watt in 1769 was not a great success. The condenser was of the surface type and was not sufficiently water- and steam-tight, and the piston leaked very seriously.

Watt became very despondent, and in his anxiety to relieve Dr. Roebuck of the financial losses that threatened he looked around for other help. This luckily came to hand in the person of Matthew Boulton, the son of a Birmingham engineer, whom Watt had met during a journey to London for the purpose of securing his patent. Thus was set up the famous partnership of Boulton and Watt that so greatly enriched the profession of engineering."

Nevertheless, it was not until 1774 that the first really successful single-acting engine was effected. The orders started to come in to the firm and Watt, with the financial anxiety removed, continued to improve the engine, receiving in 1781 a patent for a double-acting steam engine.

Steam Engine and the Urbanization of Society

... the textile revolution ... demanded that every inch of fall on every reasonably big river or stream should be harnessed for production; and so it came about that textile mills were spread out, like beads on a thread, along the rivers of the textile counties. During this phase the industrial revolution was essentially rural in its location; concentration into great industrial cities and conurbations came later, as a consequence of the wider use of the steam engine. (Cardwell, p.68)

4.4.2 Richard Trevithick and the Railway Locomotive

The final major improvement of the steam engine is due to Richard Trevithick who pioneered the use of high-pressure steam, much to the elderly Watt's disapproval. Cardwell comments:

"The advantages of high-pressure steam were obvious enough: the higher the pressure the smaller the cylinder needed to generate the same power. Moreover, the higher the pressure the less necessary the condenser became; to do without one and exhaust the steam directly into the air meant, for an engine working at an excess pressure of five atmospheres, losing only one-sixth or about 17 per cent of the total available pressure difference, while gaining enormously in simplicity, cheapness and mobility. Without the cumbersome condenser there was no real obstacle to putting a small and powerful high-pressure engine on wheels. Hence the railway locomotive was born from the supersession of Watt's low-pressure condensing engine by Trevithick's high-pressure engine. Further and incidental advantages were that the high-pressure engine, being small, might be expected to lose less heat by convection, conduction and radiation than the big low-pressure ones; and having fewer moving parts - for example no air-pump was required - it would lose less power through friction. It was therefore reasonable to expect that it would be markedly more efficient than the low-pressure engine." (Cardwell, p.84)

4.5 The Science of Heat

While the study of the steam engine provided the main impetus for the development of thermodynamics, its theoretical background - even though a very limited one - was provided by the Science of Heat. We present therefore briefly, the development of this field through the 18th century, following the major events outlined in Table 4.3.

Cardwell suggests that Fire, Air, Water, and Earth were the four Elements of the Empedoclean (5th century B.C.) classification. Other sources, however, indicate that 'Aether', and not Fire,was one of the Empedoclean Elements. Anyway, this is as much of a scientific description of

Table 4.3 Major Events in the Development of the Science of Heat

5th Century B.C.	Empedocles suggests that Fire, Air, Water, and Earth are the four Elements of the World.
17th Century	The Thermometer is introduced; The first theories of heat are suggested by Galileo and Descartes.
1727	Boerhaave proposes the axiom that 'Fire is Conserved'.
1760-1799	Black replaces Fire with Heat in Boerhaave's axiom; he also introduces the concepts of 'Specific' and of 'Latent' heat.
1779	Cleghorn presents the postulates of the Caloric theory of heat.
1798	Count Rumford shakes the foundation of the Caloric theory.

fire as the ancient Greek aphorism 'Fire, Women, and Sea'(!). It represents however, along with the conjugated qualities of 'hot' and 'cold', the extent of knowledge on the subject of heat up to the 16th century.

4.5.1 The Thermometer

The first important development towards the establishment of the science of heat was the invention of the thermometer in the beginning of the 17th century by Galileo. A friend of his recorded the following in 1638 (Hatsopoulos and Keenan, 1965): "Galileo took a glass vessel about the size of a hen's egg, fitted it to a tube the width of a straw and about two spans long; he heated the glass bulb in his hands and turned the glass upside down so that the tube dipped in the water contained in another vessel; as soon as the ball cooled down, the water rose in the tube to the height of a span above the level in the vessel; this instrument he used to investigate the degrees of heat and cold". (p.xv)

The first liquid thermometer was built in 1631 by Jean Rey a French physician and was used to take the temperature of patients. The stem was, however, open and the water evaporation effected the accuracy of reading. The first thermometer with a scaled stem was built in 1641 by the Grand Duke Ferdinard II of Tuscany, using alcohol instead of water.

The calibration of the thermometer involving the fixing of two points, the lower and upper and the divisions in the between, took quite a while. Fixed points involving cold winter- and hot summerdays; the melting of butter; and the somewhat constant temperature of cellars; among others, were proposed. Finally, in 1694 Carlo Renaldini of Padua proposed the freezing and the boiling water as fixed points. When it was realized that atmospheric pressure had an effect on these points, pressure of one atmosphere was specified. (The well-known scales of Fahrenheit and Celsius were not established until later, in 1709 and 1750 respectively.)

The introduction of the thermometer early in the 17th century, even in its crude forms, helped focus attention into two very important, but also very hazy at that time, concepts: 'Heat' and 'Hotness'.

Cardwell, for example, comments (p.2): "Unfortunately there was for a long time doubt about what it (the thermometer) actually measured. As a clinical instrument it certainly reduced the subjective element in the diagnosis of fevers, but beyond this the exact significance of its reading was uncertain and interpretation of them depended on the observer's philosophy of nature."

And Hatsopoulos and Keenan write on the same subject: "Inventors of the thermometer were trying to devise an instrument which would measure a quantity as yet undefined. The physiological sensations of hotness and coldness were involved, but they were inadequate. The thermometer would indicate the same temperature for a block of wood and a block of copper exposed to the same 'severe winter cold' even though these two blocks seemed to be of quite different coldness by the sense of touch. An equilibrium concept was doubtless involved - namely, that all bodies exposed to the same cold atmosphere would ultimately attain a uniform degree of coldness, despite contrary evidence of the senses." (This remark should remind us of the arguments in support of the Zeroth law.)

"A second concept played its part, namely, the idea that temperature was a driving potential which caused some influence to pass from one body to another of unequal hotness or coldness. Thus, when a body was taken from a warm room to the cold outdoors, its length began immediately to change. The difference in coldness between the body and its environment caused the environment to affect the body." (p.xv)

4.5.2 The Nature of Heat

Two, thus, important questions had to be answered:

1. What was responsible for the sense of 'hotness' and 'coldness'?
2. What was responsible for the observable effects? What, for example, made a copper wire increase in length when heated?

Two main schools of thought, that of Galileo and that of Descartes, were advanced for this purpose providing a very hazy answer however. Cardwell writes:

"Galileo argued, in '*Il saggiatore*' and again in the '*Discorsi*', that the sensation of 'heat' is caused by the very rapid motion of certain specific atoms.

'Heat' is thus an illusion of the senses, a product of the mental alchemy which transmutes the eternal scurrying of inert atoms of the 'real' world into our sensual world of colour and warmth, sweetness and bitterness, comfort and discomfort. There is therefore no such thing as 'heat' if there is no one present to experience it: in exactly the same way there is no such thing as a real pain at the tip of a dentist's drill, or a real tickle at the end of a feather. (The examples are from the writings of Sir Arthur Eddington and Galileo respectively.) Yet the sensation we call 'heat' is correlated with definite physical changes: bodies do expand as they seem to get 'hotter' and if they are 'hot' enough they change state, melting or vaporizing as the case may be.

Clearly, while we cannot talk about a quantity of 'hotness' (how can one quantify a subjective sensation?), we must try to account for expansion and change of state. This the Galilean atomic model can do, at least qualitatively, and if we have a measuring instrument, or thermometer, we can talk reasonably about a 'quantity of heat'. For the expression now indicates the total quantity of motion of all the specific atoms; the assumed cause of such (objective) phenomena as expansion and change of state.

The views of Rene Descartes (1596-1650) are more explicit than those of Galileo... To account for the world as we know it Descartes requires three types of particles, which are distinguished from one another by their geometrical properties only - that is, by 'extension'. These are the small 'fire particles'- some of them are infinitely small - the intermediate 'boules' and the 'gross material particles' which in aggregate constitute all material bodies. The first two sorts of particles form the all-pervading aether which circulates in a gigantic but invisible whirlpool or vortex about that large hot body, the sun. It is the motion of this vortex that carries the planets round in their circular orbits, and it is the pressure due to the aether flow round the earth - which has its own relatively small vortex - that accounts for the phenomenon we call 'weight'.

Light is a sensation caused by pressures transmitted through the aether to the retina of the eye, while the sensation of hotness is due to the vibrations of (probably) the material particles stimulating the nervous system. On the other hand the phenomena of expansion and change of state are caused by fire particles and boules penetrating between the material particles and forcing them farther and farther apart. It may even be that vibrations of fire particles and boules agitate material particles and are thus the first cause of the sensation of hotness. Dr G.R. Talbot has pointed out that later Cartesians who did not share their master's anxiety about the void, or completely empty space, saw no reason for both fire particles and boules and so lumped them together as a 'subtle fluid': a rather hazy concept which developed in the course of the eighteenth century into the doctrine of material 'caloric'." (Cardwell, p.2)

4.5.3 The Conservation of Fire

We arrive, thus, in the first half of the 18th century. In this period, in spite of some relevant to the science of heat work by Isaak Newton and Jean Bernoulli's son Daniel, the major contribution comes from Herman Boerhaave (1668-1738), professor of Medicine, Chemistry and Botany at the University of Leyden. He , one could argue, sowed the seeds for the first law.

Boerhaave was an excellent lecturer, that students from all over the civilized world came to listen to. His work was first published - in a pirated(!) edition - in 1727, and is described by Cardwell as follows (Cardwell, p.27):

"Boerhaave's lucidly expounded ideas set the pattern for his contemporaries and successors. He was basically a Cartesian and his concern was therefore with 'Fire', which he held to be distributed throughout all bodies and throughout 'space'. He maintained that Fire is the great universal agent of change, and is abundantly present even in the coldest things under the coldest conditions; for it had been shown that one could strike sparks from a flint in the depths of an arctic winter. An ultimate limit of cold can be imagined, he says, when a body would be wholly deprived of Fire, but such a state cannot be reached either by nature or by technics: it is therefore not worth bothering about it. In fact, Boerhaave's attitude throughout was one of common sense, so even when he was wrong, by later standards, he was never unreasonably or dogmatically so.

He sets out as an essential axiom the proposition that *Fire is always conserved*; he argues that it cannot be created *de novo*, as claimed by certain English philosophers, and that in all changes the total quantity remains unaltered. Now this statement was perhaps crucial, for it is very difficult to see *how a science of heat could have developed without a basic conservation principle: such a principle was plainly essential for the formulation of the concepts of quantity of heat, of specific heat and of latent heat.*" (emphasis added.)

4.5.4 The Differentiation of Heat and Temperature

In addition to the fundamental axiom of 'the conservation of fire' a second, but somewhat diffuse development, took place in the first half of the eighteenth century: the notion of 'quantity of heat' as a concept distinct from that of 'intensity or degree of heat' was introduced. Cardwell comments (p.31): "Curiously, no one claimed to be the inventor of this concept. It seems that a number of scientists started using it independently, more or less simultaneously, and without any fuss. It was a concept that led inevidently to the postulation of an absolute zero of temperature and, a little later on, to the discovery of specific and latent heats."

4.5.5 The Conservation of Heat

The third quarter of the 18th century is dominated by Joseph Black (1728-99) who was a professor of medicine and chemistry first at Glasgow and from 1764 in Edinburgh. He was an excellent teacher, but published little during his career. (The edict - aphorism? - of modern academia: 'publish or perish' had not , I guess, appeared yet!). What we know is based on notes from his lectures.

Black was influenced by Boerhaave's ideas but his philosophy was different. Cardwell contrasts the two philosophies and presents Black's contribution as follows:

"The latter (Boerhaave) had postulated Fire as the general principle of change but had been unable to bridge the gulf between familiar thermal phenomena and their postulated metaphysical cause. In contrast, Black's attitude smacked of the pragmatic positivism that sometimes characterized Newton: we do not know the cause of gravity but we can devise quantitative laws that account for the behaviour of bodies when gravity acts upon them. Accordingly, Black confessed himself agnostic on the nature of heat, and with a sound grasp of the realities of the situation ignored the elusive 'Fire' to concentrate on the operationally intelligible measures of temperature and 'quantity of heat'. He formulated the latter quite naturally, and was almost certainly unaware of the great advance that it represented. The conservation principle is maintained but is tacitly transferred from Fire to heat." (Cardwell, p.35)

The next step towards the establishment of the first law was, thus, taken.

Having formulated the concept of 'quantity of heat', Black introduced the ideas of 'specific heat' and of 'latent heat'. Black's discussion of the second item is indicative of the level of understanding of thermal phenomena at that period:

"Black gives us two examples from common experience which show how important this phenomenon of latent heat really is. A kettle of water at a temperature marginally or infinitesimally below the boiling-point would, on the common-sense view, require the addition of only a trivial amount of heat to raise its temperature above the boiling-point. But if this happened the water would be all converted into steam, technically into super-heated steam, with explosively disastrous consequences! This does not occur because a vast amount of latent heat is required if water is to be converted into steam. Again, on the common-sense view only a trivial amount of heat would serve to convert a vast snowfield just below freezing-point into (disastrous) flood water. This does not happen; snow is notoriously slow to melt and on the northern slopes of the highest Scottish mountains it lingers on through summer. Once more the doctrine of latent heat provides the explanation. In this way an advance in the science of heat gave new insights into a range of meteorological and other 'transformation' processes."(Cardwell, p.38)

At this point we should remember that James Watt had stumbled upon the phenomenon of latent heat, which was an important factor in the development of his engine. And since he was working at the same place with Black, Glasgow University, a long-lived legend developed according to which Black's discovery of the latent heat was instrumental in the development of Watt's engine. For example, as late as 1961 Charles Singer in *A Short History of Scientific Ideas* (Oxford University Press, p.300) stated: "Watt (1765) applied Black's discovery (of latent heat) to his contrivance of the separate condenser."

A fascinating account of the origin and growth of this legend is presented by Cardwell (p.41) who demonstrates rather clearly that it is a false one.

4.5.6 The Caloric Theory and its Problems

We come, hence,to the last quarter of the 18th century. The 'caloric' theory of heat, whose origin lies - as we have seen - with Descartes' 'subtle fluid', is fairly well accepted but there is some strong opposition to it. Hatsopoulos and Keenan describe the situation as follows:

"According to the caloric theory, caloric or 'matter of heat' was an all-pervading subtle fluid which could be neither created nor destroyed. This idea was in tune with its era, which was an age of subtle fluids like electricity and phlogiston. Probably no authoritative statement of the caloric theory ever existed. A fairly complete statement is contained, however, in the following postulates of William Cleghorn (1779):

1.Caloric is an elastic fluid, the particles of which repel one another strongly.

2.Particles of caloric are attracted by particles of ordinary matter in different degrees for different substances and different states of aggregation.

3.Caloric is indestructible and cannot be created.

4.It is either sensible or latent. Latent caloric is caloric which has combined chemically with particles of solid to form liquid or of liquid to form vapor.

5. Caloric has weight.

These postulates were ingeniously contrived to account for certain observed phenomena. Postulate:

(1) accounts for expansion and contraction upon heating and cooling;
(2) for variations of specific heat capacity;
(3) for calorimetry;
(4) for latent heat; and
(5) for gain in weight of certain metals when heated (calcined) in the presence of air.

The plausibility of such postulates prompted Joseph Black, a dedicated experimentalist and a sceptic regarding all theories, to say that 'such an idea of the nature of heat is the most probable of any I know'.

The gap in the armor of the caloric theory was the heating effect of friction, a phenomenon so common that it seems incredible that any theory of heat which ignored it should ever have been seriously entertained. Thus, John Locke (1772) observed: 'The axle trees of carts and coaches are often made hot, and sometimes to a degree that it sets them on fire, by the rubbing of the naves of the wheels upon them.' In an attempt to close this gap some argued that friction reduced the attraction between caloric and material particles and so freed the caloric to flow to the surroundings.

Count Rumford conducted a relentless campaign against the caloric theory. He showed, first, that caloric is weightless; second, that metal which has been subjected to friction has the same heat capacity as the native metal; third, that the ability of brass to produce heat by rubbing is inexhaustible. This last point, which he demonstrated by his famous cannon-boring experiments in his arsenal at Munich, proved to be the crucial one. In the words of Rumford, 'The source of the heat generated by friction in these experiments, appeared evidently to be inexhaustible. It is hardly necessary to add that anything which any insulated body, or system of bodies, can continue to furnish without limitation cannot possibly be a material substance'." (pp.xvi-xvii)

4.6 The Establishment of Thermodynamics

We enter thus the 19th century that saw the establishment of thermodynamics through the development of the first and second laws. But first let us summarize the state of the Science of Heat and of the Steam Engine in the early part of the 19th century.

Heat is made of fire particles (caloric), even though some disagree

Table 4.4 Major Developments Leading to the Establishment of Thermodynamics

1824	Carnot publishes his book: *Reflexions sur la Puissance Motrice du Feu*.
1834	Clapeyron presents the Carnot cycle in terms of a *P-V* diagram and develops an expression for its efficiency.
1842	Mayer publishes his theory about the conversion of heat to work.
1843-1850	Joule, through a series of experiments, establishes the convertibility of work to heat.
1849	Kelvin expresses doubts that both Carnot and Joule can be right; also introduces the term *Thermodynamics*.
1850	Clausius reconciles the findings of Carnot and of Joule and presents the analytical statement of the first law and the verbal one of the second.
1851	Kelvin accepts Joule's findings; introduces the 'Absolute Temperature' and the analytical statement of the second law for reversible processes.
1865	Clausius introduces the concept of *Entropy* and the analytical statement for the second law for irreversible (and reversible) processes.

strongly and hold that it is the 'vis viva (living force)' of the atoms. It is also conserved, i.e. the total amount of heat is constant and it can not be converted into anything else.

The steam engine has been perfected: from the embryonic Savery type to the Watt type with a separate condenser and to the high pressure type of Trevithick.

Yet, while the efficiency of water-wheels, the main source of power up to this period, can be properly determined - work produced as compared to the maximum obtainable one (Example 2.7) - the performance of the steam engine is still measured in terms of *work* ($ftlb_f$) *produced per bushel of good(!) coal*.

It is also a revolutionary period. Cardwell comments (p.190): "In France political revolution had let loose an abundance of creative talent while at the same time providing scientists with the organization, the incentives and the moral stimulus to attack the widest range of problems. In England revolution was revealed in the rise of new forms of industry wherein manufacturing processes were analyzed into their major components so that each stage could be mechanized."

In addition, the continuing economic growth demanded more energy and more efficient use of it. Energy consumption, actually, became - and still is today - the barometer of economic growth. Thus Leibig stated: *Civilization is the Economy of Power.*

This was the picture - scientific, technological, and social - in the early part of the 19th century that the two fathers of Thermodynamics: Sadi Carnot and James Prescott Joule, entered in. A new field of science and technology was established in about half a century, following the major developments presented in Table 4.4. (The 'christening' was done by Lord Kelvin in 1849).

4.7 Sadi Carnot

> *Sadi Carnot was born in 1796 when his father was just 43 years old. The young Sadi entered the Ecole Polytechnique in 1812 with the intention of becoming, like his father before him, an officer in the engineers, the 'Genie'. He saw action at Vincennes in 1814, but the long war was coming to an end and the years of peace that followed Waterloo afforded less opportunity for distinction and promotion to an army officer; especially to one whose father was a political exile. In 1820 Sadi Carnot went on half-pay and, Professor Mendoza tells us, devoted his energies to study, making the best use of the educational and research facilities of Paris. He studied principally physics and economics and, according to Professor Mendoza, spent a good deal of time visiting factories and studying the organisation and economics of various industries, becoming an expert on the commerce and industries of different European countries. The fruit of this devoted labour was a short book, some 118 pages long, which he published in 1824 under the title, 'Reflexions sur la puissance motrice du feu'.* (Cardwell, p.191)

We will start with a discussion of Carnot's motivation for his work and proceed with his arguments and conclusions. We will close with Cardwell's commentary on the fate of his work (and see what happens when one is born before his time).

4.7.1 Motivation

Carnot's main motivation came from his admiration for the steam engine and the lack of theory about its operation. Cardwell writes:

"He is impressed by the widespread utilisation of steam-power in England, and he is unstinting in his praise for the engineers - most of them British - who have developed the steam-engine up to its latest, high-pressure form. With profound insight he remarks that the steam-engine is now more important for England's existence than is her Navy - evidently his readings in economics and his comparative studies of industry and technology had not been a waste of time.

Nevertheless, in spite of all this scientific knowledge and technological development there is no general theory of the heat-engine, applicable to all conceivable forms of heat-engine, whatever their working substances or their mechanical principles. He observes:

> *Machines which are not driven by heat, those which are driven by the power of men or of animals, by a fall of water, by the movement of air, etc., can be analyzed down to their last details by mechanical theory. Every event is predictable, all possible movements are in accordance with established general principles which are applicable in all circumstances. This, of course, is characteristic of a complete theory. A similar theory is obviously required for heat engines."* (Cardwell, p.192)

4.7.2 Arguments

We discussed Carnot's cycle in Chapter 3 and demonstrated its two basic points, the need for:

* a heat reservoir and a heat sink; and
* a reversible engine.

Here now is how Cardwell presents Carnot's arguments:

"He begins by grasping and emphasizing two essential points. The first one is that if you want to use heat to generate power you must have a cold body as well as a hot one. That is to say, a 'fall of caloric' - or a temperature difference - is absolutely essential. The water-power analogy is very close at this stage of the argument: a given quantity of water will only generate power if there is a lower level to which it can flow; a tail-race as well as a penstock. For Carnot, of course, the actual motive agent in a heat-engine, the analogue to the water in a hydraulic machine, is the highly elastic fluid, caloric.

The second essential point is that for the generation of the maximum amount of power there must at no point and at no time be a useless flow of heat. In other words, heat which could cause the working substance to expand and do work must not flow uselessly from the hot body to the cold one. This means that ideally there must be no contact between bodies at different temperatures during the operation of a heat-engine; a condition which, the logician will object, would make a perfect or ideal engine quite impossible. But this objection could, with equal force, be raised against Lazare Carnot's conditions for the maximum efficiency of a hydraulic engine; that the water enter the machine without shock of impact of any sort and leave it without relative velocity (Example 2.7). Again the analogy between the working of a hydraulic engine and a heat-engine is a close one; the head or, since the one determines the other, the velocity of the water being the analogue of the temperature difference. In the case of the heat-engine Sadi Carnot answers the logician's objection by asserting that vanishingly small temperature differences are, for his purposes, sufficient to cause heat to flow without violating the prime condition that in a perfect heat-engine bodies in contact must be at the same temperature." (Cardwell, p.193)

We should notice here that Carnot had accepted the axiom of the 'conservation of heat'. Hence, when he talks about a 'fall of caloric' he means that all caloric that left the heat reservoir finds itself in the heat sink, just as all the water that falls through a hydraulic engine arrives at the bottom. Work is produced because of the 'fall' of caloric, not through conversion of it.

4.7.3 Conclusions

By using Carnot's cycle in Chapter 3 we concluded that:

a. A reversible engine has the highest efficiency which is independent of the working fluid;

b. This efficiency depends on the temperature difference (t_1 - t_2), where t_1 and t_2 refer to the heat reservoir and heat sink respectively; and

c. It also depends, for a given (t_1 - t_2) value, on the temperature level: it is higher when t_1 = 200°C and t_2 = 100°C, than when t_1 = 300°C and t_2 = 200°C.

On the first Cardwell writes: "... Carnot has no difficulty in showing that a heat-engine working according to the cycle he prescribed, and of course mechanically perfect, is the most efficient possible. For let us suppose that there is an even more efficient engine; then if it is used to drive Carnot's engine backwards so that it acts as a heat-pump, all the heat 'let down' through the more efficient engine will be restored to the source. But the latter, by definition, will deliver more power than is required to drive the heat-pump. The consequence will therefore be the generation of a net amount of power without any net 'fall' of heat and this would, in fact, mean the production of perpetual motion. Thus no heat-engine can be more efficient than one which can be reversed in the way in which Carnot prescribes. It is obvious that the nature of the working substance cannot affect this conclusion; all such engines, no matter what the working substance - air, steam, alcohol-vapour or anything else - are necessarily of exactly equal efficiency." (Cardwell, p.199)

On the second, Carnot himself states: "In accordance with the principles we have now established, we can reasonably compare the motive power of heat with that of a head of water: for both of them there is a maximum which cannot be exceeded, whatever the type of hydraulic machine and whatever the type of heat engine employed. The motive power of a head of water depends upon its height and the quantity of water; the motive power of heat depends also on the quantity of caloric and on what may be called - on what we shall call - the height of its fall, that is on the temperature difference of the bodies between which the caloric flows."

The situation with respect to the third item is more interesting for the analogy to the hydraulic engine would suggest to Carnot that the fall of caloric from 100 to 90°C and from 10°C to 0°C should produce the same effect. (Remember that the expression: $n = 1 - (T_2/T_1)$ was not developed until later). Yet he was more cautious and proceeded to demonstrate that for a given $(t_1 - t_2)$, the lower the temperature of the heat sink t_2, the higher the efficiency. (See Cardwell, p.204.)

But how could Carnot using the false axiom of the conservation of caloric arrive at the correct conclusions? For the first two, the proof is independent of the conversion of heat to work. As for the third, this is one case where two wrongs make a right. Again from Cardwell, p.205: "The conclusion was drawn from an argument based on the false assumption of the conservation of heat and from misleading experimental evidence that the specific heat of gases increases with volume. Nevertheless the conclusion happens to be correct."

4.7.4 The Acceptance of his Work

"*Reflexions* received a long and favourable notice in the 'Revue Encyclopedique' and was mentioned in the 'Bulletin des Sciences Technologiques', but for the rest it was simply ignored in France, Britain and elsewhere. When, many years later, the young and enthusiastic student William Thomson (Kelvin) tried to buy a copy of the book he found that the Paris booksellers had never heard of it, or its author; the only Carnot they knew of was the father, the 'Organiser of Victory', Lazare Carnot.

It is not difficult to understand why *Reflexions* was a failure. To the engineer, the postulation of the ideal engine and the reversible cycle must have seemed closer to the more imaginative passages in the 'Timaeus' than to the needs of power-engineering. Christian, Poncelet and others had advised French engineers to keep practical requirements always in mind and to avoid being too abstract and theoretical. Always there was the example of the Cornish engines, whose extremely high standards of performance had been achieved by careful attention to details and whose designers seemed to have made little use of abstract theory. In fact, Carnot had very little to teach the engineers of his time: the superiority of the high-pressure condensing steam-engine had been established, so had the advantages of expansive operation, and by 1824 engineers knew well enough that, all things considered, there was no better working-substance than steam. We must conclude that Carnot's message was essentially for engineers of a later generation with the resources of a more advanced technology.

On the other hand Carnot's ideas were general or cosmological in their scope and therefore, in principle, relevant to physics, chemistry and meteorology. In these fields however his mode of reasoning would, one infers, be unfamiliar and thus perhaps unacceptable to those who had had little or no experience of power-engineering. In sum, the '*Reflexions*' fell - inevitably, when one recalls its scope - between a number of specialist stools: the recognition of its technological importance lay in the future, while for the physical sciences the clarification and demonstration of its relevance called for the further, and highly original, work of Clapeyron and Kelvin (in physics), Gibbs (in chemistry) and Napier Shaw (in meteorology)." (Cardwell, pp.211-212)

4.8 Emile Clapeyron

The next major step in the development of thermodynamics belongs to Emile Clapeyron.

Emile Clapeyron (1799 - 1864) was an engineer, a graduate of the Ecole Polytechnique, whose interests seem to have been in the theory of structures and in civil engineering generally. His posthumous distinction however rests on one paper, published in 1834, in which with superb timing (as it transpired) he rescued Carnot's ideas from the oblivion into which they had fallen and re-stated them with such conviction and clarity that they aroused the interest of the perceptive minority that exists in every generation. More, he made some highly original contributions to the science - thermodynamics - which Carnot had founded, for all that he, like Carnot, subscribed to the axiom of the conservation of heat. (Cardwell, p.220)

4.8.1 Contributions

Clapeyron made three main contributions:

First, he represented the Carnot cycle on an 'indicator diagram', i.e. a pressure-volume diagram (Figure 3.1). Carnot's abstract ideas are now presented in a graphical form. (How Clapeyron came to know about this diagram, that was so closely guarded by Boulton and Watt, is still a mystery.)

Second, he developed an expression for the maximum amount of work (W) produced by the fall of an amount of caloric Q from temperature $(t+dt)$ to t, i.e. the efficiency of the Carnot cycle:

$$\frac{dW}{Q} = \frac{1}{C}\,dt \tag{4.8.1}$$

where $(1/C)$ is the maximum amount of work that a unit of heat can perform when falling through 1°C *at a temperature t*; in other words, C is a function of temperature. It was later called the 'Carnot factor'.

Third, using the Carnot cycle, he developed - as shown in Example 4.1 - the 'Clapeyron equation' for the determination of the heat of vaporization L of a pure fluid at some temperature t:

$$L = (V_2 - V_1)\frac{dP^s}{dt}\,C \tag{4.8.2}$$

where: V_1 and V_2 are the saturated molar volumes, liquid and vapor respectively; and

dP^s/dt,is the slope of the vapor pressure curve;

all evaluated at t.

The development of Eq.4.8.2 was of extreme importance for Clapeyron used it, along with available at that time vapor pressure and heat of vaporization data, to determine experimental values for C (see Example 4.2 for water). He then demonstrated using this expression that, just as Carnot anticipated, $(1/C)$ decreased with increasing temperature. (Undergraduate Students can now move to Section 4.9.)

4.8.2 The Efficiency of the Carnot Cycle

Once Clapeyron described the Carnot cycle on a P-V diagram he proceeded to determine its efficiency.

In those days the ideal gas equation had the form:

$$PV = R(t + 267)$$

with t in °C. He could thus calculate the work involved in the two isothermal steps (Figure 4.1). There was no relationship, however, available between P and V for the two adiabatic steps.

To get around this point, Clapeyron very shrewdly considered that the two temperatures differed by a very small amount dt. As a result, the two isotherms in Figure 4.1 are parallel, the segments 1-4 and 2-3 are straight lines, and the amounts of work involved in them cancel each other. Hence the work produced in the cycle is:

$$dW = \int_1^2 P\,dV + \int_3^4 P\,dV =$$

$$= R(t + dt + 267)\ln(P_1/P_2) + R(t + 267)\ln(P_3/P_4)$$

Clapeyron had also developed (Cardwell, p.222) the following expression for the total heat content of the gas:

$$Q = R(B - C\ln P)$$

where B and C depend on temperature only. Thus, the heat absorbed by the gas is:

$$Q = R(B - C\ln P_2) - R(B - C\ln P_1) = RC\ln(P_1/P_2)$$

Since lines 1-2 and 4-3 are parallel, $(P_1/P_4) = (P_2/P_3)$, or $(P_2/P_1) = (P_3/P_4)$, and we conclude after some algebra that:

$$\frac{dW}{Q} = \frac{1}{C}dt \tag{4.8.1}$$

Thus, $(1/C)$ becomes the measure of the work produced by a unit of heat falling by 1°C at a temperature t, and is a function of temperature only.

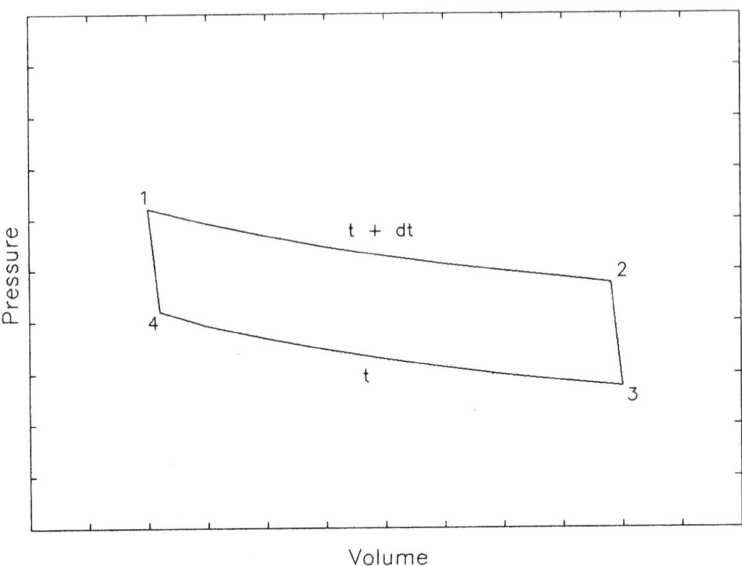

Figure 4.1 *P-V* diagram for the development of the Carnot cycle efficiency by Clapeyron.

4.8.3 Example 4.1. The Clapeyron Equation

Using one mole of a saturated liquid undergoing the following cycle (Figure 4.E.1):

1. Vaporization at constant pressure P^s from the saturation liquid volume V_1 to the saturation vapor volume V_2;
2. Heating from P^s to $P^s + dP^s$ at constant volume V_2;
3. Cooling at constant pressure to V_1;
4. Cooling at constant volume from $P^s + dP^s$ to P^s;

develop Eq. 4.8.2.

We note that:

* work produced by the cycle is: $dW = (V_2 - V_1)dP^s$;
* heat absorbed: the heat of vaporization, L.

Hence:

$$\frac{dW}{Q} = \frac{(V_2 - V_1)}{L} dP^s$$

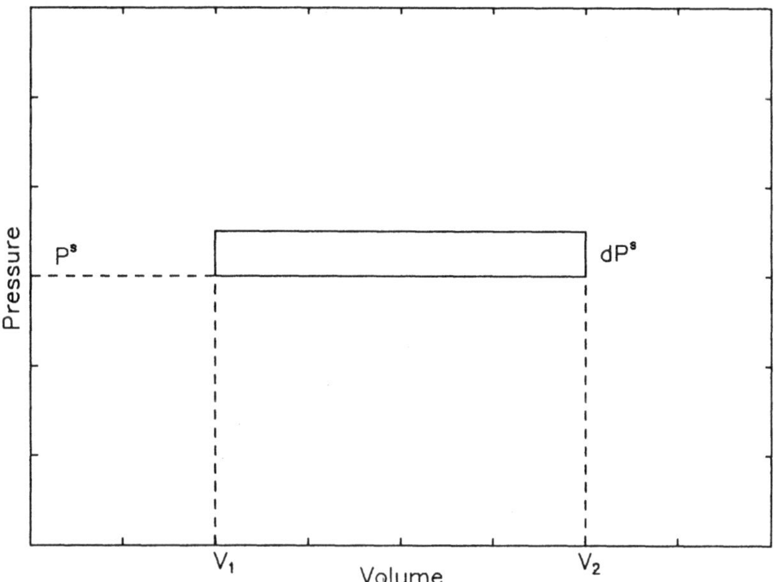

Figure 4.E.1 *P-V* diagram for the development of the Clapeyron equation.

Clapeyron then argued in his 1834 paper that the maximum amount of work produced by a unit of heat cannot, by Carnot's axiom, exceed the work that it could do if it were applied to any substance. It must therefore be given by Eq. 4.8.1. Thus:

$$\frac{(V_2 - V_1)}{L} dP^s = \frac{1}{C} dt$$

After rearrangement we obtain the Clapeyron equation.

4.8.4 Example 4.2. Evaluation of the Carnot Factor *C*

Use data from the Steam Tables, say in the interval 2°C to 80°C, to obtain numerical values for *C*:

$$C = \frac{L}{(V_2 - V_1) \dfrac{dP^s}{dt}}$$

at different temperatures.

All pertinent information and C values are presented in the top of the next page:

t (°C)	$V_2 - V_1$ (m^3/kg)	dP^s/dt (kPa/°C)	L (kJ/kg)	C	$(1/C)10^4$
2	179.90	0.050	2496.8	277.6	36.0
20	57.84	0.145	2454.3	293.7	34.1
50	12.04	0.610	2382.9	324.5	30.8
80	3.408	1.920	2308.8	352.8	28.3

Comments

1.$(1/C)$ increases with decreasing temperature just as Carnot had anticipated. But Clapeyron proved it using experimental data available at that time (water, ethanol, ether, and turpentine).

2.If we compare Clapeyron's efficiency of the Carnot cycle, Eq.4.8.1, with the one we developed in Chapter 3, Eq.3.5.3, it follows that $C = T$, i.e. the absolute temperature. (This, of course, was not done until later.) And this is why the calculated values of C are, within the accuracy of evaluating the slope (dP^s/dt), equal to the absolute temperature in Kelvin.

Carnot's somewhat abstract ideas about the heat engine and Clapeyron's quantitative interpretation of them laid the foundations for the establishment of the second law. This, however, was not done until the 1850's and 1860's by Kelvin and Clausius. In the between, the foundations for the first law were laid by Joule.

4.9 The Conversion of Work to Heat

This idea, as expected, generated tremendous opposition because it was contrary to the doctrine of the 'conservation of heat' that was deeply rooted as the basis of the science of heat at that time. Its establishment belongs to Julius Robert Mayer and, especially, to James Prescott Joule.

4.9.1 J.R. Mayer (1814-1878)

Mayer was trained as a medical doctor and was a ship-surgeon on a vessel trading to the East Indies. His first ideas about the conversion of work to heat came from his observation of the difference in venous blood color between the tropics and higher latitudes. He concluded that in the warm tropical climate, the inhabitants needed less oxygen to maintain their body's temperature, hence the bright red color of their venous blood. Mayer went on to establish a balance between energy intake, as measured by the oxygen used, and the energy consumption, determined by the heat lost by the body and by manual work.

To determine the equivalence between heat and work Mayer relied on the difference in gas heat capacities at constant pressure and at constant volume. In the case of the constant pressure heat capacity C_P, the amount of added heat not only raises the temperature of the gas, but also must account for the work of expansion of the gas against the external pressure. This work is not, of course, involved in the constant volume heat capacity C_V, and thus C_P is larger than C_V. Arguing that *all* the extra heat was converted to work, he arrived at the number of 356 kg.m per kilocalorie (664.8 $ftlb_f$/Btu), about 15% less than the accepted value today.

Mayer published his findings in 1842, but his ideas were not accepted: First, he did not prove that *all* the extra heat (for C_P) was used for work against the external pressure.

Second, the idea of a loss of caloric (heat) by conversion to work was against the aforementioned 'conservation of heat' axiom.

Finally, as Cardwell suggests (p.231), "he was outside the tradition of research and scholarship in the science and technology of heat; he stood apart from the then Establishment."

This rejection of Mayer's pioneering ideas, along with some personal problems, led to his mental collapse. He did achieve, however, the recognition he deserved due to the efforts of Clausius, recovered from his breakdown and was voted into the Royal Society in 1862.

4.9.2 J.P. Joule (1818-1889)

Joule faced also a lot of opposition. But his deep scientific insight, outstanding experimental work, and the fact that he was a member of a scientific establishment - in Manchester - helped him overcome it.

Joule carried out, as we mentioned in Chapter 2, a series of very careful experiments whereby he demonstrated that work was converted to heat; and that there was an exact relationship between work consumed and heat produced. Caloric was thus generated, leading - eventually - to the collapse of the caloric theory.

Here is how Cardwell describes Joule's work, his resolution of the heat capacities problem that Mayer faced, and the reaction of the then scientific community.

4.9.3 Early Experimental Work

"Joule was led to consider the problems of energy conversion through an early interest in magneto-electric machinery. His first papers were of a practical, engineering nature: he looked forward to the day when electric motors would replace steam-engines for driving machinery. From electric motors he graduated to a study of the generation and the heating effects of electric currents. In 1841 he discovered the important law that the heating effect of an electric current is proportional to the resistance of the circuit and the square of the current flowing.

Two years later came the very important paper: 'On the Caloric Effects of Magneto-Electricity, and on the Mechanical Value of Heat'. In this paper he demonstrated that the heat caused by the passage of an electric current is not transferred from another part of the circuit, which is correspondingly cooled, ... but is actually generated. In his own words, he showed that:

> *... the mechanical power exerted in turning a magneto-electric machine is converted into the heat evolved by the passage of the currents of induction through its coils; and, on the other hand, that the motive power of the electro-magnetic engine is obtained at the expense of the heat due to the chemical reactions of the battery by which it is worked.*

The mechanical power was provided - and measured - by the descent of a pair of weights placed, one each, in two scale pans; the latter were connected by means of fine strings passing over two pulleys to a vertical axle, round which a great length of each string was wound. As the weights descended a measured distance so the axle was rotated and drove the magneto-electric machine (dynamo).

As a result of a number of experiments Joule concluded that the mechanical value of a unit of heat was 838 ft.lbs of work expended to raise one pound of water by one degree (^oF).

These experiments, together with the conclusions drawn, were announced to the Chemical Section of the British Association meeting at Cork in 1843. They were received in silence; partly, no doubt, because some of his audience were unable to grasp the full import of what was at stake, partly because others were unwilling to throw overboard the grand principle of the conservation of heat on which the entire science had been erected, and partly because still others felt that too much was being inferred from just one set of experiments.

But Joule had already, in a postscript to this paper, transferred his attention from the electrical generation of heat to the purely frictional. By forcing water through tiny holes in a perforated cylinder, he had been able to evolve a detectable amount of heat, which compared with the work expended yielded a mechanical value for a unit of heat of 770 ft.lbs. This was in tolerable if not good agreement with the value obtained by means of the electrical experiments."

Joule's Some Other Experiments (!)

Another of his early and little known experiments was to examine the effects of electricity on human beings. His subject was a servant girl who reported her sensations as the voltage was stepped up - he used a larger battery - to the moment when she became unconscious, whereat he felt it advisable to stop the experiment. Unfortunately the young lady's opinions were not recorded. (Cardwell, 1983)

4.9.4 The Expansion/Compression of Gases and Later Work

"But if expending mechanical power on water in this way can generate heat, what of the famous phenomenon of the 'adiabatic' heating of gases? Joule's next experiment was to compress a gas in a strong copper vessel, which was placed in a large calorimeter. The walls of the calorimeter were made as impermeable to heat as was possible, and it was filled with water. By comparing the heat imparted to the water with the work performed in compressing the gas Joule found a new value for the unit of heat. This was 823 ft.lbs.

At this point Joule faced the unresolved difficulty that had made Mayer's determination of the mechanical value of heat uncertain. What proof was there that *all* the work performed on the gas had been converted into heat; alternatively, that *all* the heat produced was due to the work performed? Might it not be that in compressing the gas - in reducing its volume - some of the heat released may have been due to 'latent' heat being made 'sensible'?

These objections did not apply to the experiments made using water and a perforated piston, for water is virtually incompressible. But until they could be answered in the case of gases, determinations based on 'adiabatic' compression or expansion would be entirely uncertain.

Joule's answer to this dilemma was another and very famous experiment. ... Two large copper vessels are connected by means of a pipe containing a stopcock and are placed in a large calorimeter full of water; the walls of the calorimeter are as impermeable to heat as is possible.

One of the copper vessels is evacuated and the other is filled with dry air at a pressure of 22 atmospheres. The stopcock is opened so that the air rushes into the empty vessel, but no detectable change in the temperature of the water in the calorimeter is noticed. In other words, when air expands *without* doing work no heat is lost.

Joule now proceeded to analyze this experiment. He provided each of the two copper vessels with its own separate calorimeter and repeated the experiment. The results indicated that the calorimeter containing the evacuated vessel gained exactly the same amount of heat as the other one lost. The sum of the two quantities of heat imparted to and abstracted from the two identical calorimeters was, therefore, zero. The implication was plain: merely changing the volume of a gas does not produce or consume heat; this happens only when work is done on or by the gas. The 'latent' heat of expansion (or compression) is therefore an illusion: it is no more than the heat value of the work done by or on the gas, against or by an external force. Joule's calculation of the mechanical value of a unit of heat by measuring the amount produced when the gas is compressed to a given extent was therefore justified in principle; and so, incidentally, was that made by Mayer.

This splendid paper was rejected by the Royal Society. But it was printed in the '*Philosophical Magazine*', and in this way the scientific world learned not only of Joule's fundamental insights and confirmatory experiments but also of his penetrating criticisms of Clapeyron and Carnot. For the first time the fundamental ideas of thermodynamics were subject to the (constructive) criticism of a first-class scientific intellect:

> *I conceive that this theory ... is opposed to the recognized principles of philosophy because it leads to the conclusion that vis viva (i.e. energy) may be destroyed by an improper disposition of the apparatus: thus Mr Clapeyron draws the inference that 'the temperature of the fire being from 1000°C to 2000°C higher than that of the boiler there is an enormous loss of vis viva in the passage of the heat from the furnace to the boiler'. Believing that the power to destroy belongs to the Creator alone I affirm ... that any theory which, when carried out demands the annihilation of force, is necessarily erroneous.*

In the following year (1845) Joule read, to the British Association meeting at Cambridge, an account of a new method of determining the mechanical value of heat. A paddle-wheel, driven by the same mechanism of falling weights that he had used to provide power for his dynamo in 1843, rotated horizontally in a calorimeter full of water. A comparison of the heat generated with the work performed yielded a mechanical value of 819 ft.lbs per heat unit. But no one seemed very interested in Joule's ideas or experiments. No one, that is, in Britain; abroad he was not without honour, for when in 1847 Helmholtz published his important paper 'On the Conservation of Energy' he cited Joule's experiments in support of his arguments. In that year Joule described his improved paddle-wheel apparatus for the measurement of mechanical value of heat and this time, at the British Association meeting at Oxford, his ideas aroused much more interest. This was largely due to the keen appreciation of the young William Thomson, later Lord Kelvin.

The Oxford meeting of the British Association marked an historic stage in the history of thermodynamics: *the real beginning of the acceptance of the first law* (emphasis added). It is, of course, true that the meausurements on which Joule had had to rely for his determinations were small fractions of a degree, but he was a masterful experimentalist and he was supported by a most able instrument-maker (see Example 2.2). The implication of Joule's ideas that one could heat up water *merely by shaking it*, seemed in the context of scientific - or common sense - thought in the 1840's quite extraordinary, and the confirmation that this was indeed the case was taken as strong confirmation of his theory.

The way ahead was now, at last, clear for Joule. He had soldiered on and had won through to recognition and acceptance. In 1850 the Royal Society published his account of new and very accurate paddle-wheel experiments which gave the figure of 772 ft.lbs for the mechanical value of a unit of heat. From this time onwards Joule could extend and consolidate his theory, devise even more accurate methods for determining his constant, and find further instances of the great principle of the convertibility of energy in nature." (Cardwell, pp.232-236)

4.10 Kelvin, Clausius and the Establishment of the First, and Second Laws

4.10.1 Lord Kelvin (1824-1907)

William Thomson was knighted in 1866 for his services in connection with the great Atlantic cable and was made Lord Kelvin of Largs in 1892.

Joule won the battle but the war was not over, by far. No other than the very sympathetic to Joule - as we have seen - Kelvin, in his 1849 paper on 'Account of Carnot's Theory of the Motive Power of Heat; with Numerical Results derived from Regnault's Experiments on Steam' still has very strong doubts. Cardwell writes:

"Kelvin mentions Joule's ideas only to reject them: '... the conversion of heat, or caloric, into mechanical effect is probably impossible, certainly undiscovered'. But he then goes on to raise an important question; what happens when heat flows by conduction from a hot to a cold body?

> *When thermal agency is thus spent in conducting heat through a solid what becomes of the mechanical effect which it might produce? Nothing can be lost in the operations of nature - no energy can be destroyed. What effect then is produced in place of the mechanical effect which is lost? A perfect theory of heat imperatively demands an answer to this question; yet no answer can be given in the present state of science.*

This was, at one and the same time, an unoriginal and a very penetrating question. It was unoriginal in that the key phrase 'Nothing can be lost in the operations of nature - no energy can be destroyed' is hardly more than a rephrasing of Joule's remark: 'Believing that the power to destroy belongs to the Creator alone I affirm ... that any theory which, when carried out, demands the annihilation of force, is necessarily erroneous' (see Section 4.9.4). Therefore Kelvin had picked up the gauntlet thrown down by Joule, even if he did not know what to do with it. On the other hand he had posed the question in the framework of thought established by Carnot: that relating to the efficiency of heat-engines. We can regard the mere conduction of heat from a hot to a cold body as the operation of a heat-engine of zero efficiency; in which case we must ask what happens to the work which a more efficient engine, in exactly the same situation, would produce? In this lay the key to the further development of thermodynamics." (Cardwell, p.241)

What is fascinating about Kelvin's doubts - and demonstrates how difficult it was for the 'conservation of heat' doctrine to be abandoned - is that by the time he published this 1849 paper, he had in his own laboratory duplicated Joule's findings. In a letter of his, dated December 5, 1847, he writes:

"Perhaps before then (Friday 7) I may have succeeded in boiling water by friction. This is not very probable, however, as the machine I have made being not strong enough. In the first experiment made with it the temperature rose from 45, 46 to 57 at about the rate 1 every five minutes during a continued turning of the instrument (a very flat disc with narrow vanes turning in a thin tin box about 8 inches in diameter of which the bottom and lid were furnished with thin fixed vanes. The bearings of the paddle were entirely within the box which was full of water). In a second experiment we began with water at 98 & the temperature rose to about 99 1/2 when part of the machine gave way before the experiment could be considered quite decisive. My assistant was preparing to make an experiment today with water at 80, 90 & to go on grinding (along with another for relief) for about 4 hours, but I have not heard yet whether it went on." (Cardwell, 1983, p.56)

The debate, thus, for and against the conservation of heat went on. Cardwell comments on Kelvin:

"The arguments on both sides seemed very strong, and it is hardly surprising that in the words of Kelvin's biographer, Silvanus Thompson, 'The apparent conflict took possession of his (Kelvin's) mind and dominated his thoughts.'

There can be little doubt that at the root of Kelvin's objections to Joule's theory lay his conviction that the whole science of heat rested on what since 1783 had been the basic axiom of conservation. If that were rejected what would happen to the impressive structure of experimental knowledge and theoretical development that had been built up by the labours of men like Delaroche and Berard, Fourier, Dulong and Petit, Poisson, Victor Regnault and many others? Thus he quoted Carnot as saying, of the axiom of *conservation*, that: 'To deny it would be to overturn the whole theory of heat, in which it is the fundamental principle'. And he himself adds the comment that if the axiom of conservation is rejected: '... we meet with innumerable other difficulties - insuperable without further experimental investigation, and an entire reconstruction of the theory of heat from its foundation.'" (Cardwell, p.244)

It is also noteworthy that in this 1849 paper, Kelvin coined the word "Thermodynamics", becoming the 'godfather' of the new field.

4.10.2 R.J.E. Clausius (1822-1888)

Cardwell presents the next phase in the development of the laws of thermodynamics as follows:

"While Kelvin was mulling over these problems he was being overtaken by two other men, R.J.E. Clausius (1822-1888) and W.J.M. Rankine (1820-1872). In some ways Clausius could almost be described as a disciple, at one remove, of Kelvin. He had not read the original Carnot volume and he knew of the theory only through the writings of Clapeyron and Kelvin. Possibly this was an advantage, for he could consider the theory dispassionately, without being unduly influenced by the persuasiveness of the whole work, and in the light of Kelvin's important observations about Joule's ideas.

In any case, after comparing Kelvin's work very favorably with that done by Holtzmann, ... he takes (in his 1850 paper) as his starting point the basic problem to which Kelvin had drawn attention: what happens to the mechanical effect which is lost when heat flows from a hot to a cold body not by way of a heat engine but by simple conduction? He sees that, in fact, the position is more open, there are rather more possibilities, than Kelvin had supposed. 'On a nearer view' - and how much is contained in that modest phrase! - the new (Joule) theory is opposed not to Carnot's theory but to the assertion that no heat is expended or lost in a cyclic operation. For it is quite possible that in the production of work *both* processes take place at the same time: '... a certain portion of heat may be consumed and a further portion transmitted from a hot body to a cold one; and both portions may stand in a definite relation to the quantity of work produced.'

These words deserve to be pondered very carefully. In Clausius' view, Joule had established the consumption of heat in a cyclic process, Carnot the transmission of heat, and both phenomena are necessarily related to the work produced." (pp.244-245)

Clausius proceeds then to develop the analytical expression of the first law in the form:

$$dQ = dU + \frac{R}{J}\frac{(a+t)}{V}\,dV$$

where J is the Joule factor for converting heat to work, and a is a constant that converts t to - in today's terms - absolute temperature (Clapeyron used $a = 267$, in developing the expression for the efficiency of the Carnot cycle, Section 4.8.2).

Hence, according to Clausius, an amount of heat Q entering a gas can be broken down into two parts:

* the first accounts for changes in sensible heat (i.e. temperature) and internal work (i.e. to overcome the intermolecular forces); and

* the second accounts for external work.

Total energy is therefore conserved, not just heat, as the caloric theory had it.

Clausius then turns to Carnot's cycle and examines the issue of the maximum efficiency of the reversible engine in the light of the new theory. This leads him to the statement of the second law. Cardwell writes:

"Clausius asserts that if there were some working substance, or engine, to provide more work for the fall of a given amount of heat than another engine; or, what amounts to the same thing, were it to provide the same work for the fall of a smaller amount of heat, then if the first engine drives the second one in reverse we would have an entirely anomalous result: no net work would be done and heat would, in effect, flow from a cold to a hot body. This, he postulates, is quite impossible; and thus he proves Carnot's fundamental principle.

This, in fact, *constitutes the first historical statement of the second law of thermodynamics* (emphasis added). Carnot, we remember, went on to argue that a more efficient engine than a completely reversible one is impossible, since it would logically imply the feasibility of perpetual motion; and so indeed it would if the caloric theory, or the axiom of conservation, were true. The situation would be exactly the same as if a perfect column-of-water engine were to be driven in reverse, as a pump, by an even more perfect engine! But we are no longer dealing with fluids, subtle or gravitating; we are dealing with energy, and the extension of the argument merely leads us to the conclusion that the flow of heat from a cold to a hot body would enable us to perform work by applying the heat to a third engine, which would thus be deriving its energy from that *stored in the cold body*. Now there is, *a priori*, nothing in such an arrangement to contradict the energy principle as enunciated by Joule and Mayer and, more recently, in a very general form, by Helmholtz. All that would be happening would be that the thermal energy in the 'cold' body would be being converted into the equivalent amount of mechanical energy - and there would be nothing absurd or self-contradictory about that. It certainly does not violate the great new principle of the conservation of energy. Clausius therefore rests his proof on the flat assertion that it is impossible for heat to flow of its own accord from a cold to a hot body. It is, of course, common experience that hot bodies always tend to cool down, that you cannot boil a kettle of water by putting it on a block of ice, and so on. Now the common experience of countless generations from time immemorial is to be elevated into a scientific axiom. We shall discuss the admissibility of this procedure later". (Cardwell, pp.247-248)

Indeed Cardwell presents a fascinating discussion on the acceptance of Clausius assertion.

4.11 The Absolute Temperature and the Analytical Expression of the Second Law

4.11.1 The Absolute Temperature

Kelvin formally accepted the dynamical theory of heat in his paper of March 1851 (Cardwell, p.254). He went on to work on the relationship between heat absorbed and work produced and between heat absorbed and heat rejected.

In his 1854 paper, he accepts Joule's suggestion that the Carnot function C is simply the value of the temperature measured from -273°C, and demonstrates that:

$$\frac{Q_1}{Q_2} = \frac{T_1}{T_2} \tag{4.11.1}$$

where T is the *Absolute Temperature* defined through the Carnot cycle.

As Kelvin put it "...the absolute values of two temperatures are to one another in the proportion of the heat taken in to the heat rejected in a perfect thermo-dynamic engine working with a source and refrigerator at the higher and lower of the temperatures respectively."(Cardwell, p.258)

4.11.2 The Analytical Statement of the Second Law: Reversible Processes

Cardwell continues: "And then, in a moment of penetrating insight, Kelvin saw that this simple relationship (Eq.4.11.1) amounted to a quantitative, or mathematical statement of the second law of thermodynamics. If we consider the heat absorbed as positive and that transmitted as negative then the relationship becomes:

$$\frac{Q_1}{T_1} + \frac{Q_2}{T_2} = 0 \tag{4.11.2}$$

He is able to generalize this to include complex reversible cycles - we have come a long way from the simple expansive operation of steam-engines - and to lay down that in such complex cycles the quantities of heat are related to the temperatures (absolute) at which they are absorbed or transmitted by the linear equation:

$$Q_1(\frac{1}{T_1}) + Q_2(\frac{1}{T_2}) + Q_3(\frac{1}{T_3}) + \ldots + Q_n(\frac{1}{T_n}) = 0 \qquad (4.11.3)$$

This, Kelvin asserts, can be taken as '... the mathematical expression of the second fundamental law of the dynamical theory of heat'. The corresponding expression for the first law he writes as:

$$W + J(Q_1 + Q_2 + \ldots + Q_n) = 0$$

Thus the first law of thermodynamics expresses the equivalence between heat energy and mechanical energy, while the second law imposes a definite restriction on the ways in which the conversion from one form of energy to the other can take place. In the case of completely reversible changes the second law can be expressed mathematically by Kelvin's linear equation." (Cardwell, pp.258-259)

4.11.3 Real Life Processes and Entropy: Irreversible Processes

The two laws of thermodynamics have been thus developed, even though Kelvin's analytical statement of the second law, Eq.4.11.3, is limited to reversible processes only. What happens with real life, irreversible processes?

The analytical statement of the second law for such processes came from Clausius a few years later, in 1865. To this purpose he introduced the concept of Entropy which he defined as follows:

"We might call '*S*' the 'transformational content' of the body, just as we termed the magnitude '*U*' the 'thermal and ergonal content'. But as I hold it better to borrow terms for important magnitudes from the ancient languages so that they may be adopted unchanged in the modern languages, I propose to call the magnitude '*S*' the 'Entropy' of the body from the Greek word τροπή, transformation."

He proceeded to demonstrate that "the entropy of the universe tends to a maximum", i.e. that for a real life, irreversible process, the total entropy change is positive: $\Delta S > 0$.

It took fifteen difficult years for Clausius - from 1850 when he presented, as we saw, the verbal statement of the second law to 1865 - to arrive at the concept of entropy. It is no wonder, therefore, that we have such a hard time comprehending this term.

Cardwell presents a detailed account of Clausius' development of Entropy. It represents a fascinating reading for any one interested in this concept. Clausius' starting, probably, point is described by Cardwell as follows (p.260):

"Clausius saw clearly that the Carnot cycle represents a limiting, or ideal case. It is one end of a wide spectrum of thermal transformations, which go on in nature and indeed under man's control as part of his developed technology. In this respect the Carnot cycle is similar in kind to Galileo's intuition of the laws of falling bodies and of (circular) inertia: eliminate all imperfections - air resistance, friction, etc. - and under these ideal conditions all bodies fall with equal acceleration, or continue in uniform motion for ever."

In other words, the same way that these imperfections prevent the continuous motion of a body or the conversion of its kinetic energy completely into work, the generation of entropy - resulting also from such physical imperfections - prevents the realization of the Carnot cycle efficiency.

It should be emphasized, however, that in the first case we are referring to *complete* conversion of kinetic energy to work; in the second, to *the conversion of the part of the flowing heat into work that is prescribed by the Carnot cycle*. Complete conversion is not, of course, possible in a cyclic process.

4.12 Thermodynamic Properties of Matter

With the development of the two laws - and the publication of books such as "Die Mechanische Wärmetheorie" by Clausius in 1876 - the field of thermodynamics was established. The rigorous analysis of the conversion of thermal energy to work, of the 'motive power of heat', that started with the pioneering work of Sadi Carnot, was achieved.

Our discussion of the laws of thermodynamics in the 2nd and 3rd Chapters indicates that their application, combined with knowledge of thermophysical properties, provides for the evaluation of the feasibility of processes and, as we will see in the next Chapter, for the efficient energy utilization, i.e. the first objective of chemical engineering thermodynamics.

The theoretical framework for the accomplishment of the second and third objectives, i.e. the evaluation of thermophysical properties and of phase and chemical reaction equilibria - based of course again on the laws of thermodynamics - was provided by J.W. Gibbs, who is thus considered the father of this filed. The application of this framework to the solution of actual problems, however, has involved - and continues to do so - several investigators.

4.12.1 J.W. Gibbs (1839-1903)

> *Gibbs grew up in his native New Haven, attended Yale College and graduated in 1858, having won a number of prizes for his work in Latin and, especially, mathematics. He continued his studies in the new graduate school at Yale, and was one of the first students to be awarded a Ph.D. by an American university. Since Gibbs is now famous for the abstract and theoretical character of his scientific work, it is a little startling to learn that his doctorate was earned in engineering and that his thesis has the title, "On the Form of the Teeth of Wheels in Spur Gearing" ...*
>
> *After receiving his doctorate Gibbs was appointed a tutor in Yale College where he gave elementary instruction in Latin and physics for three years. During this period he continued to work on engineering problems, and in 1866 he obtained a patent on his design for an improved railway car brake. That same year Gibbs left New Haven for three years of study in Europe, the only time he would ever be away from his native city for an extended period...*
>
> *In 1871 Gibbs was appointed to the newly created professorship of mathematical physics at Yale, but the official notice of this appointment stated flatly that it was "without salary". The Yale administration was presumably aware that Gibbs' financial independence would make it possible for him to accept this unendowed professorship. In any case his teaching duties would be very light: he had no undergraduate courses, and very few of the students in the small graduate department were ready to tackle the courses Gibbs offered. He received no salary for the first nine years of his appointment, and seems not even to have been tempted by an offer from Bowdoin College that would have given him $1800 a year and his choice of the chairs of mathematics and physics. Only in 1880, when Gibbs was on the verge of leaving Yale to join the faculty of the appealing new Johns Hopkins University, a university devoted primarily to graduate study and research, did his home institution offer him a salary. Even though Yale offered only two thirds of what Hopkins would have paid, the advantages of remaining in his familiar environment, of being at home, convinced Gibbs to stay. His decision was surely influenced by the warm way in which his colleagues reacted to the threat of losing him. "Johns Hopkins can get on vastly better without you than we can. We cannot." These were the words of James Dwight Dana, the distinguished Yale geologist; Gibbs found it impossible to leave."* (Klein, 1982)

A very interesting account of the man and his contributions is given by Klein (1982; 1983). Very briefly:

1. In the first paper (1873): 'Graphical Methods in the Thermodynamics

of Fluids', he introduced the *fundamental equation*, which for a closed homogeneous system has the form (Chapter 9):

$$U = f(S,V)$$

Gibbs places the emphasis on the properties of the material system rather than the 'motive power of heat'. Thus the state functions U, S, and V, take precedence over the quantities that depend on the process carried out by the system, the work and heat it exchanges with the surroundings. The latter was the approach used until then, such as in the aforementioned book by Clausius. Thermodynamics, thus, *expands its domain beyond work and heat, into the description of material properties.*

In the same paper he argued in favor of representing the second law in terms of a *S-T* diagram rather than a *P-V* diagram, the one that - as we have seen - Clapeyron used about forty years earlier and was first introduced by Watt (see Problem 4.51).

2. In the second paper published three months later 'A Method of Geometrical Representation of the Thermodynamic Properties of Substances by Means of Surfaces', he arrived at an answer to a very important question at that time: the determination of the pressure at which a liquid and a vapor coexist, e.g. the pressure at which a liquid boils at a given temperature. [Maxwell arrived at the 'equality of areas' approach in 1875, and Clausius at the same, but independently, five years later (Problem 9.57)].

3. In the third, his masterpiece, paper 'On the Equilibrium of Heterogeneous Substances' (1878), he further extended the world of thermodynamics by treating chemical - and other - phenomena. He introduced the chemical potential and its use in describing phase and/or chemical equilibria.

To appreciate the latter, consider the following simple problem: what are the equilibrium compositions of a methanol-water mixture at a specified temperature and pressure?

The solution of this problem is carried out in the three-step procedure depicted in Figure 4.2:

Step I. Translation of the real problem into the abstract level of pure thermodynamics;

Step II. Solution of the problem in the abstract level;

Step III. Translation of the abstract solution into the real world.

The purpose of Step I is to define the appropriate function to be used in Step II. Gibbs introduced this function, the *chemical potential* μ_i, and provided a very simple answer for Step II: the chemical potential of each component of the equilibrium mixture in the one phase equals that in the other. He provided, thus, the solution to the theoretical part of the problem.

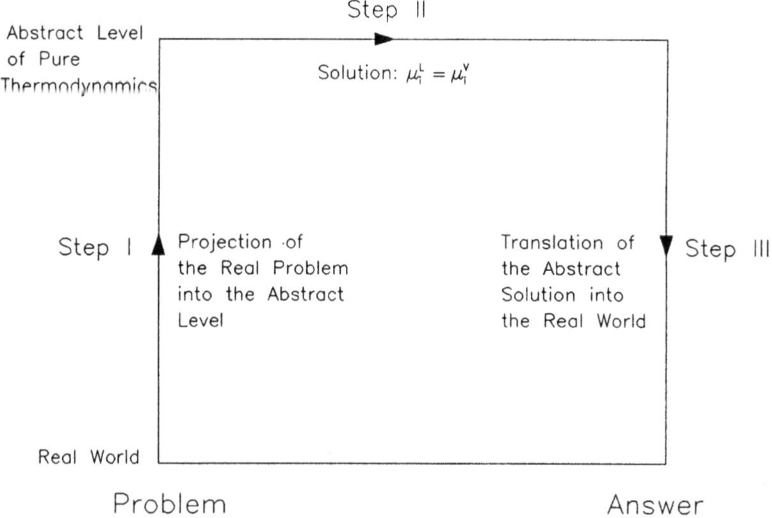

Figure 4.2 The three-step procedure for the application of thermodynamics to the solution of the phase equilibrium problem. (Prausnitz et al, 1986, p.4, reprinted by permission of Prentice-Hall, Englewood Cliffs, New Jersey.)

The really difficult step in practice is Step III. The translation of theory into the real world, i.e. the expression of the theoretical function through the measurable quantities: pressure, volume, temperature, composition, and heat capacities. While this translation was greatly facilitated by G.N. Lewis, who introduced early in this century the concepts of partial property, fugacity and ideal solution, it has been - and still is - the task of many people, who have been using two main tools to this purpose:

a. Equations of State, and
b. Activity coefficients.

In addition, the need of absolute entropy values in chemical reaction equilibrium calculations relies on Nernst's third law of thermodynamics.

4.12.2 Equations of State

The pioneer of the filed is J.D. van der Waals (1837-1923) who introduced in 1873 the first equation of state (EoS) applicable to real fluids:

$$P = \frac{RT}{V-b} - \frac{a}{V^2}$$

where V is the molar volume and a and b are positive constants, specific for each fluid. Here is how Otto Redlich (1976) presents it:

"The famous equation of van der Waals has the advantages of a simple theoretical basis and of elucidating the whole field of the gaseous and liquid states.

The two reasons for deviations from the perfect gas equation at moderate pressures had been discussed even before van der Waals. The volume available to a molecule for its motion is the total volume V reduced by the volume b occupied by all other molecules present in the same volume V. The true concentration in the freely available space is therefore $N_A/(V\text{-}b)$ rather than N_A/V. (N_A: Avogadro number). Therefore the term RT/V of the perfect gas is replaced by $RT/(V\text{-}b)$. The second reason is the intermolecular attraction (which leads to condensation at high pressures); it increases with decreasing distances between the molecules. Its effect is to pull back the molecules from the wall to the bulk. Thus it results in a reduction of the pressure; for each molecule near the wall of the vessel the effect is proportional to the concentration of the retracting molecules or to $1/V$. Since the concentration of the molecules hitting the wall during a second is also proportional to the concentration, the total effect on the pressure is proportional to its square, or to $1/V^2$.

Van der Waals not only derived these two effects but also discussed the consequences of his equation for the equilibrium between gas and liquid, for the critical state, and for the correspondence in the behavior of different substances.

Qualitatively the equation was tremendously successful. The quantitative representation of experimental data was unsatisfactory." (Redlich,p.47)

The van der Waals (vdW) EoS, and the 'corresponding states' principle that follows from it (Section 8.10.3), represents the basis on which the famous compressibility charts of Hougen, Watson, and Ragatz were developed around the middle of this century. It also provided the information necessary for Dewar and Onnes to liquify many of the so-called permanent gases. Van der Waals received the Nobel prize in Physics in 1910.

The next major improvement came from the modification of the vdW EoS by Redlich and Kwong (RK, 1949):

$$P = \frac{RT}{V-b} - \frac{a}{\sqrt{T}\,V(V+b)}$$

for which Redlich writes:

"There is no really good theoretical justification for the two changes in the denominator of the a-term but the agreement with observed data is considerably improved so that the equation is sufficient for a number of practical applications."

The RK EoS dominated the area of prediction of the volumetric behavior of fluids for about two decades. Within the same period, 1942 to 1951, the famous Benedict-Webb-Rubin (BWR) EoS was developed to meet the need for the description of the volumetric and phase equilibria behavior of hydrocarbons for the Petroleum Industry. It is a non-cubic EoS containing eight parameters per species, whose evaluation is very cumbersome (a whole floor of the M.W. Kellogg Co., where the EoS was developed, full of hand-operated calculators was used; one can imagine the noise!) The BWR EoS has been used for years, even today, for calculations involving mainly hydrocarbon systems.

The major step in the application of EoS in phase equilibrium calculations came in 1972 when G. Soave, following the pioneering work of G. Wilson (1964) and J. Joffe and D. Zudkevitch (1966), presented his famous modification of the RK EoS, that was successful in the correlation and prediction of phase equilibria in nonpolar systems. After the complicated BWR EoS, the field of EoS applications returned to the simplicity and elegance of the van der Waals form.

In the same period J. Prausnitz (1969); C. Tsonopoulos (1974, 1975); and J. O'Connell (1975), following the pioneering work of K. Pitzer in the middle 50's, presented generalized correlations for the second virial coefficient. The theoretically based virial equation introduced by K. Onnes in 1901, could be now used in the prediction of the volumetric behavior of nonpolar and often polar gases and vapors, but only at low pressures.

Question and ... Question

* *What did money contribute to thermodynamics? Or,*
* *What would Joule and Gibbs have contributed without it?*

4.12.3 Activity Coefficients

Around the same period with K. Onnes, Francois-Marie Raoult introduced his famous relationship:

$$\bar{P}_i = x_i P_i^s$$

known as Raoult's law: the partial pressure of component i over its equilibrium solution is proportional to its concentration; the proportionality constant is its vapor pressure P_i^s.

Raoult's experimental work was unsurpassed, agreeing in many instances with modern measurements to within 0.0001°C. His thermometer was considered by many antediluvian (i.e. belonging to an era before the Flood), but as Van't Hoff expressed it: "with this antediluvian thermometer the world was conquered."

Raoult's law is, of course, valid for mixtures of very similar compounds, i.e. 'ideal' solutions. For 'nonideal' solutions:

$$\bar{P}_i = \gamma_i x_i P_i^s$$

where the activity coefficient γ_i can be viewed as the 'correction factor' to Raoult's law and is a function of concentration and temperature, for low to moderate pressures.

The first theory for nonideal solutions was proposed in 1910 by J.J. van Laar who was one of van der Waals' students and, later, collaborators. His development was based on the vdW EoS and it is very elegantly described by Prausnitz et al (1986), leading to an expression for the activity coefficient involving two temperature-dependent parameters.

It took a little over half a century for the next major development to occur. G. Wilson in 1964 introduced the idea of 'local compositions' which led to an expression for γ_i bearing his name and later to the NRTL and UNIQUAC expressions by J. Prausnitz and his co-workers (1967 and 1975). The expressions have two main advantages over the van Laar model; first, their parameters have a built-in temperature dependency making them very useful in distillation applications (why?); second, they provide successful prediction of multicomponent activity coefficients from binary ones. Finally, based on the UNIQUAC model, Fredenslund and his co-workers (1977) have developed a group-contribution model, UNIFAC, that can be used for the estimation of activity coefficients for a large variety of systems.

4.12.4 The Third Law of Thermodynamics

Application of the first and second laws requires, as we have seen, the evaluation of differences of the thermodynamic properties: internal energy, enthalpy, and entropy. No absolute values of these properties are thus required.

Evaluation, on the other hand, of equilibrium conversions in chemical reactions requires absolute values of the entropy. This evaluation utilizes the Third Law (Section 1.14.1) introduced in 1906 by Walther Nernst

(1864-1941). For an interesting account of the development of this law, see Denbigh (1981) and Hatsopoulos and Keenan.

4.12.5 Other Contributors

The application of chemical engineering thermodynamics in the industrial world is the result of the experimental and theoretical contributions of a large number of individuals in addition to those already mentioned.

Among them: K. Denbigh, R.L. Dodge, P. Flory, E. Guggenheim, J. Hildebrand, O. Hougen, D. Katz, W. Kay, W. Lacey, G.N. Lewis, B. Linnhoff, K. Pitzer, I. Prigogine, R. Reid, F. Rossini, J. Rowlinson, B. Sage, T. Sherwood, and H. Van Ness, stand out. Finally, as judged by the relevant literature and international conferences, over 500 individuals are working today in the field, especially in thermophysical data and in phase equilibria evaluation.

References

Cardwell, D.S.L., 1971. *From Watt to Clausius, the Rise of Thermodynamics in the Early Industrial Age*, Cornell University Press, Ithaca, N.Y.

Cardwell, D.S.L., 1983. In *Springs of Scientific Creativity*, Aris, R., Davis, H.T., Stuewer, R.H., Eds, University of Minnesota Press, Minneapolis, MN. (Copyright 1983 by the University of Minnesota, reprinted with permission.)

Cleghorn, W., 1779. *Inaugural Dissertation at the University of Edinburgh*, (as quoted by Hatsopoulos and Keenan, op. cit.).

Denbigh, K. 1981. *The Principles of Chemical Equilibrium*, Cambridge University Press, Cambridge.

Hart, I.B., 1928. *The Great Engineers*, Books for Libraries Press, Freeport, N.Y.

Hatsopoulos, G.N., Keenan, J.H., 1965. *Principles of General Thermodynamics*, Wiley, N.Y.

Klein, M.J., 1982. *Foote Prints*, 45, **2**, 15.

Klein, M.J., 1983. In *Springs of Scientific Creativity*, op. cit.

Locke, John, 1772. *Elements of Natural Philosophy*, quoted in *The Early Development of the Concepts of Temperature and Heat*, by Duane Rotter, Harvard University Press, Cambridge, Mass., 1950.

Prausnitz, J.M., Lichtenthaler, R.N., de Azevedo, E.G., 1986. *Molecular Thermodynamics of Fluid-Phase Equilibria*, 2nd Ed., Prentice-Hall, Englewood Cliffs, N.J.

Redlich, O., 1976. *Thermodynamics: Fundamentals, Applications*, Elsevier, N.Y.

Problems

4.1 Describe the operation of the following devices:
a. Heron's Aeolopile;
b. della Porta's;
c. Branca's.

4.2 For the Savery engine:
a. describe the method of operation;
b. what is the maximum depth that the engine could raise water from?
c. what should the boiler pressure and temperature be to raise water 150 ft from its level in the mine?
d. why was the presence of air detrimental to the engine's operation?

4.3 The same as the previous Problem - parts (a) through (c) - but for the Newcomen engine.

4.4 What is the origin of the atmospheric and of the steam pressure?

4.5 Explain Watt's dilemma that: "for maximum economy the cylinder must be kept hot all the time, but for maximum power it must be cooled down once every cycle."

4.6 What was the main improvement of Watt's engine over that of Newcomen?

4.7 The same as the previous Problem, but for Trevithick's one over that of Watt.

4.8 Why would Galileo's thermometer give different readings from day to day, even when the actual temperature was the same?

4.9 With reference to Galileo's and Descartes' theories about the nature of heat:
a. what questions were they trying to answer?
b. discuss and compare the two theories.

4.10 What makes the "Principle of Conservation of Fire - and later of Heat", important in the development of thermodynamics?

4.11 Who discovered the 'latent' heat?

4.12 What do you consider as the first most important step in the development in the science of thermodynamics?

4.13 With reference to the 'Caloric' theory:
a. state it;
b. what is its origin?
c. how did it explain phase changes?
d. give the main argument against it.

4.14 In comparing the efficiency of a water-wheel to that of the Carnot engine, discuss the:

a. similarities, and

b. differences.

4.15 Discuss the most important - in your opinion - contribution that Clapeyron made.

4.16 With reference to the development of Eq.4.8.1 by Clapeyron, explain why the amounts of work involved in steps 41 and 23 cancel each other.

4.17 Using data for a fluid of your choice, other than steam, calculate values for the Carnot factor using the Clapeyron equation.

4.18 What are the main reasons - according to Cardwell - that caused Mayer's failure to convince his contemporaries about the convertibility of work to heat?

4.19 Discuss the importance of Joule's work.

4.20.a. Why did Joule have to prove that the difference between C_P and C_V for a gas is a consequence of the work involved in the change of the gas volume?

b. Describe the experiments that he used for this purpose.

c. Where else could the heat absorbed by the water come from?

d. Did he prove the same for liquids? Explain.

4.21 With reference to Joule's experiments involving the expansion of air from one container to an other, what is responsible for the observed temperature changes of the two baths?

4.22 Describe one of Joule's experiments for the conversion of work to heat, and - using what you consider reasonable data - calculate the obtained temperature increase. Any comments on the required accuracy?

4.23 Explain Joule's criticism of Carnot's and of Clapeyron's work.

4.24 How did Clausius reconcile the findings of Carnot and of Joule?

4.25 With reference to the verbal statement of the second law:

a. who made it?

b. how did he prove it?

4.26 Who presented the analytical statement of the second law, for:

a. reversible processes; and

b. real life, irreversible, processes?

4.27 Compare Carnot's and Clausius' proofs that a reversible heat engine produces the maximum amount of work for given temperatures of the heat reservoir and heat sink.

4.28 How did Clausius define entropy?

4.29 Present the major steps in the development of the first and second laws.

4.30 Discuss, briefly, the contributions of:

a. Gibbs;

b. van der Waals;

c. Raoult,

in the development of chemical engineering thermodynamics.

4.31 Discuss the origin of the name 'thermodynamics'.

4.32 Present a flow diagram that identifies the important steps in the development of chemical engineering thermodynamics.

4.51.a. Develop an expression for the thermal efficiency of a Carnot cycle using: (i) a *P-V*, and (ii) a *S-T* diagram.

b. Use the approach you consider easier to prove that for a real engine:

$$\Delta S_{total} > 0.$$

4.52 Discuss briefly the importance of Gibbs' contributions in determining the thermodynamic properties of fluids.

4.53 An equimolar liquid mixture of water with ethanol at 80°C is in equilibrium with the vapor above it, whose composition is sought.

a. Can it be determined without resorting to an experimental measurement?

b. If yes, how and what problems do you expect?

4.54 Develop the van der Waals EoS following Redlich's arguments.

4.55 Explain the practical advantages of Wilson's equation over that of van Laar.

5
Efficient Energy Utilization: Energy Conservation

As for energy conservation progress thus far, I believe only the surface has been scratched in the process industries. At present, inefficient units have been shut down. The conservation measures implemented on those units still operating are largely in the realm of housekeeping and simple additional heat recovery. Few fundamental improvements in process or equipment design have been implemented. These will be necessary for dealing with the economics of shortage in the future. (Kenney, 1984, p.ix)

5.1 Introduction

Until the early 1970's chemical engineers did not pay much attention to efficient energy utilization. Energy was very cheap, at least in the United States, and its impact on plant operating cost relatively small.

The oil embargo of 1973, however, and the resulting drastic increase in oil prices, changed the picture substantially. This is demonstrated in Figure 5.1, which shows the energy cost profile from 1972 to 1980 of a typical Gulf Coast petrochemical plant. Notice that the energy cost jumped, from about 20% of the total in 1972 to about 60% in 1980. And this in spite of the significant effects of energy conservation that yielded, as shown in Figure 5.1, savings in 1980 equal to the total operating cost in 1972.

Efficient energy utilization, i.e. energy conservation, became consequently a factor of paramount importance in the Chemical and Petroleum industries. Thus the U.S. Chemical industry committed to, and achieved, a 1/3 reduction in energy consumption by the early 1980's.

But cost - in spite of the dramatic impact demonstrated in Figure 5.1 - was not the only reason. It became also apparent - somewhat all of a sudden - that the availability of energy was not unlimited.

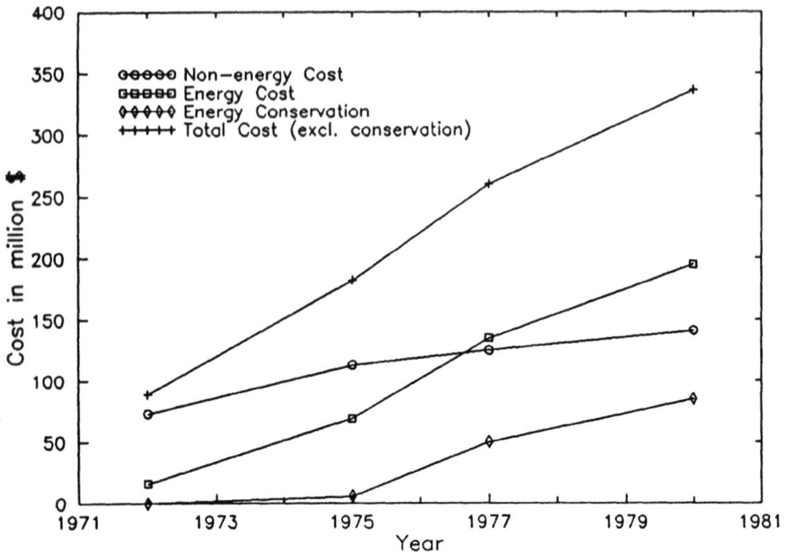

Figure 5.1 A comparison of energy-related and non-energy-related costs (excluding feed-stock costs) for a typical U.S. petrochemical plant, including the effect of energy conservation. (Information from Kenney, p.3.)

The conservation effort, however, is really in its beginning as Table 5.1 demonstrates:

* first, appreciable energy savings can be achieved using existing technology, but
* second, even with this improved operation, we are a long way from the minimum energy consumption suggested by thermodynamics. In the petroleum refining, for example, use of 1973 technology will reduce the energy consumption from 11 times to only 8 times the minimum amount required.

5.2 Objective and Approach

Efficient energy utilization - energy conservation - is a very broad and complex subject to be covered in the limited amount of space allocated here. Our objective, therefore, is to focus attention to the problem and

Table 5.1 Comparison of the Specific Fuel Consumption of Known Processes with the Theoretical Minimum for Selected U.S. Industries(a). (From Kenney, p.3.)

Process	Specific fuel consumption 1968 (MBtu/ton)	Potential specific fuel consumption using 1973 technology (MBtu/ton)	Theoretical minimum specific fuel consumption (MBtu/ton) (b)
Iron and steel manufacturing	26.5	17.2	6.0
Petroleum refining	4.4	3.3	0.4
Paper manufacturing	39.0 (c)	23.8	> -0.2 (d) < 0.1
Primary aluminum production (e)	190.0	152.0	25.2
Cement production	7.9	4.7	0.8

(a) Gyftopoulos et al (1974)
(b) Based on thermodynamic availability analysis
(c) Includes 14.5 MBtu/ton of paper produced from waste products consumed as fuel by the paper industry
(d) A negative value means that no fuel is required
(e) Does not include the effect of scrap recycling

provide the necessary background for dealing with it. More specifically, it is to:

1. Introduce the concepts that are necessary to evaluate the efficiency of a given process and, thus, determine the available room for energy conservation.
2. Demonstrate the effect that the temperature and pressure of steam - the major agent for the utilization of our main source of energy, thermal - have on its ability to produce useful work.
3. Emphasize the importance of the temperature, that an amount of thermal energy is available at, on the value of this energy.
4. Apply these concepts to a major approach towards energy conservation, Cogeneration.

We will start with establishing the 'standard of excellence' for efficient energy utilization: the concept of *ideal work*. When this is reached, no further energy conservation is possible.

We continue with a discussion of the 'usefulness' of the various forms of energy and the method for determining it through the use of a relatively new concept, *exergy*.

We proceed with a presentation of the quantitative criteria used to determine the degree of efficient energy utilization in a given process: *first and second law efficiencies*, followed by a brief description of one of the methods used to obtain increased such efficiencies in the use of thermal energy, *Cogeneration*. We demonstrate the energy savings achieved and discuss the limitations involved and the ways around them.

We continue with Kenney's commentary on the favorable economics of energy conservation, but also on the Institutional barriers that limit its wider applicability, and close the Chapter with some concluding remarks.

5.3 Ideal Work

5.3.1 Definition

For a given process, involving a specified change, there is an amount of work associated with it, referred to as *Ideal Work* $W(id)$, which is:

* the minimum one, if the process is work consuming;
* the maximum one, if the process is work producing.

To evaluate the ideal work, we set up a process that brings about the same change as the real one, whereby:

1. all operations are carried out reversibly;
2. all heat transfer - to, or from, the system - is carried out using the surroundings - reversibly, of course - as a 'heat reservoir' or as a 'heat sink' respectively.

Thus, if heat is transferred to the system then, no matter what its origin in reality is, it must come from the surroundings in calculating the ideal work. The same applies when heat is rejected by the system. A temperature of T_0 = 298 K is typically accepted for the surroundings.

Such a process is, of course, a hypothetical one. If an expansion is involved, for example, it would require a 'perfect', i.e. a completely reversible turbine; if heat is to be transferred from the system at temperature T to the surroundings (at T_0), a Carnot cycle must be employed.

The practical usefulness of such a hypothetical reversible process lies with the observation that, since it effects the same change as the real process, it can be used as a basis for evaluating the efficiency of energy use in the latter.

In the following Example we demonstrate the concept of ideal work and then proceed to develop the general formulation for its determination.

5.3.2 Example 5.1

One kg of steam at $P_1 = 60$ bar and $t_1 = 540°C$ flows through a turbine with an exhaust pressure of $P_2 = 1$ bar. Assuming that the turbine is adiabatic but irreversible, with an efficiency of 0.9 calculate the ideal work $W_s(id)$.

To calculate the ideal work, $W_s(id)$, we must first establish the state of the steam leaving the turbine, since we only know the exit pressure. To this purpose, we calculate first the reversible and adiabatic work, $W_s(RA)$.

* From the first law: $W_s(RA) = -[H_2(RA) - H_1]$
* From the Mollier diagram: $H_1 = 3515$; $H_2(RA) = 2540$, both in kJ/kg.

The real work is, thus:

$$W_s = 0.9W_s(RA) = 878 \text{ kJ}$$

leading to an exit enthalpy of $H_2 = 3515 - 878 = 2637$ kJ which, together with $P_2 = 1$ bar, specifies the state of the steam leaving the turbine.

To calculate $W_s(id)$ we must now devise a process that brings the steam from state 1 [$P_1 = 60$ bar, $t_1 = 540°C$] to state 2 [$P_2 = 1$ bar, $H_2 = 2637$ kJ] reversibly, such as the one shown in Figure 5.E.1:

1.a reversible and adiabatic expansion to $P_2 = 1$ bar, and the corresponding enthalpy H_2', and

2.a Carnot cycle, operating between the exiting steam and the surroundings, to bring the enthalpy from H_2' to H_2 at a constant pressure of $P_2 = 1$ bar.

The ideal work will then be the sum of the work involved in the two steps:

$$W_s(id) = W_s^I + W_s^{II}$$

From the calculation of H_2 we easily see that:

$$H_2' = 2540 \text{ kJ, and: } W_s^I = 975 \text{ kJ.}$$

To bring the enthalpy of the steam from H_2' to H_2 at $P = 1$ bar we must add: $Q_1 = H_2 - H_2' = 97$ kJ at a temperature of 372.8 K, since both H_2 and H_2' represent wet steam - of different quality - at $P = 1$ bar.

To this purpose, we set up a Carnot cycle that absorbs heat from the surroundings, at 298 K, consumes an amount of work W_s^{II}, and delivers $Q_1 = 97$ kJ into the steam. Hence:

$$W_s^{II} = -Q_1\,[1 - (298/372.8)] = -19.5 \text{ kJ}$$

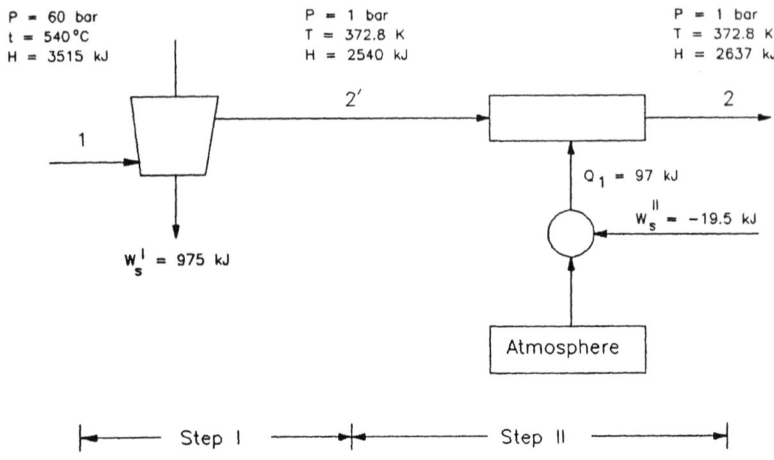

Figure 5.E.1 Evaluation of ideal work for Example 5.1.

The ideal work is, therefore:

$$W_s(id) = 975 - 19.5 = 955.5 \text{ kJ}$$

as compared to the real one of 878 kJ.

Comments:

1. Notice the difference between $W_s(RA)$ and $W_s(id)$:

* the first, represents the work obtained by expanding the steam reversibly and adiabatically to the exhaust pressure of 1 bar. The resulting state, thus, is *not* the actual final state.

* the second, represents the work that would be obtained by bringing the steam reversibly to its *actual* final state.

2. The cycle used in calculating W_s^{II}, whereby heat is transferred from a lower temperature to a higher one through the consumption of work, is referred to as a 'Heat Pump'. The same way that a pump delivers a liquid from a lower level to a higher one, a heat pump delivers heat from a lower temperature to a higher one. Your refrigerator is a heat pump.

5.3.3 Evaluation of Ideal Work: the General Expression

We will demonstrate next that evaluation of the ideal work for a given process can be accomplished using the properties of the system in its initial and final states *only*, namely without resorting each time to the development of a hypothetical reversible process, as it was done in Example 5.1.

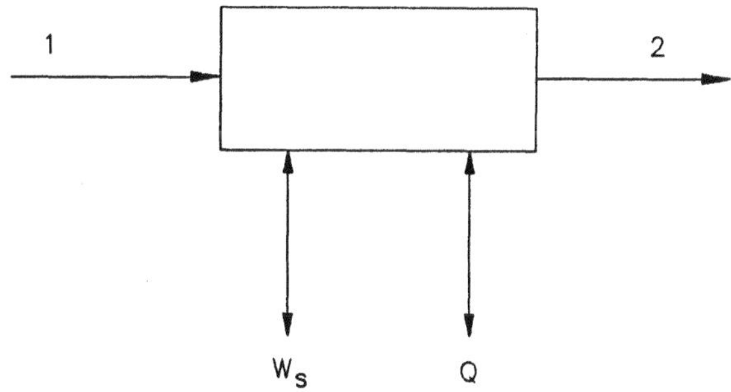

Figure 5.2 Evaluation of ideal work.

Consider a flow process that involves a change of a given stream from state 1 to state 2, as shown in Figure 5.2. An amount of heat Q and one of shaft work W_s are associated with this real process and our task is to determine the amount of ideal work, i.e. the work that would be involved if the same change - from state 1 to state 2 - were to be accomplished in a reversible fashion with the surroundings acting as heat reservoir or heat sink.

Application of the first law to this ideal (reversible) process yields:

$$W_s(id) = Q(id) - \Delta H(id) \tag{5.3.1}$$

where $W_s(id)$, $Q(id)$ and $\Delta H(id)$, refer respectively to the shaft work, amount of heat, and enthalpy change associated with the ideal process.

Since this reversible operation effects the same change in the system as the real process:

$$\Delta H(id) = H_2 - H_1 = \Delta H\ [system]$$

To develop an expression for $Q(id)$, also in terms of the system properties, we note that:

* $Q(id) = -Q(id)_{sur}$, since all heat comes from, or goes to, the surroundings;

* $Q(id)_{sur} = T_0 \Delta S(id)_{sur}$
where T_0 is the temperature of the surroundings;

* $\Delta S(id)_{sur} = -\Delta S(id) = -(S_2 - S_1) = -\Delta S[system]$, as in the case of determining $\Delta H(id)$.

Eq.5.3.1 becomes, therefore:

$$W_s(id) = T_0 (S_2 - S_1) - (H_2 - H_1) = T_0 \Delta S - \Delta H \tag{5.3.2}$$

Thus, for a flow process - the typical case in practice - involving a change of a given system from state 1 to state 2, Eq.5.3.2 allows the evaluation of the ideal work from the initial and final properties of the system only.

Let us now ask the following question: How much work is lost because of the irreversibilities present in the process? We note that:

$$W_s(lost) = W_s(id) - W_s = [T_0\,\Delta S - \Delta H] - [Q - \Delta H] = T_0\,\Delta S - Q \tag{5.3.3}$$

Notice, finally, that since $Q = -T_0 \Delta S_{sur}$, Eq.5.3.3 becomes:

$$W_s(lost) = T_0\,\Delta S + T_0\,\Delta S_{sur} = T_0\,\Delta S_{total} \tag{5.3.4}$$

Thus, the total entropy change in a given process - multiplied by the temperature of the surroundings - represents the measure of the work lost in this process as a result of the irreversibilities involved: entropy becomes the 'yardstick' of lost mechanical energy.

In the next two Examples we demonstrate the evaluation of ideal work in flow processes with Eq.5.3.2.

5.3.4 Example 5.2

Repeat Example 5.1 using Eq.5.3.2.

From the Mollier diagram:

H_1 = 3515 kJ/kg; S_1 = 7.00 kJ/kg K

H_2 = 2637 kJ/kg; S_2 = 7.27 kJ/kg K

And: $W_s(id)$ = 298(7.27 - 7.00) - (2637 - 3515) = 958 kJ, which is the same, within the accuracy of chart reading, with the answer obtained in Example 5.1.

5.3.5 Example 5.3

Calculate the minimum amount of work required for the production of 1 kg of liquid nitrogen at its normal boiling point (T = 77.4 K; P = 1 atm) in a flow process, starting with:

a. Pure nitrogen at 1 atm and 298 K;

b. Air at 1 atm and 298 K.

The following data are available for N_2 (Sussman, 1981):

State	P (atm)	T (K)	H (kcal/kg)	S (kcal/kg K)
2	1	77.4	0.0	0.0
1	1	298	102.5	0.9519

a. From Eq.5.3.2:

$$W_s(min) = [T_0(S_2 - S_1) - (H_2 - H_1)] = -181.2\,\text{kcal}$$

b. This case is closer to reality, since liquid nitrogen is produced by the liquefaction of air followed by fractionation of the obtained liquid. The difference with the previous case lies with the fact that the entropy of nitrogen in state 1 is now larger, by the entropy of mixing (see Section 3.12):

$$S_1 = 0.9519 + (R/28)\ln(1/0.79) = 0.9686\,\text{kcal/kg K}$$

and: $W_s(min) = -186.2$ kcal.

5.4 Useful Energy

All animals are equal, but some animals are more equal than others.
(*Animal Farm*, George Orwell)

All joules are equal, but some joules are more equal than others.
(*Availability Analysis*, M.V. Sussman)

5.4.1 Definition

An amount of energy that can be used to effect any type of desirable change is referred to as *useful energy*. This change could involve the raising of a weight, the heating of a fluid, the compression of a gas, the production of electricity, etc.

Let us now ask the following question: If we have one Joule of energy, can we effect all the aforementioned changes? Thus, in Example 5.1 we used 19.5 kJ of mechanical work to transfer 77.5 kJ of heat from the surroundings to the steam for a total of 97 kJ. Could the same be accomplished by using 19.5 kJ of heat?

To answer this question, let us summarize briefly our experiences with the various forms of energy in the previous Chapters.

1. All forms of energy, except heat, are completely interconvertible, at least in principle. A Joule of mechanical energy will give, thus, a Joule of electricity or potential energy and vice versa. It will also give a Joule of heat. Here, in other words: 'A Joule, is a Joule, is a Joule'.
2. When it comes to heat, however, a Joule of thermal energy will not give a Joule of mechanical energy; it is prohibited by the second law of thermodynamics (Statement A). Thus 19.5 kJ of heat cannot accomplish the task that the same amount of mechanical energy did in Example 5.1, i.e. to transfer 77.5 kJ of heat from the surroundings to the steam.

3.Rather, the amount of mechanical energy that can be obtained from a given amount of thermal energy depends on the temperature:

a.at which it is available, T, and

b.of the heat sink, T_0, into which the part that is not converted into mechanical energy will be rejected.

4.The higher the value of T and the lower the value of T_0, the larger the amount of mechanical energy produced.

5.A given amount of heat, available at a temperature T, can be transferred to a fluid that is at a lower temperature. If, however, the fluid is at a higher temperature, only a fraction of this heat can be transferred to it without, of course, the consumption of any mechanical energy. (How much? See Example 5.4). In other words, when it comes to heat: 'One Joule, is not always one Joule'. (Discrimination, unfortunately, has been practiced by Nature well before it has by Humans.)

We conclude, therefore, that:

1.One Joule of any energy, except thermal, represents one Joule of useful energy. We refer to it as *useful energy* or just *work*.

2.One Joule of thermal energy contains a smaller amount of useful energy, which is determined by the two temperatures: T, at which it is available, and T_0, of the heat sink.

3.Since the typical heat sink is the atmosphere, T_0 is fixed for practical purposes, and the amount of useful energy (work) that can be obtained from an amount of heat depends on T only. Hence, when we refer to the useful energy of a given amount of thermal energy, T_0 = 298 K is implied.

We will discuss how this qualitative picture is quantified through the concept of *exergy* after the next Example.

5.4.2 Example 5.4

One Joule of thermal energy is available at 400 K. What fraction, if any, of this Joule can be transferred to a temperature of 500 K?

We set up, first, a Carnot cycle that operates between 400 K and the atmosphere at 298 K, and produces:

$$W = 1\ [1 - (298/400)] = 0.255\ \text{J}$$

of work from this thermal energy. (This, by the way, represents the amount of useful energy associated with the given Joule of thermal energy).

We set up, next, a Carnot cycle that uses this work to absorb an amount of heat Q_0 from the surroundings and reject an amount of heat Q_1 at T_1 = 500 K, i.e. it operates as a heat pump:

$$\frac{Q_0}{Q_1} = -\frac{T_0}{T_1} \qquad \text{(From Eq.3.5.4)}$$

$$Q_1 + Q_0 = W$$

Solution of this system of two equations and two unknowns yields:

$$Q_1 = -0.631 \text{ J}$$

Thus, 0.631 J of the one Joule at 400 K can be transferred to 500 K. (Why is Q_1 negative?)

5.5 Exergy

5.5.1 Maximum Obtainable Useful Energy from a System

If the qualitative picture presented in the previous Section is to be used for practical applications, some quantitative measure of useful energy must be established.

To this purpose, let us consider the following general question: If we have a system at a temperature T and a pressure P, what is the maximum amount of useful energy that can be extracted from it? If, for example, we have one kg of steam at 60 bar and 550°C, what is the maximum useful energy that can be obtained from it?

The extraction of useful energy from a given system will involve, of course, a change to some final state characterized by the temperature T_0 and the pressure P_0. We must, therefore, specify this final state, before we can determine the obtainable useful energy.

It is apparent that if the production of work involves expansion - of the steam for example - the lower the exhaust pressure P_0, the larger the amount of useful energy produced. Or, if work is to be produced by setting up a Carnot cycle - that absorbs heat from the system and rejects part of it to a heat sink that is at a temperature T_0 - the lower the value of T_0, the more the amount of useful energy produced.

What, then, should we choose for P_0 and T_0? The obvious choices are those of the atmosphere, for: if P_0 is less than one atm, work must be consumed to create the vacuum; if T_0 is less than the atmospheric temperature, work will be required for the necessary refrigeration to cool the heat sink to this lower temperature.

5.5.2 Definition

We accept, hence: $P_0 = 1$ atm, $T_0 = 298$ K, and - since flow processes represent the typical industrial case - we introduce the following measure for maximum useful energy (work):

"The maximum amount of shaft work that can be extracted from a unit mass of matter as it flows into equilibrium with the atmosphere, is called *exergy* or *steady flow availability*, Φ." (Sussman, 1981).

The term exergy will be used here and it will be extended to include all the maximum obtainable work associated with different forms of energy (thermal, electrical, etc.). For other related terms that are used in the literature, see Kenney (p.22). For the case of a closed system, see Problem 5.57.

We discuss next how exergy is evaluated, and then how it is used in establishing the 'yardstick' of efficient energy utilization, the second law efficiency.

5.6 Evaluation of Exergy

5.6.1 Exergy of a System at T and P

To evaluate the exergy of a system at a temperature T and a pressure P, we must determine the maximum work that can be obtained by bringing it through a flow process to a final state, where its temperature is $T_0 = 298$ K and its pressure is $P_0 = 1$ atm.

If we refer, to this purpose, to our discussion of ideal work, we notice that the exergy of a system represents the ideal work obtained from it, when its final state is: $T_2 = T_0$ and $P_2 = P_0$. Hence, from Eq.5.3.2:

$$\Phi = T_0(S_0 - S) - (H_0 - H) \tag{5.6.1}$$

Evaluation of the exergy of the system, therefore, requires knowledge of the entropy and enthalpy values of the system in its actual state $[P,T]$ and in the atmospheric one $[P_0,T_0]$.

Exergy thus measures the value of a system in terms of the maximum amount of shaft work that can be extracted from it in a flow process, when its final state is the atmospheric one: $P_0 = 1$ atm and $T_0 = 298$ K.

On the other hand, the maximum amount of heat that can be extracted by bringing the system to its atmospheric state:

$$Q = \Delta H_0 = H - H_0 \qquad (5.6.2)$$

is referred to as the *heating* value. In the case of steam, for example, by converting it to water at $T_0 = 298$ K and $P_0 = 1$ atm.

In the following Examples, we evaluate the exergy and heating value of steam as a function of the temperature T and pressure P that it is available at. And, to demonstrate the practical implications of this effect, we calculate the monetary value of the steam as a function of T and P.

We proceed then with the evaluation of the exergy of thermal energy as a function of the temperature it is available at.

5.6.2 Example 5.5

One kg of steam is available at $P = 60$ bar and $T = 550$ K.

a. Calculate its exergy and heating value.

b. Assuming that the cost of electricity is: \$0.016/1000kJ, and of heat: \$0.005/1000 kJ - typical numbers for the late 70's, calculate the monetary value V of the steam, per 1000 kg, based on:

* exergy, $V(\Phi)$, and
* heating value, $V(\Delta H_0)$

a. From the Steam Tables:

$H = 3539.3$ kJ/kg; $S = 7.0285$ kJ/kg K

$H_0 = 104.8$ kJ/kg; $S_0 = 0.3670$ kJ/kg K

(Notice that the final state represents water at 25°C and 3.166 kPa, because values at 1 atm are not available. The effect of such a small pressure difference on the enthalpy and entropy values of water is negligible, however). Hence:

$$\Phi = 298\,(0.3670 - 7.0285) - (104.8 - 3539.3) = 1449.4 \text{ kJ/kg}$$

$$\Delta H_0 = 3539.3 - 104.8 = 3434.5 \text{ kJ/kg}$$

b.

$$V(\Phi) = (1449.4 \text{ kJ/kg})(\$0.016/1000\text{kJ}) \times 1000 \text{ kg} = \$23.2$$

$$V(\Delta H_0) = (3434.5 \text{ kJ/kg})(\$0.005/1000\text{kJ}) \times 1000 \text{ kg} = \$17.2$$

This difference in the steam value, resulting from the way the steam is used, is demonstrated further in the next Example.

5.6.3 Example 5.6

Using the data on electricity and heat costs of Example 5.5, plot the value of steam on the basis of exergy and heating value, in \$/1000 kg of steam, as a function of:

a. its saturation pressure, from 1 to 100 bar, and

b. its temperature at a pressure of 30 bar, from the saturation value to 800°C.

The values are calculated as in Example 5.5, and are plotted in Figures 5.E.6a and 5.E.6b. The following observations can be made:

1. Low pressure, up to about 15 bar, steam has more heating than exergy value.
2. High pressure, especially combined with high temperature, steam has an increasingly larger exergy than heating value.

Conclusion: Use high pressure steam for the production of mechanical energy by expanding it through a turbine, and then use the exhaust low pressure steam for heating purposes. (Remember this when we discuss, in Section 5.8, Cogeneration.)

5.6.4 Exergy of Thermal Energy Available at a Temperature T

To evaluate the exergy of an amount of thermal energy Q, available at a temperature T, we set up a Carnot cycle operating between T and the surroundings temperature T_0. Then:

$$\Phi = Q\left[1 - \frac{T_0}{T}\right] \tag{5.6.3}$$

The amount of thermal energy that is rejected in the atmosphere (Q - Φ) is not, of course, able to produce any useful work. It is referred to as *anergy* (from the Greek: an-ergon = no work; the unemployed, for example, is called in Greek: "anergos").

Figure 5.3 presents the amounts of exergy and anergy for one Joule of thermal energy as a function of the temperature T, at which it is available. Notice the - expected - pronounced effect of temperature on both of them.

This dependency of the exergy of thermal energy on the temperature that it is available at is, of course, reflected on its monetary value. Thus, using the data of Example 5.5, we conclude that the value of heat, per 1000 MJ:

* when based on the heating value, remains constant with T ($T > T_0$, of course), equal to $5;
* when based on exergy, increases with T. It is zero, for example at 298 K, 0.201x1000x0.016= $3.2 at 373 K, and $11.2 at 1000 K.

5.6.5 Exergy of Any Energy Except Thermal

The exergy of an amount of energy other than thermal (heat): electrical, mechanical, etc., is of course equal to this amount of energy.

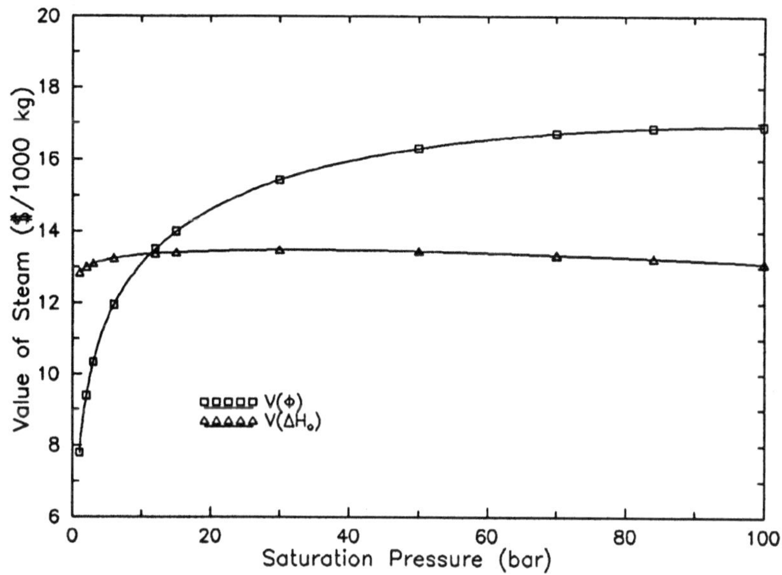

Figure 5.E.6a Value of steam, \$/1000 kg, as a function of its saturation pressure.

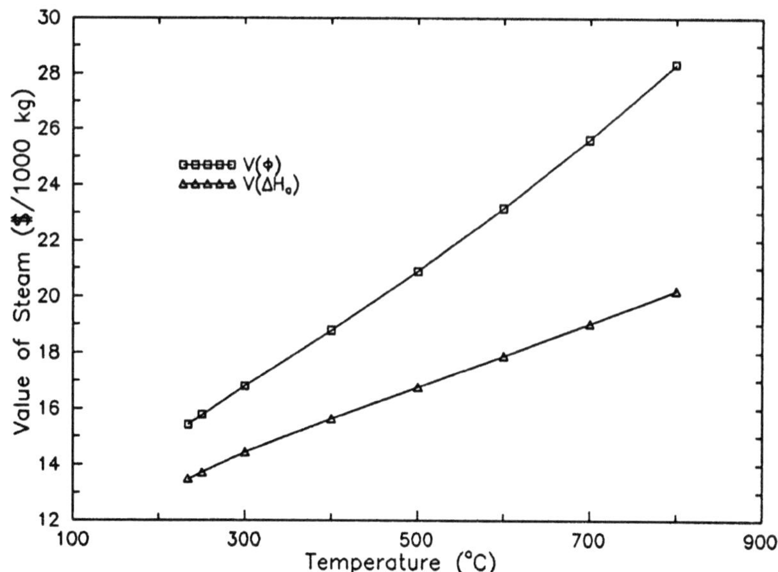

Figure 5.E.6b Value of steam, \$/1000 kg, as a function of temperature at P = 30 bar.

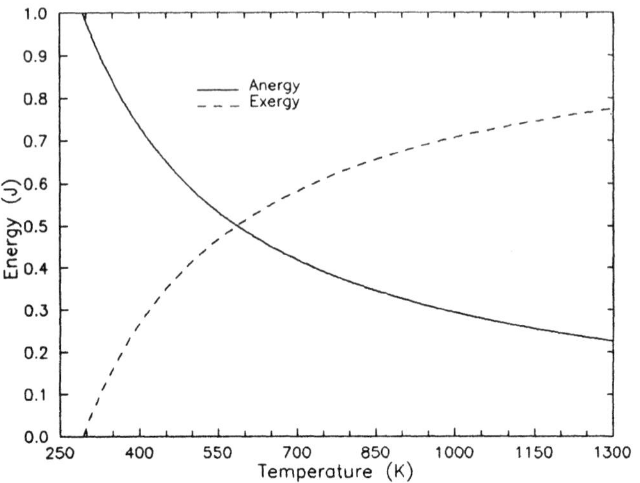

Figure 5.3 Exergy and anergy of one Joule of thermal energy as a function of the temperature it is available at.

5.6.6 Summary

1.Exergy represents the real measure - the 'yardstick' - of the ability of a system, or of a certain type of energy, to produce useful work by coming into equilibrium with the atmosphere in a flow process.

2.Ideal work, on the other hand, refers to a specified final state for the system. If this final state is the atmospheric, the ideal work is the same with the exergy of the system.

3.The exergy of heat - our main source of energy through fossil fuel combustion - is a strong function of the temperature it is available at.

4.The exergy of steam - the major fluid for the conversion of thermal energy into mechanical - is a strong function of its temperature and pressure.

5.These conclusions are summarized pictorially in Figure 5.4.

6.Exergy is not conserved, but it is continuously degraded into anergy. Thus, if one Joule of heat available at 373 K escapes to the atmosphere, its exergy goes from 0.201 J to zero.

In the next Section we will use exergy to establish the 'yardstick' for measuring the degree of efficient energy utilization in a given process, the *second law efficiency*. The *first law efficiency*, is also introduced for comparison purposes.

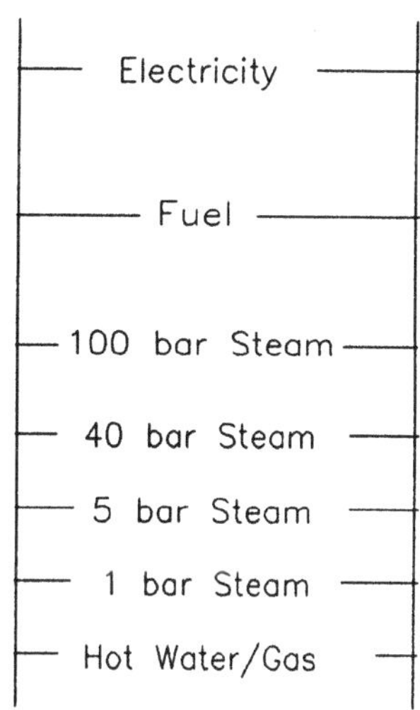

Figure 5.4 Relative energy value based on its exergy content.

5.7 First and Second Law Efficiencies

5.7.1 Definitions

Consider a flow process that involves several input and output streams. Each stream has an amount of energy and exergy associated with it. We define the first (n^I) and second (n^{II}) law efficiencies as follows:

$$n^I = \frac{\Sigma E_i}{\Sigma E_m} \tag{5.7.1}$$

$$n^{II} = \frac{\Sigma \Phi_i}{\Sigma \Phi_m} \tag{5.7.2}$$

Here, E stands for all forms of energy (thermal, mechanical, enthalpy, etc.), and the summation over i covers the *useful* output streams *only*, while the summation over m covers *all* the input streams.

Typical first and second law efficiencies are presented in Table 5.2 (Kenney, p.19). The drastically lower n^{II} values, as compared to those for n^I, demonstrate that energy degradation, i.e. the creation of anergy, occurs even in processes that are rather efficient from the first law point

Table 5.2 Comparison of First and Second Law Process Efficiencies % (a). Kenney, p.19.

Unit Operation (or Process)	First Law	Second Law
Residential heater (fuel)	60	9
Domestic water heater (fuel)	40	2-3
High-pressure steam boiler	90	50
Tobacco dryer (fuel)	40	4
Coal gasification, high Btu	55	46
Petroleum refining	~ 90	10
Steam-heated reboiler	~ 100	40
Blast furnace	76	46

(a) Gaggioli, 1980

of view.

We demonstrate the calculation of first and second law efficiencies in the following two Examples, and close the Section with some comments on their evaluation.

5.7.2 Example 5.7

Calculate the first and second law efficiencies of a boiler that generates saturated steam at 15 bar, used for process heating. Assume that:
* the temperature of the flue gases in the boiler is 1800°C;
* 10% of the fuel heating value is lost with the flue gases;
* feed water is the condensate, saturated liquid at 15 bar.

In this case we have:
* Input streams: Fuel
* Useful output streams: Heat as saturated steam at 15 bar.

Basis: 1 kJ of fuel heating value.

1.First law efficiency.

Since only 90% of the heating value is absorbed by the steam:

$$n^I = 0.9/1.0 = 0.9$$

2.Second law efficiency

* input exergy:

(1 kJ of heat at 1800°C) = 1{1 - [(273 + 25)/(273 + 1800)]} = 0.856 kJ

* output exergy:
  (1 kJ of heat at 198.3°C) = 1{1 -[(273 + 25)/(273 + 198.3)]} = 0.368 kJ

Exergy produced per kJ of fuel heat: 0.9 x 0.368 = 0.331 kJ

$$n^{II} = 0.331/0.856 = 0.387$$

Notice the drastic difference between the two efficiencies.

5.7.3 Example 5.8

Calculate the first and second law efficiencies for the power plant of Example 3.12. Assume, as in Example 5.7, that the temperature of the flue gases in the boiler is 1800°C, with a first law efficiency of 90% in fuel heating value utilization.

Energy input streams:
* fuel heat used in the boiler, Q_1
* pump work, W_p

Useful energy output streams:
* shaft work from turbine, W_t

On the basis of one kg of steam, we have from Example 3.12:

$$Q_1 = 2916.94/0.9 = 3241.0 \text{ kJ}$$
$$W_p = 3.55 \text{ kJ}$$
$$W_t = 713.7 \text{ kJ}$$

For the first law efficiency:

$$n^I = 713.7/(3241.0 + 3.55) = 0.220$$

For the second law efficiency, we must calculate - as in Example 5.7 - the exergy of the fuel heat:

$$\Phi(Q_1) = 3241.0(1 - T_0/T) = 2775.1 \text{ kJ}$$

Hence the second law efficiency is:

$$n^{II} = 713.7/(2775.1 + 3.55) = 0.257$$

5.7.4 Comments

1.Notice that our definitions of first and second law efficiencies, Eqs 5.7.1 and 5.7.2 respectively, are *pragmatic* for they consider the useful output streams only. If we consider all exit streams, then:

a.The first law efficiency will be equal to 1, since energy is conserved, which can be misleading.

b.The second law efficiency will be higher, which can be also misleading. It is recommended, however, that the enthalpies and exergies of the non-useful streams be evaluated to identify possible sources of substantial energy losses.

2.Thus, in the power plant discussed in Example 5.8, the exergy of the flue gases and of the cooling water leaving the condenser should be considered. Evaluation of the first requires knowledge of the stoichiometry of the combustion reaction and of the excess air used (see Sussman for details). Evaluation of the second, is straightforward (see Problem 5.17). In general, use of the flue gases for preheating purposes leads to low exit temperature and, consequently, low exergy losses, typically about 2% of the fuel exergy value.

3.In calculating the fuel exergy so far, we have used the heat of combustion at the temperature of the flue gases (1800°C, for example). The true fuel exergy, however, i.e. the maximum possible useful energy that can be obtained from it, is evaluated by assuming that the fuel reacts reversibly with oxygen in a so called 'fuel cell'.

We will discuss the method of evaluating this exergy - which is referred to as *chemical* exergy - when we develop the necessary background in Chapter 15 (Example 15.4). We will see that in this case it is the Gibbs free energy of the combustion reaction that is used and, as a result, the obtained chemical exergy is larger than that based on the heat of combustion.

We discuss next how improved first and second law efficiencies in the use of thermal energy can be realized through a very interesting approach, Cogeneration.

5.8 Cogeneration

5.8.1 Rationale

How can the low second law efficiency of the boiler in Example 5.7 and the low first and second law efficiencies of the power plant of Example 5.8 be improved? How, in other words, can we improve the utilization of the fuel heating value, thus effecting energy conservation?

The answer is suggested by the results of Example 5.6: High temperature and pressure steam has high exergy value and should be used for the production of mechanical work and electricity; low pressure steam has low exergy, but still high heating value, and should be used for heating purposes. This combination is called *Cogeneration*.

The following Example demonstrates the advantages of cogeneration.

5.8.2 Example 5.9

A chemical plant needs 1000 kW (1 MW) of electric power and 17282 MJ/hr (1.8 MW) of process heat as 10 bar saturated steam. Two schemes, shown in Figures 5.E.9a and 5.E.9b, are considered for this purpose.

In the 'conventional' scheme, two separate boilers are used, that produce:

* the one, high pressure steam (110 bar and 550°C), which is expanded to a pressure of 0.1 bar for power generation;
* the other, the 10 bar saturated steam.

Both boilers are efficient in the first law sense, with an efficiency of 0.90.

In the 'cogeneration' scheme, a single boiler is used generating steam at 110 bar and 550°C which flows through a turbine with an exhaust pressure of 15 bar. This steam is used then in a heat exchanger to generate the 10 bar steam for process heating. This boiler also has a first law efficiency of 0.9.

The following information is given:

* turbines are adiabatic but irreversible, with an efficiency of 0.8;
* the process steam returning to the boiler can be assumed to be saturated liquid at 10 bar;
* similarly, the water from the condenser is saturated liquid at 0.1 bar, and from the heat exchanger, at 15 bar;
* pump duties can be neglected.

The pertinent data are presented in Figures 5.E.9a and 5.E.9b.

Determine the:

a. Fuel heat requirement for the two schemes;
b. First law efficiencies; and
c. Second law efficiencies;

assuming the fuel to be methane.

Given:

Enthalpy of combustion: 212.8 kcal/mol = 49.5 MJ/kg

Chemical Exergy of CH_4: 195.5 kcal/mol = 45.5 MJ/kg (Sussman)

(Undergraduate students can proceed to the Comments at the end of the Example.)

a. Fuel Heat Requirement

1. Conventional Scheme

$$W_s = 0.80(3490 - 2120) = 1096 \text{ kJ/kg} = 1.096 \text{ MJ/kg}$$

Steam generated:

$$(1000 \text{ kW})(3600 \text{ kJ/kWh})/(1096 \text{ kJ/kg}) = 3285 \text{ kg/hr}$$

Heat absorbed in power plant boiler:

$$(3285 \text{ kg/hr})(3490 - 192) \text{ kJ/kg} = 10834 \text{ MJ/hr}$$

Heat absorbed in process steam boiler: 17282 MJ/hr

Total fuel heat:

$$(10834 + 17282)\cdot 10^6/0.9 = 31240 \text{ MJ/hr}$$

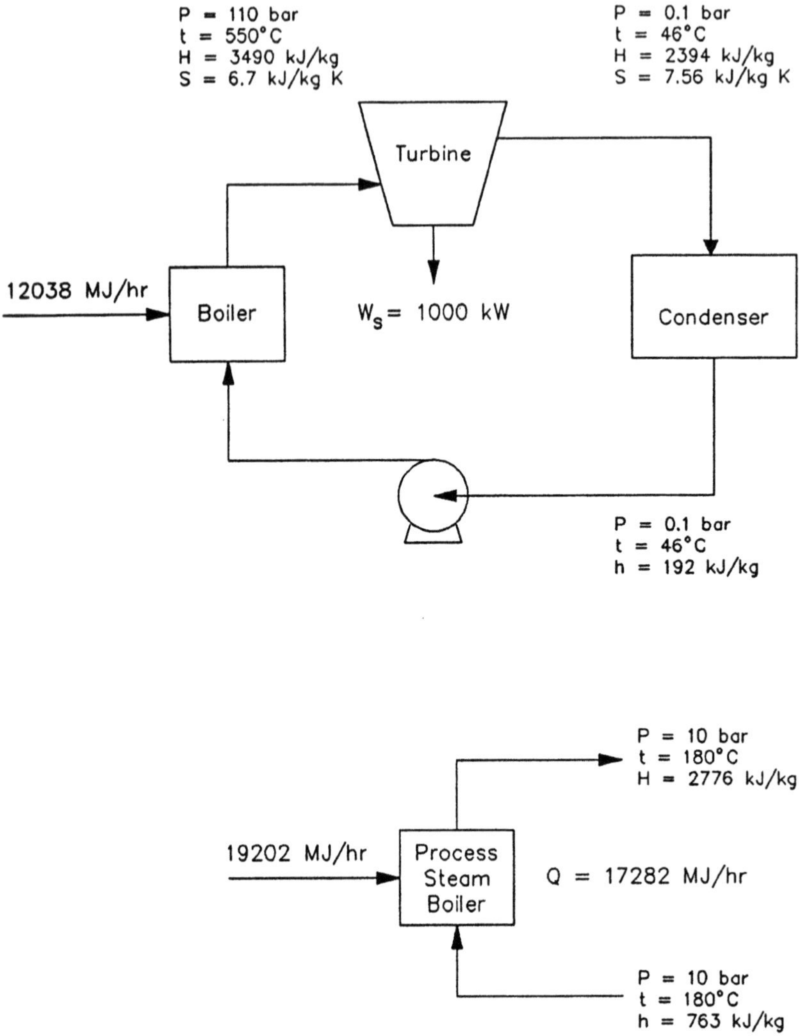

P = 110 bar
t = 550°C
H = 3490 kJ/kg
S = 6.7 kJ/kg K
P = 0.1 bar
t = 46°C
H = 2394 kJ/kg
S = 7.56 kJ/kg K
Turbine
12038 MJ/hr
Boiler
Ws = 1000 kW
Condenser
P = 0.1 bar
t = 46°C
h = 192 kJ/kg
P = 10 bar
t = 180°C
H = 2776 kJ/kg
19202 MJ/hr
Process
Steam
Boiler
Q = 17282 MJ/hr
P = 10 bar
t = 180°C
h = 763 kJ/kg

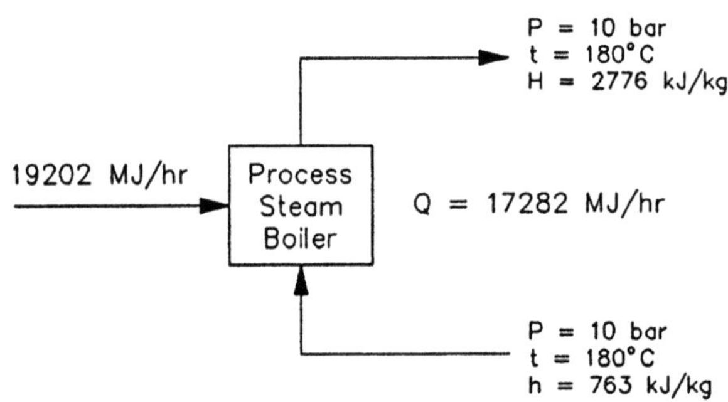

Figure 5.E.9a Conventional Scheme for Example 5.9.

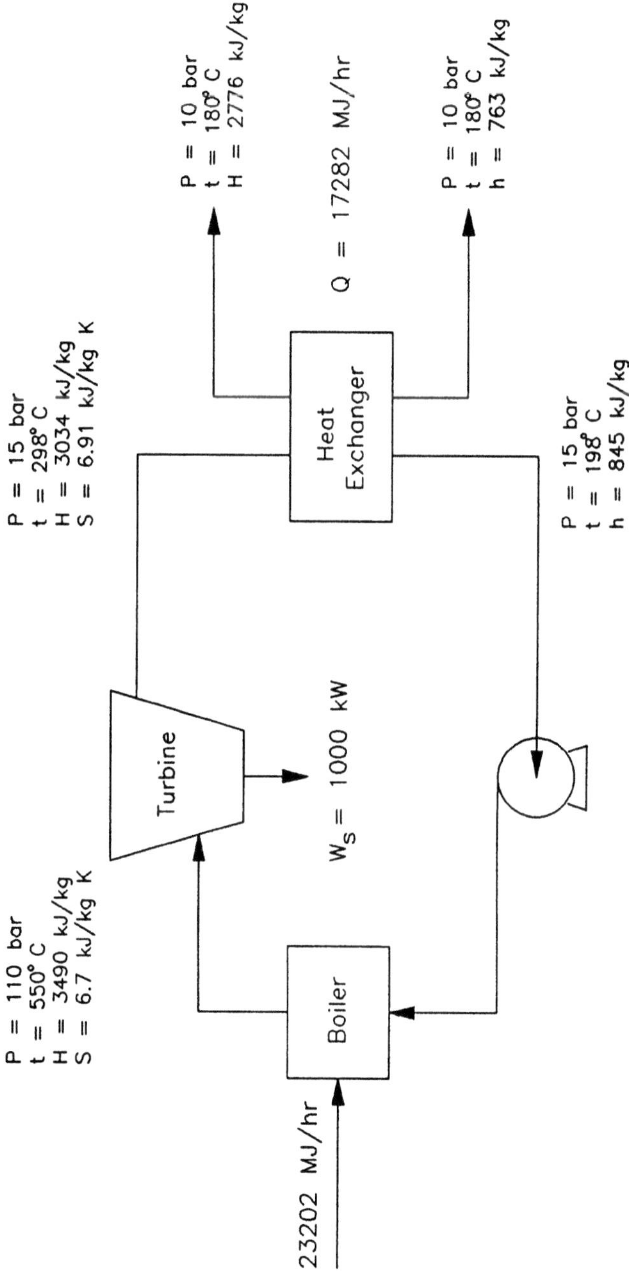

Figure 5.E.9b Cogeneration Scheme for Example 5.9.

2. Cogeneration

$$W_s = 0.80(3490 - 2920) = 456 \text{ kJ/kg} = 0.456 \text{ MJ/kg}$$

Steam generated:

$$(3600 \text{ MJ/hr})/(0.456 \text{ MJ/kg}) = 7895 \text{ kg/hr}$$

Heat absorbed in boiler:

$$(7895 \text{ kg/hr})(3490 - 845) \text{ kJ/kg} = 20882 \text{ MJ/hr}$$

Fuel heat:

$$20282/0.9 = 23202 \text{ MJ/hr}$$

Process heat:

$$(7895 \text{ kg/hr})(3034 - 845) \text{ kJ/kg} = 17282 \text{ MJ/hr (As required).}$$

b. First Law Efficiency

$$n^I \text{ (conventional)} = (3600 + 17282)/31240 = 0.67$$
$$n^I \text{ (cogeneration)} = (3600 + 17282)/23202 = 0.90$$

c. Second Law Efficiency

1. Conventional Scheme

Total fuel: (31240 MJ/hr)/(49.5 MJ/kg) = 631.1 kg/hr

Exergy (fuel) = (631.1)(45.5) = 28715 MJ/hr

Exergy (electricity) = 3600 MJ/hr

Exergy (heat at 180°C) = (17282)(1-298/453) = 5913 MJ/hr

$$n^{II} = (3600 + 5913)/28715 = 0.33$$

2. Cogeneration

Total fuel: (23202 MJ/hr)/(49.5 MJ/kg) = 468.7 kg/hr

Exergy (fuel) = (468.7)(45.5) = 21326 MJ/hr

$$n^{II} = (3600 + 5913)/21326 = 0.45$$

Comments:

1. All pertinent information is summarized in Table 5.E.9.

2. An inspection of the energy consumption suggests that cogeneration saves: (31240 - 23202) = 8038 MJ/hr of fuel heat, for a 26% fuel savings over the conventional cycle.

3. Using as a 1984 price of heat: $0.0047/MJ (Kenney, p.37), this represents savings of about $ 330,000 per year.

4. It should be noted, in addition, that the cost of the two schemes is about the same. Kenney (p.250), for example, considers two schemes similar to those used here and demonstrates that the cost (1981) is $2.8 million for the cogeneration scheme as compared to $2.9 million for the conventional one.

5. As expected the first law efficiency for the cogeneration cycle is much higher - 0.90 versus 0.67 for the conventional one - for all energy absorbed in the boiler, is utilized. There is no condenser, as in the conventional scheme. This not only saves energy, but it also eliminates a major source of thermal pollution.

6. Notice that even the cogeneration scheme, in spite of its high first law efficiency of 90%, has a second law efficiency of only 45%.

Table 5.E.9 Summary of Performance of the Two Schemes

	Conventional	Cogeneration
Energy Output (MJ/hr)		
* Electricity	3600	3600
* Process Steam	17282	17282
Total	*20882*	*20882*
Energy Input (MJ/hr)		
* Power Plant Boiler	12038	23202
* Process Steam Boiler	19202	-
Total	*31240*	*23202*
First Law Efficiency	0.67	0.90
Exergy Input (MJ/hr)		
* Power Plant Boiler	11065	21326
* Process Steam Boiler	17650	-
Total	*28715*	*21326*
Exergy Output (MJ/hr)		
* Electricity	3600	3600
* Process Steam	5913	5913
Total	*9513*	*9513*
Second Law Efficiency	0.33	0.45

7.If, however, the technology were available for using the chemical exergy directly - with an efficiency of, say, 90% - the second law efficiency would be twice that of the cogeneration scheme.

We identify the sources of these low second law efficiencies in the next Example. (Undergraduate students can proceed to Section 5.10.)

5.8.3 Example 5.10

Identify the exergy losses in the two schemes of the previous Example.

a. Conventional Scheme

1. Power plant boiler

Exergy in: 11065 MJ/hr

$\Phi(\textit{produced steam}) = T_0 (S_0 - S) - (H_0 - H)$
$= 298(0.367 - 6.7) - (104.8 - 3490) = 1498$ kJ/kg
Exergy out (steam): (3285 kg/hr)(1498 kJ/kg) = 4921 MJ/hr
Exergy loss: 11065 - 4921 = 6144 MJ/hr

2. Turbine
Exergy in: 4921 MJ/hr
Exergy out (electricity): 3600 MJ/hr
$\Phi(\textit{produced steam})$: 298(0.367 - 7.56) - (104.8 - 2394) = 146 kJ/kg
Exergy out (steam) = 480 MJ/hr
Exergy loss: 841 MJ/hr

3. Condenser
Condenser duty: (2394 - 192) = 2202 kJ/kg
Exergy loss: (3285 kg/hr)(2202 kJ/kg)(1 - 298/319) = 476 MJ/hr

4. Process Steam Boiler
Exergy loss: (17650 - 5913) = 11737 MJ/hr

b. Cogeneration

1. Boiler
Exergy in: 21326 MJ/hr
Exergy out (steam): (7895 kg/hr)(1498 kJ/kg) = 11827 MJ/hr
Exergy loss: 9499 MJ/hr

2. Turbine
Exergy in: 11827 MJ/hr
Φ(steam) = 298(0.367 - 6.91) - (104.8 - 3034) = 979 KJ/kg
Exergy out (steam) = (7895 kg/hr)(979 kJ/kg) = 7729 MJ/hr
Exergy loss: 11827 - 7729 - 3600 = 498 MJ/hr

3. Process steam
Exergy loss: 7729 - 5913 = 1816 MJ/hr

The results are tabulated in Table 5.E.10 and indicate that the major exergy losses are in the boilers.

Combustion, as carried out in boilers, is highly irreversible, even with high pressure and temperature steam produced. The old slogan "You've got a long way to go baby" is very applicable here, before really efficient use of fuel energy is achieved.

Table 5.E.10 Exergy Losses in the Two Schemes, MJ/hr

	Conventional		Cogeneration	
	loss	%	loss	%
Power Plant				
* boiler	6144	32.0	9499	80.4
* turbine	841	4.4	498	4.2
* condenser	476	2.5	-	-
Process Steam Boiler	11737	61.1	1816	15.4
Total	19198	100.0	11813	100.0

5.9 Limitations of Cogeneration

Let us now consider the following question: What will happen in the cogeneration scheme of Example 5.9 if the power requirement is double, i.e. 2000 kW, while the process heat needs remain at the same level of 17.3 GJ/hr.

Since the boiler pressure and temperature are high enough for an installation of this size, we could consider lowering the turbine exhaust pressure. But this is not possible because the resulting steam would not be hot enough to produce the 10 bar process steam.

The only solution would be to produce more steam and exhaust part of it to a low pressure. This approach, however, is not very efficient as the conventional cycle demonstrates. And since Utility companies operate at higher efficiencies, the cost of the power produced would be higher than that obtained from them.

It is apparent therefore that, once:

* the boiler pressure and temperature, and
* the process steam pressure

are specified, there is only one power to process heat ratio that can be obtained through this type of cogeneration. This ratio provides, of course, the maximum efficiency for the specified conditions.

In the typical case, the maximum pressure and temperature of the boiler are set by the manufacturer and, consequently, the ratio is determined by the turbine exhaust pressure, i.e. the process steam pressure. Typical values of this ratio are given in Table 5.3 and indicate that only low values of power to process heat ratio can be met with this type of cogeneration scheme.

Higher values, up to one, can be achieved by switching to gas turbines and still higher ones, by using diesel engines. For details, see Kenney.

5.10 Energy Conservation: The Favorable Economics and the Institutional Barriers

Kenney presents a very eloquent and demonstrative discussion of this topic. He writes (pp.245-248):

Table 5.3 Ratios of Power to Process Heat (a) for Maximum Efficiency as a Function of the Turbine Exhaust Pressure

Inlet steam: P = 110 bar; t = 550°C; H = 3490 kJ/kg
Turbine: adiabatic with an efficiency of 0.8

P(bar)	Ratio
15	0.21
30	0.15
40	0.12

(a)Process heat: Exhaust steam enthalpy - Saturated liquid enthalpy

"The usual expectation in the process industries is that improved energy efficiency requires increased investment. Much of this expectation is derived from the 'fix-up' nature of early conservation activities, from a lack of understanding of the complete plant energy system, and from a deep-seated conviction that nothing comes free. However, from the more fundamental view point of the second law of thermodynamics, one might expect to be able to reduce both the capital investment of a plant and its inefficiency at the same time. This comes from the knowledge that the second law efficiency of most industrial processes is very low and from the feeling that if less fuel is burned on a given site, then surely less hardware needs to be installed to handle the products of combustion.

Various energy policy studies performed in the aftermath of the 1973 oil embargo showed that on a global scale the required investment for efficient facilities was less than that for a number of conventional processes and their associated energy supply systems. Gyftopoulos and Widmer (1980) pointed out that the industrial sector of the U.S. economy operates at an overall second law efficiency of about 13%, leaving considerable room for improvement. These authors and co-workers also showed that approximately 25% of the projected 1985 energy usage in the manufacturing sector of the U.S. economy could be saved through conservation measures with capital costs on the order of $125 billion. By comparison, they estimated that to produce the same amount of energy from grass roots sources would require a total investment of about $175 billion. They quote an American Gas Association report that indicates that, apart from the Middle East, the current capital investment required to provide new supplies of energy is very high. For example, each new source of petroleum or natural gas equivalent to 456 kB/d costs between $5 and $10 billion in investment. A similar amount of 'synthetic' gas or liquid fuel derived from coal would be even more costly, on the order of $15 billion. Although equivalent new coal supplies are still obtainable at a capital cost of about $2 billion, environmental restrictions at the mine and the user's site and the investment required for transportation equipment add significantly to this cost.

The capital cost for industrial energy is highest when the energy is provided in the form of electricity. In its 1977 annual report, ThermoElectron Corporation summarized its studies of the capital cost of central station power systems compared to industrial cogeneration systems for electricity production....... The total investment for central station power systems using coal or oil is on the order of $50 to $60 billion/$10^{15}$ Btu·yr (one quad) of electricity. This includes transmission and distribution facilities, fuel production facilities, and the majority of the costs of the generating plant itself. Industrial cogeneration, on the other hand, requires only 50 to 65% of this investment, because the total investment is shared with onsite users of process heat that would otherwise need a separate supply. Obviously, the nation would be far better off if industrial cogeneration investments were substituted for grass roots fuel production and central station power generation.

These global studies point out the economic potential of energy efficiency in the United States. Although other countries may be more productive in the use of energy in the industrial sector, appreciable potential exists for them as well. The practical difficulty with implementation of global studies is that institutional barriers play an important role in society, but are ignored in the studies. Industrial managers function within an assigned area and work very hard to maximize the profit impact of their sector. They get no credit for savings achieved by others and have little time to spend in persuading other parts of their (or other) organizations to go into a cooperative venture in the energy field. In spite of the global potential for energy conservation ideas specific to their own area (see Table 5.4 for a few examples), the perceived rewards are not high enough to merit action. Table 5.4 shows that the total capital cost for new supplies often exceeds that for conservation measures. Still, much of the capital costs considered for new supplies would have to be spent outside the industrial plant gate. Therein lies part of the problem for the industrial energy conserver.

Typical of this situation was the case of a plant manager who proposed to install several million dollars of cogeneration equipment powered solely by waste heat. This investment would have replaced at least $5 million of investment in a central station power plant and would have provided a 22% return on investment at an electricity price of 2.6 c/kWh.

Despite the fact that the return was more than twice the company's average return on assets, top management rejected the proposal. The reasons given were two-fold:

1. the plant management was not faced with the $5 million investment for power generation facilities;
2. the management had set a 30% return on the investment hurdle rate for investments not related to the production of product.

By company practice the manager was forced to use the average cost of electricity rather than the replacement value in evaluating his project, and could not take credit for investment savings by others. As a result his project, however beneficial to the nation as a whole, was turned down.

Table 5.4 Examples of the Cost Effectiveness of Conservation Measures(a). Kenney, p.247.

Conservation step	Energy savings (MBtu/h)	1975 capital ($\$10^6$)	Annual quad fuel ($\$10^9$)	Capital for comparable new supply
Place recuperators on the steel reheating furnaces (170 ton/h)	100	480	0.7	10xCons. (b)
Optimize the crude product train (100,000 B/d)	23	1000	4.7	1.5xCons.
Add feed water heaters on the styrene plant furnace (400 ton/d)	4	17	0.6	4-12xCons.
Generate steam from the flue gas of NH_3 reformer (1000 ton/d)	11	130	1.1	2-6xCons.

(a) Gyftopoulos and Widmer
(b) Cons. stands for the capital required for the conservation step.

Thus, the institutional nature of the industrial organization presents additional problems for the energy conservation engineer that are not faced by global policymakers."

5.11 Concluding Remarks

1.Energy conservation has a very long way to go before the theoretical minimum level of energy consumption is reached. Efficiency, thus, in energy utilization of the chemical and petroleum industries, of direct interest to chemical engineers, is at a very low level of 10-20%.

2.There are three main reasons for this low efficiency:

a.The necessary technology is not always available;

b.Lack of understanding of the second law efficiency, and of the fact that it can be very low, even when the first law one is very high;

c.Institutional: objection to change combined with the deeply rooted belief that "nothing comes free".

3.Yet, some change is taking place. Basta and Price(1985), for example, comment: "A favorable regulatory climate (in the U.S.), combined with investment returns at 20% or better, is making cogeneration the hottest construction item in the chemical process industries." They report that cogeneration certificates rose over 100% between 1979 and 1984: from about $11 \cdot 10^9$ megawatts in 1979, to nearly $25 \cdot 10^9$ megawatts in 1984.

4.While cogeneration is a major method for realizing increased efficiency in energy utilization, a more effective method is to consider a "total plant energy integration" (cogeneration considerations would of course be part of it). Using this approach in a major consumer oriented chemical company, Steinmetz and Chaney (1985) report energy savings of $4.5 million per year with rates of return in excess of 30%.

5.As mentioned in the Introduction, energy conservation is a very complex subject. We only scratched the surface here. For a detailed discussion of the subject, the excellent book of Kenney: *Energy Conservation in the Process Industries* is a 'must'.

The references given in the end of this Chapter should also be consulted for their content itself, as well as for the references given in them. Finally, for further understanding of exergy calculations, and for some interesting examples, the monograph of Sussman: *Availability (Exergy) Analysis*, is recommended.

References

Basta, N., Price, W., August 5, 1985. *Chemical Engineering*, 14.

Boland, D., Hill, J.C., Hindmarsh, E., Williams, P.J., 1983. AIChE Meeting, Washington, D.C. (and ICI Tensa Technology, Tracerco Corp., P.O. Box 192, South Houston, Texas).

Flower, J.R., Linnhoff, B., 1979. *Computers and Chemical Engineering*, 3, 283.

Gaggioli, R., November, 1980. *Second Law Analysis for Process and Energy Engineering*, AIChE Meeting.

Gyftopoulos, E.P., Lazarides, L.J., Widmer, T.F., 1974. *Potential Fuel Effectiveness in Industry*, Ballinger, Mass.

Gyftopoulos, E.P., Widmer, T.F., 1980. *Energy Conservation*, Am. Chem. Soc. Symposium Series 122, *Thermodynamics: Second Law Analysis*, Washington, D.C.

Haywood, R.W., 1974. *J. of Mechanical Engineering Science*, **16**, 160; 258.

Kenney, W.F., 1984. *Energy Conservation in the Process Industries*, Academic Press, Orlando, Florida.

Lamaeda, T., Harada, T., Shirako, K., 1979. *Computers and Chemical Engineering*, **3**, 273.

Moser, F., Schnitzer, H., 1985. *Heat Pumps in Industry*, Elsevier Science Publishing Co, New York.

Peterson, J.F., Mann, W.L., Oct. 14, 1985. *Chemical Engineering*, 62.

Riekert, L., 1982. *Energy: Money, Materials and Engineering*, I. Chem. E. Symposium Series, No 78, London.

Steinmetz, F.J., Chaney, M.O., July 1985. *Chem. Eng. Progress*, 27.

Sussman, M.V., 1981. *Availability (Exergy) Analysis*, Mulliken House, Lexington, Mass.

ThermoElectron Corporation, 1977 Annual Report. Waltham, Mass.

Townsend, D.W., Linnhoff, B., 1982. *The Chemical Engineer*, Issue No 378, 91.

Problems

5.1 Discuss the factor(s) that made energy conservation such an important subject after the early 1970's.

5.2 With reference to ideal work:

a. define it;

b. discuss the approach used in determining it;

c. how does it differ from the reversible work?

d. what is its importance in practice?

5.3 In Section 5.3.3, why is $\Delta S(id) = \Delta S[system]$?

5.4 One mole of air at 1 atm and 25°C is compressed to 5 atm in a flow process. The compressor is adiabatic with an efficiency of 0.8:

a. determine the temperature of the compressed air and the minimum work required;

b. develop a reversible process that brings the air to its final state, and calculate the work involved.

Assume an average heat capacity in the temperature range involved equal to 29.3 J/mol K.

5.5 What is the minimum work required to dry one mole of air at 1 atm and 25°C that contains 2%(mole) moisture?

5.6 With reference to Example 5.3:

a. what accounts for the difference in minimum work in the two cases?

b. what conclusion do you reach with respect to the energy consumption for liquefaction versus that for separation?

5.7 One kg of wet steam at 15 bar is throttled to 1.3 bar and 110°C. Determine:

a. the quality of the steam before throttling;

b. the ideal work;

c. the lost work.

5.8 Dry saturated steam at 1 bar is compressed adiabatically in a flow process to 4 bar. The actual enthalpy change is 20% greater than its value if the compression to 4 bar were carried out reversibly. Calculate:

a. the temperature of the steam after the compression; and

b. the lost work per kg of steam.

5.9 In a flow process a stream of water at 25°C is mixed with superheated steam at 7 bar and 260°C to produce dry saturated steam at 100°C. Assuming that 100 kg/hr of superheated steam are used and no heat losses occur to the surroundings, calculate:

a. the kg/hr of water used; and

b. the lost work.

5.10 Propane gas enters a heat exchanger at the rate of 1 mol/s at 400°C and 5 atm and leaves at 40°C and 1 atm. The cooling medium is water that enters the exchanger at 20°C and leaves at 40°C. The average heat capacity of propane in this range is 100 J/mol K. Calculate:

a. the entropy change of propane;

b. the entropy change of water; and

c. the lost work.

5.11.a. What is useful energy?

b. what contains more exergy, one Joule of:

* mechanical energy,
* heat, or
* kinetic energy?

c. the same as (b), but for useful energy.

5.12 What is the difference between ideal work and exergy?

5.13 One Joule of heat is available at 200°C. What is the maximum fraction of it that can be transferred to a heat reservoir being at:

a. 100°C

b. 300°C

without the consumption of any work?

5.14 It is recommended that the coils in the back of your refrigerator be kept clean from dust and away from the wall. Why? Be specific.

5.15 With reference to Problem 5.4, calculate the first and second law efficiencies.

5.16 Steam at 300 psia and 540°F expands through an adiabatic throttling valve to 100 psia:

a. what is the enthalpy and temperature of the exiting steam?

b. what are the first and second law efficiencies of the process?

5.17 The exhaust steam from the turbine of a power plant, at 0.3 bar and a quality of 5%, is condensed to saturated liquid with cooling water entering at 25 °C. If the cooling water leaves at 50°C (higher temperatures are undesirable to avoid excessive corrosion), calculate its exergy per kg of steam condensed.

5.18 Steam at 540°C and 20 bar expands through a turbine to 1.5 bar. If the turbine is adiabatic with an efficiency of 0.9, calculate the first and second law efficiencies.

5.19 With reference to Example 5.8, explain the higher second law efficiency as compared to the first law one.

5.20 The maximum heating requirement for a house is 10^8 Joule/hr in the coldest period, when the temperature outside is -20°C. The house is to be maintained at 25°C. Two heating systems are considered for this purpose:

a. hot air generated by burning natural gas, whose cost is \$5/10^9 Joules. The furnace has a first law efficiency of 0.5.

b. a heat pump that, as we saw in Example 5.1, 'pumps' heat from the lower atmospheric temperature into the house. In this case, the operating fluid will absorb heat at -30°C and reject heat in the house at 35°C.

Which of the two systems would you recommend on the basis of energy cost?

Given:

* Heat absorption and rejection take place at constant, for practical purposes, temperature since they involve a constant pressure phase change of the fluid.

* A 10°C difference, 'the temperature approach' as it is referred to - between the operating fluid and the atmosphere or the house - is needed to provide the driving force for the required heat transfer.

* Assume that the compressor that drives this heat pump is adiabatic with an efficiency of 0.75, but that - to compensate for insulation losses - only the reversible work appears in the hot fluid.

* Electricity cost: \$15/10^9 Joules.

5.21 If the maximum cooling requirement for the house in the previous Problem during the hottest period of the summer is $1.5 \cdot 10^8$ J/hr, with an outside temperature of 40°C and an inside temperature of 28°C, will the motor power - based on the winter operation - suffice?

5.22 With reference to cogeneration:

a. what is it?

b. give an example where you could use it.

5.23.a. What are the main objectives of this Chapter?

b. Summarize the important items that you have learned.

c. Give an Example in each case.

5.51 Develop the expression for ideal work in a non-flow process.

5.52 Assuming that the boiler in Example 5.7 operates with methane fuel, and using the pertinent data from Example 5.9, calculate the second law efficiency.

5.53 For the power plant of Example 5.8, assuming that the fuel is methane and using the heat of combustion and chemical exergy data of Example 5.9:

a. calculate the second law efficiency of the whole cycle;

b. identify the exergy losses in the cycle;

c. present a list of recommendations for improving the cycle's second law efficiency.

5.54 Example 5.10 demonstrates that major exergy losses occur in the boilers.

a. What method(s) would you recommend for improvement?

b. Are there any problems associated with your recommendations?

5.55 Discuss the limitations of cogeneration.

5.56 A chemical plant needs 800 kW of power. Two approaches are considered:

1. a conventional cycle that operates as follows: steam at 100 bar and 550°C expands through an adiabatic turbine with an efficiency of 0.75 to an exhaust pressure of 0.1 bar.

2. a cogeneration cycle, since the plant has need for process steam at 12 bar, to be generated with the exhaust steam at 15 bar.

a. How much process steam will be generated?

b. What are the first and and second law efficiencies of the two approaches?

c. Identify the exergy losses in them.

Use the same fuel and boiler first law efficiencies of Example 5.9, and neglect pump duties.

5.57 Show that the availability - to do useful work - of a closed system is given by: $(U - U_0) + P_0(V - V_0) - T_0(S - S_0)$, where the subscript $_0$ denotes the surroundings.

5.58 Prepare a report on fuel cells: current status and future expectations.

6
Second Law and Entropy: A Broader View

The Second Law:

* *the premier law of all Sciences* (Einstein)
* *the supreme metaphysical law of the entire Universe* (Sir Arthur Edington)
* *the supreme law of Nature that governs everything we do* (Rifkin)

6.1 Introduction

We saw in Section 5.3.3 that any process that brings about a total increase of entropy ΔS results in the loss of an amount of useful energy equal to $T_0 \Delta S$, where T_0 is the temperature of the atmosphere. Notice that this energy, while it is still around us, is lost for ever with respect to its ability to provide useful work such as, for example, to produce mechanical work.

But we also know that all real life processes are, by their very nature, irreversible causing an increase in total entropy and, consequently, leading continuously to loss of useful energy. Thus, according to Table 5.2, we only use between 10% and 50% of the useful energy of the fuels, our main source of energy.

While this percentage will be increasing with technological advances, we should keep in mind that it can never reach 100%, for a degree of irreversibility is a practical requirement for all real life processes. A reversible heat exchanger, for example, must operate with a vanishingly small temperature difference, hence requiring an infinite area for heat transfer.

The net effect, thus, of the second law is a continuous:

* increase of entropy; and
* dissipation of useful energy.

6.2 Objective and Approach

Our main objective in this Chapter is to examine, albeit briefly, the broader implications of this dissipation process dictated by the second law. More important, to demonstrate the paramount impact that this dissipation has had - and will have - on the energy supply and, consequently, *on the way of life on this planet*. After all, *we are educating the technologists and technological leaders of tomorrow, who can play a major role in slowing this process down.*

We will start by suggesting that this dissipation process does not only apply to energy, but also to materials (minerals, for example), leading to the generation of *'material-entropy'* and the *'Fourth'* law of thermodynamics. The terms energy and entropy can be viewed, hence, in a broader sense, that includes materials as well.

We proceed then to consider the historical implications of the second law and suggest, following Rifkin (1980), that the occurrence of major *'Entropy Watersheds'* - whereby all available useful energy was exhausted - are responsible for the two main transitions in world history: the Agricultural and the Industrial Revolutions.

This interpretation of history will naturally lead to the very obvious question: "Is there a third Entropy Watershed coming, bringing a new Revolution"?

To answer this question - of immense obviously importance - we discuss first the three main issues that will determine the answer:

a. Energy demand in the future;
b. Energy supply in the future; and
c. Associated, to energy supply, cost and environmental problems.

We proceed then to consider the aforementioned question and close with some concluding remarks.

6.3 Material Dissipation and the 'Fourth' Law

Dissipation, unfortunately, does not only apply to energy, but to materials as well. We cannot, for example, recover the rubber particles that come off the car tires on the highway; or the top soil that is blown away by the wind; or all metals used in our technological society (even recycling provides, typically, well below 50% recovery for most metals). Thus, even though total matter is conserved, the amount of useful matter is continuously decreasing, just like useful energy is. Such observations actually led Georgescu-Roegen (1971; Rifkin) to propose a 'Fourth' law which proclaims for matter what the second law does for energy. Thus an isolated system, instead of tending toward heat death, where all energy is unavailable, tends toward chaos, where all matter is unavailable.

We turn next to the continuous energy dissipation - and the corresponding entropy increase - and examine its implications on the development of mankind, from the very early days to the future.

6.4 Second Law and World History

Our presentation of this rather unexpected relationship will begin by establishing first the concept of a 'common thread' behind the world history.

We will then proceed to suggest, following Rifkin, that the second law can be such a common thread by demonstrating, for example, how the two main transitions in world history, the Agricultural and the Industrial Revolutions, can be explained in terms of major entropy watersheds.

6.4.1 Common Threads in World History

> *His claim to universality rests on discovering a consistent human nature immanent in history that will allow an intelligent reader, guided by a wise and diligent historian, to interpret the future.* (Kagan on Thucydides, the historian of the Peloponnesian War.)

The purpose of studying History is not just to accumulate facts and dates, but rather to identify some 'common threads' that will help humanity understand the past and, hopefully, shape the future in the right direction.

This point is clearly underscored by Donald Kagan, Yale professor of History and Classics, in his article *The Changing World of World Histories*. (The New York Times, Book Review, p.1, Nov. 11, 1984). He writes:

"Perhaps the most striking difference between these modern historians and their forerunners in the long tradition of universal histories is not in their concern or approaches but in their lack of clearly stated purposes and programs. Why do they write world histories, and what can the reader hope to learn from them? They neither ask nor try to answer such questions, tacitly rejecting the age-old practice of seeking in the history of mankind the answers to larger religious and philosophical questions".

After a brief, but illuminating discussion of how past historians have dealt with these issues, he closes his article as follows:

"What we need are fuller more detailed and more systematic comparisons between societies. The new world historians should try to answer those truly interesting questions to which only a broad view of history can hope to supply the response:

* What are the enduring elements of human nature?
* How broad and deep is that complex of human traits we call by that name?
* How great, on the other hand, is that part of our lives freely open to our own will?
* To what extent was Herodotus right in saying, "Custom is king"?
* Or Thucydides, who sought "an exact knowledge of the past as an aid to the interpretation of the future, which in the course of human things must resemble if it does not reflect it"?

These are the important questions world historians must address, for whatever the proportions may be, a vital part of man's destiny is in his own hands, and he needs a broad and accurate picture of where he has been if he is to decide where he can and should go."

The search for such 'common threads' should not be surprising to us, for it was exactly this search for the common threads behind the physical phenomena that led to the development of the laws of thermodynamics. The identification, thus, of the 'common threads' behind world history represents, in essence, an attempt to identify the 'Laws of History'.

Rifkin suggests that the second law - the 'Entropy Law' as he refers to it - is not only a common thread behind physical phenomena, but also behind world history, as very briefly discussed next.

6.4.2 Entropy Watersheds and the Main Transitions in World History

To support the contention that the second law does represent a common thread behind world history we will use, as an example, its effect on the two main transitions:

* the Agricultural Revolution; and
* the Industrial Revolution.

In each case, we will demonstrate that the underlying cause for these transitions was the fact that the existing society exhausted all available energy, i.e. it reached an *Entropy Watershed.*

a. The Agricultural Revolution

We all know that our early ancestors were hunters and food gatherers. Evidence based on tribes that survived into the last centuries in isolated areas of America, Australia, and Africa suggests that they lived a rather pleasant life, probably much better than that of millions of people in the world today: they seem to have sufficient food and the effort required to gather it was only about a few hours per day (Stavrianos, 1984). Here is how J. Bargert, a Jesuit who lived between 1750 and 1767 with some California Indians, described their life: "They are generally strong, daring and healthier than thousands of other people that live daily with the best meals that are prepared by the skilful Paris chefs ... There is no doubt that they live a much happier life than the civilized people of Europe." (Smithsonian Institute, 1864.)

Why did these ancestors of ours change such a pleasant way of life? What brought about the Agricultural Revolution?

In fertile areas, one or two people needed about a square mile to support themselves; in desert or cold weather ones, 20 to 30 square miles were needed per person (Stavrianos). As the world population increased with the passage of thousands of years, consumption of food - the main energy source of the people at that time - grew faster than nature could provide for. Humanity, thus, reached its *first major entropy watershed.* A new energy source was needed: *Agriculture provided it.* There is no doubt that people preferred the 'free style' of the hunting life to the hard work of cultivating the earth. But they had no choice. It was a matter of survival.

b. The Industrial Revolution

Farm land becomes, thus, essential to humans as the world enters the agricultural period, while forests supply their needs in wood, and both are plentifully available. With the passage of time, however, two main factors change this picture leading to the industrial revolution.

First, technological developments utilize progressively more and more wood rendering it, by the 19th century, essential to the existence of society in Europe, just like oil is today. Mumford (1934) describes its importance as follows: "The carpenters' tools were of wood but for the last cutting edge; the rake, the oxyoke, the cart, the wagon, were of wood; so was the wash tub in the bathhouse; so was the bucket and so was the broom; so in certain parts of Europe were the poor man's shoes. Wood served the farmer and the textile worker; the loom and the spinning wheel, the oil presses and the wine presses, and even a hundred years after the printing press was invented, it was still made of wood. The very pipes that carried water in the cities were often treetrunks; so were the cylinders of pumps ... the ships of course were made of wood and ... the principal machines of industry were likewise made of wood."

Second, population growth, accelerated by the technological developments in the 17th and 18th century, bring the world population (Figure 6.1) from 5 million in the beginning of the agricultural period to 1000 million by the 19th century (up by a factor of 200!).

The problem becomes apparent. Increased population requires more farm land leading to increased deforestation which, combined with the also increasing need for wood, led to an acute shortage of the latter. Thus comes the *Industrial Revolution* with coal - and a little later oil - as the new energy sources, and the steam - and later other - engines the tools for harnessing them.

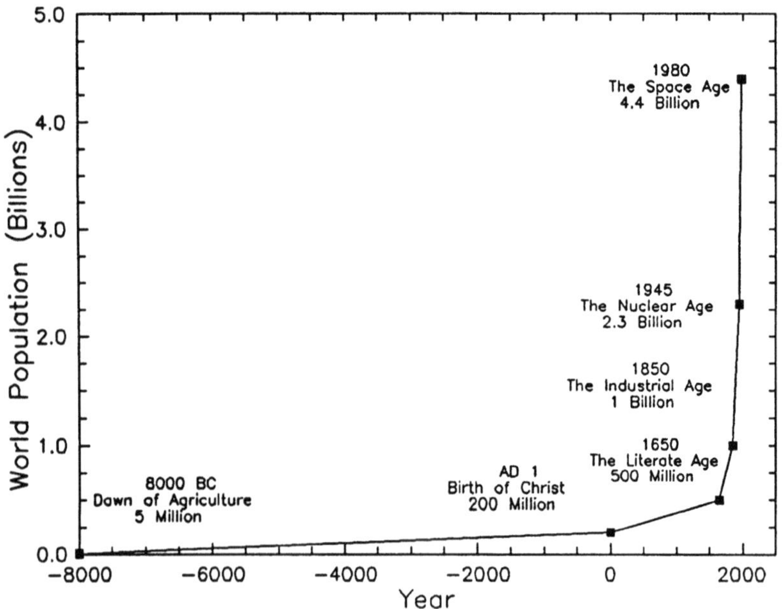

Figure 6.1 World population growth.

It should be noted, however, as we contemplate the possibility of an energy shortage developing today, that this replacement of the energy source was not a simple task.

First, coal was much more difficult to extract and process than wood (and oil is still more difficult). The problems with opening the mines, ventilating them, and moving the coal to the surface and then to the consumers, were further exacerbated by flooding for, as mining advanced deeper into the ground, the water table was reached. (This is exactly the problem that the Savery steam engine was intended to solve, as we saw in Chapter 4, thus its name: the miner's friend.)

Second, it was a very slow process as the aforementioned difficulties suggest. It was not until 1700 that coal replaced wood in England and still later in western Europe.

Finally, it effected drastically the quality of people's life creating hardship and misery as eloquently discussed by Rifkin. *The cultures of Europe had been thoroughly integrated into a wood based existence, like into an oil one today, and the changeover necessitated the radical uprooting of an entire way of life.*

Many will argue, and correctly so, that other factors played an important role in bringing about the Industrial Era besides the energy crisis. But the major role of the latter cannot be overlooked, neither can the hardships that people suffered, and both should be taken into account as we - naturally - ask the question: "*Is there a third Entropy Watershed coming?*"

Before we consider this question, we examine first the factors that would determine the answer to it.

6.5 Energy Demand and Supply in the Future

6.5.1 The Energy Crisis of the Early 1970's: Was it a Warning?

During the Oil Crisis of the early 1970's half-mile to mile long lines in gasoline stations were common in the U.S. and shootouts were reported. The Government was printing coupons to ration gasoline and the President was pleading with the citizens to set the thermostats low (68°F) during the winter and high (78°F) during the summer. The Government was also setting performance standards for new automobiles, in terms of miles per gallon, while the petroleum and chemical industries were setting energy conservation standards. And the price of oil increased drastically from $3 to $35 per barrel from 1970 to 1981. In other countries alternative day or weekend driving was imposed.

Today, there are no lines in the gasoline stations. Conservation, the high price of oil, and a mild world recession, have reduced the consumption of oil leading to an 'oil glut' with prices reaching the $18-20 per barrel (1990).

But is the crisis really over? Or have we just given a shot in the arm to a person suffering from an incurable disease? Is, in other words, a new Entropy Watershed coming, with the oil crisis of the 1970's serving just as a warning?

To answer this question, let us examine the energy demand and supply for the near future, year 2000, and the not so near future: 2025 and 2050. The required to this purpose projection is, as expected, difficult, but it does provide a qualitative - at least - picture of the energy demand and supply for this period.

To demonstrate the difficulties involved with such projections and, especially, the problems involved in meeting the demand of energy in the future, we consider first just the oil demand/supply picture for one country, the U.S., and for a short time range only, up to the year 2000.

The global demand/supply picture up to the year 2050, based on the excellent study of Edmonds and Reilly (1985), is then discussed, while problems associated with meeting this demand are considered in the next Section.

6.5.2 The U.S. Oil Demand and Supply to the Year 2000

Figure 6.2 presents the total oil demand for the U.S. projected to year 2000 by the Exxon corporation (Kenney, 1984). Notice the uncertainty in projections demonstrated by the pre-1973 curve; and to appreciate the problems involved in meeting this demand, consider the supply picture presented in Figure 6.3. Two main uncertainties appear:

First, it is projected that by the year 2000, about 20% of all the oil will be synthetic (synfuels). Recent (1984-85), however, actions by the U.S. Government - spurred by the oil glut - have brought the Synfuels program to a halt, for all practical purposes.

Second, will all the oil, projected to be imported, be available - in spite of the late 1980's glut - or will political developments, such as those experienced in the 1973 Embargo, and more recently in the Gulf War, lead

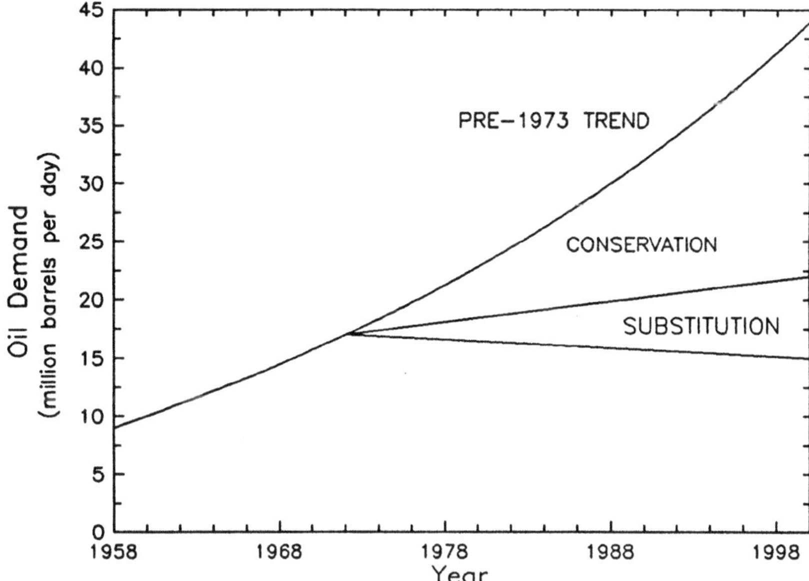

Figure 6.2 A projection of U.S. oil demand to the year 2000 (information from Kenney, p.6).

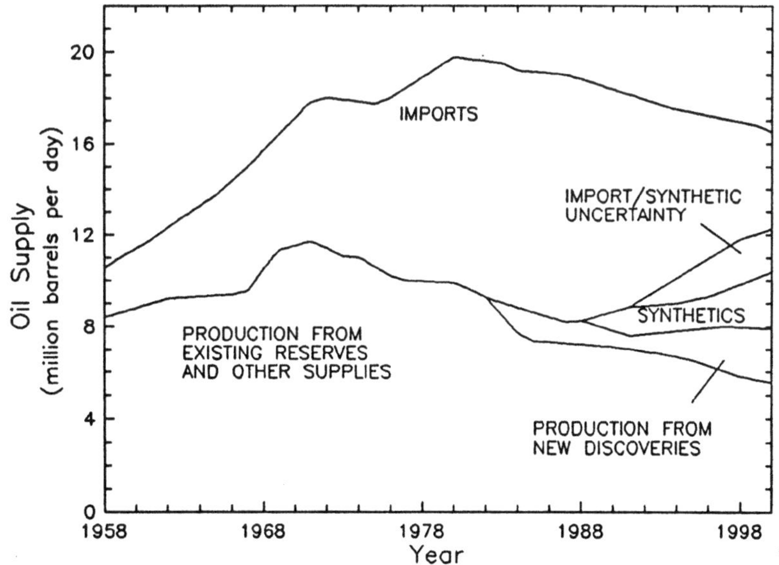

Figure 6.3 A projection of U.S. oil supply by source to the year 2000 (information from Kenney, p.7).

to new problems? Kenney, thus, asks: "Will there be adequate supplies once the current recession is over?"

The problem is further exacerbated by the fact that the United States, with only 6% of the world population, consumed in 1985 about 1/3 of the total world oil production (excluding the Soviet Union).

Let us now consider the future energy demand/supply picture on a global basis.

6.5.3 Global Energy Demand and Supply in the Future

Projection of energy demand and supply for the whole world, especially to the distant future, is of course very difficult and uncertain considering the aforediscussed case for the demand and supply of oil in the United States to year 2000. Edmonds and Reilly, however, present an extensive and well documented description of this demand/supply picture to the year 2050 that can be viewed, at least, as a 'plausible scenario'.

a. Some Qualitative Observations

Before we consider their findings, however, the following observations should help us understand the forces that will shape one part of the picture, the demand:

1. The aspiration of the Third World (Latin America, Africa, Asia) is, of course, to reach the standards of living of the advanced western societies.
2. To accomplish these standards of living the U.S., as we have seen, consumes 1/3 of the world oil production with only 6% of the population.
3. Sweden and Germany, however, have similar standards of living with only one-half the per capita energy consumption of the U.S.
4. Using this consumption as a basis, and assuming a 1986 population of 3 billion for the Third World and 0.25 for the U.S., we can easily see that its oil demand alone would be about: (3/0.25)(1/2) = 6 times that of the U.S.
5. The problem is further exacerbated by the exponential population growth, Figure 6.1. To put things in perspective, notice that it took:

 * 2 million years for the world to reach 1 billion people;
 * 100 years to reach 2.4 billion; and
 * 35 years to reach 4.4 billion.

And the projections are for about 6 billion by 2000 and 8 billion by 2050 (Edmonds and Reilly).

One cannot help but ask: *where will all this energy come from?* or, *will the Third World accept its impoverished state of existence for ever?* Let us now turn to the more quantitative picture of Edmonds and Reilly.

b. A Plausible Quantitative Picture

Figure 6.4 presents the projected global energy demand and supply picture presented by Edmonds and Reilly as a result of their 'scenario'. Here is how they introduce (p.277) their findings:

"If the scenario we have constructed comes to pass, global energy supply and demand configurations will have undergone a dramatic evolution by the year 2050. Global reliance on conventional oil and gas is likely to fall to less than 10 percent of total energy needs, down from two-thirds in 1975. The United States, the world's largest oil importer in 1975, could well be exporting oil in 2050. Coal will exceed oil as the most important traded fuel. Coal will become the dominant primary energy source, but nearly three-quarters of all coal produced in 2050 will be transformed into liquids, gases, or electricity (35 percent will be converted to synthetic oil and gas, and approximately 40 percent will be burned by electric utilities). Despite the fact that conventional oil becomes a mi-

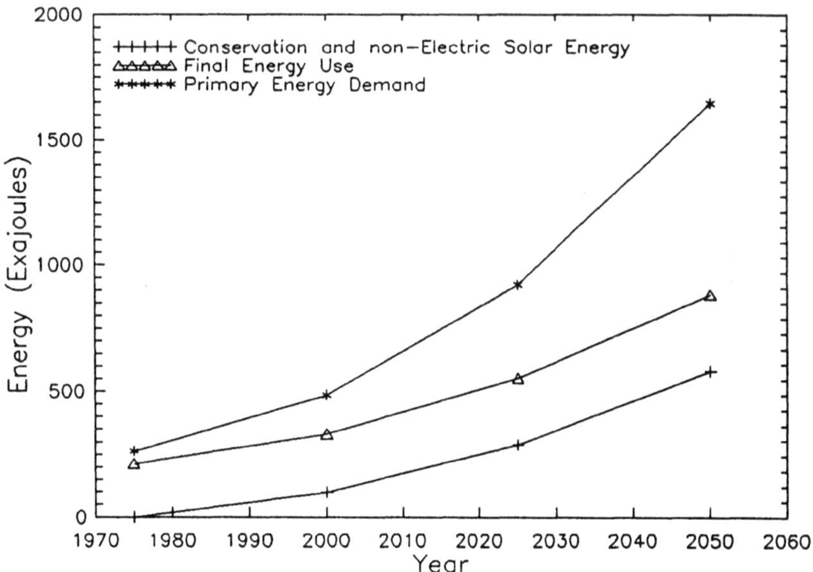

Figure 6.4 Projected: primary energy demand (*); final energy use (Δ); conservation and non-electric solar (+). (Information from Edmonds and Reilly, p.282 and 284.)

nor source of energy by 2050, liquids remain important, accounting for just under a quarter of demand. The continued use of liquids is made possible by shale oil production and synfuel conversions from coal, which emerge as major energy technologies after the turn of the century.

Under our base-line scenario, the rate of growth of energy demand between 1975 and 2050 slows dramatically from pre-1975 rates. Whereas global energy demand grew at an average rate of 5 percent per year between 1950 and 1975, growth rates are expected to decline to less than half that rate between 2025 and 2050. The reasons are slower rates of population and GNP growth, and higher energy prices. The share of energy demand accounted for by the developed OECD regions falls dramatically over the period, from 54 percent in 1975 to 36 percent in 2050. In contrast, the developing nations increase their relative shares of energy consumption from 11 percent to 35 percent. The global primary energy/GNP (E/GNP) ratio remains relatively stable despite major contributions by conservation and direct end-use solar applications (which save almost a third of all energy consumed by 2050). This stability results from relatively rapid economic development in third world nations where E/GNP ratios rise, counterbalanced by declining E/GNP ratios in the developed world.

Finally, projected energy price increases are expected to be dampened over time. The largest jump in price occurs between 1975 and 2000, when the price of oil increases by 75 percent. This, however, implies that between 1981 and 2000 the oil market will remain soft, with real prices of oil possibly declining from their 1981 peak. Oil prices (in real terms) increase by 36 percent between 2000 and 2025 and are nearly steady after 2025."

Comments:

1.According to their scenario, supply meets demand. Thus, when they talk about demand they also imply that the necessary supply will be there.

2.Conventional oil and gas: oil available using recovery techniques known up to 1980. Thus, oil from: tertiary oil recovery; coal; shales; etc., is referred to as 'unconventional' oil.

3.GNP (Gross National Product): the sum of products and services of a Nation.

4.OECD (Organization for Economic Cooperation and Development): U.S., Canada, and West European Countries.

5.Primary Energy Sources: raw energy source, before converted to usable (secondary) energy. Coal, thus, is a primary source; electricity produced from it, a secondary.

6.Nuclear power plays an increasingly significant role: from 8% of the energy supply in the year 2000, to 17% in 2025, to 22% in 2050.

7.Conservation and non-electric use of solar energy (hot water, heating, passive solar heating or cooling through building design, etc.) play an increasingly major role as shown in Figure 6.4.

6.6 Meeting the Energy Demand in the Future: Problems

The findings of Edmonds and Reilly, indicate that in order that the future energy demand of the world is met, their scenario has to rely on:

* Nuclear (conventional and breeder) energy
* Coal (directly and through synfuels)
* Oil and Gas (conventional and unconventional)
* Extensive Use of Solar Energy and Conservation

Let us take a very brief look at some of the major problems associated with achieving these goals. We will assume, in the interest of space, that all the required technological developments and breakthroughs will be realized and concentrate on two major areas: Cost and Environmental problems.

6.6.1 Cost

To just get a picture of the magnitude of the problem consider the following examples:

1.A major coal-fired cogeneration project (conservation) will cost several billion dollars (Kenney).

2.The cost of a typical nuclear power plant is over $4 billion (1985 basis).

3.To reach the "1 million barrels per day of oil from coal" goal set in the late 1970's by the U.S. Government, Schuyten (1979) estimated a total cost of 300 billion dollars. (This figure includes not only plant cost but also the 'hidden' ones: pipelines for fuel transport and beefing up the transportation system for the shipment of over 100 billion tons per year of additional coal).

To place this number in perspective, notice that $300 billion represents about 30% of the total U.S. budget in the middle 1980's. And all this to provide only about 6% of the 1985 oil consumption for the U.S.

6.6.2 Environmental

We will concentrate here on three major such problems:
* Carbon dioxide
* Acid rain
* Nuclear

a. Carbon Dioxide

This problem has generated, until very recently, the least amount of concern among the three, yet it may turn out to be the most serious one because of the *Greenhouse Effect*. Edmonds and Reilly write:

"Energy consumption and the environment have been closely linked in the past and are likely to be even more closely associated in the future. Environmental concerns range over the entire energy system from problems associated with mining and extraction to those associated with transformation and consumption. The problems of acid rain, land reclamation, water pollution, and nuclear waste disposal have become all too familiar. Carbon dioxide (CO_2) is a less familiar, but no less ominous problem looming on the long-term horizon.

Carbon dioxide is an important by-product of the combustion of fossil fuels. It is a nontoxic, colorless gas with a faintly pungent odor and acid taste. Since carbon dioxide is nontoxic and does not appear to be associated with the formation of other undesirable compounds, it is not commonly thought of as a pollutant. But CO_2 plays an important role in the determination of the global climate. The presence of CO_2 in the atmosphere produces a 'greenhouse effect', allowing incoming sunlight to penetrate, but trapping heat radiated back from earth. Man's ability to significantly affect CO_2 levels through the use of fossil fuels gives rise to the possibility of an unprecedented global problem.

The CO_2 question is a unique environmental question in a number of ways. Unlike other environmental issues in which the problem is restricted to a region, CO_2 mixes rapidly with the earth's atmosphere, yielding a common concentration. Thus, all regions of the world could experience global warming, though it is expected to be greatest in the earth's polar latitudes and least in the equatorial regions. In addition, changes in the earth's temperature are expected to trigger changes in climatological patterns. There will be both winners and losers in this shift, but there is little certainty as to how the gains and losses will be distributed. Finally, there may be discontinuous changes brought about by global warming. Because the polar regions will likely experience the greatest warming, temperature increases may result in a warming sufficient to melt the Antarctic ice shelf which, sitting atop the south polar land mass, could add enough water to the earth's oceans to raise sea levels an average of 25 feet. Such an event would alter the world's coastlines, leaving most of the world's great port cities below sea level.

Preindustrial concentrations of carbon dioxide are estimated to have been in the range of 270 to 295 ppm. By 1980 the concentration had increased to 339 ppm. It is widely argued that at atmospheric concentrations of 600 ppm or higher, global temperature and climate changes would almost certainly become significant and be cause for concern. However, climatic change in response to CO_2 concentrations significantly less than 600 ppm may present problems of adaptation for human societies. Human activities which produce CO_2 have been increasing. These include forest clearing, cement manufacture, and the combustion of fossil fuels." (Edmonds and Reilly, pp.265-266)

They proceed to demonstrate that the 'doubling time' - the time required for carbon dioxide levels to reach the aforementioned level of 600 ppm, double the preindustrial level - will come around the year 2050. Even with an unrealistic worldwide 100% tax on coal, and a proportionately lower tax on other fuels according to their carbon content, the doubling time will be only postponed by 15 years.

Edmonds and Reilly conclude (p.275): "Because it is so far in the future, affecting our children's lives more directly than our own, the CO_2 question has an unreal aura about it. It lacks the impact of an oil embargo. Yet the question is real and while the work we present here indicates that major man-made climate changes may not be waiting just beyond the turn of the century, we do find that CO_2 concentrations will almost certainly double within 100 years; the magnitude of the temperature and climate change in response to projected increases in atmospheric CO_2 and the associated effects on human societies remain uncertain."

To appreciate, finally, the magnitude of the problem note that:

1. Current (1990) global emissions are about 5.5 billion tons per year of carbon in the form of carbon dioxide.

2. Other gases generated in large amounts (CH_4, N_2O, CFCs, etc.) contribute significantly to the greenhouse effect.

Temperature Rises at 600 ppm CO_2 Level

* *midlatitudes: 3-6 °C*

* *near the poles: 9-12 °C*

(U.S. National Academy of Sciences, 1977)

b. Acid Rain

While the increase of CO_2 levels in the atmosphere has attracted relatively little attention until very recently, the presence of the toxic gases, sulfur dioxide and oxides of nitrogen, has caused significant concern. Their impact is much more noticeable. Here is how O'Sullivan describes the problem:

"The acid rain issue is at once simple and extremely complex. Its simplicity rests on the nature of the chemicals involved, essentially sulfur dioxide and sulfates, nitrogen oxides (generally referred to as NO_x), and ozone produced in the atmosphere from nitrogen oxides and hydrocarbons under the influence of ultraviolet light.

That atmospheric acidity adversely affects a variety of ecosystems - from forests in central Europe and Scandinavia to lakes in Scandinavia and Scotland - can hardly be refuted. But the complexity of the subject is reflected in the many theories, some of them conflicting, that have been proposed to explain its mode of action.

Apart from circumstantial evidence, no one really knows precisely the chemical and physiological pathways involved. However, much more is known about the acidification of lakes from acid deposition than is known about forest damage from air pollution. Meanwhile, the issue has been pushed into the forefront of the international political arena, spurred by economic, health and aesthetic considerations, and to some extent by emotive journalism. The term 'acid rain' itself is something of a misnomer. It was coined more than a century ago by Robert Angus Smith when making a detailed study of the atmospheric conditions in the industrialized area of Manchester, England, and the term has since taken hold.

Acid deposition is the more accurate and scientifically preferred term. It is true that sulfur dioxide undergoes oxidation to the trioxide, which in turn converts to sulfuric acid in clouds, fogs, and mists. On the other hand, dry deposition of acidic aerosols and particulates is deemed to be just as damaging, especially as far as the adverse effects on trees and historic buildings are concerned. Moreover, ozone and heavy metals, which contribute to ecosystem damage, are not themselves acidic.

Increase in the level of man-made air pollutants over Europe, and the concomitant rise in the intensity of acid deposition, can be traced back to the industrial revolution. As recently as 30 years or so ago, air quality over heavily industrialized centers had deteriorated to the point where women's hose were known to disintegrate during rainy spells because the level of acidity was high enough to attack the man-made fibers.

Widespread burning of coal both for power generation and domestic heating also led to development of severe episodes of smoke haze combined with weather related fog. London's 'pea-soupers' are legendary. In 1952, some 4000 died, due largely to inhalation of smoke particles on which sulfur dioxide was absorbed. A similar incident there in 1962 brought death to about 800. So bad were conditions that an internationally acclaimed collection of pine trees at the Botanical Gardens of Kew had to be moved out of London to ensure their survival." (Reprinted with permission from *European Concern About Acid Rain is Growing*, D.A. O'Sullivan, Chem. & Eng. News, January 28, 1985, 63(4), Copyright 1985 American Chemical Society.)

To get an idea about the magnitude of the problem - but also its international character - notice, for example, that:

* the United Kingdom has an average monthly deposition of 847,000 metric tons of sulfur with only 20% coming from outside, while
* Norway has 255,000 tons per month of sulfur deposition with 92%(!) coming from outside (O'Sullivan).

The problem is, of course, no less serious in the U.S. and Canada. Living in New York City is equivalent to smoking one pack of cigarettes a day. And the acidification of the lakes in the North East and Canada, with main source the SO_2 and NO_x gases coming from the West through the westerly winds, is well known.

Finally, in spite of the world-wide concern about this problem, very little is really done and the main reason is cost. Thus the Edison Electric Institute (EEI) indicates that the "cost of an acid rain mitigation scheme (in the U.S.) ... could be $5.6 billion annually. EEI pegs job losses at 40,000 to 50,000." (Chemical & Engineering News, Nov. 25, 1985, p.23.)

1910: Swiss Oppose the Automobile.

GENEVA - The idea that the French Automobile Club or any other body may be able to influence the Swiss into relaxing regulations in this country as regards automobiles is one little likely to be materialized. The disposition prevails here further to stiffen automobile regulations. That is easily understood in this land where the tourist is everything. The tourist complains that the automobile is an abomination which makes the roads impassable, raises dust and gives out a vile smell. Automobiles are not admitted in the Canton of Grisons. There is much the same feeling in other cantons. The general sentiment appears to be that the automobile is a doubtful adjunct to the general prosperity of Switzerland. (International Herald Tribune, Sept. 28, 1910. In IHT, Sept. 28, 1986)

c. Nuclear

Our perception of the environmental problems associated with nuclear energy is shaped mainly by the safety of nuclear power plants, especially after the Three Mile Island accident and, much more, after the Chernobyl one.

It could be argued of course that, considering the projected need for nuclear energy, the world will have to learn to live with such uncertainties. After all, thousands of people die on the highways every year, about 50,000 in the U.S. alone, but more and more automobiles are built.

The most serious problem of nuclear energy, however, may turn out to be the 'radioactive wastes' of nuclear plants. Some of their constituents have a half-life of 80,000(!) years. They also include large amounts of Plutonium, which some people fear could be used for the production of bombs by terrorists, since a great deal of the required technology is available on library shelves. In addition, it is extremely poisonous.

Harvey Brooks (1980) of Harvard University, head of the U.S. National Academy of Sciences committee on nuclear waste disposal, sums up the picture as follows: "I would predict that should nuclear energy ultimately prove to be socially unacceptable, it will be primarily because of the public's perception of the waste disposal problem."

It should be noted, in closing, that our discussion of the energy demand/supply picture and the associated problems has been, in the interest of space, extremely brief. Our purpose, however, is not to create familiarity with such a complex issue. Rather, to generate awareness of it and, hopefully, to 'whet' your appetite for further reading. *This is the least that can be expected from the technologists and technological leaders of tomorrow.*

Let us now turn to the very important question:

6.7 So, Is There a Third Entropy Watershed Coming?

Our brief account of the global energy demand for the future - and the associated problems with meeting it, would probably force a 'realist' to answer: 'YES'. Let us then consider briefly the following two major issues associated with such an answer:

a. the magnitude of the problem, and

b. life after this watershed.

We will close with some encouraging - perhaps - comments.

6.7.1 The Magnitude of the Problem

This will be determined to a great extent by the answers to the following questions:

a. Will the predicted population growth occur, or will the continuously increasing awareness of the problem help ameliorate it? For example, the population projection of Edmonds and Reilly in 1985 for the year 2050 is significantly smaller than those made in the 1970's.

b. Will all the necessary technologies be developed in time?

c. Where will the vast amounts of money needed to implement these new technologies be found considering, especially, the low economic growth of the world today?

d. Will non-electric solar energy use and conservation play the significant role predicted by Edmonds and Reilly (Figure 6.4)?

e. Will all the materials needed in the 'scenario' of Edmonds and Reilly be available, and to what degree will the 'fourth' law effect their supply and cost?

f. Will the problem of nuclear safety and waste disposal be resolved in time, reversing thus the current trend, especially in the U.S., away from nuclear energy?

g. What - and this may turn out to be the crucial issue - will be the actual effects of the continuously increasing concentration of CO_2 in the atmosphere?

These, and the answers to unforeseen questions, will determine the magnitude of the problem associated with a third entropy watershed.

Let us now turn to the second issue.

6.7.2 Life After This Entropy Watershed

Rifkin, one of the 'realists', gives a fascinating and lengthy vision of this post-industrial 'low-entropy era' as he calls it, which is guided by "the physical limits we face - the limits of our planet's resources and the limits we must impose on the use of technology". He predicts drastic changes in our way of life and deeply held values such as:

* Material wealth and consumption: 'more is less' becomes the way of life, not just a slogan.
* Work ethic: work is intended to fulfil a person, not lead to material wealth.
* Education: conservation becomes its underlying principle.
* Urban life: labor-intensive agriculture will replace slowly the current

energy-intensive one leading to the disappearance of the megalopolis, with 'large' cities returning to the preindustrial level of 50,000 to 100,000 inhabitants.

* and so on.

There is no question that Rifkin's reading of History is somewhat simplistic, his predictions of the consequences of the 'Third Entropy Watershed', being rather (hopefully!) exaggerated. But the changes in people's life during the transition to the Industrial Era were no less severe. Anyway, this is his approach in trying to attract attention to very serious problems. Here is, for example, how R.M. Baum (Chemical and Engineering News, February 3, 1986) concludes his critique of Rifkin's book on Biotechnology: "He is, undoubtedly, a fanatic. His reading of philosophy and the history of science, at least as evidenced in *Declaration of a Heretic* (his biotechnology book), is superficial at best."

But Baum concludes: "But I do not think he can be dismissed. I wish he could be."

Rifkin, however, is not the only realist. Following are the comments of Dr Lave, an economist and environmental expert at Carnegie-Mellon University, as quoted by P. Shaberoff (1988): "Major changes in daily life would be necessary, he said. 'We get to drive and fly less, and use a lot less fuel for space conditioning. Our houses can't be as warm in winter or as cold in summer. A lot of materials we use such as plastics and aluminum are really congealed energy. We are going to have to be less of a throwaway society'."

Additional support for this rather bleak picture comes from the 'State of the World 1987' (Worldwatch Institute, Washington, D.C., as quoted in the New York Times, Feb. 15, 1987, p.4): "Efforts to improve living standards are themselves beginning to threaten the health of the global economy. *The very notion of progress begs for redefinition in light of the intolerable consequences as a result of its pursuit. A sustainable society satisfies its needs without diminishing the prospects of the next generation. ... By many measures, contemporary society fails to meet this criterion.*" (Emphasis added.)

A Positive Note

> *"It is conceivable however that life may have a larger role to play than we have yet imagined. Life may succeed against all of the odds in molding the universe to its own purpose. And the design of the inanimate universe may not be as detached from the potentialities of life and intelligence as scientists of the twentieth century have tended to suppose."* (Dyson, 1971)

6.7.3 Comments

Rifkin and the other realists could of course turn out to be wrong. It is very doubtful, for example, that the world crisis predicted for the year 2000 by Meadows and Meadows (1972) in the *Limits of Growth* will be realized. (It may be, of course, that it will be just postponed).

In addition, the greenhouse effect has started to generate interest on the Government level. Thus, from Shaberoff again: "Delegates to a recent conference on world climate change, convened in Toronto by the Canadian government, called for a 20 percent cut in global emissions of carbon dioxide by the year 2005, with the ultimate goal of reducing emissions by 50 percent. The world's population is now sending about 5.5 billion tons of carbon dioxide into the atmosphere each year, about half of which is absorbed by the oceans and forests. Thus a 50 percent cut in emissions would stabilize the amount of the gas in the atmosphere."

And from Chemical and Engineering News, May 22, 1989, p.5: "After a week of controversy, President Bush has reversed his earlier wait-and-see stance and committed his Administration to working toward an international agreement on global warming. The scope and importance of the issue are so great, says White House chief of staff John H. Sununu, that it is essential for the U.S. to exercise a leadership role.

Specifically, the U.S. will host an international workshop this fall that will try to reach a consensus on the scientific, legal, technical, and economic issues surrounding the question of how to respond to climate change. Actual negotiation of an international treaty that would control emission of greenhouse gases would begin about a year later under the Administration's plan."

It remains of course to be seen if this continuously increasing reaction by people and governments will avert the greenhouse effect or just delay it. For it is very possible that the energy requirements associated with sustaining the standards of living in the developed countries and - more important - with improving those in the Third World ones, could prevent this reaction from leading to any substantial reduction in CO_2 emissions.

6.8 Concluding Remarks

Few of us would ever suspect that this big 'enigma' of thermodynamics, Entropy, would have such broad implications. And we have only 'scratched the surface' here, for the Second Law and Entropy seem to make their presence felt in a variety of areas. It has been used thus:

* in information theory (see, among others, Tribus, 1981);
* to support the existence of God (Inge, 1934), an inference against which argued the Russian physicist Bazarov using the dialectic materialism of Engels (1964);
* to support the evolutionary Heraclitean philosophy of nature, as contrasted to the static Eleatic one supported by the first law (an elegant and fascinating article, Astarita, 1977);
* in economics (Georgescu-Roegen, 1971; Daly, 1977). The latter argues for example that growth, when measured by the GNP (Gross National Product), may be misleading: a rise in GNP may reflect the *increased cost of production* - since the richer and more accessible resources are the first to be used - rather than *increased level of production.*

For an extensive review of entropy involvement in a variety of subjects including Social Sciences, Literature, Art etc., see Kyle (1988).

A most fascinating application of the second law, however, involves the concept of time and the associated cosmological implications discussed briefly below.

Classical mechanics, quantum mechanics, and relativity theory, do not preclude the reversibility of processes, since their equations are symmetric in time. On the other hand, application of the second law to an isolated system yields, since there is no entropy change in the surroundings:

$$\Delta S_{sys} > 0$$

Entropy increase, therefore, reflects the spontaneous evolution of the system, in other words the *arrow of time*, as Eddington called it. For an isolated system, the future is the direction of increasing entropy.

If we now consider the whole Universe a closed system by definition, the second law transcends the steam engine and other technological applications and acquires a cosmological stature. Thus, Clausius stated the second law as: *The entropy of the Universe tends towards a maximum.* No temperature differences will be present, resulting in what is referred to as the *heat death* of the Universe. [For interesting discussions of time in connection with the second law, see: Kyle; Prigogine and Stengers (1984); Fraser (1982); and Denbigh (1981).]

Notice, in closing, the ultimate application of the Carnot cycle depicted in Figure 6.5. Nothing could give more meaning to Socrates' famous statement: "It is a wise man who knows, that he knows not".

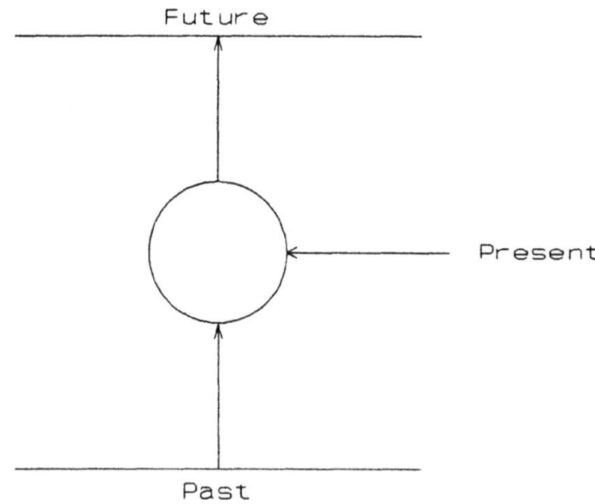

Figure 6.5 The ultimate application of the Carnot cycle.

References

Astarita, G., 1977. *Ind. Eng. Chem. Fundam.*, **16**, 138.

Bazarov, I.P., 1964. *Thermodynamics*, McMillan, New York.

Brooks, H., 1980. As quoted by Rifkin, op. cit., p.109.

Daly, Herman E., 1977. *Steady-State Economics*, W.H. Freeman & Co., San Francisco.

Denbigh, K.G., 1981. *Three Concepts of Time*, Springer-Verlag, Berlin.

Dyson, F., 1971, *Scientific American*, **225**, 50.

Edmonds, J., Reilly, J.M., 1985. *Global Energy, Assessing the Future*, Oxford University Press, Oxford.

Eubank, Ph.T., 1989. *Chemical Engineering Thermodynamics: Some Pedagogic Tricks*, Texas A&M University, College Station, Texas.

Fraser, J.T., 1982. *The Genesis and Evolution of Time*, University of Massachusetts Press, Amherst, Mass.

Georgescu-Roegen, N., 1971. *The Entropy Law and the Economic Process*, Harvard University Press, Cambridge, Mass.

Inge, W.R., 1934. *God and the Astronomers*, Longmans, Green & Co., London.

Kenney, W.F., 1984. *Energy Conservation in the Process Industries*, Academic Press, Orlando, Florida.

Kyle, B.G., *Chem. Eng. Educ.*, Spring 1988.

Meadows, D., Meadows, D., 1972. *The Limits of Growth*, University Books, New York.

Mumford, L., 1934. As quoted by Rifkin, op. cit., p.71.
Prigogine, I., Stengers, I., 1984. *Order Out of Chaos*, Bantam Books, N.Y.
Rifkin, J., 1980. *Entropy*, Viking Press, New York.
Shaberoff, P., 1988. *International Herald Tribune*, July 21.
Schuyten, P.J., 1979. *New York Times*, July 15, Sec. 3, p.1.
Smithsonian Institute, Annual Report, Washington, D.C., 1864, p.352.
Stavrianos, L., 1984. *History of the Human Race*, G. Kiotis Press, Athens, (in Greek).
Tribus, M., *CHEMTECH*, Oct. 1981, 634.

Problems

6.1 What is the 'Fourth' law? What are some of its implications?
6.2.a. What are the main Entropy Watersheds?
  b. How are they used in explaining the major transitions in the world history?
6.3 With reference to Synthetic Fuels:
  a. what are they?
  b. what are the main obstacles in their availability?
6.4 Discuss the advantages and disadvantages of the late 1980's oil glut and resulting lower oil prices.
6.5 With reference to the energy supply in the future:
  a. what will be the main source(s)?
  b. discuss some of the problems associated with this supply.
6.6 Why addition of 'final energy use' plus 'conservation and non-electric solar energy' does not give the 'primary energy demand' in Figure 6.4?
6.7 With reference to the greenhouse effect:
  a. what is it?
  b. what problems could it lead to?
6.8 Do you believe that a third Entropy Watershed is coming? Present the main arguments in support of your position.
6.9 Does the 'entropic' interpretation of world history meet any of the criteria that Kagan (Section 6.4.1) sets to this purpose?
6.10 Discuss the relationship between entropy and time.
6.11 Assuming that the total solar flux to the earth is $1.7 \cdot 10^{14}$ kW and the energy produced from fossil fuels is $9 \cdot 10^{6}$ kW (Eubank, 1989):
  a. calculate the fraction of entropy generated by the latter;
  b. discuss any implications concerning the energy supply in the world.
Assume: T(sun) = 5900 K; T(earth, average) = 275 K.

Thoughts

("Confusion on Confusion: A Portrait of Entropy")

Mood: Just Thinking

Here I am. I went through the second law; the definition of *S*; the Carnot cycle and its relationship to *S*; the molecular interpretation of *S*; exergy and *S*; I even read the 'broader' view of *S*.

But I still do not know what *S* is.

Mood: Philosophical

I know what *S* is. She is the Goddess of Thermodynamics. *Goddess Entropy*. She keeps an eye on how well we, Humans, use our Resources and *measures* it. She even gets technical some times:

"You got low efficiency, you dummies", she says.

"Don't blame us", I hear us say. "We are only *irreversible*. And you made us so, with your darn appetite to grow bigger and bigger."

Man, am I confused.

Mood: Emotional

I hear her again:

"You know", she says, "I measure your *personal efficiency* too. How much you give to others as compared to how much you take from them. And you got low efficiency again, you *self-centered* bastards."

"Look who is talking", I hear us say. "You are the greediest thing I have ever seen. You must be the *original* capitalist."

Man, am I mixed up.

Mood: Sober

"Entropy? Who ever understood a woman, let alone a Goddess! I have never heard of her. *She is just S*."

7
Intermolecular Forces

7.1 Introduction

We have concluded, thus far, the discussion of the first objective of chemical engineering thermodynamics: *the evaluation of the feasibility of processes and the efficient energy utilization in them.* Before we proceed with the second one, *the evaluation of thermophysical properties of fluids*, we consider next *intermolecular forces* because they are useful in understanding the behavior of fluids as well as *phase equilibrium*, which is part of the third objective.

Intermolecular forces are present everywhere. If it were not for the attractive forces to hold the molecules together, there would be no solids and liquids; the kinetic energy of the molecules would spread them all over the available space. Everything, including ourselves, would be in the gaseous state. And if it were not for the repulsive forces, solids and liquids would not resist compression.

The importance of intermolecular forces in explaining the thermodynamic behavior of fluids will be evident in the ensuing Chapters. We have already used them, however, in Example 1.5, to explain the observation that the deviation of steam from ideal gas behavior:

a. increases with increasing density at constant temperature; and
b. decreases with increasing temperature at constant density.

7.2 Objective and Approach

Our objective in this Chapter is to develop an understanding of:

1. The origin and types of intermolecular forces;

2. Their relative magnitude, using approximate expressions for its determination; and

3. The degree to which knowledge of intermolecular forces can help predict - qualitatively and quantitatively - macroscopic behavior.

We will start with a discussion of the relationship between intermolecular forces and molecular potential energy and then proceed to identify the origin and types of these forces.

Following a somewhat arbitrary - but convenient - classification of the attractive forces into *physical* and *chemical* ones, we will discuss them separately. In each case, through theoretical considerations and an Example, we will try to identify the relationship between these forces and macroscopic behavior.

Undergraduate students should be exposed, at the very least, to the qualitative considerations, Sections 7.4, 7.5, and 7.7.1.

7.3 Intermolecular Forces and Potential Energy

The presence of forces among the molecules of a system gives rise to their potential energy which, along with their kinetic energy, comprise the internal energy of the system. The potential energy Γ between two molecules, whose centers are at a distance r, is related to the intermolecular force F by:

$$F = -\frac{d\Gamma}{dr} \tag{7.3.1}$$

and it represents the amount of work involved in separating the two molecules from distance r to infinity. In the case of attractive forces work must be done on the system and, consequently, the potential energy is negative, while the opposite is the case for repulsive forces.

Equation 7.3.1 is applicable when the potential energy is a function of the distance between the centers of the molecules only, which is the case for small symmetric molecules (Ar, Xe, CH_4, etc.).

In the more general case:

$$F(r,\theta,...) = -\nabla\Gamma(r,\theta,...) \tag{7.3.2}$$

where ∇ is the gradient, and θ,....are the additional parameters needed to specify the potential energy function (for a detailed discussion, see Gray and Gubbins, 1984).

7.4 Origin and Types of Intermolecular Forces

Electrostatic (*Coulombic*) interactions among molecules give rise to intermolecular forces, that are classified as *Attractive* and *Repulsive* and their types are shown pictorially below:

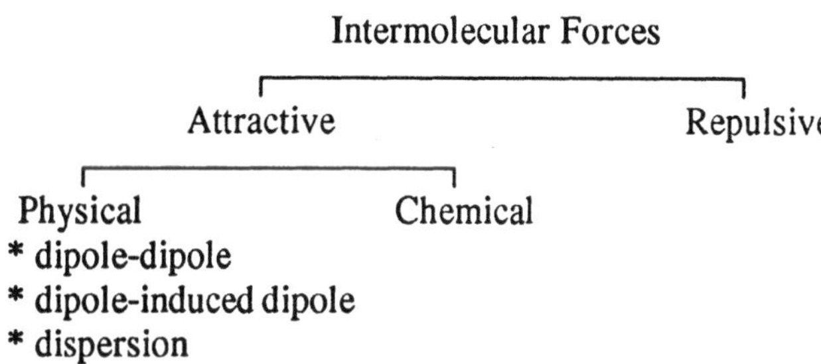

7.4.1 Attractive Forces

Their origin lies with electrostatic interactions between opposite charges. But, if molecules are electrically neutral, how can there be positive and negative charges?

To answer this question, consider the schematic diagram of a water molecule shown in Figure 7.1. We know from chemistry that because of the electronegative character of the oxygen atom, the two electrons that it shares with each hydrogen have the tendency to spend more time with the oxygen than with the hydrogen atoms.

As a result, water ends up with a net negative charge of (-*e*), whose center lies close to the center of the oxygen atom; and with a net positive charge of (+*e*), whose center lies in the middle of the distance between the two hydrogen atoms. We refer to such a molecule as a *dipole*, with a *dipole moment*:

$$\mu = el \tag{7.4.1}$$

where l is the distance between the centers of the two charges. The common unit for the dipole moment is the Debye = 10^{-18} esu cm.

All asymmetric nonhydrocarbon molecules (ketones, alcohols, ethers, etc.) are dipoles since the centers of the positive and negative charges,

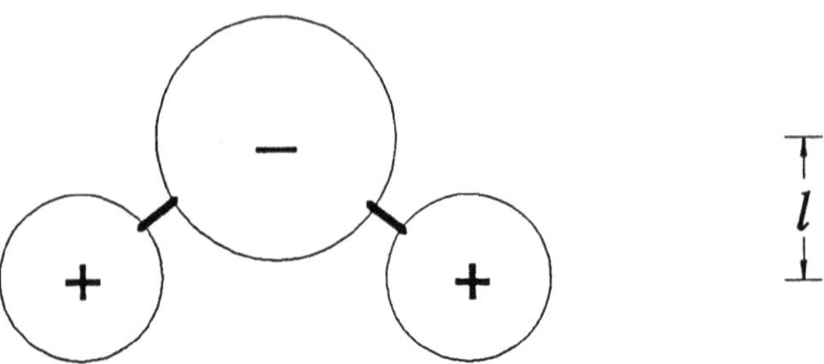

Figure 7.1 The water dipole.

because of their asymmetry, do not coincide. Such molecules are referred to as *polar*, while those that do not have a dipole moment, as *nonpolar*.

On the basis of the strength of the attractive forces among molecules it is convenient to classify them, somewhat arbitrarily, into two main types:

* physical (weaker); and
* chemical (stronger).

In the second group the interaction is so strong that physical aggregates are formed. The most typical example is hydrogen bonding where, in the case of acetic acid for example, dimers are formed.

7.4.2 Repulsive Forces

If the distance r between two molecules i and j becomes very small, their electron clouds start to overlap leading to strong repulsive forces. Such forces are not well understood and, while theoretical considerations suggest that the repulsive potential should be an exponential function of the intermolecular distance (Reed and Gubbins, 1973), the following expression is typically used as more convenient:

$$\Gamma = \frac{A}{r^n} \tag{7.4.2}$$

where A is a positive constant and n is usually between 8 and 16 (Prausnitz et al, 1986).

We will discuss next the physical attractive forces, and the resulting intermolecular potentials, and then consider the chemical ones.

7.5 Physical Attractive Forces

The physical attractive forces are classified as follows (Section 7.4):
* dipole-dipole;
* dipole-induced dipole;
* dispersion (among instantaneous dipoles).

7.5.1 Dipole-Dipole Forces

If two polar molecules *i* and *j* are placed in the vicinity of each other, attractive and repulsive forces - depending on their relative orientation - will develop.

This relative orientation is influenced by two opposing effects. On the one hand, the electric field of the molecules tends to line them up leading to strong attractions. On the other hand, their kinetic energy tends to toss them around in a random fashion. As a result, as the temperature increases - and, hence, their kinetic energy - the attraction becomes progressively smaller. (Remember that this is how we explained in Example 1.5 the observation that, when compared at the same density, steam deviation from ideal gas behavior was determined by its temperature: the higher temperature gave the lower deviation.)

The average potential energy between two dipoles *i* and *j* at a fixed distance *r*, averaged over all orientations, is given by (Prausnitz et al):

$$\bar{\Gamma}_{ij} = -\frac{2}{3}\frac{\mu_i^2\mu_j^2}{kTr^6} + \ldots\ldots \qquad (7.5.1)$$

where *k* is the Boltzmann constant and the additional terms include higher orders of $(1/kT)$. Eq.7.5.1 is consistent with the aforementioned observation of decreasing attractive energy with increasing temperature.

Dipole moment values are given by Reid et al (1987) and values for some selected compounds in Table 7.1.

7.5.2 Dipole-Induced Dipole Forces

Consider a nonpolar molecule *i* in the vicinity of a polar molecule *j*. The latter generates an electric field which causes a displacement of the electrons of molecule *i* and hence *induces* a dipole on it. The resulting force

Table 7.1 Dipole Moments, Polarizabilities, and First Ionization Potentials for Selected Compounds (From Prausnitz et al, pp. 52, 55, 58)

Compound	μ (debye)	a ($cm^3 x 10^{25}$)	I (eV)
CH_4	0.0	26.0	13.0
H_2	0.0	7.9	15.4
c-C_6H_{12}	0.0	109.0	11.0
Cl_2	0.0	46.1	13.2
CCl_4	0.0	105.0	11.0
CO_2	0.0	26.5	13.7
$CHCl_3$	1.1	82.3	11.5
NH_3	1.5	22.6	11.5
CH_3OH	1.7	32.3	10.8
H_2O	1.8	15.9	12.6
$(CH_3)_2CO$	2.9	63.3	10.1

is always attractive and the corresponding average energy, first calculated by Debye, is given by:

$$\bar{\Gamma}_{ij} = -\frac{a_i \mu_j^2}{r^6} \tag{7.5.2}$$

where a_i is called the *polarizability* of molecule i and reflects the ease with which its electrons can be displaced by an electric field. For asymmetric molecules the polarizability is not constant but it varies with the orientation of the molecule relative to the field. Average values are given by Prausnitz et al and for some selected compounds in Table 7.1.

If both molecules i and j are polar, they induce a dipole on each other and the average potential energy is given by:

$$\bar{\Gamma}_{ij} = -\frac{a_j \mu_i^2 + a_i \mu_j^2}{r^6} \tag{7.5.3}$$

7.5.3 Dispersion Forces

We can see, so far, why a system composed, for example, of acetone molecules or a mixture of acetone and methane molecules should experience attractive forces. But what about a system of methane molecules? Why should there be attractive forces there?

Or, simply, why should methane or argon exist in the liquid phase? What kind of attractive forces keep the molecules in this phase?

The answer was provided in 1930 by London who explained the nature of these interactions as follows: "*If one were to take an instantaneous photograph of a molecule at any time, one would find various configurations of nuclei and electrons, showing in general dipole moments.*"

These instantaneous dipoles, whose dipole moments are rapidly changing in magnitude and direction and average to zero over a finite time period, lead to interaction among the molecules.

Using quantum mechanics and some simplifying assumptions London developed an expression for the energy of interaction between two spherically symmetric molecules i and j at large distances:

$$\Gamma_{ij} = -\frac{3}{2}\frac{a_i a_j}{r^6}\left[\frac{h\nu_{0_i}\, h\nu_{0_j}}{h\nu_{0_i} + h\nu_{0_j}}\right] \tag{7.5.4}$$

where h is the Planck constant and ν_0 is a characteristic for each molecule electronic frequency in the unexcited state. This frequency is related to the variation of the index of refraction of the substance with light and, thus, the name *dispersion* for the attractive forces between nonpolar molecules. They are also called *London* forces. For a given molecule, the product $h\nu_0$ is very close to its first ionization potential I, which represents the work required to remove the first electron from the molecule. Eq.7.5.4 becomes, thus:

$$\Gamma_{ij} = -\frac{3}{2}\frac{a_i a_j}{r^6}\left[\frac{I_i I_j}{I_i + I_j}\right] \tag{7.5.5}$$

In the case where i and j are the same:

$$\Gamma_{ii} = -\frac{3}{4}\frac{a_i^2}{r^6} I_i \tag{7.5.6}$$

Values for the first ionization potential are given by Prausnitz et al and for some selected compounds in Table 7.1.

7.5.4 Example 7.1 Comparison of Attractive Forces

We notice that the potential energy for all physical attractive forces is of the type:

$$\Gamma = -B/r^6$$

B is a positive constant for each type of forces except for the dipole-dipole ones, where it is also a function of the reciprocal temperature. It represents, thus, a measure of the potential energy at a given distance.

With this background, calculate *B* values for the different types of forces exerted in the following three pairs of molecules at 273 K:
a. cyclohexane-cyclohexane
b. ammonia-ammonia
c. cyclohexane-ammonia

a. Since the dipole moment is equal to zero, the *B* values for dipole-dipole (*Bdd*) and dipole-induced dipole (*Bdi*) interactions are equal to zero. For the dispersion forces (*Bds*) using:
$a = 109 \cdot 10^{-25} cm^3$; $I = 11$ eV; and 1 eV $= 1.602 \cdot 10^{-19}$ J
$Bds = 1570 \cdot 10^{-67}$ J cm^6

b. Using:
$\mu = 1.5$ Debye $= 1.5 \cdot 10^{-18} (erg\ cm^3)^{1/2}$
$a = 22.6 \cdot 10^{-25} cm^3$; $I = 11.5$ eV; $k = 1.380 \cdot 10^{-16}$ erg/K
$Bdd = 89.6 \cdot 10^{-67}$ J cm^6
$Bdi = 10.2 \cdot 10^{-67}$ J cm^6
$Bds = 70.6 \cdot 10^{-67}$ J cm^6

c. Using parameter values from the previous cases:
$Bdd = 0.0$; $Bdi = 24.5 \cdot 10^{-67}$ J cm^6
$Bds = 332.8 \cdot 10^{-67}$ J cm^6

7.5.5 Comments

1. *B* values for several pairs of molecules are presented in Table 7.2, calculated as in Example 7.1, and indicate that:

a. the dispersion forces are relatively large;

b. the induction forces, in both polar-nonpolar and polar-polar pairs, are relatively small;

c. dipole-dipole forces become significant only when $\mu > 1.0$ Debye, for a pure polar compound or mixtures of such compounds. This is apparent from Eq. 7.5.1, where μ is raised to a total power of four.

2. Eqs 7.5.1 through 7.5.6 are approximations and are useful for comparison purposes only.

3. In certain molecules, such as CO_2 for example, charge concentrations in four locations within the molecule give rise to quadrupoles and additional intermolecular forces. They are smaller, but not insignificant.

We consider next the analytical description of intermolecular potentials and then proceed with a discussion of chemical forces.

Table 7.2 Comparison of Intermolecular Forces in Terms of the Values of the Parameter *B* (Example 7.1) at 273 K.

Molecules		*Bdd*	*Bdi*	*Bds*
1	2		($J\ cm^6 x 10^{67}$)	
c-C_6H_{12}	c-C_6H_{12}	0.0	0.0	1570.0
CH_4	CH_4	0.0	0.0	105.6
CO_2	CO_2	0.0	0.0	115.6
$CHCl_3$	$CHCl_3$	25.9	19.9	935.9
NH_3	NH_3	89.6	10.2	70.6
CH_3OH	CH_3OH	147.7	18.7	135.4
H_2O	H_2O	185.7	10.3	38.3
$(CH_3)_2CO$	$(CH_3)_2CO$	1251.0	106.5	486.2
c-C_6H_{12}	NH_3	0.0	24.5	332.8
c-C_6H_{12}	$(CH_3)_2CO$	0.0	91.7	873.0
CH_4	$(CH_3)_2CO$	0.0	21.9	224.8
H_2O	$(CH_3)_2CO$	482.0	33.9	135.6

7.6 Intermolecular Potentials

7.6.1 The Mie Potential

To account for both repulsive and attractive forces between two molecules, Mie proposed the following expression for the intermolecular potential:

$$\Gamma = \Gamma_{repulsive} + \Gamma_{attractive} = \frac{A}{r^n} - \frac{B}{r^m} \tag{7.6.1}$$

where A, B, n, and m are positive constants and r is the intermolecular distance.

Our discussion so far suggests a value of $m = 6$. In addition, since the repulsive potential becomes, as we saw, very strong at small distances only, where electron cloud overlapping occurs, n should be larger than m.

A typical intermolecular potential as a function of the distance r of the centers of two molecules is shown in Figure 7.2. The value of r where $\Gamma = 0$ is usually depicted by σ, while the minimum energy - and the distance where it occurs, by $-\epsilon$ and r^* respectively. Notice that r^* represents the equilibrium distance for a pair of molecules.

7.6.2 The Lennard-Jones Potential

If we set $n = 12$ and $m = 6$ in Eq.7.6.1 we obtain the well known and extensively studied Lennard-Jones (LJ), or 12-6, potential. This potential is usually written in the reduced form:

$$\frac{\Gamma}{\epsilon} = 4\left[\left(\frac{\sigma}{r}\right)^{12} - \left(\frac{\sigma}{r}\right)^{6}\right] \tag{7.6.2}$$

which is easily obtained from Eq.7.6.1 (Problem 7.8).

To obtain numerical values for the parameters ϵ and σ the LJ potential must be related to some macroscopic property, typically: vapor phase volumetric behavior, as expressed through the second virial coefficient; and transport properties, such as viscosity and diffusivity. Values for several compounds are given by Reid et al.

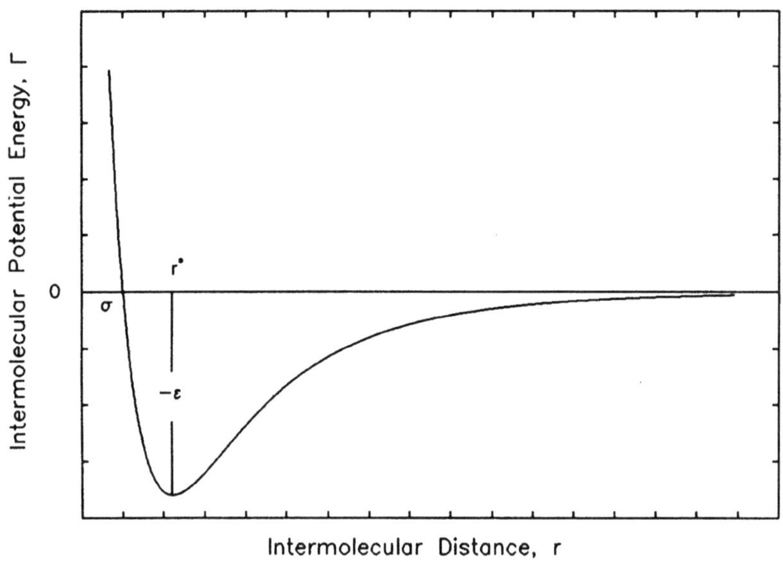

Figure 7.2 Intermolecular potential energy as a function of the distance between the centers of two molecules.

7.6.3 Comments

1.Can we expect the LJ, or the more general Mie potential, to be applicable to all types of molecules?

Eq.7.6.1 expresses the intermolecular potential as a function of the distance between the centers of the two molecules but not of their relative orientation. It should, hence, be applicable to molecules whose force fields are spherical. Strictly speaking, this would include only the heavy noble gases: argon, krypton, and xenon. But other nonpolar (or slightly polar) molecules containing a small number of atoms such as oxygen, nitrogen, or carbon monoxide, fall approximately in this category. Pitzer refers to them as *simple* molecules.

2.Molecules that deviate from this behavior cannot be expected to follow Eq.7.6.1. For large and/or polar molecules, their relative orientation plays an important role in their intermolecular potential.

3.The Lennard-Jones potential represents an approximation, even for simple molecules. As a result, different sets of parameter values are obtained from second virial coefficient data and from transport properties data; and the discrepancy increases, for the aforementioned reasons, as the deviation from the simple molecule concept becomes larger (Hirschfelder et al, 1954).

4.Prausnitz et al suggest that for many practical applications the Lennard-Jones potential is entirely adequate, but not for very accurate calculations. They report that even for argon, Michels et al (1958) have shown that their highly accurate second virial coefficient data in the range -140 °C to 150°C cannot be fitted to within experimental error with one set of parameters.

5.Eq.7.6.2 indicates that the intermolecular potential, when reduced by a characteristic energy parameter ϵ, is a universal function of the distance, reduced by a characteristic parameter σ.

This represents the molecular basis for the *Corresponding States Principle*, whose practical importance will become apparent in the next Chapter.

6.It appears, therefore, that application of intermolecular potentials for quantitative answers is limited to very small molecules only. Recent statistical mechanics studies, however, using molecular simulations, appear more promising for other types of molecules as well. (See, for example, Gubbins, 1983).

Qualitative insight, on the other hand, can be often achieved as Examples 7.1 and 7.2 demonstrate, and as we will see in the ensuing Chapters.

7.The LJ potential is not, in spite of its apparent simplicity, an easy expression to use in statistical mechanics calculations. Thus, the simpler square-well potential, defined by:

$$\begin{aligned} \Gamma(r) &= \infty \quad \text{for } r \le \sigma \\ \Gamma(r) &= -\epsilon \quad \text{for } \sigma < r \le R\sigma \\ \Gamma(r) &= 0 \quad \text{for } r > R\sigma \end{aligned} \tag{7.6.3}$$

where $R\sigma$ and ϵ are the *width* and *depth* of the potential well respectively, is often used. Eq.7.6.3 represents a very crude approximation of the LJ potential, but its simplicity facilitates greatly the numerical calculations, at least for preliminary purposes.

8.For a realistic description of intermolecular forces more complex models are used as discussed, for example, by Prausnitz et al and, in a more comprehensive fashion, by Reed and Gubbins (1973) and by Gray and Gubbins (1984).

7.6.4 Example 7.2. The Origin of the 'Geometric Mean' Rule

Show that the attractive potential between two unlike nonpolar molecules 1 and 2 is given by the geometric mean of their attractive potentials.

From London's formula for dispersion forces, Eqs 7.5.5 and 7.5.6:

$$\Gamma_{11} = -\frac{3}{4} a_1^2 \frac{I_1}{r^6}$$

$$\Gamma_{22} = -\frac{3}{4} a_2^2 \frac{I_2}{r^6}$$

$$\Gamma_{12} = -\frac{3}{2}(a_1 a_2)\left(\frac{I_1 I_2}{I_1 + I_2}\right)\frac{1}{r^6}$$

Notice that:

1.The attractive potential is a stronger function of the polarizability than of the ionization potential; and

2.Values for the ionization potential are, as suggested by Table 7.1, fairly close to each other, especially when compared to the polarizability ones. If we neglect, therefore, these small differences in the I values, we obtain:

$$\Gamma_{11} = \beta \frac{a_1^2}{r^6}$$

$$\Gamma_{22} = \beta \frac{a_2^2}{r^6}$$

$$\Gamma_{12} = \beta \frac{a_1 a_2}{r^6}$$

where β is a constant. It follows, then, that:

$$\Gamma_{12} = (\Gamma_{11} \Gamma_{22})^{1/2} \quad (7.6.4)$$

Comments:

1. Eq.7.6.4 provides the theoretical basis for the so-called 'geometric mean' rule approximation used in corresponding states calculations involving nonpolar mixtures.

2. As the dissimilarity between the components of the mixture increases, however, such as in the systems methane-decane or methane-carbon tetrachloride for example, this approximation fails.

7.7 Chemical Forces

7.7.1 Hydrogen Bonding

Consider the physical properties of the two isomeric compounds, ethanol and dimethyl ether, presented in Table 7.3. Since the dipole moment values are so close, how can we explain the much higher normal boiling point temperature of ethanol? Or, the more than double amount of energy needed to vaporize one mole of ethanol as compared to one of dimethyl ether? In other words, what kind of intermolecular forces are responsible for these differences?

To answer these questions we notice that, as discussed in Section 7.4.1, oxygen in the hydroxyl group has the tendency to 'steal' the electron from the hydrogen atom rather than share it, along with one of its own, with the latter.

Table 7.3 Physical Properties of two Isomers: Ethanol and Dimethyl Ether

Property	Ethanol	Di-me Ether
Dipole Moment (Debye)	1.7	1.3
Normal Boiling Point (T_b, °C)	78.3	-24.9
Heat of Vaporization at T_b (kJ/mol)	42.6	18.6

As a result, we have an oxygen atom that has more than its share of electrons and, consequently, can act as an *electron donor* or a *proton acceptor*; and a hydrogen atom, that has less than its share of electrons and can act as an *electron acceptor* or a *proton donor*.

When, now, ethanol molecules come close, the electron donor and acceptor properties of the oxygen and hydrogen atoms in the hydroxyl groups lead to the formation of complexes, such as those shown in Figure 7.3, that can involve up to several molecules.

This tendency of atoms to saturate their electronic structure leads, of course, to the formation of chemical bonds and new species. The aggregates formed here, however, are weaker but their origin is the same, hence the name *chemical* forces. The difference is basically the strength of the created bond: the bond strength for most hydrogen bonds lies in the region of 8 to 40 kJ/mol, whereas the usual covalent bond strength is in the region 200 to 400 kJ/mol. As a result, the hydrogen bond is broken rather easily and it is for this reason that hydrogen bonding effects usually decrease at higher temperatures, where the kinetic energy of the molecules is sufficient to break the loose bonds (Prausnitz et al).

Molecules, thus, containing a hydrogen atom linked to an electronegative one - such as acids, alcohols, and amines - have a strong tendency to form aggregates with each other or with other molecules containing accessible electronegative atoms.

When the aggregates formed contain molecules of the same type, such as in the aforementioned case of ethanol, the phenomenon is referred to as *association*; when they contain molecules of different type, such as ethanol and water ones for example, as *solvation*.

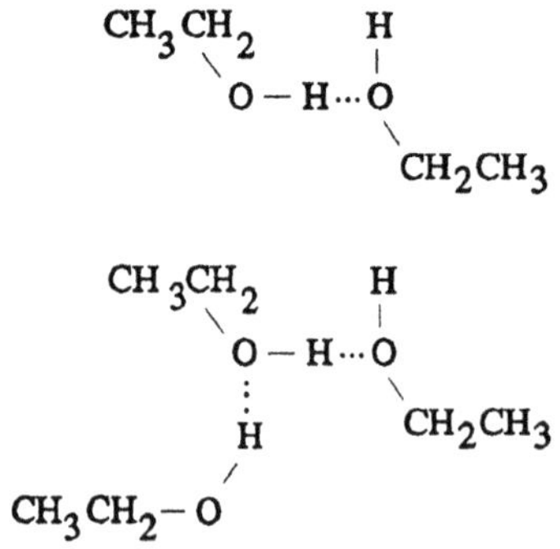

```
CH3CH2        H
      \       |
       O - H···O
                \
                 CH2CH3

CH3CH2        H
      \       |
       O - H···O
       :        \
       H         CH2CH3
      /
CH3CH2-O
```

Figure 7.3 Ethanol aggregates.

7.7.2 Weaker Forces

While hydrogen bonding represents the case of strongest and most common chemical forces, weaker associations formed between electron donors and electron acceptors are also encountered in solutions. For example, Weimer and Prausnitz (1966) demonstrated experimentally that aromatic compounds form weak complexes with a variety of polar solvents, the latter acting as electron acceptors and the former as electron donors. They observed no complex formation with saturated hydrocarbons.

7.7.3 Applications

There are several cases where chemical effects are very important. Some typical examples are (Prausnitz et al):

* the solubility of silver chloride in water is very small, but it increases significantly with the addition of some ammonia due to the formation of a silver-ammonia complex;
* acetone is a much better solvent for acetylene than for ethylene because, unlike ethylene, acetylene can form hydrogen bonding with acetone;
* the well known industrial processes for the absorption of carbon dioxide in amines, where carbon dioxide (acidic) forms a complex with the amine (basic);
* the aforementioned tendency of polar solvents to form complexes with unsaturated hydrocarbons, but not with saturated ones, represents the basis for various commercial separation processes in the petroleum industry, including the well known Edeleance and Udex processes.

Several experimental results demonstrating the importance of hydrogen bonding are discussed by Prausnitz et al. For a more detailed study, see Pimentel and McLellan (1960).

7.7.4 Example 7.3

Methanol is mixed with n-hexane:

a. is the process exothermic or endothermic?

b. what will the shape of the curve look like if we were to plot the heat of mixing - per mole of methanol - versus the methanol mole fraction at constant temperature and pressure?

a. The mixing of methanol with an other solvent requires the breaking of the methanol aggregates which requires energy. Since hexane does not solvate with methanol no new aggregates, which would release energy, will be formed. The process is, consequently, endothermic.

b. When the first drop of methanol is added to hexane, all its aggregates must break down because in the resulting solution the methanol molecules are completely surrounded by hexane molecules. The heat of mixing thus, per mole of methanol, assumes its largest value. This phenomenon becomes less pronounced, but still very important, with the next drop of methanol.

When, however, we reach the point where the amount of methanol in the solution becomes significant, so that the addition of more methanol does not necessitate the breaking of a large number of aggregates, the effect becomes small.

We would expect, therefore, the heat of mixing - on a per mole of methanol basis - to be very large at very low methanol mole fraction (x_M), decline sharply with increasing values of x_M, and then approach zero at higher x_M values.

This is indeed the case, as the experimental data in Figure 7.E.3 indicate. The results for cyclohexane-hexane are also shown to demonstrate the large impact of the hydrogen bonding.

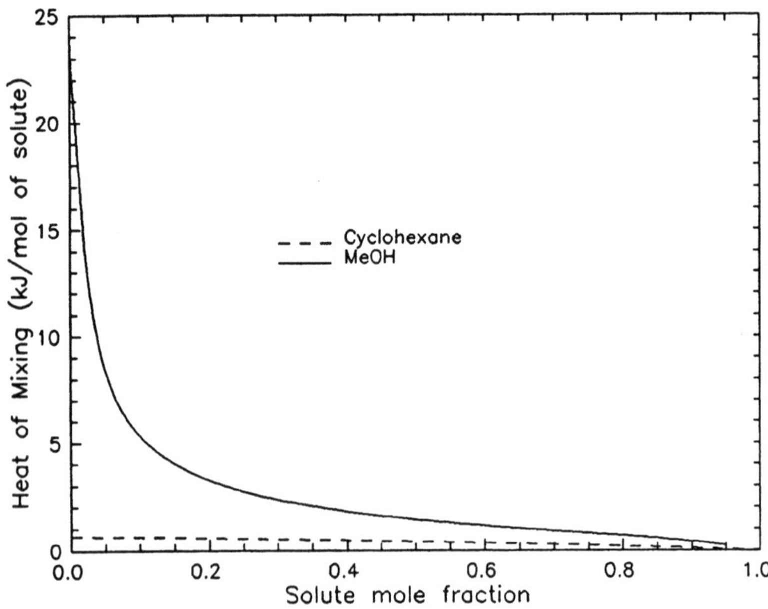

Figure 7.E.3 The effect of hydrogen bonding: heat of mixing with n-hexane (solvent) at 313 K for two chemically different solutes: methanol, that demonstrates strong hydrogen bonding and cyclohexane, that does not.

7.8 Concluding Remarks

1.Electrostatic (coulombic) interactions among molecules give rise to intermolecular forces, repulsive and attractive. The former are dominant at short intermolecular distances, the latter at larger ones (Figure 7.2).

2.Statistical mechanics provides the relationship between macroscopic thermodynamic properties and molecular behavior. The degree to which quantitative results will be obtained for realistic systems, however, will depend on the progress in understanding, and expressing quantitatively, intermolecular forces (see Chapters 16 and 17).

3.Use of intermolecular forces for quantitative results is thus limited mostly to simple systems so far.

4.A quantitative description of chemical forces through pure component properties, even to the limited degree that was the case with physical forces, is not possible yet. Results from experimental measurements are usually correlated using association constants (see, for example, Weimer and Prausnitz).

5.In spite of our limited understanding of intermolecular forces, qualitative answers are often possible, such as in the case of separation processes. For a review, for example, of the use of intermolecular forces - both physical and chemical - in screening potential solvents for extractive distillation, and for some applications to separation processes, see this author (1969 and 1972).

References

Gray, C.G., Gubbins, K.E., 1984. *Theory of Molecular Fluids*, Oxford University Press, London.

Gubbins, K.E., 1983. *J. Am. Chem. Soc.*, **87**, 4573.

Happel, J., Cornell, P.W., Eastman, D., Fowle, M.J., Porter, C.A., Schutte, A.H., 1946. *Trans. AIChE*, **42**, 189.

Hirschfelder, J.O., Curtiss, C.F., Bird, R.B., 1954. *Molecular Theory of Gases and Liquids*, Wiley, New York.

Michels, A.J., Levelt, M., De Graaf, W., 1958. *Physica*, **24**, 659.

Mikolaj, P.G., Pings, G., 1967. *J. Chem. Phys.*, **46**, 1401.

Pimentel, G.C., McLellan, A.L., 1960. *The Hydrogen Bond*, W. H. Freeman & Co., San Francisco.

Prausnitz, J.M., Lichtenthaler, R.N., de Azevedo, E.G., 1986. *Molecular Thermodynamics of Fluid-Phase Equilibria*, 2nd Edition, Prentice-Hall, Englewood Cliffs, N.J. (The information in Table 7.1 and Problems 7.13, 7.14 and 7.16 is reprinted by permission of Prentice-Hall.)

Reed, T.M., Gubbins, K., 1973. *Applied Statistical Mechanics*, McGraw-Hill, New York.

Reid, R.C., Prausnitz, J.M., Poling, B.E., 1987. *The Properties of Gases and Liquids*, McGraw-Hill, New York.

Tassios, D.P., 1969. *Chem. Eng.*, **76**, 3, 118.

Tassios, D.P., 1972. Ed., *Extractive and Azeotropic Distillation, Advances in Chemistry Series*, No 115, Am. Chem. Soc., Washington, D.C.

Weimer, R.F., Prausnitz, J.M., 1966. *Spectrochim. Acta*, **22**, 77.

Problems

7.1 What type of interactions exist between a molecule of methane and one of water? Put them in order of decreasing magnitude.

7.2 Estimate the attractive forces between a molecule of carbon tetrachloride and one of methanol.

7.3 Discuss qualitatively, i.e. without using Eq.7.5.1, why at high temperatures dipole-dipole interactions become negligible.

7.4 Using the data in the Steam Tables demonstrate the effect of temperature on the deviation of steam from ideal gas behavior.

7.5 With reference to Eq.7.6.1, explain why:

a. $m = 6$

b. $n > m$

7.6 What are the limitations of the LJ potential?

7.7 Discuss the successes and failures in using the geometric mean rule with equations of state. Give some specific examples.

7.8 For the Lennard-Jones potential show that:

a. the parameters in Eq. 7.6.1 are given by: $A = 4\epsilon\sigma^{12}$; $B = 4\epsilon\sigma^{6}$, which lead to Eq.7.6.2;

b. $r^* = 2^{1/6}\sigma$

7.9 What is the origin of hydrogen bonding?

7.10 What is the difference between association and solvation?

7.11 Estimate the enthalpy of formation of hydrogen bonding per mole of hydrogen bond using data of your choice.

7.12.a. Mikolaj and Pings (1967) report, on the basis of the experimental radial distribution function for argon, that the equilibrium distance for the liquid state at -125°C is equal to about 1.6σ. How is this reconciled with Problem 7.8b?

b. What is the radial distribution function?

7.13.a. What is meant by the terms 'electron affinity' and 'ionization potential'?

b. How are these concepts related to the Lewis definition of acids and bases?

c. In the separation of aromatics from paraffins by extraction, why is liquid SO_2 a better solvent than liquid NH_3?

7.14 At 30°C, the solubilities of ethane and acetylene in n-octane are about the same. However, at the same temperature, the solubility of acetylene in dimethyl formamide is very much larger than that of ethane. How do you explain this in terms of intermolecular forces?

7.15.a. What is the purpose of extractive distillation?

b. What role do intermolecular forces play in it?

c. What type of solvents would you consider for the separation of butanes from butenes? [For actual cases, see Happel et al (1946).]

7.16 Sketch qualitatively a plot of the compressibility factor z vs the mole fraction for the following mixtures:

a. dimethylamine-H_2; and

b. dimethylamine-HCl.

Briefly explain your reasoning.

Note. Problems 7.13, 7.14, and 7.16 are taken from Prausnitz et al (pp.91,92).

8
Physical Properties of Pure Fluids

8.1 Introduction

We will present the second main objective of Chemical Engineering Thermodynamics, the *Evaluation of Thermophysical Properties* (Section 1.14.2), by considering *Pure Fluids*, in this and the ensuing two Chapters; and *Mixtures*, in Chapters 11 through 15. (Actually Chapters 13 and 14 deal with Phase Equilibrium and Chapter 15 with Chemical Equilibrium, subjects whose study represents the third objective of Chemical Engineering Thermodynamics. Their separate consideration, thus, is intended to emphasize their importance in our profession.)

We start by discussing in this Chapter the *Physical Properties* of pure fluids:

* Volumetric: Pressure-Volume-Temperature (*PVT*) behavior, and
* Thermal: Enthalpies of Vaporization and Heat Capacities.

Values of these properties are essential in chemical engineering calculations (see Table 1.4 for example), such as the sizing of pipes and vessels, the evaluation of duties in heat exchangers, etc. In addition, they are required information in the determination of the *Thermodynamic Properties* of pure fluids, enthalpies, entropies, etc., that are considered in the next Chapter; and of mixture properties.

The importance of thermophysical properties has not become noticeable in our discussion so far because, for simplicity reasons:

* first, we assumed ideal gas behavior in all calculations involving gases; and
* second, we worked with steam, where - mainly due to its importance in the conversion of thermal energy into mechanical - the extensive information contained in the Steam Tables is available.

We know, of course, that gases behave as ideal ones in the limit of zero pressure, and that other compounds - besides steam - are important in chemical engineering calculations.

8.2 Objective and Approach

Experimental pressure-volume-temperature and thermal property data are available in the literature for a limited number of compounds only. In practice, therefore:

a. The operating conditions of temperature and pressure may be different than those of the experimental data, if available;

b. Considering the vast number of compounds of interest in chemical engineering applications, limited - or no - experimental data may be available in the typical case.

Our objective, therefore, is to develop familiarity with:

1. The experimental volumetric and thermal behavior of pure fluids and some of the corresponding data sources;
2. Correlations that can be used for extra- or interpolation purposes; and, especially,
3. Estimation methods.

We will start with a description of the volumetric behavior of pure fluids, by considering first gases and then vapors and liquids. (The difference between gases and vapors is that a gas cannot be condensed when compressed isothermally; a vapor, can. The two terms are often used interchangeably, however.)

We proceed, then, to discuss some common characteristics in the *PVT* behavior of all pure fluids, which are helpful in its quantitative description and estimation that follows.

For the latter we present the two main approaches used:

a. the Corresponding States Principle, and

b. Equations of State

We will close the Chapter with a discussion of vapor pressures and the two thermal properties:

* enthalpies of vaporization; and
* heat capacities.

<u>Note.</u> Thermal properties are classified by some authors as thermodynamic rather than physical properties. To be consistent with our classification, however, that thermodynamic properties are derived ones and physical measurable ones, we discuss thermal properties in this Chapter.

8.3 The Volumetric Behavior of Pure Gases

8.3.1 The Origin of Fluid Pressure

Before we consider the *PVT* behavior of pure gases let us pose the following question: what do we really mean when we say that the pressure of a gas, say i-butane, is equal to some specified value?

To this purpose, let us consider the following mental experiment. A transparent cylinder, with a cross-sectional area of 10 cm^2, is fitted with a piston and immersed into a constant temperature bath at 420 K. The piston has a plate, on which we can place weights, and a volume indicator calibrated to read cm^3. We assume, for simplicity, that the piston is weightless and frictionless and that the outside pressure is zero.

We now place one mole (58.124 grs) of i-butane in the cylinder, and the necessary weights on the plate, so that the volume becomes equal to some specified value, say 171 cm^3. We wait until thermal equilibrium is reached with the bath, adjusting the weights so that the volume of 171 cm^3 is maintained. The final weight is $5.75 \cdot 10^3$ N, which - when divided by the piston area - gives a pressure of 57.5 bar.

What is pushing the piston upwards, balancing the weights? We know that the i-butane molecules, all 6.023×10^{23} of them, are moving continuously in all directions striking the internal surface of the cylinder and the piston. The resulting force, divided by the area, represents the gas pressure and, since it balances that of the weights, it is equal to 57.5 bar.

If we now increase the bath temperature, and thus that of i-butane, the kinetic energy of the molecules and, consequently, the force exerted by them when they strike the piston surface will also increase. Thus, if the volume is to be maintained constant, more weight must be added on the plate: the pressure of the gas has increased.

Atmospheric Pressure, Premature Births, and Shrinking (!)

Jeanette Madea of Pompano Beach, Fla., saw a newspaper report of a slight increase in premature births at hospitals in Long Island, New York, when Hurricane Gloria shot through last month. One of the spokesmen was quoted as follows: "It's a well-known hurricane phenomenon .... When the air pressure drops suddenly it may put enough external pressure on a woman to start labor a little early .... The pressure can be incredible .... I've heard if you measure yourself during a hurricane you'll be shorter because the air is pressing you down."
(C&EN, p.72, October 28, 1985)

8.3.2 Pressure - Volume Isotherms

We now change the weight on the plate, corresponding to a pressure P_1, wait until thermal equilibrium is reached, and record the corresponding value of the volume V_1. We repeat the measurement several times generating a list of P_i, and the corresponding V_i, values.

Such data for i-butane are shown in Figure 8.1. The resulting curve is referred to as an isotherm (from the Greek: "iso", the same; and "therme", heat).

8.4 The Volumetric Behavior of Pure Vapors and Liquids

8.4.1 Pressure - Volume Isotherms

We now repeat the measurements at a lower bath temperature of 400 K. Starting with a low pressure and increasing it gradually, Figure 8.1, we notice that when we reach the value of 31.885 bar, the volume indicator reads 483 cm^3 and a very light dew of liquid droplets is formed.

Further reduction of volume does not require an increase of pressure: the piston moves downwards, but the pressure remains constant. We notice, however, that more and more liquid is formed. We eventually reach an indicator reading of 170 cm^3, where all the i-butane is liquid except for a few bubbles. Note also that in this two-phase region ($L+V$) the relative amounts of liquid and vapor - for a mixture of specified volume - are given by the well known inverse lever rule.

Once all vapor becomes liquid, we can increase the pressure again. We notice, however, that large increases are needed now to effect even a small volume change. The molecules in the liquid phase are very close to each other, and attempts to bring them closer must overcome the very large, as we saw in the previous Chapter, repulsive forces.

Thus an increase in pressure from 31.885 bar to 669.5 bar, i.e. by a factor of over 20, leads to an increase in density from 341.9 kg/m^3 to 560 kg/m^3, i.e. by only 64%. This is why we often refer to liquids as *incompressible*.

This volumetric behavior involving single- and two-phase regions occurs at all subcritical (Section 8.4.3) temperatures as shown in Figure 8.1.

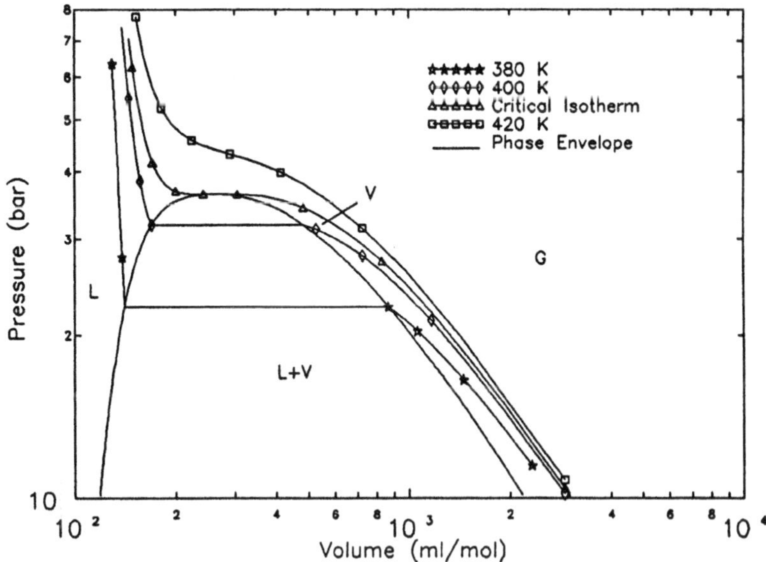

Figure 8.1 P-V isotherms and the two-phase region for i-butane (Goodwin and Haynes, 1982). L: Liquid; V: Vapor; G: Gas.

8.4.2 Terminology

a.Dew Point

We saw that as the pressure of the vapor at 400 K was increased to the value of 31.885 bar, a very light dew of liquid droplets appeared. This pressure is referred to as the *Dew Point pressure* of i-butane at 400 K.

Notice that if our experiment involved cooling of gaseous i-butane at the constant pressure of 31.885 bar, the first dew would appear at 400 K. This represents the dew point *temperature* of i-butane at 31.885 bar.

b.Bubble Point

Assume next that we start our experiment with liquid at 400 K and a high pressure - say 100 bar - and proceed to reduce the pressure slowly. According, now, to the discussion in Section 8.4.1, when the pressure of 31.885 bar is reached, the first vapor bubbles will appear. This is referred to as the *Bubble Point* pressure of i-butane at 400 K. And following the same line of reasoning, as with the dew point temperature, 400 K represents the bubble point temperature of i-butane at 31.885 bar.

c. Saturation

When the pressure of the i-butane vapor increases to 31.885 bar, at 400 K, the vapor becomes saturated with molecules that want to be in the liquid phase, thus the appearance of the first droplets. It is referred to, therefore, as *saturated vapor.*

Similarly, when the pressure of the liquid is reduced to 31.885 bar, also at 400 K, the liquid becomes saturated with molecules that want to be in the vapor phase, thus the appearance of the first bubbles. It is referred to as *saturated liquid.*

The values of pressure, temperature, and vapor and liquid volumes at this point, represent the saturation values and are designated with the superscript '*s*'; thus for i-butane vapor at 400 K $V^s(vapor)$ = 0.483 l/mol; and $V^s(liquid)$ = 0.170 l/mol. Finally, the locus of saturated liquids and vapors is referred to as the *saturation* or *phase envelope* (Figure 8.1).

d. Equilibrium

Notice that at the dew as well as at the bubble point two phases: a vapor and a liquid, coexist. The same occurs, at 400 K, in the volume range from 0.483 to 0.170 l/mol (Figure 8.1). The amounts of the two phases of course vary, from all vapor at the dew point, to all liquid at the bubble point. These coexisting phases are said to be in *equilibrium.*

e. Vapor Pressure and Boiling Point Temperature

The pressure over a liquid in equilibrium with its vapor is called the *saturation* or *vapor* pressure; and the temperature, the *saturation* or *boiling point* temperature.

f. Comments

1. Notice that if a vapor and a liquid are in equilibrium, then:
   * the vapor is saturated, and at its dew point;
   * the liquid is saturated, and at its bubble point;
   * the pressure is the vapor pressure; and
   * the temperature is the boiling point temperature.
2. If the pressure is equal to 1 atm, the corresponding boiling point temperature is referred to as the *normal* boiling point temperature, T_b.
3. At a given pressure, a liquid that is at a temperature below the boiling point value that corresponds to this pressure, is called *subcooled.*

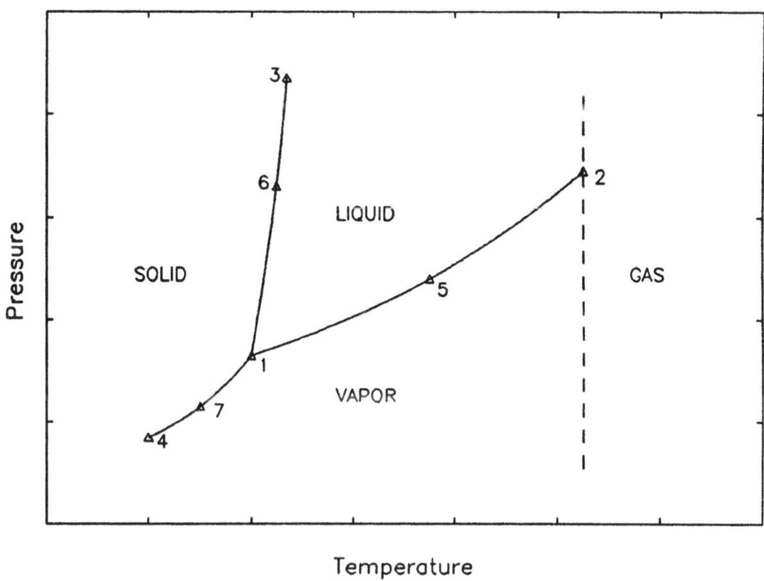

Figure 8.2 The phase diagram for a pure compound.

4. At a given pressure, a vapor that is at a temperature above the saturation value that corresponds to this pressure, is called *superheated.*

8.4.3 Phase Diagrams

Figure 8.1 represents a *P-V phase diagram*. An other useful type of phase diagram is a *P-T* one presented in Figure 8.2.

Line 12 represents the coexistence of vapor and liquid phases and it is referred to as the *vapor pressure* curve. Point 5 indicates, thus, that the vapor pressure of the liquid at the temperature T_5 is equal to P_5 which we depict, as we have seen, as P_5^s. Notice that increasing temperatures lead, as expected, to increasing vapor pressure values. Beyond the temperature of point 2, however, the kinetic energy of the molecules is too high to allow them to form a liquid phase.

We refer to point 2 as the *critical point*, and the corresponding values of temperature, pressure and volume, as the critical ones depicted by T_c, P_c and V_c respectively. Notice that according to our classification, a vapor at a temperature higher than T_c, is a gas.

Line 13 represents the coexistence of solid and liquid phases and is called the *fusion* curve. Point 6, thus, depicts the melting point pressure P_6 at the temperature T_6.

Line 14 represents the vapor-solid coexistence - or *sublimation* - curve. Point 7, thus, depicts the vapor pressure of the solid at the temperature T_7. Finally, point 1 represents the coexistence of all three phases: vapor, liquid and solid. It is called, appropriately, the *triple point* T_t.

8.5 Volumetric Behavior Common to all Pure Fluids

The volumetric behavior for a number of pure gases and vapors - often up to high pressures - has been studied, but only for few liquids, because of the difficulties arising from their very limited compressibility.

The following observations reflect the actual behavior of pure vapors and gases:

a. They approach ideal gas behavior as the pressure approaches zero, i.e:

$$\lim_{P \to 0} (PV) = RT \quad (T = Constant)$$

b. More generally, they approach ideal gas behavior as the density tends to zero.

c. At the critical point (*cp*), the critical isotherm shows a horizontal inflection point (see Figure 8.1). Hence:

$$\left(\frac{\partial P}{\partial V}\right)_{cp} = 0 \tag{8.5.1}$$

$$\left(\frac{\partial^2 P}{\partial V^2}\right)_{cp} = 0 \tag{8.5.2}$$

d. As the critical point is approached, the molar volume of the liquid phase approaches that of the vapor (see Figure 8.1), and this applies to all properties of the two phases.

We define, therefore, the *critical point* as that pressure and temperature where the liquid and vapor properties - molar volumes, enthalpies, etc. - become identical. The critical temperature also represents the temperature above which a pure fluid cannot exist as a liquid. As we will see in Chapter 14, however, this does not necessarily apply to a mixture of fluids.

8.6 The Volumetric Behavior of Pure Fluids: Quantitative Treatment

8.6.1 The Compressibility Factor

The *PVT* behavior of pure fluids is typically expressed in terms of the *compressibility factor z* :

$$z = \frac{PV}{RT} \tag{8.6.1}$$

where V is the molar volume. Notice that, since for an ideal gas $z = 1.0$, the compressibility factor can be viewed as a measure of the deviation of the fluid from ideal gas behavior (but see also item (d) below).

Two sets of z values for i-butane, at 380 K and 420 K, are plotted versus pressure in Figure 8.3. Let us examine these two curves in terms of our understanding of intermolecular forces:

a. At very low pressures, the molecules are very far from each other and experience, thus, no intermolecular forces. As a result, the fluid appro-

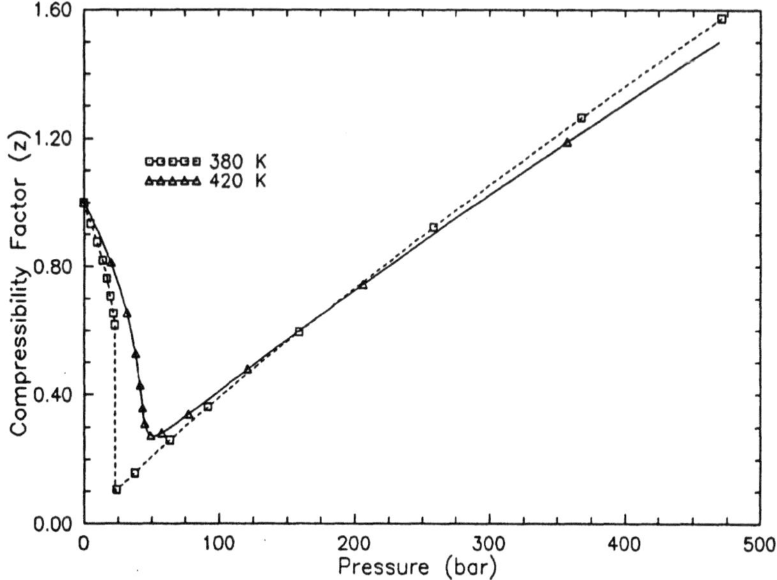

Figure 8.3 Compressibility factors for i-butane (Goodwin and Haynes).

aches ideal gas behavior and the compressibility factor approaches the value of one.

b. As the pressure increases, the molecules are forced closer to each other and the attractive forces become dominant. As a result, their volume becomes smaller, than it would be at the same conditions if there were no intermolecular forces, and the compressibility factor becomes smaller than one.

c. At very high pressures, the opposite occurs. The molecules are forced very close to each other, the repulsive forces become dominant, and z becomes larger than one.

d. There is a high pressure value where $z = 1$. This does not, of course, correspond to ideal gas behavior.

8.6.2 Determination of the PVT Behavior

Experimental determination of *PVT* data is time-consuming and costly. Such data, therefore, are available for a small number of compounds only, mostly light hydrocarbons and some inorganic gases (Section 8.5).

Limited data - basically critical properties and some vapor pressures - are available, however, for a rather large number of compounds. The most comprehensive compilations of such data are given by Reid et al (1987); and by Daubert and Danner (1985,1986). Critical property values for some selected compounds are presented in Appendix C.

In the typical case, therefore, the volumetric behavior of a fluid is estimated using the available critical property values. This task is accomplished through two approaches:

a. the Corresponding States Principle (CSP); and

b. Equations of State (EoS).

8.7 The Corresponding States Principle

8.7.1 The Two-Parameter Version

Figure 8.4 presents a plot of the compressibility factor z versus pressure - at the same temperature - for two pure gases A and B, consisting of small molecules such as argon and methane. The difference in the z values, at a given pressure, is apparent.

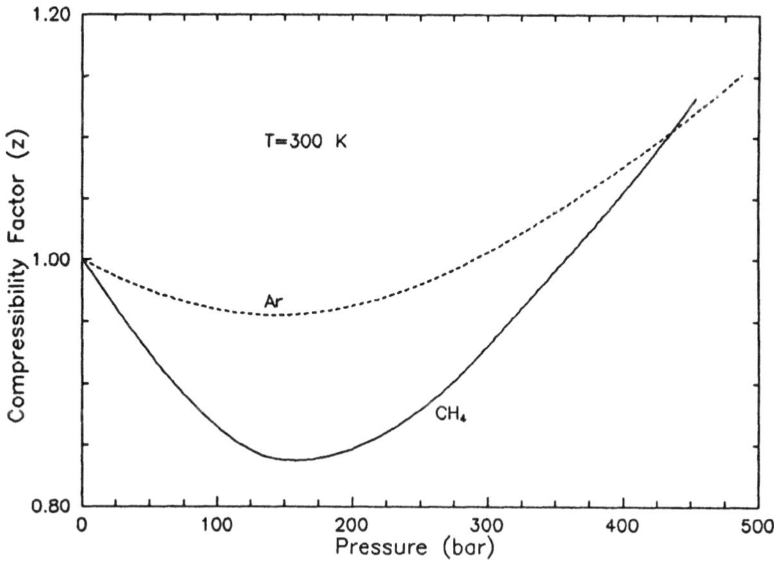

Figure 8.4 Variation of z with pressure for argon and methane at 300 K.

We plot next in Figure 8.5 the compressibility factors for the same gases, but this time versus the *reduced pressure*: $P_r = P/P_c$, and at the same value of the *reduced temperature:* $T_r = T/T_c$, say equal to 1.05. We notice that in this case, the points for both gases fall on the same practically curve, i.e. they have the same value of z at the same values of P_r and T_r (this is why only the curve for methane is shown). We refer to fluids that are at the same P_r and T_r values, i.e. at the same distance from the critical point, as being in *Corresponding States*.

Observations such as the above have led to the following statement, known as the *Two-Parameter Corresponding States Principle* (*CSP*): *When compared at the same T_r and P_r values, all fluids demonstrate the same deviation from ideal gas behavior.*

Hence, z is a function of T_r and P_r only:

$$z = f(T_r, P_r) \tag{8.7.1}$$

The practical implications of this principle are obvious: establishment of the volumetric behavior of one fluid, say Ar, from experimental measurements, suffices for the prediction of such behavior for all other fluids, for which no experimental data are available.

8.7.2 The Three-Parameter Version

Let us consider next the case where species A and B are dissimilar, such as methane (C_1) and n-decane (nC_{10}). A plot of z versus P_r at $T_r = 1.05$ is shown for both of them in Figure 8.5 and indicates a difference in z values for the same values of the reduced pressure. Obviously, the two-parameter CSP does not apply here.

Furthermore, the expression depicted by Eq.8.7.1 indicates that all compounds should have the same value of the compressibility factor at the critical point, z_c. Indeed: $z_c(\mathrm{Ar}) = 0.291$ and $z_c(CH_4) = 0.288$, but $z_c(nC_{10}) = 0.230$. Actually, z_c values cover the range of about 0.23 to 0.30, as Table 8.1 demonstrates.

On the positive side, however, notice that the differences in the z values between C_1 and nC_{10} - Figure 8.5 - are not that large, suggesting that the addition of one more parameter might rectify the problem.

Three such parameters have been proposed through the years: the *critical compressibility factor*; the *Riedel factor*; and the *acentric factor omega* (ω). The last one, proposed by Pitzer in 1955, has found the widest application and is defined as follows:

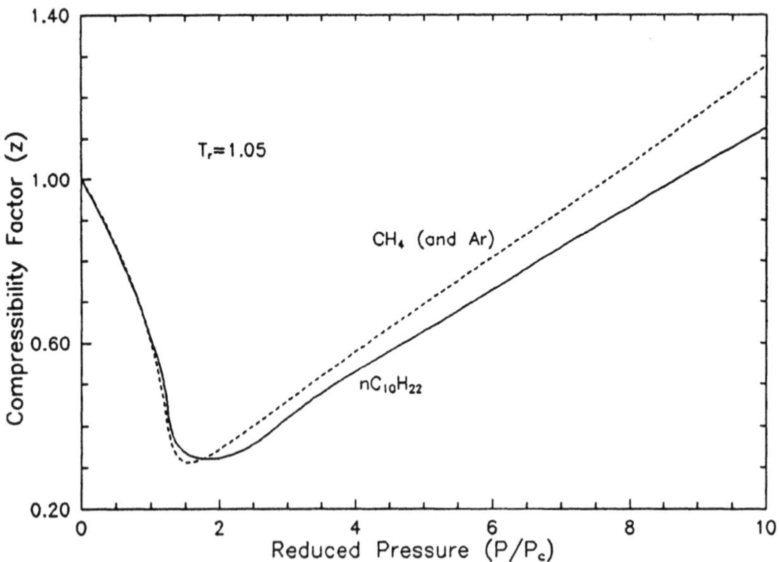

Figure 8.5 Variation of z with reduced pressure for methane (and argon) and n-decane at $T_r = 1.05$.

Table 8.1 Typical Values for the Critical Compressibility Factor

z_c	Representative Compounds
0.22 - 0.24	H_2O, Acetone, Methanol
0.24 - 0.26	NH_3, Ethers, Alcohols
0.26 - 0.28	About 2/3 of compounds, mostly hydrocarbons
0.28 - 0.30	O_2, N_2, CO, H_2S, CH_4, C_2H_6, Ar, Ne

$$\omega = -\log_{10} P_r^s(\text{at } T_r = 0.7) - 1.000 \tag{8.7.2}$$

Hence, only the vapor pressure at the reduced temperature of 0.7 is needed to establish the value of ω.

It should be noticed that accurate vapor pressure data in this temperature range are more readily available than z_c values, which are difficult to establish experimentally. This is the main advantage of using ω as a third parameter. Values of ω for a variety of compounds are given by Reid et al and by Daubert and Danner; and for some selected ones, in Appendix C.

Eq.8.7.1 should, therefore, be written as:

$$z = f(T_r, P_r, \omega) \tag{8.7.3}$$

and represents the *Three-Parameter Corresponding States Principle*: *Fluids demonstrate the same deviation from ideal gas behavior at the same distance from the critical point, provided that they also have the same acentric factor.*

For the functional dependency of z on T_r , P_r and ω, Pitzer suggested the following power series in the acentric factor:

$$z = z^{(0)} + \omega z^{(1)} + \ldots \tag{8.7.4}$$

where $z^{(0)}$, $z^{(1)}$, ... is each a function of T_r and P_r, with the first two terms being sufficient in practice.

Using compressibility factor data for small spherical molecules, such as Ar and CH_4 - which Pitzer refers to as 'simple' fluids and have ω values equal, or very close, to zero - tables for $z^{(0)}$ in terms of T_r and P_r were developed by Pitzer and his coworkers (1955); and from data for compounds with $\omega > 0$, tables for $z^{(1)}$, again in terms of T_r and P_r.

More extensive tables, that also provide somewhat better results, have been developed by Lee and Kesler (1975) and are presented in Appendix D. Their use is demonstrated in Example 8.1.

8.7.3 Example 8.1

Estimate the compressibility factor of n-butane at 425.2 K and 4.4586 MPa. (Exp. value: 0.2095). Given: T_c = 425.18 K, P_c = 3.7969 MPa - hence: T_r = 1.0, P_r = 1.175 - and, ω = 0.1931.

From Appendix D, $z^{(0)}$ = 0.2320 and $z^{(1)}$ = -0.0643, both obtained by linear interpolation. Thus:

$z = 0.2320 + 0.1931(-0.0643) = 0.2196$, with an error of 4.8%.

Note: In the neighborhood of the critical point, $z^{(0)}$ and $z^{(1)}$ are strong functions of P_r and judgement should be used in obtaining the appropriate values, especially for $z^{(0)}$.

In this case, for example, plots of $z^{(0)}$ and of $z^{(1)}$ versus P_r suggest that the 'best' approach would be to average the interpolated between P_r = 1.0 and 1.2 values, with those obtained by extrapolation from P_r = 1.2 to 1.5. Using this method:

$z^{(0)} = (0.2320 + 0.2208)/2 = 0.2264$; $z^{(1)} = -(0.0643 + 0.0902)/2 = -0.0772$; and, z = 0.2115, with an error of 1%.

8.7.4 Comments

1.On the acentric factor, Pitzer states: "The third parameter is required because the intermolecular force in complex molecules is a sum of interactions between the various parts of the molecules - not just their centers - hence the name 'acentric factor' is suggested". This subject is further discussed, for graduate students, in Example 8.2.

2.The acentric factor represents, therefore, a measure of the acentricity (non-sphericity) of the molecule. It is, therefore, equal to zero for monoatomic gases, very small for methane, and increases with molecular weight. It is also effected by the polar character of the molecule. It reflects, thus, the geometry and polarity of the molecule.

3.The overall performance of the Lee and Kesler tables for the compressibility factor of gases, vapors, and liquids indicates errors of - typically - 1-2 % in the T_r range 0.3 to 4 and the P_r range 0 to 10. Larger errors can be expected close to the critical point.

4.These tables are not intended to be applicable to strongly polar fluids, though they are often so used with surprising accuracy, except at low temperatures near the saturated vapor region (Reid et al).

5.The corresponding states approach is not only applicable to the estimation of the volumetric behavior of fluids but, as we will see in the next Chapter, to all properties that depend on intermolecular forces.

We consider, after Example 8.2, the second approach to the prediction of the *PVT* behavior of fluids, Equations of State.

8.7.5 Example 8.2. The Molecular Theory of the Corresponding States Principle (Graduate Students)

We mentioned in Chapter 7 that Eq.7.6.2:

$$\Gamma/\epsilon = f(r/\sigma) \tag{7.6.2}$$

represents the basis for the molecular theory of corresponding states.

The complete set of assumptions behind this theory is given by Prausnitz (1969):

1. The potential energy between two molecules depends only on the distance between them and not on their mutual orientation; that is, the force fields of the molecules are spherically symmetric.
2. The potential energy, reduced by a characteristic energy, can be written as a universal function of the intermolecular distance, reduced by a characteristic length, i.e. Eq.7.6.2.
3. Classical (rather than quantum) statistical mechanics is applicable.
4. The potential energy of a system of molecules is the sum of the potential energies of all possible pairs of molecules.

It is apparent that Eq.7.6.2 is equivalent to the two-parameter corresponding states principle, i.e. Eq.8.7.1.

If the molecules are not small and spherical, the force field around them is not spherical either and, consequently, the distance between the centers of the molecules is not sufficient to describe the intermolecular potential. A third, at least, parameter - such as the acentric factor - is required for this purpose.

Notice that this applies not only to nonspherical molecules, such as water and n-pentane, but also to large molecules that happen to be spherical, such as neo-pentane: the intermolecular potential between two neo-pentane molecules is determined by the distance between the methyl groups, not the two central carbon atoms.

8.8 Equations of State: General Remarks

8.8.1 The Purpose of an Equation of State

An analytical expression that relates the volumetric properties of a fluid, such as:

$$P = f(V, T) \tag{8.8.1}$$

or,

$$V = f(P, T) \tag{8.8.2}$$

is referred to as an *Equation of State* (EoS). It is called so for - in principle at least - it completely describes the state of a given system:

* liquid, vapor, solid, or mixtures of them; and
* pure compounds or mixtures of them.

Such equations are apparently very useful tools for both, correlation and prediction purposes.

8.8.2 Classification of Equations of State

We can classify EoS as follows:

* ideal gas
* virial
* cubic
* noncubic

The ideal gas EoS, based on the assumption of zero intermolecular forces and negligible molecular volume, has already been discussed in Chapter 2. We will start, hence, with the virial EoS and then concentrate on cubic EoS, which are becoming more and more popular because of their simplicity and good performance. From the noncubic EoS, we will only mention - very briefly - the Benedict-Webb-Rubin (BWR) one.

8.9 The Virial Equation of State

8.9.1 The Expression

The virial EoS is derived from a series expansion in powers of $(1/V)$ of the compressibility factor z, at some specified temperature, around its value as $P \rightarrow 0$, i.e. around $z = 1$, and has the following form:

$$z = 1 + \frac{B}{V} + \frac{C}{V^2} + \frac{D}{V^3} + \dots \tag{8.9.1}$$

where V is the molar volume.

The coefficients B, C, D, ... are therefore functions of temperature only - and also of composition for mixtures. They can be related to the intermolecular potential through Statistical Mechanics and reflect: B, two-; C, three-; D, four-; etc., body interactions. This is why they are referred to as 'virial' coefficients (second, third, etc.)

Eq.8.9.1 is known as the volume series of the virial equation. A pressure series expansion for the compressibility factor:

$$z = 1 + B'P + C'P^2 + D'P^3 + \dots \tag{8.9.2}$$

is also used. B', C', etc., are again functions of temperature and composition only.

It can be easily shown (Example 8.3) that the two sets of coefficients B, C, ..., and B', C', ..., are interrelated. Thus, for the second and third ones:

$$B' = \frac{B}{RT} \tag{8.9.3}$$

$$C' = \frac{C - B^2}{(RT)^2} \tag{8.9.4}$$

8.9.2 Evaluation of Virial Coefficients

Virial coefficients can be evaluated through Statistical Mechanics, using expressions for the intermolecular potential that are not, unfortunately, available in the typical case. Virial coefficients are available, therefore, in practice from two sources:

a. experimental, from *PVT* data, and

b. estimated, from empirical correlations, in the absence of *PVT* data.

a. Experimental Virial Coefficients

Determination of virial coefficients from *PVT* data is discussed in Example 8.4, which should be of interest to Graduate Students only. It demonstrates, however, that the reliability of their values fails considerably beyond the second one. This is why experimental values for third virial coefficients are scarce, and practically non-existing for higher ones.

As a result, the virial equation is typically used 'truncated' after the second term. We will concentrate, therefore, our attention on the second coefficient only.

Second virial coefficients are strong functions of temperature, especially at lower temperatures, as demonstrated with the typical values for methane, butane, and methanol, shown in Figure 8.6.

For a qualitative understanding of this temperature effect, we solve the volume series, Eq.8.9.1, truncated after B, for the pressure:

$$P = \frac{RT}{V} + B\frac{RT}{V^2}$$

Notice that the first term represents the ideal gas pressure, i.e. the pressure in the absence of intermolecular forces, at the same temperature T and volume V as those of the real gas.

Turning to the second term, we notice that at low temperatures the long-range attractive forces dominate and tend to reduce the gas pressure as compared to the ideal gas value at the same T and V. Thus, the second virial coefficient assumes negative values.

As the temperature increases, however, so does the kinetic energy of the molecules and, consequently, the number of collisions among them. As a result, the short-range repulsive forces become increasingly important and B becomes less negative until, at some temperature - known as the *Boyle* temperature - they balance out the attractive forces, and $B = 0$.

At still higher temperatures, the repulsive forces dominate, the gas pressure becomes larger than that of the ideal gas - at the same T and V - and B becomes positive.

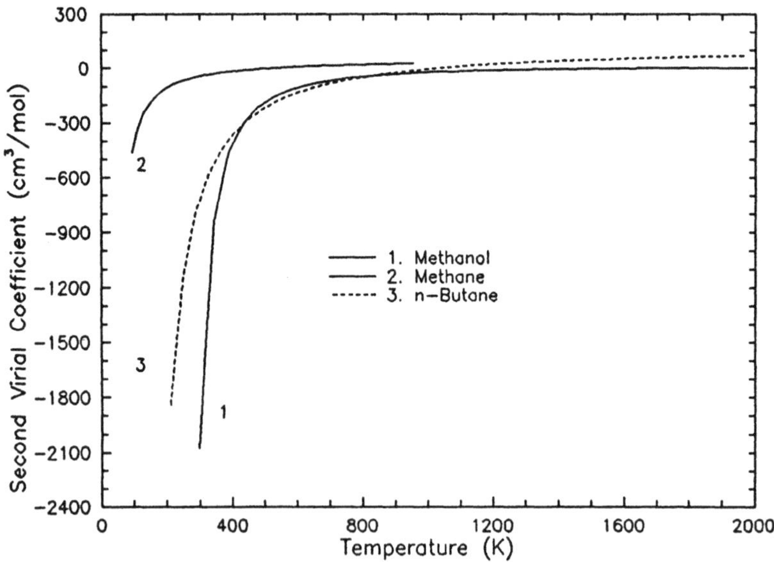

Figure 8.6 Typical values of the second virial coefficient (Daubert & Danner).

The largest compilation of B values is given by Dymond and Smith (1969), while Daubert and Danner present correlations of experimental B values as a function of temperature for a large number of pure fluids.

b. Estimation of Second Virial Coefficients

Two empirical correlations for the estimation of second virial coefficients find extensive use: the one was developed by Tsonopoulos (1974; 1975), and the other by Hayden and O'Connell (1975).

The first is simpler and provides comparable results with the second. A computer subroutine is available, however, for the latter, along with an extensive data bank of required parameters (Prausnitz et al, 1980).

Estimation of B values is discussed in Examples 8.5 and 8.6, to which Undergraduate Students should proceed.

8.9.3 Example 8.3. The Relationship Among the Coefficients of the Volume and Pressure Series Expansions

Solving Eq. 8.9.1 for P we obtain:

$$P = z\frac{RT}{V} = \frac{RT}{V} + \frac{BRT}{V^2} + \frac{CRT}{V^3} + \ldots \tag{A}$$

Introduction of Eq. (A) into Eq. 8.9.2 yields:

$$z = 1 + \frac{B'RT}{V} + \frac{B'BRT}{V^2} + \frac{B'CRT}{V^3} + \ldots +$$

$$+ C'\left[\frac{(RT)^2}{V^2} + \text{terms involving } (\frac{1}{V^n}),\ n>2\right] + \ldots$$

$$= 1 + \frac{B'RT}{V} + \frac{B'BRT + C'(RT)^2}{V^2} + \text{ terms involving } (\frac{1}{V^n}),\ n>2$$

Comparison of the last equation with Eq. 8.9.1 yields Eqs 8.9.3 and 8.9.4.

8.9.4 Example 8.4

a. For the virial equation show that:

$$B = \lim_{\rho \to 0}\left(\frac{z-1}{\rho}\right) \quad ; \quad C = \lim_{\rho \to 0}\left(\frac{z-1-B\rho}{\rho^2}\right)$$

where ρ is the molar density.

b. Using the density and compressibility factor data for steam presented in Table 8.E.4, calculate the B and C values for steam at 200°C.

a. Written in terms of density, the volume series of the virial equation becomes:

$$z = 1 + B\rho + C\rho^2 + D\rho^3 + \ldots.$$

and: $$B = \frac{(z-1)}{\rho} - C\rho - D\rho^2 - \ldots \quad ; \quad C = \frac{(z-1-B\rho)}{\rho^2} - D\rho - \ldots$$

Hence: $$B = \lim_{\rho \to 0}\left(\frac{z-1}{\rho}\right) \quad ; \quad C = \lim_{\rho \to 0}\left(\frac{z-1-B\rho}{\rho^2}\right)$$

Table 8.E.4 Compressibility Factors, Molar Densities, and F_b and F_c Values for Steam at 200°C

P(kPa)	z	ρ(mol/l)	F_b	F_c
1	1.0000	0.0003	-0.1939	29.0460
10	0.9993	0.0025	-0.2894	-34.6309
20	0.9983	0.0051	-0.3244	-24.1791
30	0.9985	0.0076	-0.2013	-0.0049
40	0.9979	0.0102	-0.2017	-0.0409
50	0.9974	0.0127	-0.2015	-0.0183
100	0.9948	0.0256	-0.2026	-0.0535
150	0.9922	0.0384	-0.2027	-0.0361
200	0.9896	0.0514	-0.2033	-0.0391
250	0.9869	0.0644	-0.2038	-0.0391
300	0.9842	0.0775	-0.2042	-0.0381
350	0.9814	0.0907	-0.2047	-0.0378
400	0.9787	0.1039	-0.2053	-0.0382
450	0.9759	0.1172	-0.2056	-0.0371
500	0.9731	0.1306	-0.2062	-0.0375
550	0.9702	0.1441	-0.2068	-0.0380
600	0.9673	0.1577	-0.2073	-0.0380
650	0.9644	0.1713	-0.2078	-0.0379
700	0.9615	0.1851	-0.2083	-0.0377
750	0.9584	0.1989	-0.2089	-0.0382
800	0.9554	0.2129	-0.2093	-0.0378
850	0.9524	0.2269	-0.2099	-0.0381
900	0.9493	0.2410	-0.2104	-0.0378
950	0.9462	0.2552	-0.2109	-0.0377
1000	0.9430	0.2696	-0.2114	-0.0374

b. Values of F_b = $[(z-1)/\rho]$ are presented in Table 8.E.4 and are plotted versus density in Figure 8.E.4a. Notice the scattering at low densities (the first three points fall outside the scale of the plot), which creates uncertainty in the extrapolation to $\rho = 0$. A value of B = -0.201 l/mol is accepted, which is in reasonable agreement with the value of -203.0±1.9 cm^3/mol given by Eubank et al (1988) using their very accurate PVT measurements.

Using B = -0.201 l/mol, the values of the quantity $F_c = [(z-1-B\rho)/\rho^2]$ presented in Table 8.E.4 and plotted versus ρ in Figure 8.E.4b are obtained. The uncertainty here is more pronounced (the first three points fall again outside the scale). A value of C = -38 cm^6/mol^2 is accepted, but rather arbitrarily. (Eubank et al: C = -31.8±6 cm^6/mol^2.)

Notice, finally, that the PVT values in the Steam Tables have been developed by fitting a very large number of reliable experimental data to an accurate equation of state. The uncertainty, therefore, in the two coefficients reflects their sensitivity to low density data, which increases from B to C and so on. This also explains the uncertainty in the recommended values for B by Daubert and Danner; and the fact that values for third virial coefficients are reported in the literature for very few compounds.

8.9.5 Example 8.5. Estimation of Second Virial Coefficients: Nonpolar Compounds

Tsonopoulos (1974; 1975) proposed the following Pitzer-type expression for the estimation of second virial coefficients of nonpolar compounds:

$$\frac{BP_c}{RT_c} = f^{(0)} + \omega f^{(1)} \tag{A}$$

$$f^{(0)} = 0.1445 - \frac{0.330}{T_r} - \frac{0.1385}{T_r^{\,2}} - \frac{0.0121}{T_r^{\,3}} - \frac{0.000607}{T_r^{\,8}}$$

$$f^{(1)} = 0.0637 + \frac{0.331}{T_r^{\,2}} - \frac{0.423}{T_r^{\,3}} - \frac{0.008}{T_r^{\,8}}$$

A simpler expression is given by Abbott in Smith and Van Ness (1975):

$$\frac{BP_c}{RT_c} = B^{(0)} + \omega B^{(1)} \tag{B}$$

$$B^{(0)} = 0.083 - \frac{0.422}{T_r^{1.6}}$$

$$B^{(1)} = 0.139 - \frac{0.172}{T_r^{4.2}}$$

Calculate B values, using these two expressions, for n-butane (nC_4) from 250 K

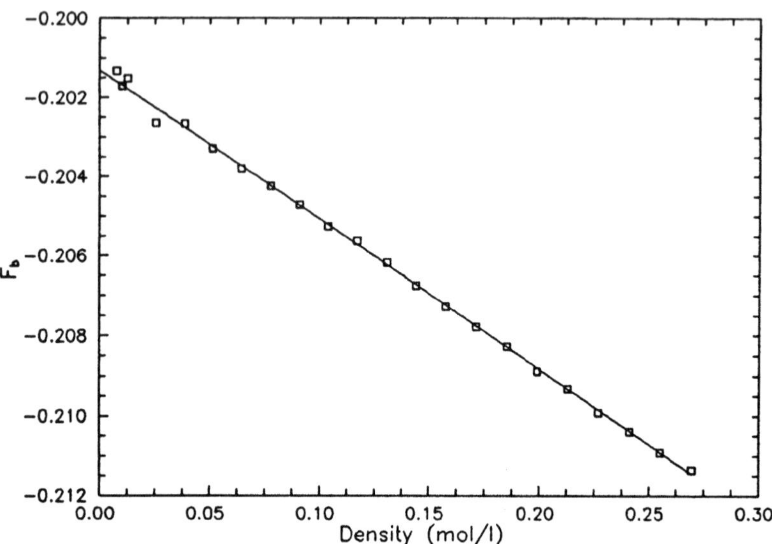

Figure 8.E.4a Evaluation of the second virial coefficient for steam at 200°C.

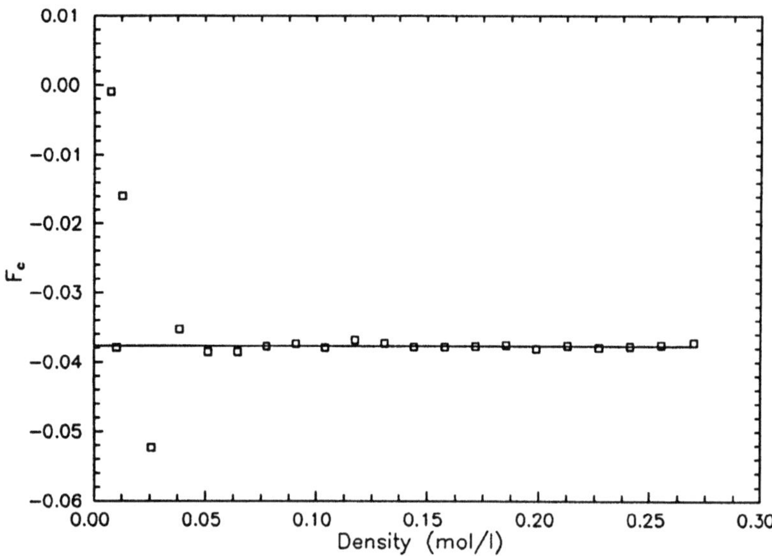

Figure 8.E.4b Evaluation of the third virial coefficient for steam at 200°C.

to 560 K, and compare them to the recommended values by Dymond (1985) presented in Table 8.E.5.

Given: T_c = 425.18 K; P_c = 3.7969·10^6 Pa; ω = 0.1931.

The following observations can be made from the prediction results, presented in Table 8.E.5 and obtained with the Program PVT presented in Appendix E.

1. The Tsonopoulos correlation gives results that are within the estimated uncertainty given by Dymond. For example, Dymond gives a value of B = -1170 ± 30 cm^3/mol at 250 K, while the predicted value is -1147 cm^3/mol.

2. The correlation of Abbott gives very good results considering its simplicity, being a little inferior to the Tsonopoulos correlation at the low T_r values only.

3. Prediction results for methane with the Abbott correlation are also presented in Table 8.E.5. Their good quality for the whole T_r range suggests that, since for methane ω = 0.008, the poorer predictions of the method for nC_4 at low T_r values are due to the $B^{(1)}$ term.

4. Notice, finally, that both the Tsonopoulos and the Abbott correlations represent an application of the three-parameter corresponding states theory to the second virial coefficient.

Table 8.E.5 Prediction of Second Virial Coefficients

	n-Butane				Methane			
		B^*					B^*	
T(K)	T_r	Expt.	Tson.	Abbott	T(K)	T_r	Expt.	Abbott
250	0.588	-1170	-1147	-1104	110	0.557	-330	-327
260	0.611	-1050	-1026	-1005	120	0.630	-273	-280
270	0.635	-950	-927	-918	130	0.682	-235	-242
280	0.659	-862	-843	-843	140	0.735	-207	-211
300	0.706	-722	-710	-718	160	0.839	-161	-164
320	0.753	-620	-609	-619	180	0.944	-129	-131
340	0.800	-535	-530	-539	200	1.049	-105	-106
360	0.847	-472	-465	-473	240	1.259	-75	-72
380	0.894	-419	-412	-418	280	1.469	-51	-50
400	0.941	-370	-367	-371	320	1.679	-34	-34
420	0.988	-332	-329	-331	360	1.889	-23	-23
440	1.035	-298	-296	-296	400	2.099	-15	-15
470	1.105	-256	-254	-253	440	2.308	-8.5	-9.0
500	1.176	-219	-220	-217	480	2.518	-2.5	-4.1
530	1.246	-188	-191	-186	520	2.728	2.0	-0.1
560	1.317	-164	-166	-160	560	2.938	5.5	3.2
					600	3.148	8.5	5.9

(*) B in cm^3/mol

8.9.6 Example 8.6. Estimation of Second Virial Coefficients: Polar Compounds

For polar and especially hydrogen bonded compounds, we cannot expect the three-parameter corresponding states correlation for B to give successful results. Indeed, Tsonopoulos adds to Eq.(A) of Example 8.5 one more term:

$$\frac{BP_c}{RT_c} = f^{(0)} + \omega f^{(1)} + f^{(2)} \tag{A}$$

where:

$$f^{(2)} = \frac{a}{T_r^6} - \frac{b}{T_r^8} \tag{B}$$

Eq.(A) provides excellent representation of existing second virial coefficient data but its prediction capabilities are limited. For non-hydrogen bonded compounds b assumes the value of zero, and an approximate expression for a is given in the 1974 article for ketones and ethers. Values of a and b for some hydrogen bonded compounds are also given.

8.9.7 Comments

1.The virial equation is applicable to the vapor and gaseous phases only (why?).

2.In the typical application, one needs to calculate the volume from the known values of the pressure and temperature, and Eq.8.9.2 is more convenient for this purpose.

3.A comparison of the two expansions for the compressibility factor indicates, however, that when truncated after C, the volume series provides better results than the pressure series; but when truncated after B, they provide comparable results, with the pressure series actually doing better at higher densities due to compensation of errors (Chueh and Prausnitz, 1967).

4.As a rough rule the virial equation, when truncated after B, is valid in the pressure range (Prausnitz et al):

$$P \leq \frac{T}{2}\frac{P_c}{T_c} = T_r\frac{P_c}{2} \tag{8.9.5}$$

5.Both the Tsonopoulos and Hayden-O'Connell correlations provide good predictions for second virial coefficients, especially for nonpolar compounds.

6.A main advantage of the virial equation, as compared to the other EoS, is that there is an exact relationship between $B(mixture)$ and the B values of the mixture components and their pairs, as we will see in Chapter 11. It finds, thus, extensive application in the description of the vapor phase nonideality in distillation or gas absorption design up to moderate pressures.

8.10 Cubic Equations of State

8.10.1 Development of a Cubic Equation of State

The most rigorous way of developing an equation of state would be through Statistical Mechanics. Unfortunately the required to this purpose information is, as we will see in Chapter 17, rather limited. Equations of state are developed, therefore, in a rather empirical - or, at best, semi-empirical - fashion.

The most popular equations of state are the cubic ones and are developed taking into account the following considerations:

1.The equation of state should reflect the intermolecular forces, repulsive and attractive, that play a major role in determining the volumetric behavior of a fluid. Thus, a typical cubic equation of state is of the form:

$$P = P_R + P_A \tag{8.10.1}$$

where the first term accounts for the repulsive forces, and the second for the attractive ones.

2.We should expect that the repulsive forces would have a positive contribution to pressure ($P_R > 0$), and the attractive forces a negative one ($P_A < 0$). This is demonstrated with the first equation of state proposed, that of van der Waals (vdW):

$$P = \frac{RT}{V-b} - \frac{a}{V^2} \tag{8.10.2}$$

where, b and a are positive constants and V is the molar volume.

3.Following the principle of corresponding states, the equation of state should be able to provide reasonably good results for nonpolar substances with two, and still better, with three adjustable parameters. Indeed, the typical cubic EoS contain two or three parameters. For very accurate results, however, non-cubic equations of state with a large number of parameters are used, such as the Benedict-Webb-Rubin one.

4.The equation of state should meet the experimentally observed behavior of pure fluids at the critical point, i.e. Eqs 8.5.1 and 8.5.2.

The last observation is used to express the equation of state parameters in terms of the critical properties - as shown in the next Example for the vdW EoS; and in Example 8.8, to demonstrate the relationship between this EoS and the corresponding states principle.

8.10.2 Example 8.7

Develop expressions for a and b in the van der Walls equation of state in terms of P_c and T_c.

From Eqs 8.10.2, 8.5.1, and 8.5.2, we have:

$$\frac{RT_c}{(V_c-b)} - \frac{a}{V_c^2} = P_c \tag{A}$$

$$-\frac{RT_c}{(V_c-b)^2} + \frac{2a}{V_c^3} = 0 \tag{B}$$

$$\frac{2RT_c}{(V_c-b)^3} - \frac{6a}{V_c^4} = 0 \tag{C}$$

From Eqs (B) and (C):

$$V_c = 3b \tag{D}$$

Introduction of Eq.(D) into Eqs (A) and (B) results in a system of two equations, involving P_c and T_c only, with two unknowns, a and b. Solution of this system yields:

$$b = \frac{RT_c}{8P_c} \tag{E}$$

$$a = \frac{27}{64}\frac{(RT_c)^2}{P_c} \tag{F}$$

8.10.3 Example 8.8

Show that the van der Waals equation of state represents an analytical expression of the two-parameter corresponding states principle.

Eq.8.10.2 can be rearranged to the following cubic in volume expression:

$$V^3 - \left(b + \frac{RT}{P}\right)V^2 + \frac{a}{P}V - \frac{ab}{P} = 0 \tag{A}$$

Introduction of $V = (zRT/P)$, and of b and a from Eqs (E) and (F) of Example 8.7, yields after algebraic manipulation the following equation:

$$z^3 - \left(1 + \frac{P_r}{8T_r}\right)z^2 + \left(\frac{27}{64}\frac{P_r}{T_r^2}\right)z - \frac{27}{512}\frac{P_r^2}{T_r^3} = 0 \tag{B}$$

Comments

1. Because the van der Waals EoS is a cubic one when solved for volume, it is referred to as a 'cubic' EoS. The same name applies to a large category of EoS that are also cubic in volume.

2. According to Eq.(B) all compounds have the same z value at the same reduced temperature and pressure and, consequently, the vdW EoS in its 'generalized' form, represents an analytical expression of the two-parameter corresponding states principle.

3. The same applies to all cubic EoS that contain two parameters, while those containing three represent an analytical expression of the three-parameter CSP.

4. The vdW EoS does not provide an accurate description of the volumetric behavior of fluids. Improved results are obtained with the cubic EoS discussed next.

8.10.4 The Redlich-Kwong Equation of State

The early purpose of equations of state was the description of the volumetric behavior of fluids, especially in the vapor phase. The most popular and successful among them was that proposed by Redlich and Kwong (RK, 1949):

$$P = \frac{RT}{V-b} - \frac{a}{T^{0.5}V(V+b)} \tag{8.10.3}$$

Using the same approach as with the vdW EoS (Example 8.7), we arrive at the following expressions for a and b:

$$a = 0.42748\frac{R^2T_c^{2.5}}{P_c} \tag{8.10.4a}$$

$$b = 0.08664\frac{RT_c}{P_c} \tag{8.10.4b}$$

We explore the prediction of the volumetric behavior of a pure fluid with the RK EoS in the next Example.

8.10.5 Example 8.9

Using the RK EoS, calculate:

a. the molar volume of i-butane (iC_4) at 420 K and 2 MPa. (Exp. value: 1411.2 cm^3/mol.)

b. the saturated vapor and liquid molar volumes of iC_4 at 380 K and 2.25 MPa. (Exp. values: $V^s(l)$ = 140.8 cm^3/mol; $V^s(v)$ = 866.1 cm^3/mol.)

From Eqs 8.10.4a and 8.10.4b (T_c = 408.1 K; P_c = 36.48 bar):

$a = 2.725 \cdot 10^7$ MPa $K^{0.5}cm^6/mol^2$ and b = 80.58 cm^3/mol.

Introduction of these values into Eq.8.10.3 and solution of the resulting cubic equation gives:

* T = 420 K; V = 1404.5 cm^3/mol
* T = 380 K; V_1 = 916.1; V_2 = 313.8; V_3 = 174.0; cm^3/mol

Graphical solution is shown in Figure 8.E.9.

The value of 1404.5 cm^3/mol at 420 K is very close to the experimental value with an error of -0.5%.

Of the three volume values at the subcritical temperature of 380 K, two are of interest:

* the largest one that represents the volume of the vapor, with an error of 5.8%; and
* the smallest one that corresponds to the liquid volume, with a substantial error of 23.7%.

8.10.6 The SRK, PR, and vdW-711 Equations of State

Following the pioneering work of Wilson (1964), Zudkevitch and Joffe (1966), and Soave (1972), cubic EoS have become progressively the main tool for phase equilibrium calculations, especially for nonpolar systems. With this objective in mind, the most important requirement for cubic EoS becomes their ability to accurately represent:

a. Vapor pressures, and

b. Saturated liquid volumes,

with the first being the most important.

We will discuss the development and performance, with respect to these objectives, of cubic EoS in Chapter 10. Here we will present three of the simplest ones, the:

* Soave-Redlich-Kwong (SRK, Soave, 1972);
* Peng-Robinson (PR, Peng and Robinson, 1976); and

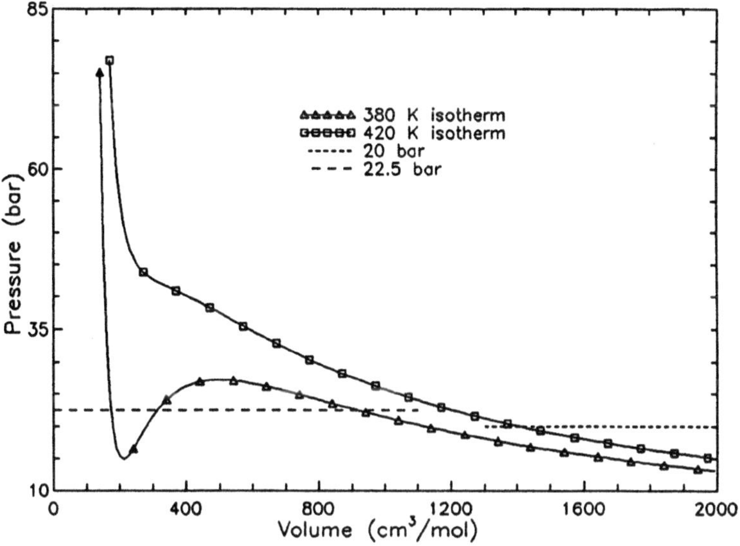

Figure 8.E.9 Graphical solution for Example 8.9 using the RK EoS.

* van der Waals-711 (vdW-711), developed by this author and coworkers (Salerno et al, 1986; Watson et al, 1986; Czerwienski et al, 1988);

and discuss their performance in the estimation of the *PVT* behavior of pure fluids, that we are concerned with in this Chapter.

The three EoS are presented in Table 8.3. Their main characteristic is that, following Soave, the parameter a in the attractive term is made a function of T_r and ω:

$$a = a_c[1+m(1-\sqrt{T_r})]^2 \tag{8.10.5}$$

$$m = d_0 + d_1\omega + d_2\omega^2 \tag{8.10.6}$$

where a_c is the value of a at the critical point calculated, along with the value of b, following the same procedure as with the vdW EoS (Example 8.7). All necessary information using these three EoS is included in Table 8.3.

8.10.7 Example 8.10

Repeat Example 8.9 using the SRK, PR, and vdW-711 EoS (ω = 0.176).

Table 8.3 The SRK, PR, and vdW-711 Equations of State

SRK $$P = \frac{RT}{V-b} - \frac{a}{V(V+b)}$$

PR $$P = \frac{RT}{V-b} - \frac{a}{V(V+b)+b(V-b)}$$

vdW-711 $$P = \frac{RT}{V+t-b} - \frac{a}{(V+t)^2}$$

$$a = \alpha a_c \qquad a_c = a_0 \frac{(RT_c)^2}{P_c}$$

$$\alpha = [1 + m(1 - T_r^{0.5})]^2$$

$$m = d_0 + d_1 \omega + d_2 \omega^2$$

$$b = b_0 \frac{RT_c}{P_c}$$

	SRK[1]	PR	vdW-711
a_0:	0.42748	0.45724	27/64
b_0:	0.08664	0.07780	1/8
d_0:	0.48508	0.37464	0.48553
d_1:	1.55171	1.54226	1.62400
d_2:	-0.15613	-0.26992	-0.21884

$$t = t_0 + (t_c - t_0)\exp[\beta|(1 - T_r)|]$$

$$t_0 = \frac{RT_c}{P_c}(0.0348 + 0.0937\,\omega - 0.1661\,\omega^2 + 0.1250\,\omega^3)$$

$$t_c = \frac{RT_c}{P_c}\left(\frac{3}{8} - z_c\right)$$

$$z_c = 0.2890 - 0.0701\,\omega - 0.0207\,\omega^2$$

$$\beta = -7.35 - 24.52\,\omega + 9.20\,\omega^2$$

[1] Values for d_0, d_1, and d_2 from Graboski and Daubert (1978)

Table 8.E.10 Summary of Results for Example 8.10

T	P	V	% Error in V		
K	MPa	cm^3/mol	SRK	PR	vdW-711
420	2.00	1411.2	-0.2	-2.5	-2.9
380	2.25	866.1	3.4	-0.3	0.5
380	2.25	140.8	17.8	4.3	-1.8

The results, obtained using the Program PVT of Appendix E, are summarized in Table 8.E.10.

8.10.8 Comments

1. At supercritical temperatures, all cubic EoS when solved for volume have one real root and two complex ones. [Two-parameter cubic EoS may, however, yield three real roots in the supercritical region (Gomez-Nieto, 1979; Lawal, 1987), but only the largest of them has a physical meaning.]
2. At temperatures below the critical, cubic EoS have one or three real roots: in the latter case, the largest corresponds to the vapor phase; the smallest to the liquid. The intermediate has no physical meaning.
3. This is better demonstrated in Figure 8.E.9, where plots of P versus V - generated by the RK EoS for iC_4 using the a and b values of Example 8.9 - are presented at 420 K and 380 K. Notice that at the supercritical temperature of 420 K only one root is possible; but three, at the subcritical T = 380 K.
4. Cubic EoS, where the attractive parameter a is not a function of temperature, such as the RK and the original vdW, fail to give reliable predictions of vapor pressures.
5. On the other hand, where a is made an appropriate function of temperature, for example in the SRK, PR, and vdW-711 EoS, very good predictions of vapor pressures are obtained for nonpolar compounds as we will see in Chapter 10.
6. The SRK, PR and vdW-711 EoS, provide also good predictions of saturation vapor volumes for nonpolar compounds as Example 8.10 demonstrates. For saturated liquid volumes, however, only the vdW-711 EoS provides consistently good results because it includes the 'translation' factor t. This is also discussed in Chapter 10.
7. A two-parameter cubic EoS gives the same z_c value for all compounds. For example, the vdW EoS gives z_c = 0.375, and the PR, z_c = 0.307. (Is this necessarily so? See Problem 8.52.) This may not, how-

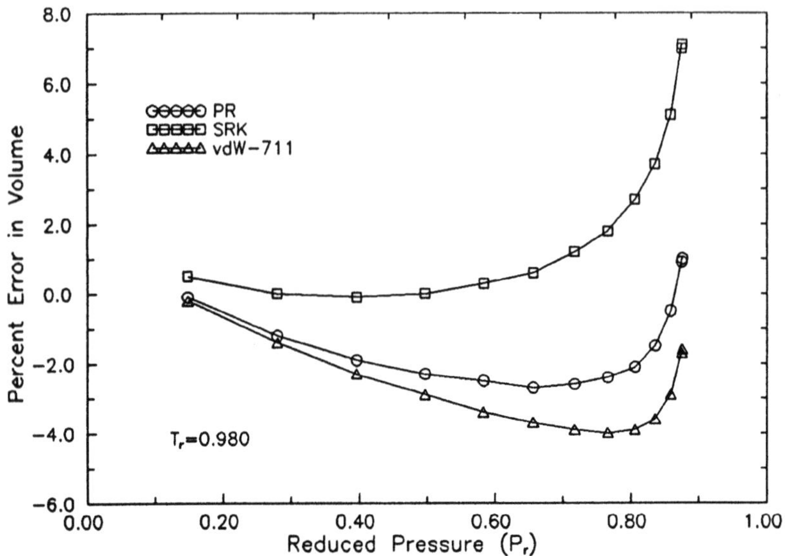

Figure 8.7a Prediction of molar volumes for i-butane at T_r = 0.980.

ever, be the case for three-parameter EoS. For the vdW-711 EoS, thus, z_c is an input quantity and is reproduced at the critical point.

8.Typical predictions of vapor and gas phase volumetric behavior with cubic EoS for nonpolar compounds are presented in Figure 8.7: at T_r < 1.0 (Figure 8.7a), reasonably accurate results are obtained; at T_r equal to or just over 1.0 (Figure 8.7b), large errors are observed at P_r > 1.0, which decline with increasing T_r values (Figure 8.7c). (Exp. data from Goodwin and Haynes.)

9.The large errors at T_r equal to, or just over 1.0, and pressures larger than the critical (Figure 8.7b), are due to the shape of the *PVT* isotherms in this range, as suggested by that of the critical one in Figure 8.1. Notice the sensitivity of volume to pressure in this range.

10.For a comprehensive discussion of the limitations of cubic EoS see Martin (1967) and Abbott (1973).

11.For some comments on the development of the van der Waals and Redlich-Kwong equations of state, see Chapter 4.

Note.The results for Figure 8.7 were obtained using the Program PVT of Appendix E.

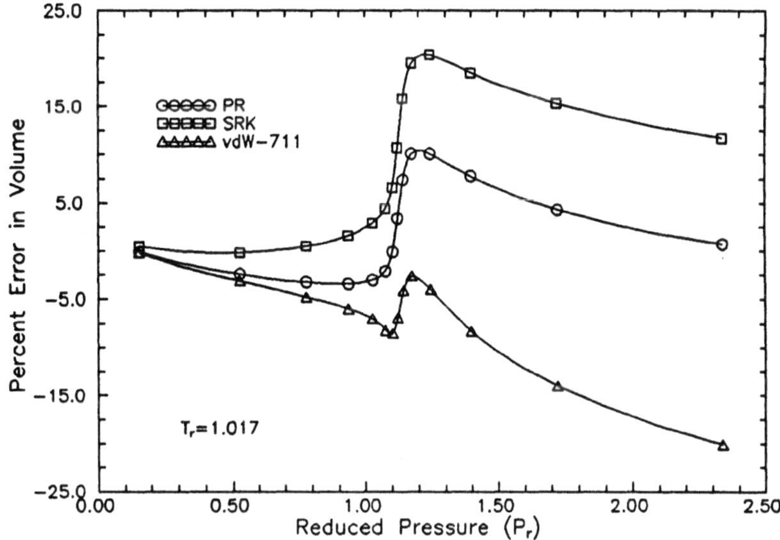

Figure 8.7b Prediction of molar volumes for i-butane at T_r = 1.017.

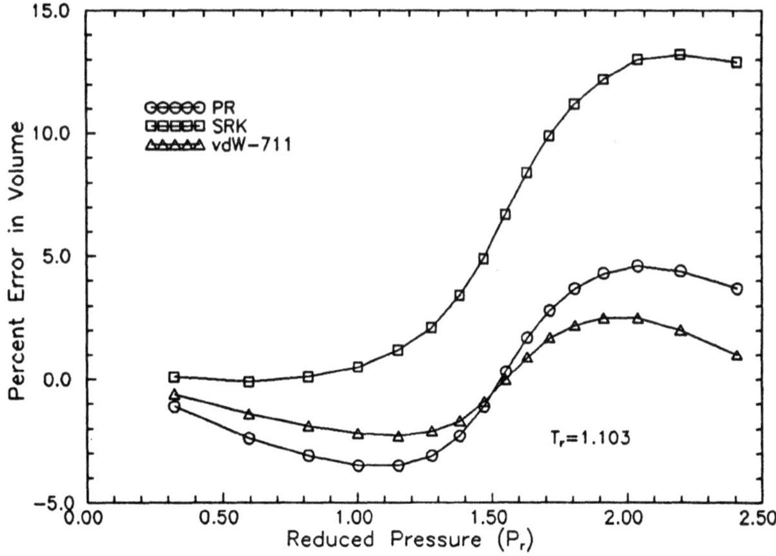

Figure 8.7c Prediction of molar volumes for i-butane at T_r = 1.103.

8.11 The Benedict-Webb-Rubin (BWR) Equation of State

One of the oldest, but still used, non-cubic equations of state is that proposed by Benedict, Webb and Rubin (1940; 1942; 1951). It contains eight parameters that are determined from experimental data. Parameter values for several compounds are given by Reid et al.

The BWR EoS provides excellent description of the volumetric behavior of pure fluids, but has two main drawbacks:

First, extensive *PVT* data are needed for the evaluation of the eight parameters. Such data, as we have mentioned, are available for a few compounds only. A generalized form, developed by Lee and Kesler is available, however. (This form, by the way, has been used for the development of the Lee-Kesler tables used in combination with Pitzer's corresponding states approach.)

Second, solution of the equation for volume is cumbersome and computer-time consuming.

8.12 Recommendations for the Estimation of the Volumetric Behavior of Pure Fluids

1.Nonpolar and weakly polar vapors and gases

a.Below the critical temperature, very good results are obtained with the cubic EoS discussed here, especially with the PR and vdW-711; and within the range of Eq.8.9.5, with the virial equation truncated after *B*. Typical errors are around 1-3%.

b.At the critical and somewhat higher temperatures and for supercritical pressures, all cubic EoS give unreliable results. In this range, the corresponding states approach of Pitzer with the Lee-Kesler tables, or the generalized BWR EoS, should be used.

c.As the temperature increases beyond the critical, so does the pressure range for which these cubic EoS give reasonable accuracy.

2.Polar vapors and gases

a.In the range of applicability of the virial equation truncated after *B*, good results are obtained depending, of course, on the accuracy of the estimated second virial coefficient values.

b. Outside this range, the modified for polar compounds vdW-711 EoS (Androulakis et al, 1989) and, for very high pressures, the Pitzer correlation with the Lee-Kesler tables, should be used.

3. Liquids

The most successful correlation is that of Hankinson and Thomson (1979) which provides excellent predictions of liquid densities. The authors report an average absolute error of 0.375% for 4508 data points on 190 different nonpolar and polar compounds and in the range $0.25 < T_r < 0.98$. The vdW-711 EoS, original and modified for polar compounds, provides also very good results for saturated liquid volumes.

8.13 Vapor Pressures

Of all physical property measurements, those dealing with vapor pressures are the most typical. Such information is available for a large number of commonly encountered compounds either as measured values or as correlations of them (Daubert and Danner; Reid et al).

For less common compounds, however, limited - or no - vapor pressure data are available. We discuss, therefore, next the correlation of a small number of data points for inter- and extra-polation purposes; and a corresponding states method for estimation purposes.

8.13.1 Correlation of Vapor Pressure Data

The starting point for this purpose is the Clapeyron equation, developed in Chapter 4 (commonly referred to as the Clausius-Clapeyron equation):

$$\frac{dP^s}{dT} = \frac{\Delta H_V}{T\Delta V} = \frac{\Delta H_V}{[(RT^2/P^s)\,\Delta z]} \tag{8.13.1}$$

where, dP^s/dT: slope of the vapor pressure curve;

ΔH_V: enthalpy of vaporization per mole;

ΔV: difference between saturated vapor and liquid molar volumes;

Δz: difference between saturated vapor and liquid compressibility factors.

All quantities are evaluated at the temperature T.

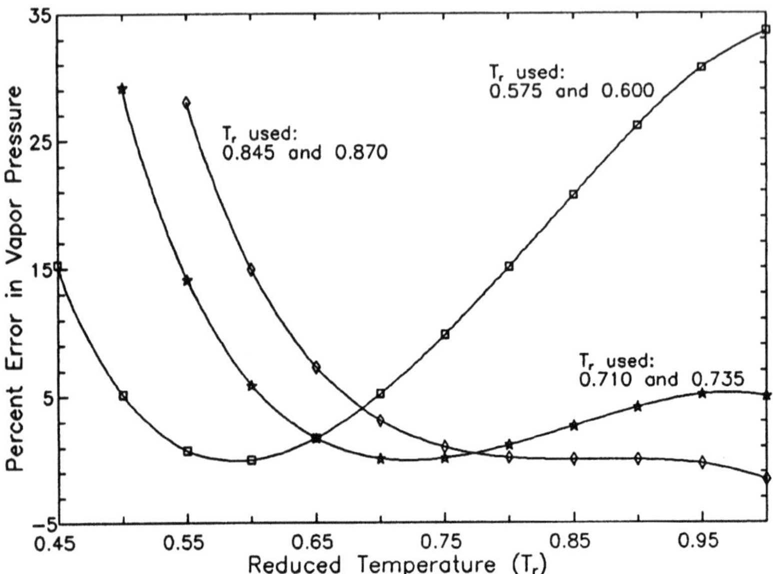

Figure 8.8. Prediction of vapor pressure for nC_7 using $lnP_r = a\text{-}b/T_r$ with a and b calculated at the designated T_r values.

If we make the simple assumption that the group $(\Delta H_V/\Delta z)$ is independent of temperature, then:

$$\ln P^s = a - \frac{b}{T} \tag{8.13.2}$$

where, a is the constant of integration, and $b = \Delta H_V/R\Delta z$.

Eq.8.13.2 suggests that if a limited number of vapor pressure data is available, a linear plot of $\ln P^s$ versus $(1/T)$ can be used for inter- and extrapolation purposes.

Surprisingly, considering the assumption involved, Eq.8.13.2 provides a fairly good approximation for small temperature ranges. Except near the critical point, ΔH_V and Δz are both weak functions of temperature and, since both decrease with an increase in temperature, there is a compensatory effect.

The temperature range must be small, however, as demonstrated in Figure 8.8 for the case of n-heptane (nC_7), where the values of the parameters a and b are evaluated using vapor pressure data at two close T_r values: extrapolation of ± 0.05 outside this T_r range - about ± 25°C for n-heptane - can lead to rapidly increasing errors.(See also Problem 8.34.)

A better way to use a limited number of experimental data, however, is to fit them to the Antoine equation:

$$\ln P^s = A - \frac{B}{C+T} \tag{8.13.3}$$

When $C = 0$, Eq.8.13.3 becomes Eq.8.13.2. Values of A, B, and C, obtained by fitting experimental vapor pressure data, are given - among others - by Reid et al in their 1974 edition. Values for some selected compounds are given in Appendix C. Notice, however, that while the temperature range of applicability for the Antoine equation is larger than that for Eq.8.13.2, it is still limited.

8.13.2 Estimation of Vapor Pressures

The Pitzer three-parameter corresponding states expression:

$$\ln(P^s/P_c) = f^{(0)} + \omega f^{(1)} \tag{8.13.4}$$

with $f^{(0)}$ and $f^{(1)}$ expressed in analytical form by Lee and Kesler:

$$f^{(0)} = 5.92714 - 6.09648/T_r - 1.28862 \ln T_r + 0.169347\, T_r^6$$

$$f^{(1)} = 15.2518 - 15.6875/T_r - 13.4721 \ln T_r + 0.43577\, T_r^6$$

gives reasonably accurate estimates of vapor pressures.

8.13.3 Comments

1.Vapor pressure data tend to be, in general, accurate at pressures above 0.1 atm. At lower pressures the uncertainty increases. Thus at 1 mm Hg and below, errors of 10% and higher are typical. For a method of screening a large number of vapor pressure data and correlating them with T_r see Wilsak and Thodos (1984); and for the latter only, Reid et al.

2.Equation 8.13.4 is applicable, in principle, to nonpolar compounds and gives very good results as demonstrated in Table 8.7 for n-butane. Notice that good results are obtained down to very low pressures.

3.But even for polar compounds reasonable estimates are obtained, however, as demonstrated with acetone (μ = 2.8 Debye), also included in Table 8.7. Notice that the accuracy improves drastically at higher temperatures where the impact of dipole-dipole interactions diminishes, as we have seen in Chapter 7.

Table 8.7 Prediction of Vapor Pressures Using the Corresponding States Principle, Eq.8.13.4

n-Butane			Acetone		
T (K)	$P^{s(1)}$ (atm)	Error (%)	T (K)	$P^{s(2)}$ (atm)	Error (%)
149.98	$0.86 \cdot 10^{-4}$	4.1	259.2	0.0421	-11.3
199.98	0.0192	-1.7	273.4	0.0937	-9.2
219.98	0.0770	-1.0	290.1	0.2124	-6.4
270.37	0.9173	0.9	320.5	0.7347	-2.6
327.59	5.5076	1.3	390.3	5.5803	1.4
383.15	18.25	0.6	446.4	17.447	0.7
409.98	29.11	0.6	470.6	26.276	0.2

(1) Salerno et al

(2) Ambrose et al (1974)

4.In obtaining the results of Table 8.7 the ω values of Appendix C were used. Reid et al suggest that ω should be calculated from an empirical correlation.

5.Vapor pressure predictions can also be obtained through equations of state. Very good results are obtained for nonpolar compounds with the SRK, PR and vdW-711 EoS discussed here, provided that accurate values for T_c, P_c and ω are available (see Chapter 10).

8.14 Enthalpies of Vaporization

The amount of energy required to convert, at a given temperature, an amount of saturated liquid into saturated vapor is called the *enthalpy* - or *heat - of vaporization*, ΔH_V:

$$\Delta H_V = H_{(saturated\ vapor)} - H_{(saturated\ liquid)} \tag{8.14.1}$$

For a pure compound, ΔH_V is a function of temperature as indicated by the two typical cases shown in Figure 8.9. It decreases with increasing temperature, the decrease becoming more pronounced as the critical point is approached.

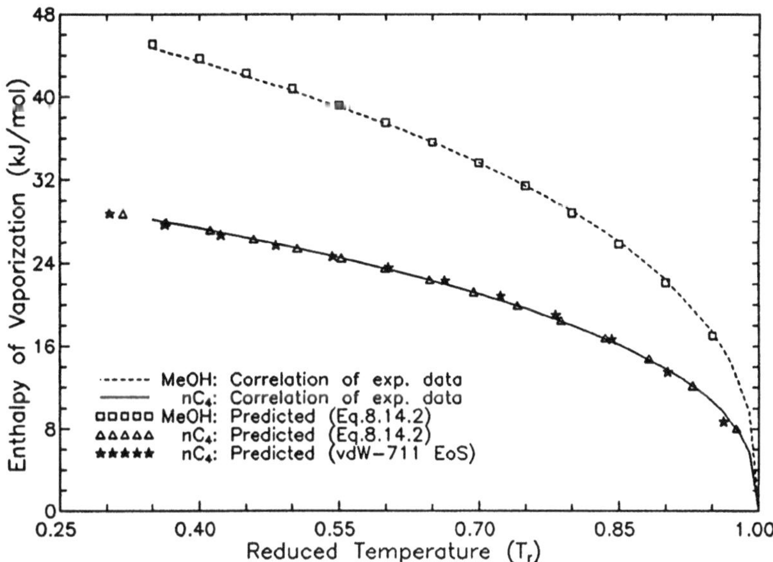

Figure 8.9 Experimental and predicted enthalpies of vaporization.

8.14.1 Determination and Prediction

Enthalpies of vaporization are obtained with two approaches:

* experimental measurements
* empirical correlations

Experimental values are determined either directly from calorimetric measurements, or indirectly through the Clausius-Clapeyron equation, using experimental data for the vapor pressure and the saturated molar volumes. Correlations of experimental data as a function of temperature for a large number of compounds is given by Daubert and Danner.

Prediction of enthalpies of vaporization is carried out with equations of state that give successful predictions of vapor pressures, such as the SRK, PR and vdW-711 ones; and with empirical correlations.

In the first approach, the EoS is used to evaluate the required information in the Clausius-Clapeyron equation. Very good results are obtained, as demonstrated with the typical ones for n-butane from the vdW-711 EoS in Figure 8.9. In the second approach empirical correlations are used which are also based on the Clausius-Clapeyron equation, with most of the difficulty involved in estimating the difference in saturation volumes.

For details see Reid et al.

A commonly encountered problem is the evaluation of the enthalpy of vaporization at some temperature T from its value at a given temperature, typically the normal boiling point.

The rigorous approach to this problem would call for use of Eq.8.13.1, a rather cumbersome task. Empirically, the commonly used expression for the estimation of the enthalpy of vaporization at some temperature T_1 from its value at temperature T_2 is that of Watson:

$$\frac{\Delta H_V(T_{r_1})}{\Delta H_V(T_{r_2})} = \left(\frac{1-T_{r_1}}{1-T_{r_2}}\right)^n \tag{8.14.2}$$

Even though n varies from compound to compound, a value of 0.38 gives satisfactory results between the normal boiling point, T_b, and the critical. Below T_b, ΔH_V increases more rapidly with decreasing temperature than predicted with a constant n. This variation of n with the reduced temperature T_r, at low T_r values, appears to be more serious with polar compounds than with nonpolar ones.

8.14.2 Example 8.11

The following enthalpies of vaporization are given at $T_r = 0.7$:
* methanol: 33.6 kJ/mol;
* n-butane: 21.0 kJ/mol.

Estimate their enthalpy of vaporization values as function of T_r.

The results obtained with Eq.8.14.2 are shown in Figure 8.9. Notice that very good results are obtained with $n = 0.38$ for the whole T_r range, which is rather surprising considering that at T_b: $T_r(nC_4) = 0.64$; $T_r(\text{methanol}) = 0.66$.

8.15 Heat Capacities

As discussed in Section 1.13, we define heat capacity as the amount of thermal energy required to raise the temperature of a given amount of substance by one degree:

C_P, if the addition of thermal energy takes place at constant pressure;

C_V, if the addition takes place at constant volume.

The difference, however, is significant for gases only.

8.15.1 Determination

Heat capacities are determined from calorimetric measurements. For gases, however, they are more often determined through a combination of Statistical Mechanics and spectroscopic measurements at atmospheric pressure. The reported values are designated as C_P^{o} and can be used as ideal gas heat capacities, C_P^{*}, since gases approach such behavior at atmospheric pressure.

Heat capacities are weak functions of pressure, especially for solids and liquids; but strong ones of temperature, as demonstrated in Figure 8.10 with the data for propane (see also Figure 2.4). Only liquid heat capacities are weak functions of temperature, below T_r of about 0.7.

Heat capacities - solid, liquid, and ideal gas - as functions of temperature are given by Daubert and Danner; for ideal gases as a function of temperature and for liquids at one temperature by Reid et al; and in the JANAF Tables (1971) and by Wicks and Block (1963) for a very large number of inorganic compounds.

Values for some selected compounds: gases, liquids, and solids, are presented in Appendix C.

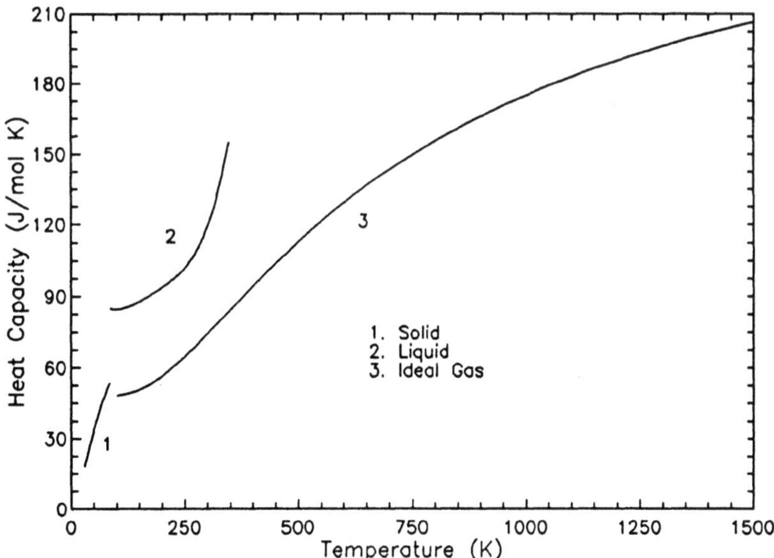

Figure 8.10 Heat capacities for propane (Daubert and Danner).

8.15.2 Example 8.12

Propane is compressed reversibly and adiabatically in a flow process from 1 bar and 20°C, to 10 bar.

a. Assuming that propane behaves as an ideal gas in this pressure range, calculate the work required per mole.

b. Provide an estimate of the error introduced by this assumption.

a. From the first law for a flow process:

$$Q = W_s + \Delta H$$

Since $Q = 0$, $W_s = -\Delta H = -\int C_P^* dT$

where: $C_P^* = a + bT + cT^2 + dT^3$, with values for the constants given in Appendix C, and the integration extends between the inlet temperature T_1 and the outlet one T_2. Hence:

$$\Delta H = a(T_2 - T_1) + (b/2)(T_2^2 - T_1^2) + (c/3)(T_2^3 - T_1^3) + (d/4)(T_2^4 - T_1^4)$$

To determine T_2 we notice that, since the process is reversible and adiabatic, the total entropy change of the propane from state 1 to state 2 is zero:

$$\Delta S = S_2 - S_1 = \int (C_P^*/T)dT - R\ln(P_2/P_1) = a\ln(T_2/T_1) + b(T_2 - T_1) +$$
$$+ (c/2)(T_2^2 - T_1^2) + (d/3)(T_2^3 - T_1^3) - R\ln(P_2/P_1) = 0$$

By trial and error $T_2 = 371$ K, and:

$$W_s = -6352 \text{ J}$$

b. At $P = 10$ bar and $T = 371$ K, we can use the virial equation, truncated after B, to determine the degree of non-ideality. From the Abbott correlation: $B^{(0)} = -0.336$; $B^{(1)} = -0.030$; and, $B = -246$ cm^3/mol. Thus:

$$z = 1 + BP_2/RT_2 = 0.92$$

Since at P_1 and T_1, $z = 1.0$, the error should be less than about 5%.

References

Abbott, M.M., 1973. *AIChE J.*, **19**, 596.

Ambrose, D., Spake, C.H.S., Townsend, R., 1974. *J. Chem. Thermodynamics*, **6**, 693.

Androulakis, I.P., Kalospiros, N.S., Tassios, D.P., 1989. *Fluid Phase Equilibria*, **45**, 135.

Benedict, M., Webb, G.B., Rubin, L.C.: *J. Chem. Phys.*, 1940, **8**, 334; 1942, **10**, 747; *Chem. Eng. Prog.*, 1951, **47**, 419.

Chueh, P.L., Prausnitz, J.M., 1967. *AIChE J.*, **13**, 1099.

Czerwienski, G.J., Tomasula, P., Tassios, D., 1988. *Fluid Phase Equilibria*, **42**, 63.

Daubert, T.E., Danner, R.P., 1985; 1986. *Data Compilation Tables of Properties of Pure Compounds*, AIChE, New York.
Dymond, J.H., 1985. *The Second International IUPAC Workshop*, Paris.
Dymond, J.H., Smith, E.B., 1969. *The Virial Coefficients of Gases*, Clarendon Press, Oxford.
Eubank, P.T., Joffrion, L.L., Patel, M.R., Warowny, W., 1988. *J. Chem. Thermodynamics*, **20**, 1009.
Gomez-Nieto, M., 1979. *Ind. Eng. Chem. Fundam.*, **18**, 187.
Goodwin, R.D., Haynes, W.M., 1982. *NBS Technical Note 1051*, National Bureau of Standards, Boulder, Colorado.
Graboski, M.S., Daubert, T.E., 1978. *Ind. Eng. Chem. Process Des. Dev.*, **17**, 443.
Hankinson, R.W., Thomson, G.H., 1979. *AIChE J.*, **25**, 653.
Hayden, J.C., O'Connell, J.P., 1975. *Ind. Eng. Chem. Process Des. Dev.*, **14**, 209.
JANAF Thermochemical Tables, 2nd Ed., NSRDS-NBS 37, 1971.
Lawal, A.S., 1987. *Ind. Eng. Chem. Res.*, **26**, 859.
Lee, B.J., Kesler, M.G., 1975. *AIChE J.*, **21**, 510.
Martin, J.J., 1967. *Ind. Eng. Chem.*, **59**, 12, 34.
Peng, D.Y., Robinson, D.B., 1976. *Ind. Eng. Chem. Fundam.*, **15**, 59.
Pitzer, K.S., Lippman, D.Z., Curl Jr, R.F., Huggins, C.M., Petersen, D.E., 1955. *J. Am. Chem. Soc.*, **77**, 3433.
Prausnitz, J.M., 1969. *Molecular Thermodynamics of Fluid-Phase Equilibria*, Prentice-Hall, Englewood Cliffs, New Jersey.
Prausnitz, J.M., Anderson, T.F., Grens, E.A., Eckert, C.A., Hsieh, R., O'Connell, J.P., 1980. *Computer Calculations for Multicomponent Vapor-Liquid and Liquid-Liquid Equilibria*, Prentice-Hall, Englewood Cliffs, New Jersey.
Redlich, O., Kwong, J.N.S., 1949. *Chem. Rev.*, **44**, 233.
Reid, R.C., Prausnitz, J.M., Poling, B.E., 1987. *The Properties of Gases and Liquids*, McGraw-Hill, New York.
Salerno, S., Cascella, M., May, D., Watson, P., Tassios, D., 1986. *Fluid Phase Equilibria*, **27**, 15.
Smith, J. M., Van Ness, H.C., 1975. *Introduction to Chemical Engineering Thermodynamics*, 3rd Ed., McGraw-Hill, New York.
Soave, G., 1972. *Chem. Eng. Sci.*, **27**, 1197.
Tsonopoulos, C., 1974; 1975. *AIChE J.*, **20**, 263; **21**, 287.
Watson, P., Cascella, M., May, D., Salerno, S., Tassios, D., 1986. *Fluid Phase Equilibria*, **27**, 35.
Wicks, C.E., Block, F.E., 1963. *Bulletin 605*, Bureau of Mines, USA.
Wilsak, R. A., Thodos, G., 1984. *Ind. Eng. Chem. Fundam.*, **23**, 75.
Wilson, G. M., 1964. In *International Advances in Cryogenic Engineering*, **11**, 392, Plenum Press, New York.
Zudkevitch, D., Joffe, J., 1966. *AIChE Meeting*, New Orleans.

Problems

8.1 If a molecule of n-butane strikes the piston mentioned in Section 8.3.1 with a velocity of $2x10^4$ m/s and at an angle of 45 degrees, what is the force exerted on the piston in Newtons?

8.2 Describe an experimental procedure that you would use to determine, for water at 100°C, the following:

* bubble point
* dew point
* saturation pressure
* saturated liquid and vapor molar volumes.

8.3 What does the triple point of water represent? What are the values of temperature and pressure at this point?

8.4 With reference to the critical point:

a. define it;

b. is there a difference between that of a pure compound and that of a mixture?

8.5 Using the Steam Tables, prepare a *P-V* diagram for water that includes at least two isotherms as well as the saturation envelope.

8.6 What is the difference between:

a. a saturated liquid and one at its bubble point?

b. a saturated vapor and one at its dew point?

c. vapor pressure and saturation pressure?

8.7.a. What is a subcooled liquid and a superheated vapor?

b. What is the normal boiling point?

8.8.a. State the Corresponding States Principle;

b. Explain its importance;

c. Give two specific cases where you would use it.

8.9 With reference to the acentric factor:

a. what is its meaning?

b. what does it depend on?

c. what is it used for?

8.10.a. Using the following compressibility factor data for N_2 [Lunbeck, Michels, and Walkers, *Appl. Sci. Res.*, A3, 197, (1952)] at 0°C, calculate the z value for propane at 400 atm and the appropriate temperature.

P(atm)	0	5	10	100	200	300	400	600
z	1.000	0.998	0.996	0.984	1.036	1.134	1.256	1.524

b. Repeat using the Pitzer approach.

c. Which method should give the better results?

8.11 How do you explain z values larger than one?

8.12 If you wanted to develop your own corresponding states tables for:

a. the compressibility factor, and

b. the second virial coefficient,

what data - and for what type of compounds - would you need?

8.13 Explain - in terms of intermolecular forces - the effect of temperature on the second virial coefficient of methanol as compared to that of n-butane (Figure 8.6).

8.14 Give the minimum information needed to estimate the second virial coefficient for a:

a. nonpolar compound;

b. polar compound;

c. associated compound.

8.15.a. Why is the virial equation not applicable to liquids?

b. What is the Boyle temperature?

8.16 Estimate the second virial coefficient for methanol as a function of temperature and comment on its accuracy.

8.17.a. What pressure-volume-temperature relationships do we rely on to develop expressions for a and b in the RK EoS?

b. Develop these expressions.

8.18 Discuss the factors - and their importance - that must be considered in developing a cubic EoS.

8.19.a. Show that the RK EoS represents an analytical expression of the two-parameter CSP.

b. Would this expression give the same value of z, at the same T_r and P_r, as the corresponding expression for the vdW EoS? Explain.

8.20 Why do attractive forces reduce the pressure of a gas as compared to an ideal gas at the same temperature and volume?

8.21 Calculate the molar volume of SO_2 at 157.5°C and 100 atm using:

a. the principle of corresponding states

b. the virial equation truncated after B (both forms)

c. the virial equation truncated after C (both forms)

d. same as (b), but with B calculated from the Tsonopoulos correlation (using the form you consider appropriate)

e. the Redlich-Kwong EoS

f. the SRK EoS

g. the PR EoS

h. the vdW-711 EoS

Tabulate your results, present the % errors from the experimental value, and comment on the obtained accuracy.

Given: B = -159 cm^3/mol; C = 9000 cm^6/mol^2;

Experimental value: V = 77.4 cm^3/mol;

8.22 Compare the RK, SRK, PR, and vdW-711 EoS with respect to their ability to predict:

a. vapor pressures

b. molar volumes (vapor and liquid)

Consider both, nonpolar and polar compounds.

8.23 In Figure 8.3, at T=380 K and a pressure of 2.25 MPa, two values for the compressibility factor, 0.62 and 0.10, are shown for i-butane:

a.why are there two values?

b.estimate them using the method you consider best.

8.24 With reference to the estimation of the volumetric behavior of a pure fluid (gas or liquid):

a.what is the minimum information required?

b.which method do you recommend in each case?

c.what accuracy do you expect?

Consider polar and nonpolar compounds, low and high pressures, and present your answers in a tabular form.

8.25 What is the density of Ar at 25°C and 100 atm? Comment on the accuracy of your answer.

8.26 To what pressure should a cylinder with a volume 0.5 m^3 be filled to store 100 kg of ethane at 25°C? Estimate the accuracy of your answer.

8.27 Natural gas is to be transported from a field in Texas to a major market in the Midwest. The gas, which is essentially pure methane, enters the pipeline at a rate of 70 lb_m/s at a pressure of 3000 psia and a temperature of 65°F. The pipeline has a 12 in inside diameter. Calculate the initial velocity in ft/s:

a.assuming methane to behave as an ideal gas; and

b.using an appropriate EoS.

8.28 One mole of water, inside a container of volume 1124 cm^3, is heated to 500°C. Determine the pressure and compare it to that obtained from the Steam Tables.

8.29.a. Determine the temperature at which 20 lb_m of ammonia stored in a volume of 10 ft^3 will exert a pressure of 700 psia.

b.Comment on the accuracy of your answer.

8.30 Estimate the vapor pressure of n-hexane and acetone using the corresponding states principle and compare the results with those obtained with the Antoine equation.

8.31 Calculate the vapor pressure of water, from the normal boiling point to the critical, using the Antoine equation and discuss the accuracy of the obtained results.

8.32 Starting with the Clausius-Clapeyron equation:

a.show that under certain assumptions the heat of vaporization of a fluid can be evaluated from knowledge of the vapor pressure as a function of temperature only;

b.state these assumptions and the range of their validity.

Hint. Notice that at low pressures, the saturated vapor volume is much larger than that of the saturated liquid.

8.33 Using the result from the previous Problem, and the vapor pressure data in the Steam Tables:

a.evaluate the enthalpy of vaporization of water in the range 90 - 250°C;

b.plot the % error versus T_r and comment on the obtained accuracy.

8.34 To examine the effect of the accuracy of the experimental vapor pressure data - used for interpolation and extrapolation purposes - on the quality of the obtained predictions, consider the P^s data for water given below:

t(°C)	P^s(kPa) I	P^s(kPa) II
40	7.375	7.228
60	19.92	19.92
80	47.46	48.31

Column I presents the actual data from the Steam Tables, while in column II errors of -2% and +2% have been imposed in the first and third data points respectively. Using the data in the two columns:

a. estimate the vapor pressure in the range 0°C to 150°C;

b. plot the error % versus t;

c. comment on the accuracy of the obtained results.

8.35 Using the results from the previous Problem, calculate enthalpies of vaporization for water in the same temperature range and compare them with the experimental values. Use the hint of Problem 8.32.

8.36 Using the enthalpy of vaporization of water at the normal boiling point, estimate its values from 0°C to the critical and comment on their accuracy.

8.37 If the compressor in Example 8.12 is adiabatic but irreversible with an efficiency of 0.75, what will the final temperature be? What would the anticipated accuracy be in this case?

8.38 Identify three equations that you consider:

a. thermodynamic,

b. empirical,

and explain your reasoning.

8.39 With reference to this Chapter:

a. present the most important in your opinion items that you have learned;

b. give in each case an application.

8.51.a. Show that $\lim_{P \to 0}(PV) = RT$ [T = constant]

b. How could you use this to determine the values of R and of the absolute temperature from several sets of experimental isothermal PVt data? (t is in °C, V in cm^3/mol.)

c. How does this approach differ from that given in Chapter 2?

8.52.a. Show that for the van der Waals EoS two more sets of expressions for a and b, in addition to those given by Eqs. (E) and (F) of Example 8.7, can be obtained.

b. Discuss the performance of these three sets (see, J.J. Martin).

8.53.a. Show, by carrying out a Taylor (McLaurin) series expansion for the compressibility factor:

$$z = \frac{P}{\rho RT}$$

around $\rho = 0$, that: $B = (dz/d\rho)$ at $\rho = 0$, where ρ is the molar density, and B the second virial coefficient.

b. What does this mean for the Boyle temperature?

8.54.a. Show that second virial coefficients can be obtained from a cubic EoS.

b. Calculate B values for n-butane, using the SRK and vdW-711 EoS, as a function of temperature and compare them to the experimental value given in Table 8.E.5.

8.55 Using the Steam Tables, determine the second and third virial coefficients of water at 300°C by using:

a. the method presented in Example 8.4;

b. a plot of $[(z-1)V]$ versus $(1/V)$. (What assumption is involved here?)

Comment on the accuracy of the two approaches.

8.56 Develop the expression for the efficiency of the Carnot cycle, Eq.3.5.3, using a real gas instead of the ideal one used there.

8.57.a. Present a step-by-step procedure for the evaluation of the enthalpy of vaporization of a pure compound with the SRK EoS.

b. Apply it to a compound of your choice and compare the obtained results with the experimental values.

The price of optimism is blood sweat and tears. (Winston Churchill)
Ditto, for understanding thermodynamics. (!)

9
Thermodynamic Properties of Pure Fluids

9.1 Introduction

In the previous Chapter we discussed the evaluation of physical, i.e. experimentally determinable, properties of pure fluids: pressure-volume-temperature (*PVT*) relationships and heat capacities. In this Chapter we turn to thermodynamic properties, i.e. properties that cannot be measured directly. Rather, they are determined indirectly by using relationships that express them in terms of physical properties. These thermodynamic properties are:

* internal energy
* entropy
* enthalpy
* Helmholtz free energy
* Gibbs free energy (and the associated properties: chemical potential and fugacity).

We will concentrate our attention to the most important among them for chemical engineering applications: entropy, enthalpy, and fugacity.

We must emphasize again that thermodynamic properties are not measurable quantities. They are concepts, mathematical constructs in a strict sense, created for thermodynamics; and their values are determined from experimentally measurable quantities through relationships developed by thermodynamics. Yet, these constructs are essential for the solution of everyday chemical engineering problems: from the shaft work required for the compression of a gas (entropy and enthalpy), to the design of a distillation tower (fugacity and enthalpy).

9.2 Objective and Approach

The main objective of this Chapter is to develop familiarity with the methods used for the evaluation and prediction of thermodynamic properties of pure fluids:

a. from experimental data, or

b. through estimation techniques, whenever such data are not available, which represents the typical case.

We will start with the development of the *fundamental equations* - the name given by J.W. Gibbs - because they contain all information needed for the complete description of the thermodynamic state of a system.

Unfortunately, the appropriate analytical expressions for the fundamental equations are not available. We turn therefore to a postulate, resulting from our empirical observations and known as the *fundamental postulate*, that states: *The macroscopic properties of a homogeneous fluid on a molar basis can be expressed in terms of its pressure, temperature and composition.*

Using this postulate, we develop a framework of expressions for the evaluation of thermodynamic properties of fluids from *PVT* and ideal gas heat capacity data; and, since in the typical case *PVT* data are not available, estimation techniques are emphasized.

We proceed then to discuss the concepts of chemical potential and fugacity, essential for Phase and Chemical Reaction Equilibrium calculations as we will see in later Chapters. We close with a summary of our observations on the evaluation of thermophysical properties of pure fluids.

9.3 The Fundamental Equations: Closed Systems

If we are to determine thermodynamic properties from the physical ones, the appropriate relationships between these two types of properties must be established. These relationships are referred to as fundamental equations, for they contain all the information needed for the complete description of the state of a system.

We will start our discussion with closed systems, i.e. systems that do not exchange mass with the surroundings and, consequently, their mass is constant, and then consider - in the next Section - open systems, i.e. systems of variable mass.

Because the approach for developing the fundamental equations is basically the same for mixtures and pure compounds, we will discuss the more general case of mixtures.

9.3.1 The Fundamental Equation for the Internal Energy

Consider a closed homogeneous system which experiences the transfer of a small amount of heat dQ and an amount of volumetric work dW. Then, according to the first law, the corresponding change in its internal energy dU is given by:

$$dU = dQ - dW \tag{9.3.1}$$

If these changes are reversible, then:

$$dW = PdV \quad ; \quad dQ = TdS$$

where P, T, V and S refer to the pressure, temperature, volume, and entropy of the system respectively.

Eq.9.3.1 becomes, thus:

$$dU = TdS - PdV \tag{9.3.2}$$

Eq.9.3.2, even though developed for a reversible change, involves state functions only. It applies, therefore, to a closed system for a reversible or irreversible change, provided that the initial and final states are equilibrium states. Notice, however, that the first and second terms on the right-hand side refer to heat transfer and work for reversible changes only.

Eq.9.3.2 indicates that the internal energy of the system is a function of its entropy and volume, i.e.:

$$U = f(S, V) \tag{9.3.3}$$

If the analytical relationship between U and S and V is available, then specification of the values of S and V establishes, obviously, the value of U; and, as we will demonstrate next, it also establishes the values of the system temperature and pressure.

Consider a differential change in S and V. The corresponding change in U, according to Eq.9.3.3 is:

$$dU = \left(\frac{\partial U}{\partial S}\right)_V dS + \left(\frac{\partial U}{\partial V}\right)_S dV \tag{9.3.4}$$

Comparison of Eqs 9.3.4 and 9.3.2 indicates that:

$$T = \left(\frac{\partial U}{\partial S}\right)_V \tag{9.3.5}$$

$$P = -\left(\frac{\partial U}{\partial V}\right)_S \tag{9.3.6}$$

Thus, the functional relationship indicated by Eq.9.3.3 establishes - in addition to the value of the internal energy U - the values of the temperature T and pressure P of the system, once its entropy and volume are specified.

The system enthalpy is also specified from the definition:

$$H = U + PV \tag{9.3.7}$$

So are the other two energy properties mentioned in Chapter 1: the Helmholtz free energy, defined by:

$$A = U - ST \tag{9.3.8}$$

and the Gibbs free energy, defined by:

$$G = U + PV - ST \tag{9.3.9}$$

We conclude, therefore, that availability of the functional relationship: $U = f(S,V)$, Eq.9.3.3, provides a complete description of the thermodynamic state of the system, i.e. it is a *fundamental equation*.

The undesirable characteristic of Eq.9.3.3, however, is that one of its independent variables, the entropy, is not a measurable quantity. We turn, therefore, to the other energy expressions: enthalpy H, Helmholtz free energy A, and Gibbs free energy G, and examine:

a. if they are also fundamental equations, and
b. if so, whether their independent variables are measurable quantities?

9.3.2 The Fundamental Equation for *H*, *A*, and *G*

The enthalpy differential, according to its definition, Eq.9.3.7, is:

$$dH = dU + PdV + VdP$$

Replacement of the sum: $dU + PdV$, through Eq.9.3.2, yields:

$$dH = TdS + VdP \tag{9.3.10}$$

which indicates that:

$$H = f(S,P) \tag{9.3.11}$$

A differential change in S and P leads to the following change in H according to Eq.9.3.11:

$$dH = \left(\frac{\partial H}{\partial S}\right)_P dS + \left(\frac{\partial H}{\partial P}\right)_S dP \quad (9.3.12)$$

and comparison of Eq.9.3.12 with Eq.9.3.10 indicates that:

$$T = \left(\frac{\partial H}{\partial S}\right)_P \quad (9.3.13)$$

$$V = \left(\frac{\partial H}{\partial P}\right)_S \quad (9.3.14)$$

Evaluation of U, A, and G is straightforward from Eqs 9.3.7, 9.3.8 and 9.3.9 respectively. Eq.9.3.11, thus, combined with Eqs 9.3.13 and 9.3.14, does provide complete description of the system once S and P are specified. It is, therefore, also a fundamental equation. One of its independent variables, however, is again the non-measurable entropy.

Following the same procedure for the Helmholtz free energy, we find that:

$$dA = -PdV - SdT \quad (9.3.15)$$

$$A = f(V,T) \quad (9.3.16)$$

$$P = -\left(\frac{\partial A}{\partial V}\right)_T \quad (9.3.17)$$

$$S = -\left(\frac{\partial A}{\partial T}\right)_V \quad (9.3.18)$$

Similarly, for the Gibbs free energy:

$$dG = VdP - SdT \quad (9.3.19)$$

$$G = f(T,P) \quad (9.3.20)$$

$$V = \left(\frac{\partial G}{\partial P}\right)_T \quad (9.3.21)$$

$$S = -\left(\frac{\partial G}{\partial T}\right)_P \quad (9.3.22)$$

It is apparent that Eqs 9.3.16 and 9.3.20 are also fundamental equations, i.e. they contain the same information as Eq.9.3.3. (For a mathematical demonstration of this equivalency, using Legendre transformations, see Modell and Reid, 1974). Furthermore, their independent variables are measurable quantities in both cases.

In the following Example 9.1 we demonstrate the importance of fundamental equations, and then consider open systems.

9.3.3 Example 9.1

a. Show that if for a homogeneous closed system A is known as a function of T and V, as dictated by Eq.9.3.16:

$$A = f(T, V) \tag{A}$$

then all thermodynamic quantities can be evaluated.

b. Show that this is not the case if A is known as a function of T and P:

$$A = f(T, P) \tag{B}$$

a. Introduction of Eq.(A) into Eqs 9.3.17 and 9.3.18 yields expressions for P and S and, consequently, all other quantities can be evaluated. For example:

$$G = U + PV - TS = A + PV$$

b. Let us now turn to Eq.(B) and use it to evaluate, for example, the volume V at some temperature T. From Eq.9.3.17:

$$V = \int \frac{-dA}{P} + c$$

Determination, thus, of the volume is not possible because of the presence of the unknown constant c. Eq.(B) is not, therefore, a fundamental equation.

9.4 The Fundamental Equations: Open Systems

In the case of an open homogeneous system, i.e. one that can exchange mass with its surroundings, the number of moles present is not constant. Eq.9.3.3 becomes, therefore:

$$U = f(V, S, N_1, N_2, \ldots\ldots, N_k) \tag{9.4.1}$$

where k is the number of components of the system. The system's internal energy is now a function of its entropy S, volume V, and the number of moles of the constituent components, N_1, N_2, ..., N_k.

From this relationship we obtain the following expression for the differential change in U resulting from differential changes in S, V, and N_i:

$$dU = \left(\frac{\partial U}{\partial S}\right)_{V,N_i} dS + \left(\frac{\partial U}{\partial V}\right)_{S,N_i} dV + \sum \left(\frac{\partial U}{\partial N_i}\right)_{V,S,N_j} dN_i \quad (9.4.2)$$

where i and j assume all values: 1, 2, ..., k, except, of course, that j cannot be i.

We notice that the first two derivatives, taken while holding constant the mole number of all components, correspond to those for a closed system. It follows from Eqs 9.3.5 and 9.3.6 respectively, that:

$$T = \left(\frac{\partial U}{\partial S}\right)_{V,N_i} \quad (9.4.3)$$

$$P = -\left(\frac{\partial U}{\partial V}\right)_{S,N_i} \quad (9.4.4)$$

Eq.9.4.2 can be thus written as:

$$dU = TdS - PdV + \Sigma\mu_i dN_i \quad (9.4.5)$$

where:

$$\mu_i = \left(\frac{\partial U}{\partial N_i}\right)_{V,S,N_j} \quad (9.4.6)$$

Notice that the multiplier of the first differential in Eq.9.4.5, the temperature, is the potential for heat transfer; and that of the second one, the pressure, is the potential for P-V work. By analogy, the multiplier of each of the remaining differentials must be also a potential. It is indeed the potential for mass transfer from one phase to an other, or for the feasibility of chemical reactions in a mixture of compounds, and it is called, appropriately, the *Chemical Potential* of component i.

It is apparent that Eq.9.4.1, combined with Eqs 9.4.3, 9.4.4 and 9.4.6, provides complete description of the thermodynamic state of an open homogeneous system, i.e. it is a fundamental equation.

Following the same approach for the enthalpy, Helmholtz free energy and Gibbs free energy, we obtain:

$$H = f(S,P,N_1,N_2,.....,N_k) \quad (9.4.7)$$

$$dH = TdS + VdP + \Sigma\mu_i dN_i \tag{9.4.8}$$

where:

$$\mu_i = \left(\frac{\partial H}{\partial N_i}\right)_{S,P,N_j} \tag{9.4.9}$$

$$A = f(V,T,N_1,N_2,\ldots\ldots,N_k) \tag{9.4.10}$$

$$dA = -PdV - SdT + \Sigma\mu_i dN_i \tag{9.4.11}$$

where:

$$\mu_i = \left(\frac{\partial A}{\partial N_i}\right)_{V,T,N_j} \tag{9.4.12}$$

$$G = f(T,P,N_1,N_2,\ldots\ldots,N_k) \tag{9.4.13}$$

$$dG = -SdT + VdP + \Sigma\mu_i dN_i \tag{9.4.14}$$

where:

$$\mu_i = \left(\frac{\partial G}{\partial N_i}\right)_{T,P,N_j} \tag{9.4.15}$$

Comments

1. It can be easily seen that Eqs 9.4.7, 9.4.10 and 9.4.13 are fundamental equations.

2. Eqs 9.4.5, 9.4.8, 9.4.11, and 9.4.14 indicate that:

$$\mu_i = \left(\frac{\partial U}{\partial N_i}\right)_{V,S,N_j} = \left(\frac{\partial H}{\partial N_i}\right)_{S,P,N_j} = \left(\frac{\partial A}{\partial N_i}\right)_{V,T,N_j} = \left(\frac{\partial G}{\partial N_i}\right)_{T,P,N_j} \tag{9.4.16}$$

i.e. the chemical potential has the same value in all cases.

3. The chemical potential, in terms of the Gibbs free energy, represents the basic property for phase and/or chemical reaction equilibrium calculations.

With this background, let us now proceed with the main objective of this Chapter: the evaluation of thermodynamic properties of pure fluids.

9.5 Evaluation of Thermodynamic Properties of Pure Fluids

9.5.1 The Problem

It is apparent from our discussion so far that if any of the fundamental equations, Eqs 9.4.1, 9.4.7, 9.4.10 and 9.4.13, is available in an analytical form, all properties can be evaluated; and for the last two, from experimentally measurable quantities.

Thermodynamics unfortunately establishes only the differential Eqs 9.4.5, 9.4.8, 9.4.11, and 9.4.14, which are referred to as *fundamental property relations* (Smith and Van Ness, 1987); but not the appropriate analytical expressions, which require consideration of the molecular behavior of matter and represent the objective of Statistical Mechanics.

In practice, however, due to the limitations in the applicability of Statistical Mechanics to real fluids, that will be considered in Chapters 16 and 17, evaluation of thermodynamic properties is carried out with the approach discussed next.

9.5.2 The Approach

The approach used is based on the following two observations:

1.In the typical case, we are interested in the difference in values of a thermodynamic property, and not in its absolute ones. In calculating, for example, the amount of energy required to heat a gas in a flow process we only need the enthalpy difference between the initial and final states, not the absolute values.

2.Experience suggests that: *the macroscopic properties of a homogeneous fluid, on a molar basis, can be expressed as functions of temperature, pressure, and composition.* Thus, for enthalpy and entropy:

$$H = f(T,P,x_1,x_2,\ldots\ldots,x_k) \tag{9.5.1}$$

$$S = f(T,P,x_1,x_2,\ldots\ldots,x_k) \tag{9.5.2}$$

The above statement is referred to by Van Ness (1964) as the *fundamental postulate* for it is derived solely from our empirical observations. Such observations, for example, indicate that the density of steam at a given temperature and pressure is always the same, no matter how the steam arrived at this state; and the same would apply for a mixture of nitrogen and methane at a specified temperature, pressure, and composition.

It should be noted, however, that these observations apply to *simple* systems, i.e. systems that are not subject to significant external effects, such as strong magnetic or electric fields, shear stresses, and strong surface tension, as in the case of droplets and bubbles. Our postulate, therefore, applies to such simple systems which represent, of course, the typical case.

For a pure fluid, Eqs 9.5.1 and 9.5.2 become:

$$H = f(T,P) \tag{9.5.3}$$

$$S = f(T,P) \tag{9.5.4}$$

To evaluate the enthalpy and entropy differences for one mole of a pure fluid, undergoing a change from state $1(P_1,T_1)$ to state $2(P_2,T_2)$, we notice that for differential changes in T and P the corresponding changes in H and S according to Eqs 9.5.3 and 9.5.4 are:

$$dH = \left(\frac{\partial H}{\partial T}\right)_P dT + \left(\frac{\partial H}{\partial P}\right)_T dP \tag{9.5.5}$$

$$dS = \left(\frac{\partial S}{\partial T}\right)_P dT + \left(\frac{\partial S}{\partial P}\right)_T dP \tag{9.5.6}$$

We can thus determine the enthalpy and entropy differences:

$$\Delta H = H_2 - H_1$$
$$\Delta S = S_2 - S_1$$

by integrating Eqs 9.5.5 and 9.5.6 between the appropriate limits of temperature and pressure, provided of course that we can express the derivatives involved in these expressions through measurable quantities. To this purpose we note that:

1.From the definition of heat capacity at constant pressure (Eq.1.13.1):

$$\left(\frac{\partial H}{\partial T}\right)_P = C_P$$

and the observation that for a constant pressure reversible process:

$$TdS = dQ = dU + PdV = dU + d(PV) = dH$$

it follows that:

$$\left(\frac{\partial S}{\partial T}\right)_P = \frac{C_P}{T}$$

and therefore Eqs 9.5.5 and 9.5.6 become:

$$dH = C_P dT + \left(\frac{\partial H}{\partial P}\right)_T dP \tag{9.5.7}$$

$$dS = \frac{C_P}{T} dT + \left(\frac{\partial S}{\partial P}\right)_T dP \tag{9.5.8}$$

2.For the derivative: $\left(\frac{\partial H}{\partial P}\right)_T$ we obtain from Eq.9.3.10:

$$\frac{dH}{dP} = T\frac{dS}{dP} + V$$

and at constant temperature:

$$\left(\frac{\partial H}{\partial P}\right)_T = T\left(\frac{\partial S}{\partial P}\right)_T + V \tag{9.5.9}$$

Introduction of Eq.9.5.9 into Eq.9.5.7 yields:

$$dH = C_P dT + \left[T\left(\frac{\partial S}{\partial P}\right)_T + V\right] dP \tag{9.5.10}$$

3.The task hence of expressing the derivatives in Eqs 9.5.5 and 9.5.6 in terms of measurable quantities would be completed if the derivative:

$$D = \left(\frac{\partial S}{\partial P}\right)_T \tag{9.5.11}$$

could be expressed in terms of such. This is accomplished by using the Maxwell relations discussed in the next Section.

4.The heat capacity C_P involved here is that of the real fluid, and not that in the ideal gas state, C_P^*, and is consequently a function of temperature and pressure. Even though $C_P(T,P)$ can be determined experimentally, it is $C_P^*(T)$ that is available typically through low pressure calorimetric or spectroscopic measurements.

To get around this problem we make use of the so called *Departure* functions, which are discussed after the aforementioned Maxwell relations.

9.6 The Maxwell Relations

Consider the continuous function: $F = f(x,y)$. For a differential change in x and y we have:

$$dF = \left(\frac{\partial F}{\partial x}\right)_y dx + \left(\frac{\partial F}{\partial y}\right)_x dy = N dx + M dy \tag{9.6.1}$$

where:

$$N = \left(\frac{\partial F}{\partial x}\right)_y \quad ; \quad M = \left(\frac{\partial F}{\partial y}\right)_x$$

In addition, since:

$$\left(\frac{\partial N}{\partial y}\right)_x = \frac{\partial^2 F}{\partial x \cdot \partial y} \quad ; \quad \left(\frac{\partial M}{\partial x}\right)_y = \frac{\partial^2 F}{\partial y \cdot \partial x}$$

and the order of differentiation in mixed second derivatives is immaterial, we obtain:

$$\left(\frac{\partial N}{\partial y}\right)_x = \left(\frac{\partial M}{\partial x}\right)_y \tag{9.6.2}$$

Applying Eq.9.6.2 to Eqs 9.3.2, 9.3.10, 9.3.15, and 9.3.19 respectively, we obtain:

$$\left(\frac{\partial T}{\partial V}\right)_S = -\left(\frac{\partial P}{\partial S}\right)_V \tag{9.6.3}$$

$$\left(\frac{\partial T}{\partial P}\right)_S = \left(\frac{\partial V}{\partial S}\right)_P \tag{9.6.4}$$

$$\left(\frac{\partial P}{\partial T}\right)_V = \left(\frac{\partial S}{\partial V}\right)_T \tag{9.6.5}$$

$$\left(\frac{\partial V}{\partial T}\right)_P = -\left(\frac{\partial S}{\partial P}\right)_T \tag{9.6.6}$$

Eqs 9.6.3 through 9.6.6 are known as the Maxwell relations.

Returning now to the problem of expressing the derivative D, Eq. 9.5.11, in terms of measurable quantities, we notice that this is accomplished through Eq.9.6.6. Hence, Eqs 9.5.10 and 9.5.8 become:

$$dH = C_P dT + \left[-T\left(\frac{\partial V}{\partial T}\right)_P + V \right] dP \tag{9.6.7}$$

$$dS = \frac{C_P}{T} dT - \left(\frac{\partial V}{\partial T}\right)_P dP \tag{9.6.8}$$

and the task of expressing the derivatives in Eqs 9.5.5 and 9.5.6 in terms of measurable quantities has been accomplished.

9.7 Departure Functions

Having expressed the derivatives in Eqs 9.5.5 and 9.5.6 in terms of measurable quantities, obtaining Eqs 9.6.7 and 9.6.8, we turn to the second problem involved in the determination of enthalpy and entropy differences: the fact that $C_P^* = f(T)$ and not $C_P = f(T,P)$ is typically available. To this purpose, we use the following approach (Reid et al, 1987).

Let L be the value of some thermodynamic property of a pure fluid at a given pressure P and temperature T, and $L°$ the value of L at the same temperature T and some reference pressure $P°$, where the fluid is assumed to behave as an ideal gas. For example, the volume of the fluid in this reference state is given by:

$$V° = \frac{RT}{P°} \tag{9.7.1}$$

We can now determine the difference: $L_2 - L_1 = L(P_2,T_2) - L(P_1,T_1)$ using the following approach.

9.7.1 The Approach

Since L is a path independent property, we can use the scheme depicted in Figure 9.1 to determine $L_2 - L_1$:

$$L_2 - L_1 = (L_2 - L_2°) + (L_2° - L_1°) + (L_1° - L_1) \tag{9.7.2}$$

The difference, thus, $(L_2 - L_1)$ - for example $(H_2 - H_1)$ - is broken into two contributions describing:

a.the one, the effect of deviation of the fluid behavior from that of an ideal gas at the final and initial conditions, $(L_2 - L_2°)$ and $(L_1 - L_1°)$, and

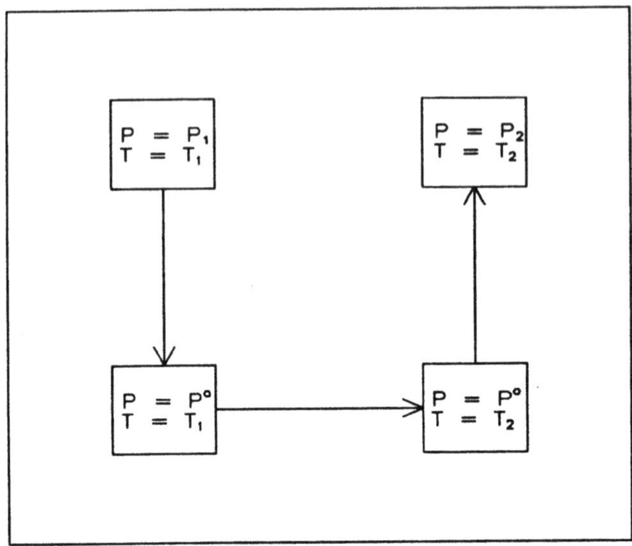

Figure 9.1 Scheme for the evaluation of the difference in thermodynamic properties between states (P_1,T_1) and (P_2,T_2).

b. the other, the effect of temperature on the fluid in its ideal gas state, $L_2^{\circ} - L_1^{\circ}$, determined through the ideal gas properties of the fluid.

The difference $(L^{\circ} - L)$ or $(L - L^{\circ})$ is referred to as the *Departure* function of the fluid at the given temperature T and pressure P.

Two choices for P° are typical: $P^{\circ} = P$ and $P^{\circ} = 1$ atm. We will use here $P^{\circ} = P$, and consequently the departure function represents the difference between:

* the value of L at a hypothetical state, L°, where the fluid behaves as an ideal gas at the given P and T, and
* the actual value of L at P and T.

Before we proceed with the evaluation of departure functions, and to develop a picture of their magnitudes as functions of pressure, we calculate enthalpy and entropy departure values in the next two Examples for i-butane.

9.7.2 Example 9.2

Table 9.E.2 presents enthalpies of i-butane(iC_4) as functions of pressure at two temperatures: 400 K and 700 K. Estimate the enthalpy departure values as functions of pressure.

Table 9.E.2 Enthalpy and Entropy for i-Butane as a Function of Pressure (Goodwin and Haynes, 1982)

P(MPa)	H(J/mol)		S(J/mol K)	
	400 K	700 K	400 K	700 K
0.01	52933	100619	346.9	434.2
0.05	52875	100591	333.4	420.8
0.101325	52809	100560	327.5	414.9
0.3	52578	100460	318.1	405.8
0.5	52354	100376	313.5	401.4
1.0	51764	100191	306.7	395.5
1.6	50922	99989	301.2	391.4
2.6	48889	99652	293.1	386.9
2.8	48274	99585	291.2	386.3
3.0	47470	99518	288.9	385.6
3.2 [(1)]	46187		285.4	
3.64		99304		383.7
4.0		99183		382.8
5.0		98850		380.6
7.5		98034		376.2
10.0		97261		372.9
14.0		96177		368.8
20.0		95013		364.4
70.0		94463		350.6

[(1)] At 400.2 K, the saturation temperature.

If we assume that iC_4 behaves, for practical purposes, as an ideal gas at P = 0.01 MPa, then:

$$H^\circ - H = H(P = 0.01\text{MPa}) - H(P)$$

since pressure does not effect the ideal gas enthalpy, i.e. $H^\circ(P) = H^\circ(P = 0.01$ MPa). The obtained results are shown as a function of pressure in Figure 9.E.2.

Comments

1. Table 9.E.2 indicates that enthalpy decreases with increasing pressure at constant temperature (see Problem 9.53).

2. The effect of pressure on the enthalpy departure is, as expected, much more pronounced at the lower temperature and it becomes still more so as the saturation temperature is approached.

3. The effect of pressure on enthalpy - at constant temperature - is much smaller (in absolute values) than that of temperature at constant pressure, but not insignificant.

4. Notice that the first case reflects the effect of pressure, and thus density, on the potential energy, i.e. on the magnitude of the intermolecular forces, while the second one, the effect of temperature on the internal (translational, rotational, and vibrational) energy of the molecules.

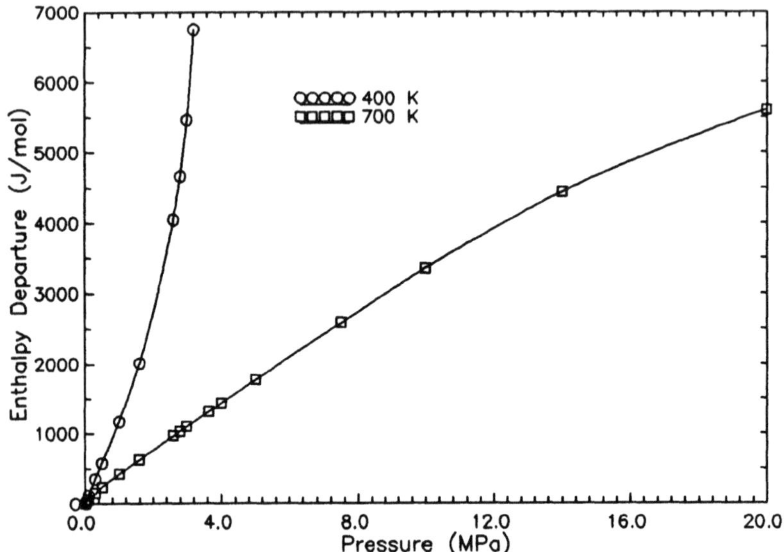

Figure 9.E.2 Variation of $(H^{\circ} - H)$ with pressure for i-butane.

9.7.3 Example 9.3

From the entropy data for iC_4 in Table 9.E.2 determine the entropy departure as a function of pressure at 400 K and 700 K.

To determine the departure function: $S^{\circ} - S = S^{\circ}(P) - S(P)$ at a specified temperature, we note that:

$$S^{\circ}(P) - S(P) = [S^{\circ}(P) - S(0.01)] + [S(0.01) - S(P)] \tag{A}$$

where $S(0.01)$ represents the entropy at a pressure of 0.01 MPa.

If we assume again that iC_4 behaves as an ideal gas at 0.01 MPa, then from Eq.3.12.5:

$$S^{\circ}(P) - S(0.01) = -R\ln(P/0.01) \tag{B}$$

Introduction of Eq.(B) into Eq.(A) yields:

$$S^{\circ}(P) - S(P) = S^{\circ} - S = [S(0.01) - S(P)] - R\ln(P/0.01)$$

A plot of $(S^{\circ} - S)$ at 400 K and 700 K versus pressure is presented in Figure 9.E.3. Notice that, as with the enthalpy departure, the deviation from ideal gas behavior is much more pronounced at the lower temperature, especially as the saturation pressure is approached.

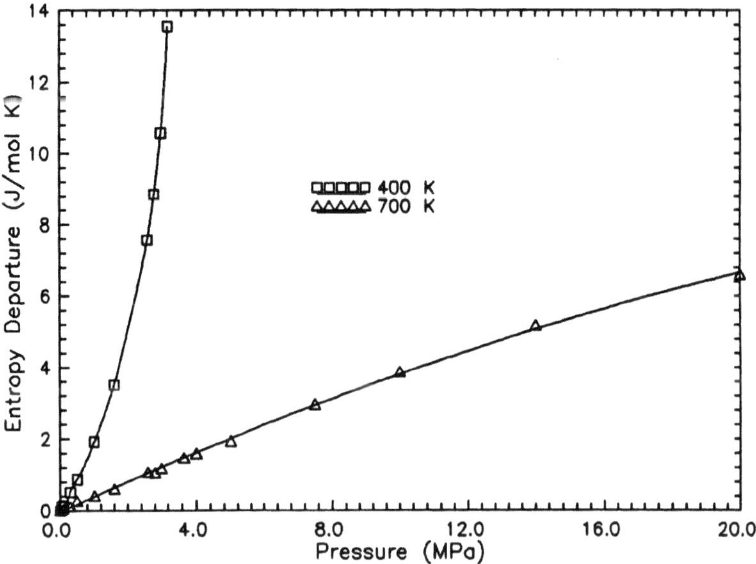

Figure 9.E.3 Variation of ($S° - S$) with pressure for i-butane.

9.7.4 Evaluation of Departure Functions

Examination of Eqs 9.3.2, 9.3.10, 9.3.15, and 9.3.19 indicates that evaluation of isothermal changes respectively for U, H, A, and G in terms of PVT data - directly or through an equation of state - can be accomplished readily for the last two only. Thus:

$$A° - A = -\int_{V}^{V°} P dV \tag{9.7.3}$$

$$G° - G = \int_{P}^{P°} V dP \tag{9.7.4}$$

Since most equations of state (EoS) are in the form $P = f(V,T)$, rather than $V = f(P,T)$, we choose Eq.9.7.3 and, to eliminate the problem of integrating between an ideal gas and a real gas state, we break the integral into two parts:

$$A° - A = -\int_{V}^{\infty} P dV - \int_{\infty}^{V°} P dV = -\int_{V}^{\infty} P dV - \int_{\infty}^{V°} \frac{RT}{V} dV$$

since the second integral covers the ideal gas region.

Finally, to avoid the difficulties involved with the infinity limit, we add to and subtract from the right-hand side the integral:

$$\int_V^\infty \frac{RT}{V} dV$$

Then, since also $(V/V^\circ) = z$:

$$A^\circ - A = -\int_V^\infty \left(P - \frac{RT}{V}\right) dV + RT\ln z \tag{9.7.5}$$

The other departure functions can be now easily developed:

$$S^\circ - S = -\left[\frac{\partial(A^\circ - A)}{\partial T}\right]_V = \int_V^\infty \left[\left(\frac{\partial P}{\partial T}\right)_V - \frac{R}{V}\right] dV - R\ln z \tag{9.7.6}$$

$$H^\circ - H = (A^\circ - A) + T(S^\circ - S) - RT(z - 1) \tag{9.7.7}$$

$$U^\circ - U = (A^\circ - A) + T(S^\circ - S) \tag{9.7.8}$$

$$G^\circ - G = (A^\circ - A) - RT(z - 1) \tag{9.7.9}$$

Evaluation of departure functions from experimental *PVT* data can be carried out graphically using these expressions. This approach, however, involves evaluation of derivatives from plots of experimental data which introduces substantial uncertainties. In the typical case, therefore, the *PVT* data are fitted to an accurate EoS to facilitate the calculations. Or, in the absence of such data, some generalized EoS can be used.

9.7.5 Evaluation of Enthalpy and Entropy Differences

Our objective of evaluating enthalpy and entropy differences for a homogeneous phase undergoing a change from state $1(P_1,T_1)$ to state $2(P_2,T_2)$, through the measurable quantities:

* *PVT*; and
* C_P^*

has been, thus, accomplished:

$$H_2 - H_1 = -(H_2^\circ - H_2) + \int_{T_1}^{T_2} C_P^* dT + (H_1^\circ - H_1) \tag{9.7.10}$$

$$S_2 - S_1 = -(S_2^\circ - S_2) + \int_{T_1}^{T_2} \frac{C_P^*}{T} dT + R\ln\left(\frac{P_1}{P_2}\right) + (S_1^\circ - S_1) \tag{9.7.11}$$

9.7.6 Development of Enthalpy and Entropy Tables and Charts

Since we are interested in enthalpy and entropy differences, the values of these properties presented in Tables and Charts are not the absolute ones but relative to an arbitrarily chosen set of values at some reference point. In the Steam Tables, for example, the H and S values are relative to a zero value arbitrarily assigned to saturated water at 273.16 K.

Once the reference point is set, enthalpy and entropy values are determined by using the following information:

1. Pressure-Volume-Temperature, *PVT*, data to evaluate the parameters in an accurate equation of state, through which the departure function values will be obtained;
2. Ideal gas heat capacities; and
3. Vapor pressure as a function of temperature to establish:
   * the phase boundary, and
   * in combination with saturated vapor and liquid volumes, the enthalpies and entropies of vaporization, if they are not available from direct measurements.

For details see, among others, Goodwin and Haynes (1982) who used this approach to develop extensive tables of thermodynamic data for i-butane. (From which the values presented in Table 9.E.2 were obtained).

In the typical case, the extensive *PVT* data needed for the evaluation of the departure function values through an accurate EoS are not available. We resort, therefore, to the estimation techniques discussed next.

9.8 Estimation of Departure Functions

9.8.1 The Methods

Estimation of departure function values is carried out by using either:

* Pitzer's corresponding states approach, or
* equations of state (EoS).

Using the corresponding states principle, Pitzer proposed that:

$$\left[\frac{(S^\circ - S)}{R}\right] = \left[\frac{(S^\circ - S)}{R}\right]^{(0)} + \omega\left[\frac{(S^\circ - S)}{R}\right]^{(1)} \qquad (9.8.1)$$

$$\left[\frac{(H^\circ - H)}{RT_c}\right] = \left[\frac{(H^\circ - H)}{RT_c}\right]^{(0)} + \omega\left[\frac{(H^\circ - H)}{RT_c}\right]^{(1)} \qquad (9.8.2)$$

For best results the Lee and Kesler (1975) Tables presented in Appendix D should be used.

Use of the SRK EoS (Table 8.3) for the Helmholtz free energy departure, Eq.9.7.5, yields (see Example 9.4):

$$A^\circ - A = RT\ln\frac{V-b}{V^\circ} + \frac{a}{b}\ln\frac{V+b}{V} \qquad (9.8.3)$$

Introduction of Eq.9.8.3 into Eq.9.7.6 yields:

$$S^\circ - S = -R\ln\frac{V-b}{V^\circ} + \frac{ma_c}{b}\frac{[1+m(1-T_r^{0.5})]}{T_c T_r^{0.5}}\ln\frac{V+b}{V} \qquad (9.8.4)$$

Introduction of Eqs 9.8.3 and 9.8.4 into Eq.9.7.7 yields:

$$H^\circ - H = \left[\frac{a}{b} + \frac{ma_c T_r^{0.5}}{b}\left[1+m(1-T_r^{0.5})\right]\right]\ln\frac{V+b}{V} - RT(z-1) \qquad (9.8.5)$$

Similar expressions are obtained with other EoS.

Development, finally, of the departure expressions for the internal energy ($U^\circ - U$), and the Gibbs free energy ($G^\circ - G$), is straightforward through Eqs 9.7.8 and 9.7.9.

9.8.2 Example 9.4

Develop the Helmholtz free energy departure function using the SRK EoS.

Introduction of the SRK EoS into Eq.9.7.5 yields:

$$A^\circ - A = -\int_V^\infty\left[\frac{RT}{V-b} - \frac{a}{V(V+b)} - \frac{RT}{V}\right]dV + RT\ln\left(\frac{V}{V^\circ}\right)$$

$$= RT\ln\frac{V-b}{V} + \frac{a}{b}\ln\frac{V+b}{V} + RT\ln\left(\frac{V}{V^\circ}\right)$$

$$= RT\ln\frac{V-b}{V^\circ} + \frac{a}{b}\ln\frac{V+b}{V}$$

i.e. Eq.9.8.3.

9.8.3 Example 9.5

In calculating enthalpy and entropy departures for i-butane in Examples 9.2 and 9.3, we assumed ideal gas behavior at 0.01 MPa (0.1 bar). Strictly speaking, of course, ideal gas behavior is approached only as $P \rightarrow 0$. Evaluate the corrections to the $(H^\circ - H)$ and $(S^\circ - S)$ values at 400 K required as a result of this assumption. Given: z(0.01 MPa) = 0.9979.

a. Enthalpy

Assuming that the SRK EoS provides an accurate description of the nonideal behavior of iC_4 from 0 to 0.01 MPa, and using T_c = 408.1K; P_c = 3.648 MPa; and ω = 0.176, we have:

$a_c = 1.349 \cdot 10^6$ MPa cm^6/(mol)^2

b = 80.58 cm^3/mol

m = 0.7533, α = 1.007

From Eq.9.8.5: $[H^\circ - H(P = 0.01 \text{ MPa})] = 14.0$ J/mol.

b. Entropy

Similarly, using Eq.9.8.4: $[S^\circ - S(P = 0.01 \text{ MPa})] = 0.027$ J/mol K.

The $(H^\circ - H)$ values in Figure 9.E.2 must be increased, therefore, by 14 J/mol; and the $(S^\circ - S)$ values of Figure 9.E.3 by 0.027 J/mol K, both very small quantities, as it should be expected.

9.8.4 Example 9.6

Estimate the entropy and enthalpy departures for i-butane at 400K and 2.80 MPa, using the Pitzer correlation and the SRK EoS, and compare them to the experimental values.

Using the Pitzer correlation with T_r = 0.980, and P_r = 0.768:

$$[(S^\circ - S)/R]^{(0)} = 0.908$$
$$[(S^\circ - S)/R]^{(1)} = 1.183$$

and: $(S^\circ - S)$ = 9.28 J/mol K.

$$[(H^\circ - H)/RT_c]^{(0)} = 1.196$$
$$[(H^\circ - H)/RT_c]^{(1)} = 1.235$$

and: $(H^\circ - H)$ = 4796 J/mol.

From Examples 9.2, 9.3, and 9.5, the experimental values for the two departure functions are:

$$(S^\circ - S) = 8.87 + 0.03 = 8.90 \text{ J/mol K}$$
$$(H^\circ - H) = 4659 + 14 = 4673 \text{ J/mol;}$$

which indicate errors of 4.3% and 2.6% in the predicted values respectively.

For the SRK EoS, we must first calculate the volume at 400 K and 2.8 MPa obtaining: V = 733.5 cm^3/mol; and the ideal gas volume at the same conditions: $V°$ = 1187.7 cm^3/mol.

Use of these values yields:

$(S° - S)$ = 8.25 J/mol K; and $(H° - H)$ = 4353 J/mol

with errors of -7.3% and -6.8% respectively.

9.8.5 Comments on the Estimation of Entropy and Enthalpy Departures

1. The Pitzer corresponding states approach, with the Lee-Kesler Tables of Appendix D, and the cubic equations of state of Table 8.3 give reasonable accuracy for nonpolar/weakly polar compounds.
2. Typical results with the Pitzer approach and the SRK EoS are presented in Table 9.1 and similar ones are obtained with the PR and vdW-711 EoS. For the latter $t = t_0$ should be used to avoid the pronounced and distorting effect that (dt/dT) has on enthalpy and entropy departure values close to the critical point (Androulakis et al, 1989).
3. At very high pressures the Pitzer approach should be preferred over the aforementioned, and other, cubic EoS.
4. For polar systems, the virial EoS truncated after B gives reasonable estimates of departure values within the range of applicability of the equation, provided that the required temperature dependency of B is available from experimental data. Use of an empirical correlation for B could be made for nonpolar/weakly polar compounds only.
5. For polar systems at higher pressures, the Pitzer correlation should be used, but the uncertainty increases as the polarity of the system increases.

Table 9.1 Estimation of Entropy and Enthalpy Departure Values for i-Butane

T	P	$(S° - S)^*$	% Error		$(H° - H)^*$	% Error	
K	MPa	J/mol K	Pit.	SRK	J/mol	Pit.	SRK
400	2.8	8.90	4.3	-7.3	4673	2.6	-6.8
440	4.0	8.69	10.5	-1.5	5046	7.1	-0.2
700	20.0	6.63	5.7	4.7	5606	1.4	-4.4

(*) Calculated from the data of Goodwin and Haynes.

6.For the estimation of departure functions for pure liquids, see Reid et al.

7.The departure functions are also used for mixtures, provided that L and L^o refer to the same composition, and the above comments apply to them as well.

8.Our discussion so far has dealt with the estimation of the nonideal gas behavior contribution to $(L_2 - L_1)$, i.e. $(L_2^o - L_2)$ and $(L_1^o - L_1)$. For $(L_2^o - L_1^o)$ the required ideal gas heat capacity depends on the internal molecular structure (e.g. bond lengths, vibration frequencies, and configuration of the constituent atoms) and, unfortunately, cannot be estimated through generalized correlations, such as those used for $(L^o - L)$, which reflect the intermolecular forces.

9.C_P^* values are available for a large number of compounds, however, mostly from spectroscopic measurements (see Section 8.15). Estimation, based on group contribution techniques, is also possible but for a limited number of compounds (Reid et al, 1987).

We proceed now with a discussion of two other very important thermodynamic properties: chemical potential and fugacity.

9.9 The Chemical Potential

The chemical potential is the key property in the most important application of chemical engineering thermodynamics, chemical and phase equilibrium. The ease of separation, for example, of methanol from its mixture with water by distillation is determined by the relationship between the concentration of methanol in the liquid and vapor phases, which - as Gibbs showed (Section 4.12) - is dictated by the chemical potential of methanol in the two phases.

We saw in Section 9.4, Eq.9.4.16, that the chemical potential is defined in terms of all four energy functions. Since, however, we are interested in chemical and in phase equilibrium reached at constant temperature and pressure, we use the one through the Gibbs free energy. For a pure compound - of interest in this Chapter - Eq.9.4.15 becomes:

$$\mu = \left(\frac{\partial G}{\partial N}\right)_{T,P} \tag{9.9.1}$$

Thus, the chemical potential is the Gibbs free energy per mole. It follows then from Eq.9.3.19 that:

$$d\mu = -SdT + VdP \tag{9.9.2}$$

where S and V represent the molar entropy and volume respectively.

To evaluate the chemical potential at some temperature T and pressure P, $\mu(T,P)$, we integrate Eq.9.9.2:

$$\mu(T,P) - \mu(T',P') = -\int_{T'}^{T} SdT + \int_{P'}^{P} VdP \tag{9.9.3}$$

where $\mu(T',P')$ is the chemical potential at some reference pressure P' and temperature T'. Hence $\mu(T,P)$ can only be evaluated relatively to its value at some other state.

This uncertainty is made worse by the following observation. Assume, for simplicity, that we:

a. choose as a reference temperature that of the system, i.e. $T' = T$; and
b. assign some finite value, say 10 J/mol, to the chemical potential at T and an arbitrarily chosen reference pressure, say $P' = 1$ atm;

and proceed to evaluate the chemical potential at the temperature $T = T'$ as a function of the pressure P, for $P < 1$ atm. Since in this range we can assume ideal gas behavior, Eq.9.9.3 gives:

$$\mu(T,P) = 10 + RT\ln(P/1)$$

We note that as P becomes progressively smaller, the chemical potential assumes large negative values, approaching negative infinity as P approaches zero.

Conversely, if we assign a value of zero to the chemical potential at T as $P \rightarrow 0$, then very large values will be obtained for it at higher pressures.

9.10 Fugacity

9.10.1 Definition

To avoid the aforementioned uncertainties associated with the evaluation of the chemical potential G.N. Lewis introduced the concept of fugacity in 1901.

Consider first the change in chemical potential of a pure ideal gas from pressure P_1 to pressure P_2 at a constant temperature T. From Eq. 9.9.2:

$$\mu_2 - \mu_1 = \mu(P_2,T) - \mu(P_1,T) = RT\ln(P_2/P_1) \tag{9.10.1}$$

This expression indicates that, for an ideal gas, the isothermal change in chemical potential can be determined through its initial and final pressure.

Turning to real systems, Lewis introduced a new function called *fugacity (f)*, such that the isothermal change in the chemical potential for any compound in any system (gas, liquid, or solid) is given by an expression similar to Eq.9.10.1:

$$\mu_2 - \mu_1 = \mu(P_2,T) - \mu(P_1,T) = RT\ln(f_2/f_1) \qquad (9.10.2)$$

Obviously, μ_i and f_i refer to the same state.

Consider now the ratio of the fugacity of a pure compound at a pressure P divided by P. This ratio is called the *fugacity coefficient*:

$$\phi = \frac{f}{P} \qquad (9.10.3)$$

Since ideal gas behavior is approached as P approaches zero:

$$\lim_{P\to 0} \phi = \lim_{P\to 0} \frac{f}{P} = 1 \qquad (9.10.4)$$

Eq.9.10.4 along with the constant temperature differential form of Eq. 9.10.2:

$$d\mu = RTd\ln f \qquad (9.10.5)$$

provide the complete definition of fugacity for a pure species.

Usually we calculate the fugacity through the fugacity coefficient. Before, however, we discuss how the fugacity coefficient is calculated, let us consider a physical interpretation of fugacity.

9.10.2 A Physical Interpretation

Fugacity has units of pressure as Eq.9.10.4 indicates, and a comparison of Eqs 9.10.1 and 9.10.2 suggests. Beyond this, however, its physical meaning becomes elusive.

Webster's Dictionary defines 'fugacity' as the "quality of being fugacious" and 'fugacious' as "fleeing, or apt to flee; passing quickly away; ephemeral". (Fugacity is indeed ephemeral; one forgets it very fast!).

Consider, now, the case of liquid water at 100°C and 1 atm, in equilibrium with its vapor, i.e. saturated water. According to Webster's definition of fugacity we would expect that, due to the equilibrium between the two phases, the molecules of the liquid are as "apt to flee" their phase as are the molecules of the vapor phase. (And this is physically the case: vapor-liquid equilibrium is not static, but dynamic, with molecules - equal in number - fleeing continuously from one phase to the other.) The fugacity of the liquid should be, thus, equal to that of the vapor.

We will demonstrate in Chapter 12, through a combination of the first and second laws, that this also represents a thermodynamic requirement:

$$f^l = f^v \tag{9.10.6}$$

where the superscripts 'l' and 'v' indicate saturated liquid and vapor phases respectively. Hence, at least in the case of phase equilibrium, fugacity may be viewed as the 'escaping tendency' of the molecules in each phase.

9.11 Evaluation of Fugacities from Experimental Data

9.11.1 The Mathematical Formulation

Combination of Eqs 9.10.5 and 9.9.2 yields, at a constant temperature T:

$$RTd\ln f = VdP \tag{9.11.1}$$

For an ideal gas, at the same temperature, Eq.9.11.1 becomes:

$$RTd\ln P = \frac{RT}{P}dP \tag{9.11.2}$$

Subtraction of Eq.9.11.2 from Eq.9.11.1 yields:

$$RTd\ln\left(\frac{f}{P}\right) = RTd\ln\phi = \left(V - \frac{RT}{P}\right)dP \tag{9.11.3}$$

and:

$$\ln\phi = \frac{1}{RT}\int_0^P\left(V - \frac{RT}{P}\right)dP \tag{9.11.4}$$

Eq.9.11.4 can be used along with experimental PVT data for the evaluation of fugacity coefficients through graphical integration. As with the evaluation of enthalpy and entropy departures, however, the integration is carried out by first fitting the PVT data to an accurate equation of state. And since such EoS express $P = f(V,T)$, rather than $V = f(P,T)$, direct use of Eq.9.11.4 is not possible.

To circumvent this problem, let us develop an expression for the difference in the chemical potential of a pure substance between its:
* real state at T and P, and
* ideal gas state at the same conditions,
through the corresponding difference in fugacities. From Eq.9.10.5:

$$\mu(T,P) - \mu^\circ(T,P) = RT\ln\left(\frac{f}{P}\right) = RT\ln\phi$$

We notice that, since according to Eq.9.9.1 the chemical potential equals the molar Gibbs free energy, the left-hand side of the above equation equals the negative of the departure function for the latter. Thus:

$$G^\circ - G = -RT\ln\phi \qquad (9.11.5)$$

From Eqs 9.7.9 and 9.7.5:

$$\ln\phi = (z-1) - \ln z - \frac{1}{RT}\int_\infty^V \left(P - \frac{RT}{V}\right) dV \qquad (9.11.6)$$

Evaluation of fugacities from experimental PVT data can be thus accomplished with two approaches:
a. with Eq.9.11.6, using the data directly, i.e. graphically, or better, after they have been fitted with an accurate EoS; and
b. with Eq.9.11.5, using enthalpy and entropy values generated from the fitted PVT data, since:

$$RT\ln\phi = (G - G^\circ) = (H - H^\circ) - T(S - S^\circ) \qquad (9.11.7)$$

9.11.2 Evaluation of Fugacities with an Equation of State

We will demonstrate this approach with the virial EoS. Use of its volume series form truncated after C into Eq.9.11.6 yields:

$$\ln\phi = -\ln z + \frac{2B}{V} + \frac{3C}{2V^2} \qquad (9.11.8)$$

while use of the pressure series, directly in this case into Eq.9.11.4, yields:

$$\ln\phi = \frac{BP}{RT} + \frac{C - B^2}{2}\left(\frac{P}{RT}\right)^2 \qquad (9.11.9)$$

As with the case of volume calculations, Eq.9.11.8 performs better than Eq.9.11.9. But when truncated after B, the expression:

$$\ln\phi = \frac{BP}{RT} \tag{9.11.10}$$

performs better than the corresponding expression obtained from Eq. 9.11.8 (Van Ness).

Evaluation of fugacity coefficients from experimental data is demonstrated in the following Examples.

9.11.3 Example 9.7

Using the Steam Tables, determine the fugacity of steam at 300°C and 70 bar.

Two approaches are possible:

a. Use of the enthalpy and entropy values; and

b. Graphically, using the *PVT* data directly.

The first one will be used here, since it is the easiest and should provide good accuracy, at least comparable to the second.

From the Steam Tables at 300°C and 70 bar (by interpolation):

$$H = 2839.3\text{ kJ/kg};\quad S = 5.9327\text{ kJ/kg K};$$

and at $P = 0.1$ bar and 300°C:

$$H = 3076.6\text{ kJ/kg};\quad S = 9.2820\text{ kJ/kg K}.$$

Since we can reasonably expect ideal gas behavior at the latter conditions:

$H^\circ = 3076.6$ kJ/kg

$S^\circ = 9.2820 + (R/18)\ln(0.1/70) = 6.2561$ kJ/kg K (see Example 9.3)

Hence, from Eq.9.11.7:

$$\ln(f/P) = 18\,[(2839.3 - 3076.6) - 573.15(5.9327 - 6.2561)]/[(8.314)(573.15)]$$
$$= -0.196$$

and $f = 57.6$ bar.

9.11.4 Example 9.8

Goodwin and Haynes give the following value for the second virial coefficient of i-butane (iC_4) at 410 K: $B = -324.6\text{ cm}^3/\text{mol}$. Calculate the fugacity coefficient of iC_4 as a function of pressure at 410 K and compare it to the recommended values of Goodwin and Haynes presented in Table 9.E.8, and arrived at by using enthalpy and entropy values.

From Eq.9.11.10: $\ln\phi = BP/RT = -0.0952P(\text{MPa})$. The results are presented in Table 9.E.8 and indicate that:

a. Within the range of applicability of the virial EoS truncated after B: $P \leq (T/2)(P_c/T_c)$, i.e. below about 1.8 MPa, good results are obtained. For still better accuracy, the third virial coefficient must be used.

b. At higher pressures the results become progressively poorer, as expected.

Table 9.E.8 Performance of the Virial Equation in the Evaluation of Fugacity Coefficients for i-Butane at 410 K as a Function of Pressure. (Experimental Values from Goodwin and Haynes, GH).

	ϕ		
P (MPa)	GH	Virial	% Error
0.101325	0.983	0.990	0.7
0.5	0.948	0.954	0.6
1.0	0.892	0.909	1.9
1.4	0.858	0.875	2.0
1.8	0.824	0.843	2.3
2.2	0.789	0.811	2.8
2.6	0.755	0.781	3.4
3.0	0.719	0.752	4.5
3.4	0.683	0.723	5.9
3.8	0.641	0.696	8.6
4.2	0.594	0.670	12.8
4.6	0.553	0.645	16.6
5.0	0.519	0.621	19.7
6.0	0.453	0.565	24.7
7.0	0.405	0.514	26.7
8.0	0.370	0.467	26.2
10.0	0.322	0.386	19.9
12.0	0.290	0.319	10.0
14.0	0.269	0.264	-1.9
16.0	0.254	0.218	-14.2
20.0	0.235	0.149	-36.6

9.11.5 Liquid Phase Fugacities

Our discussion about fugacities has focused, so far, on vapors and gases. In addition, since the fugacity of a saturated liquid is equal to that of the saturated vapor in equilibrium with it (Eq.9.10.6), our discussion applies also to saturated liquids as well.

We will consider next how the fugacity of a subcooled liquid, i.e. one where the pressure on the liquid is higher than its vapor pressure at the given temperature ($P > P^s$), is determined.

We start again with Eq.9.11.4 and break the integral into two parts:

$$\ln\phi = \frac{1}{RT}\int_0^{P^s}\left(V_v - \frac{RT}{P}\right)dP + \frac{1}{RT}\int_{P^s}^{P}\left(V_l - \frac{RT}{P}\right)dP$$

where the superscript 's' stands for saturation, and the subscripts 'v' and 'l' refer to vapor and liquid phases respectively.

Notice that the first integral represents the fugacity coefficient of the saturated vapor and, consequently:

$$\ln\phi = \ln\phi^s + \int_{P^s}^{P} \frac{V_l}{RT}\, dP + \ln\left(\frac{P^s}{P}\right) \qquad (9.11.11)$$

Eq.9.11.11 gives, after some algebra, the expression for the liquid phase fugacity, f_l:

$$f_l = \phi^s P^s (Pe) \qquad (9.11.12)$$

where:

$$Pe = \exp \int_{P^s}^{P} \frac{V_l}{RT}\, dP \qquad (9.11.13)$$

is known as the *Poynting* effect, and represents the effect of pressure on the fugacity of a pure liquid, from saturation to system pressure.

Evaluation hence of the liquid phase fugacity at a given temperature T and pressure P requires:

a. the value of the fugacity coefficient of the saturated vapor at the temperature T, and

b. evaluation of the integral in Eq.9.11.13, either graphically or through an equation of state applicable to the liquid region.

As we have seen, however, liquid volumes are fairly insensitive to pressure - except close to the critical point - and V_l can be replaced by its average value $V_l(avg)$ in the range P^s to P. Under this assumption, Eq. 9.11.12, becomes:

$$f_l = \phi^s P^s \exp\left[\frac{V_l(avg)(P - P^s)}{RT}\right] \qquad (9.11.14)$$

Evaluation of liquid phase fugacity is considered in the next Example.

9.11.6 Example 9.9

Determine the fugacity of i-butane at 310.9 K and 10MPa. Given (Goodwin and Haynes): P^s = 0.5 MPa; V_l (0.5 MPa) = 108.8 cm^3/mol; V_l (10 MPa) = 104.8 cm^3/mol; ϕ^s = 0.8725.

From Eq.9.11.14, using V_l (avg) = 106.8 cm^3/mol, we obtain f_l = 0.646 MPa, which is in agreement with the value given by Goodwin and Haynes of 0.647 MPa

Notice the insensitivity of the liquid molar volume to pressure: a twenty-fold increase in pressure has an effect of less than 4% on the volume; but the significant effect of pressure on the liquid phase fugacity as indicated by the value of Pe = 1.48.

9.12 Estimation of Fugacities

9.12.1 The Methods

Estimation of fugacity coefficients - and hence fugacities - can be accomplished with two approaches:

a. The Pitzer corresponding states principle; and

b. Equations of state.

According to Pitzer's method:

$$\log_{10}\left(\frac{f}{P}\right) = \log_{10}\left(\frac{f}{P}\right)^{(0)} + \omega \log_{10}\left(\frac{f}{P}\right)^{(1)} \tag{9.12.1}$$

and best results are obtained using the Lee and Kesler Tables presented in Appendix D.

Estimation of fugacities with the virial EoS truncated after B uses Eq. 9.11.10, with second virial coefficient values obtained from the empirical correlations of Tsonopoulos or of Hayden and O'Connell (Chapter 8).

Use of the SRK EoS into Eq.9.11.6 leads to:

$$\ln\phi = (z-1) - \ln(z-B) - \frac{A}{B}\ln\left(\frac{z+B}{z}\right) \tag{9.12.2}$$

where: $A = aP/(RT)^2$; $B = bP/RT$

Results with this approach are presented in the next Example.

9.12.2 Example 9.10

Estimate the fugacity coefficient of i-butane at 410K and the pressures shown in Table 9.E.10 using the SRK EoS. Compare your results with the experimental values presented in Table 9.E.8.

Table 9.E.10 Performance of the SRK EoS in the Estimation of the Compressibility Factor and Fugacity Coefficient for i-Butane at 410 K as a Function of Pressure. Values for z and ϕ from Goodwin and Haynes.

P (MPa)	z	% Error	ϕ	% Error
2	0.787	-0.1	0.806	1.9
3	0.626	1.1	0.719	2.0
4	0.220	22.7	0.617	2.4
5	0.237	17.7	0.519	3.5
6	0.269	14.9	0.453	4.2
10	0.404	9.7	0.322	6.2
20	0.727	4.8	0.235	9.2

The calculation of fugacity coefficient using Eq.9.12.2 requires the value of z obtained from the SRK EoS. To this purpose, it is more convenient to write the SRK EoS in the form:

$$z^3 - z^2 + (A - B - B^2)z - AB = 0; \; A = aP/(RT)^2 \text{ and } B = bP/RT$$

Thus at 2 MPa: $A = 0.231$ and $B = 0.047$, which gives $z = 0.786$, and $\phi = 0.821$.

Calculated fugacity coefficients for iC_4 at 410 K as a function of pressure are presented in Table 9.E.10, along with the corresponding deviations from the experimental values. The errors for the compressibility factor are also included for comparison purposes. (See Problem 9.24.)

9.12.3 Comments

1.For nonpolar/weakly polar compounds the SRK EoS provides good estimates up to moderate pressures, as demonstrated with the typical - for such compounds - results presented in Table 9.E.10. Even at the higher pressures, $P > 6$ MPa (60 bar), the errors are not that high. Similar results should be expected with the PR and vdW-711 EoS.

2.For nonpolar and polar compounds the virial equation truncated after B, with the latter obtained from empirical correlations, gives reasonable accuracy within the range of its applicability.

3.At high pressures the Pitzer correlation should be used. The results become poorer, however, with increasing polarity.

4.The importance of fugacity in solving chemical engineering problems will become apparent when we discuss phase and chemical reaction equilibrium. We consider, for example, in the next Chapter a simple such application, the use of fugacities in calculating vapor pressures through cubic EoS.

9.13 Summary

We can summarize our discussion on the evaluation of thermophysical properties of pure fluids, presented in the last two Chapters, as follows:

1.In the typical case, only limited experimental information is available - mostly the critical properties and acentric factor, and the ideal gas heat capacity as a function of temperature. We must, consequently, resort to some estimation (predictive) technique.

2.At low pressures, the virial equation truncated after B, with the second virial coefficient values predicted by the Tsonopoulos or the Hayden-O'Connell correlations, gives the best results. Cubic EoS are also successful, but for nonpolar/weakly polar compounds only.

3.At higher pressures and for nonpolar compounds, cubic EoS can be used except close to the critical temperature. More reliable results in this region can be obtained with Pitzer's corresponding states approach, combined with the Lee-Kesler Tables, which should be also used at high pressures. [While the Tables cannot be easily adapted for computer use, the generalized BWR EoS, from which these Tables (Appendix D) were developed, can be used for this purpose.]

4.The Pitzer-Lee-Kesler approach can be also used for polar compounds at higher pressures. The accuracy, however, decreases with increasing polarity and the presence of hydrogen bonding.

5.Finally, before an estimation approach is used, all available sources of data should be examined.

References

Androulakis, J., Kalospiros, N., Tassios, D., 1989. *Fluid Phase Equilibria*, **45**, 135.

Goodwin, R.D., Haynes, W. M., 1982. *Thermophysical Properties of Isobutane from 114 K to 700 K at Pressures to 70 MPa*, Nat. Bur. Standards (U.S.), Washington, D.C.

Lee, B.I., Kesler, M.G., 1975. *AIChE J.*, **21**, 510.

Modell, M., Reid, R.C., 1974. *Thermodynamics and its Applications*, Prentice-Hall, Englewood Cliffs, N.J.

Reid, R.C., Prausnitz, J.M., Poling, B.E., 1987. *The Properties of Gases and*

Liquids, 4th Ed., McGraw-Hill, New York.
Smith, J.M., Van Ness, H.C., 1987. *Introduction to Chemical Engineering Thermodynamics*, 3rd Ed., McGraw-Hill, New York.
Van Ness, H.C., 1964. *Classical Thermodynamics of Non-Electrolyte Solutions*, McMillan, New York.

Problems

9.1 What is the importance of the fundamental equations?
9.2 Is the expression: $I = U + TS$, where I is some energy, a fundamental equation? Justify your answer.
9.3 Why is the chemical potential called so?
9.4 With reference to the fundamental postulate:
  a. state it;
  b. why is it called a postulate?
  c. what is its importance?
9.5 With reference to the Maxwell relations:
  a. what is their mathematical origin?
  b. give an application of them;
  c. develop Eq.9.6.6.
9.6.a. What is the purpose of the departure functions?
  b. Show that evaluation of the departure functions for H and S suffices for the determination of those for U, A, and G.
9.7 With reference to the Steam Tables:
  a. are the enthalpy and entropy values absolute or relative (to what)?
  b. does this create any problems in chemical engineering calculations?
9.8 With reference to the estimation of the enthalpy difference between two gaseous states of a pure compound:
  a. what is the minimum information needed?
  b. what method(s) do you recommend?
  c. what accuracy do you expect as a function of pressure?
Consider: (i) nonpolar; and (ii) polar compounds.
9.9.a. Show that use of the virial equation in evaluating departure functions gives:

$$A^\circ - A = \frac{-BRT}{V} + RT \ln\left(\frac{V}{V^\circ}\right)$$

$$S^\circ - S = \frac{BR}{V} + \frac{RT}{V}\frac{dB}{dT} - R \ln\left(\frac{V}{V^\circ}\right)$$

b. Calculate the enthalpy and entropy departure values for i-butane at 700 K and 2.80 MPa and compare them with those obtained from the data in Table 9.E.2.

9.10 At 0.01 MPa and 350 K the enthalpy of i-butane is 47044 J/mol. Calculate the enthalpy at 2 MPa and 400 K (exp. value: 50238 J/mol).

9.11 Saturated steam at 1 bar is compressed to 30 bar in a flow process. If the compressor is adiabatic and irreversible with an efficiency of 0.75:

a. how much work is required per kg of steam?

b. what is the temperature of the steam leaving the compressor?

c. what is the second law efficiency of the process?

9.12 Propane at 1 bar and 20°C is compressed to 10 bar in a flow process. If the compressor is reversible and adiabatic, how much work is required per mole of propane? Compare your answer to that obtained in Example 8.12.

9.13 With reference to the previous Problem, if the compressor is adiabatic but irreversible with an efficiency of 0.75:

a. how much work is required?

b. what is the temperature of the gas leaving the compressor?

c. what is the lost work?

9.14 Methane is compressed isothermally at 300 K from 1 bar to 10 bar in a nonflow process. If the compressor is irreversible with an efficiency of 0.75, calculate on a per mole basis:

a. the work required;

b. the amount of heat exchanged with the surroundings.

Given: B=-161 cm^3/mol.

9.15 The enthalpy and entropy of saturated liquid water at 273.16 K are set arbitrarily equal to zero.

a. Calculate the enthalpy and entropy of saturated vapor and liquid at 10 bar and compare them with the values given in the Steam Tables.

b. How would you use the extensive *PVT* and vapor pressure data available for water in carrying out the calculations in part (a)?

9.16 Wet steam at 11 bar and with a quality of 95.0% is throttled until it becomes dry and saturated. What are the final pressure and temperature of the steam?

9.17 Estimate the enthalpy difference between saturated steam at 10 bar and at 100 bar. Compare your value with the experimental one.

9.18 Between 0 and 80°C the cubical expansion of liquid water is given by:

$$V = V_0\,[1\text{-}5.3255\cdot 10^{-6}t + 7.61153\cdot 10^{-7}t^2 - 4.37217\cdot 10^{-9}t^3 + 1.64322\cdot 10^{-11}t^4]$$

where V_0 is the volume at 0°C (cm^3/mol) and t in °C.

Calculate the entropy change in J/mol K, when liquid water is compressed from 0 to 50 atm at 50°C. Neglect the pressure effect on volume.

9.19 For a certain fluid the functional relationship between enthalpy and temperature and pressure is given by:

$$H = aT + bP$$

where a and b are constants. Show that for this fluid the Joule-Thomson coefficient: $(\partial T/\partial P)_H$ is a constant, and express it in terms of a and b.

9.20.a. What is the minimum information needed for the estimation of the fugacity of a pure vapor or liquid?

b. What method(s) do you recommend?

c. What accuracy would you expect as a function of pressure?

Consider: (i) nonpolar; and (ii) polar compounds.

9.21 Estimate the fugacity coefficient of n-butane at 340 K and 0.7 MPa using the virial, SRK, and Pitzer methods and compare them with the recommended value of 0.862.

9.22 The vapor pressure of n-butane at 357.0 K is 1.1 MPa. Estimate the fugacity of:

a. saturated liquid at this temperature.

b. liquid at this temperature and 2MPa.

9.23 Estimate the fugacity of methanol at 380 K and 0.25 MPa. What accuracy do you expect?

9.24 Explain the difference in errors in the prediction of ϕ and z of Example 9.10 for iC_4.

9.25 If the compressibility factor of a gas is equal to 1 at some temperature and pressure, will its fugacity coefficient be also equal to 1 at the same conditions? Explain.

9.26 With reference to this Chapter:

a. what is the main objective?

b. write a short essay describing the important items that you have learned, including some applications.

9.51 Starting with the fact that for a constant composition gas: $S = f(T,P)$, show that for the isentropic change of an ideal gas:

$$PV^{\gamma} = \text{constant } (\gamma = C_P^*/C_V^*).$$

9.52 Show that the departure values for H and U do not depend on the value of the reference pressure $P°$, but those for S, A, and G, do.

9.53 Examine the variation of enthalpy with pressure, at constant temperature, for a pure gas.

9.54 Show that for a gas that obeys the Berthelot EoS:

$$P = \frac{RT}{V-b} - \frac{a}{TV^2}$$

$$\left(\frac{\partial U}{\partial V}\right)_T = \frac{2a}{TV^2} \; ; \; \left(\frac{\partial S}{\partial V}\right)_T = \frac{R}{V-b} + \frac{a}{(TV)^2} \; ; \; \left(\frac{\partial C_V}{\partial V}\right)_T = \frac{-2a}{(TV)^2}$$

9.55 Show that use of the van der Waals EoS leads to the following expression for the fugacity coefficient:

$$\ln\phi = \ln\frac{RT}{P(V-b)} + \frac{b}{V-b} - \frac{2a}{VRT}$$

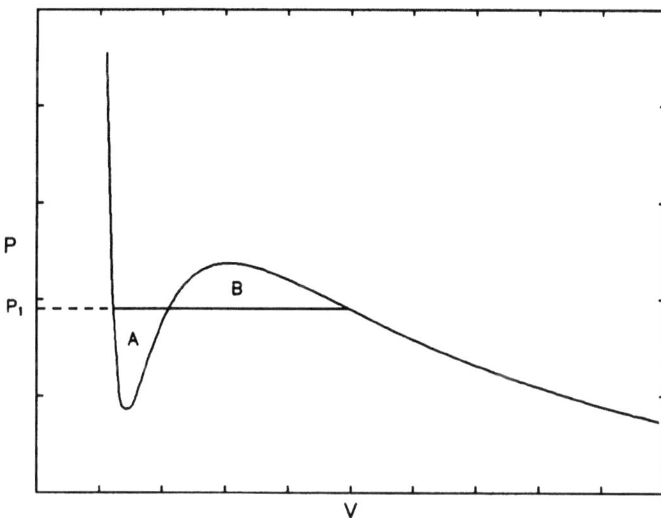

Figure 9.P.1 A pressure-volume isotherm for a pure fluid at a temperature T_1 generated from an equation of state.

9.56 Starting with Eq.9.10.6 present a flow diagram that demonstrates how you would calculate the vapor pressure of a pure fluid at some temperature T using a cubic EoS.

9.57 The isotherm ($T = T_1$) in Figure 9.P.1 is generated by some equation of state for a pure fluid. Show that if: Area A = Area B, then P_1 represents the vapor pressure of this fluid at T_1.

9.58 We have seen that corresponding states methods give progressively poorer results in the prediction of thermophysical properties as the polarity of the fluid increases. How do you explain this in terms of intermolecular forces?

9.59.a. Show that use of an EoS combined with C_P^* data can provide a complete description of the properties of a pure fluid.

b. Is this in conflict with Section 9.5.1?

9.60 Develop Eq.9.4.16.

10
Cubic Equations of State

10.1 Introduction

Fluid-phase equilibria: vapor-liquid, gas-liquid, and liquid-liquid, represent a most important application of chemical engineering thermodynamics in the petroleum and chemical industries. Quantitative description and/or prediction of such equilibria is essential in the design of distillation, extraction, and gas absorption units.

Cubic Equations of State (EoS) are progressively becoming the main tool for phase equilibria calculations and, even though they are - so far - successful for nonpolar/weakly polar systems only, it will not be long before they can handle polar systems as well. The Soave-Redlich-Kwong (SRK, Soave, 1972) and the PR (Peng and Robinson, 1976) EoS - modifications of the first EoS proposed, that of van der Waals (vdW) - are the most commonly used among them.

For the successful description of phase equilibria EoS must meet the following requirements:

1. Provide reliable prediction of vapor pressures of pure compounds over a wide temperature range;
2. Reflect a three-parameter corresponding states principle, i.e. predict successfully vapor pressures for nonpolar/weakly polar compounds where T_c, P_c, and ω values are available;
3. Provide reliable values of the saturated vapor and liquid volumes, especially the of latter;
4. Provide successful correlation and prediction of phase equilibria.

Application of cubic EoS in phase equilibria calculations of mixtures will be discussed in Chapter 14. Here, we will concentrate on pure fluids only.

This Chapter is intended for Graduate Students. Undergraduates should read through and, if time allows, should go into the details and do the problems.

10.2 Objective and Approach

The main objective of this Chapter is to:

1. Demonstrate the methodology used in developing cubic EoS for phase equilibria calculations; and
2. Explore their capabilities and limitations with respect to the aforementioned requirements (Section 10.1) with emphasis on pure compounds.

To develop the appropriate background we will begin by addressing the following question: Given an EoS, how is the vapor pressure of a pure fluid evaluated at a specified temperature?

The results will suggest that the traditional cubic EoS - such as the vdW and Redlich-Kwong (RK) - fail to provide satisfactory values for vapor pressures. We proceed, thus, to discuss how this problem is dealt with - using the vdW EoS as a starting point - presenting the approach that is typical for this type of cubic EoS, such as the SRK and PR ones.

We turn next to the second requirement - mentioned in the Introduction - by outlining the procedure whereby this modified vdW EoS can be generalized to provide prediction of vapor pressures for nonpolar compounds, just as the SRK and PR are, from knowledge of the critical properties and acentric factor values.

We proceed then with the third requirement and demonstrate how this vdW EoS is modified further, to provide successful predictions of saturated liquid molar volumes. This will lead to the vdW-711 EoS, which is typical of cubic EoS as far as the first two requirements are concerned, and somewhat novel with respect to the third one.

We follow with some comments on the capabilities and limitations of these cubic EoS and close with the concluding remarks.

10.3 Evaluation of Vapor Pressures with an Equation of State

10.3.1 The Problem

The problem can be stated as follows: Given an EoS, evaluate the vapor pressure of a pure fluid at a specified temperature, i.e. the pressure where a vapor and a liquid phase coexist in equilibrium with each other.

In the previous Chapter we stated - and will prove it in Chapter 12 - that the thermodynamic requirement for equilibrium between two phases is that their fugacities be equal. Thus, for vapor-liquid equilibrium of pure compounds:

$$f^l = f^v \tag{9.10.6}$$

where the superscripts l and v stand for saturated liquid and vapor phase respectively.

The problem, therefore, becomes: find the pressure where - at the specified temperature - the two fugacities, calculated through the given EoS, are equal.

We demonstrate the method for accomplishing this task in the next Example using the RK EoS.

10.3.2 Example 10.1

Calculate the vapor pressure of benzene at 373.15 K using the RK EoS (experimental value: 1.801 bar). Given: T_c = 562.16 K; P_c = 48.98 bar.

The fugacity coefficient of a pure compound according to the RK EoS is given by:

$$\ln\frac{f}{P} = \ln\phi = (z-1) - \frac{a}{bRT^{1.5}}\ln\frac{V+b}{V} + \ln\frac{V}{V-b} - \ln z \tag{A}$$

where V is the molar volume and z the compressibility factor, and (Eq.8.10.4):

$$a = 0.42748\,\frac{R^2 T_c^{2.5}}{P_c} \quad ; \quad b = 0.08664\,\frac{RT_c}{P_c}$$

For the calculation of the compressibility factor it is more convenient to write the RK EoS in the following form:

$$z^3 - z^2 + (A - B - B^2)z - AB = 0 \tag{B}$$

where:

$$A = 0.42748\,\frac{P_r}{T_r^{2.5}} \quad ; \quad B = 0.08664\,\frac{P_r}{T_r}$$

Eqs (A) and (B), combined with Eq.9.10.6, suggest the following trial-and-error procedure for the determination of the vapor pressure:

Step 1. Assume a pressure P.
Step 2. Solve Eq.(B) for the vapor and liquid phase compressibility factors, z^l and z^v.
Step 3. Using these values, calculate f^v and f^l through Eq.(A).

Step 4. Check if these values satisfy Eq.9.10.6 within some tolerance e, say, $e \leq 0.005$ (relative).
Step 5. If the answer is positive, the assumed pressure P represents the vapor pressure; if not, adjust P and return to Step 2.

We now apply this procedure to the case of benzene at 373.15 K.

Step 1. Assume $P = 2.635$ bar.
Step 2. Using $A = 0.0641$ and $B = 0.0070$ in Eq. (B):
$z^v = 0.9398, z^l = 0.9437 \cdot 10^{-2}$, and hence:
$V^v = 11.0649, V^l = 0.1111$, both in l/mol.
Step 3. From Eq.(A): $f^v = 2.4852$ bar; $f^l = 2.5327$ bar.
Step 4. $e = |f^v - f^l| / f^v = 0.019 > 0.005$.
Step 5. Generate a new pressure using the ratio of liquid to vapor fugacity (why, see Problem 10.13):
$P_{new} = P_{old}(f^l/f^v) = 2.685$ bar.
This pressure value leads to $f^v = 2.5295$ bar and $f^l = 2.5339$ bar, which satisfy the requirement: $e \leq 0.005$, and the calculated vapor pressure of benzene at 373.15 K, according to the RK EoS, is 2.685 bar.

10.3.3 Comments

1. The approach described in Example 10.1, applies to all EoS.
2. The obtained value for the vapor pressure of benzene at 373.15 K of 2.685 bar is in substantial error, nearly 50% higher than the experimental one. Poor results would be also obtained with the vdW EoS.

We consider next how a cubic EoS is adjusted to eliminate this very serious problem by using the vdW EoS as a starting point. This approach is similar - in principle - to that used with other cubic EoS of this type, such as the SRK and PR ones.

10.4 Development of a Cubic Equation of State for Vapor Pressure Prediction

10.4.1 The Approach

To correct this failure of the traditional cubic EoS - such as the RK, vdW, and others - to provide reliable predictions of vapor pressures, we start by reversing the problem addressed in Example 10.1: If the vapor pressure value, at a given temperature, is available from experimental measurements, how can an EoS - for example the vdW - be adjusted to represent this value?

To this purpose, we write the vdW EoS as follows:

$$P = \frac{RT}{V-b} - \frac{\alpha a_c}{V^2} \qquad (10.4.1)$$

where b and a_c are given by Eqs (E) and (F) of Example 8.7 respectively.

Next, following the procedure of Example 10.1 - using the given values of temperature and vapor pressure this time - we adjust the value of alpha (α) until the fugacities of the vapor and liquid phases become equal (within e).

Typical results are presented in Table 10.1 for n-butane and demonstrate that alpha is a function of temperature, not equal to one, as the original equation assumes.

10.4.2 The Dependency of Alpha on Temperature

All this exercise will not be of much practical use unless the functional dependency of alpha on temperature is established, so that interpolation and extrapolation from a limited number of data points can be effected for a given compound.

Indeed, after a few trials using the data of Table 10.1 we arrive at the following relationship between alpha and the reduced temperature:

Table 10.1 Values of Alpha as a Function of the Reduced Temperature for n-Butane

T (K)	T_r	P (bar)	Alpha	m
149.98	0.3527	$0.2318 \cdot 10^{-5}$	1.7539	0.7987
179.98	0.4233	$0.3369 \cdot 10^{-2}$	1.6348	0.7974
212.65	0.5001	$0.4834 \cdot 10^{-1}$	1.5230	0.7995
249.98	0.5879	0.3915	1.4091	0.8019
272.65	0.6412	1.0136	1.3467	0.8054
319.98	0.7525	4.5747	1.2280	0.8160
379.98	0.8937	17.3888	1.0966	0.8635

$$\alpha^{1/2} = 1 + m(1 - \sqrt{T_r}) \tag{10.4.2}$$

where $m = 0.8004$. Notice, that the functional relationship of Eq.10.4.2 is of the same type as that introduced first by Soave in his famous modification of the RK EoS, and also used by Peng and Robinson (see Table 8.3).

10.4.3 Generalization of the EoS

The establishment of Eq.10.4.2 allows for the inter- and extrapolation of existing data for a given compound. To effect, however, prediction of vapor pressures for other compounds a method for predicting the slope m for them is needed.

To this purpose we note that, since two-parameter cubic EoS reflect a two-parameter corresponding states principle (Chapter 8), a relationship between the parameter m and the acentric factor ω should be possible, at least for nonpolar compounds.

Following Soave the m values for the C_1 to C_{16} n-alkanes, obtained using a reliable data base of vapor pressures (Salerno et al, 1986), were correlated (Watson et al, 1986) very satisfactorily to ω with Eq.8.10.6:

$$m = d_0 + d_1\omega + d_2\omega^2 \tag{8.10.6}$$

where the values of d_0, d_1, and d_2 are given in Table 8.3.

We consider, next, the third requirement to be met by these EoS, the prediction of saturation liquid volumes, and then return to discuss vapor pressure predictions.

10.5 Prediction of Saturated Liquid Volumes

10.5.1 The Problem

This form of the vdW EoS (Eq.10.4.1 combined with Eq.10.4.2), the SRK, and PR EoS, while providing reliable predictions of vapor pressures for nonpolar compounds, lead to poor results for saturated liquid volumes. This is demonstrated with the results for the C_1 to C_{20} n-alkanes, obtained with the PR EoS and shown in Figure 10.1, where the average absolute error (AAE) is defined by:

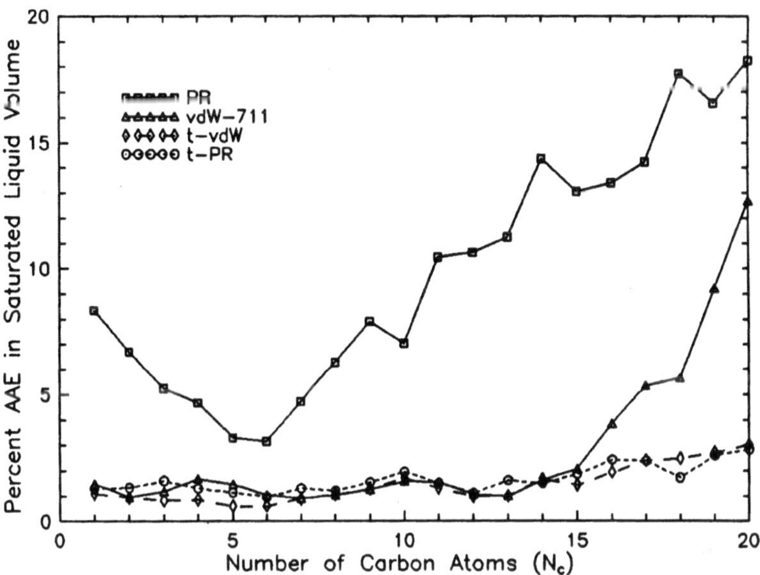

Figure 10.1 Prediction of saturated liquid volumes for the C_1 - C_{20} n-alkanes. (Experimental data from the correlations of such data by Daubert and Danner, 1985, 1986.)

$$AAE = \frac{1}{n}\sum \frac{|Q_{cal}-Q_{exp}|}{Q_{exp}} \tag{10.5.1}$$

Q is here the saturated liquid volume, the subscript *cal* stands for calculated and *exp* for experimental values, and the summation is over the n data points, here thirty, equidistanced on a T_r basis, from the triple to the critical point. The results are poorer with the SRK EoS and still poorer with the vdW EoS.

10.5.2 Methodology and Results

To correct for this failure, we take advantage of the suggestion of Peneloux et al (1982) according to which the vapor pressure calculated by an EoS does not change if the volume is 'translated' by some amount t. Thus, in the case of the vdW EoS, Eq.10.4.1 provides the same value of vapor pressure as its translated form:

$$P = \frac{RT}{(V+t)-b} - \frac{\alpha a_c}{(V+t)^2} \tag{10.5.2}$$

Using experimental saturated liquid molar volume data for a given compound, values of t can be calculated as a function of the reduced temperature. Typical results are presented in Figure 10.2 for some n-alkanes and indicate that the translation t:

a. is approximately constant up to about T_r = 0.7 to 0.8;
b. increases as the critical point is approached; and
c. increases with increasing ω values.

Using the translation values for the C_1 to C_{16} n-alkanes, a generalized expression for t as a function of T_r and ω was developed and is given in Table 8.3. The obtained modified vdW EoS - Eq.10.5.2 - referred to as the vdW-711 EoS, is also presented in this Table. Details are given by Watson et al.

The quality of the obtained saturated liquid volume predictions is demonstrated in Figure 10.1. Notice that very good results are obtained except for high ω values (large n-alkanes).

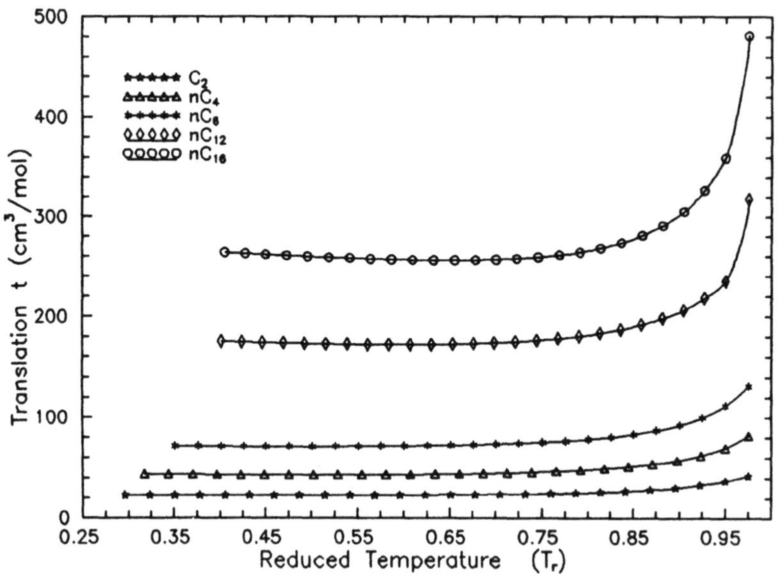

Figure 10.2 Variation of the translation t with T_r for the vdW EoS. (Lines for identification only.)

10.6 Comments

1.The quadratic expressions for m as a function of ω presented in Table 8.3 have a maximum at the ω value shown below:

Equation	ω
SRK	4.97
PR	2.86
vdW-711	3.71

To avoid this maximum, of importance to the petroleum industry where hydrocarbons with very high molecular weight are encountered, a cubic dependency of m on ω has been proposed (Soave, 1979, for the SRK; Robinson and Peng, 1978, for the PR; Magoulas, 1990, for the vdW-711):

$$m = d_0 + d_1\omega + d_2\omega^2 + d_3\omega^3 \tag{10.6.1}$$

Values for the parameters d_0, d_1, d_2 and d_3 are presented in Table 10.2.

2.To extend the good predictions of saturated liquid molar volumes of the vdW-711 EoS to high molecular weight hydrocarbons, Magoulas proposes a similar expression as that of Table 8.3 for the translation t:

$$t = t_0 + (t_c - t_0)\exp(\beta\,|1 - T_r|) \tag{10.6.2}$$

$$t_0 = \frac{RT_c}{P_c}(k_0 + k_1\,\omega + k_2\,\omega^2 + k_3\,\omega^3 + k_4\,\omega^4) \tag{10.6.3}$$

$$\beta = l_0 + l_1\,\omega \tag{10.6.4}$$

$$t_c = \frac{RT_c}{P_c}(z_c' - z_c) \tag{10.6.5}$$

with the parameter values given in Table 10.2. z_c is the experimental compressibility factor which, when not available can be estimated through ω using the expression of Table 8.3; and z_c' is the compressibility factor value given by the original van der Waals EoS and equal to 3/8.

Equation 10.5.2 with the new expression for m, Eq.10.6.1., and the new parameter values for t_0 and β, is referred to as the translated van der Waals (t-vdW) EoS.

3.To improve PR's vapor pressure predictions, and also avoid the need for two m expressions, see Table 10.2, Magoulas proposes:

$$m = d_0 + d_1\omega + d_2\omega^2 + d_3\omega^3 + d_4\omega^4 \tag{10.6.6}$$

with the parameter values given in Table 10.2.

To improve, in addition, the saturated liquid volume predictions of this PR EoS, he presents a translated form:

$$P = \frac{RT}{V+t-b} - \frac{\alpha a_c}{(V+t)(V+t+b)+b(V+t-b)} \quad (10.6.7)$$

where t is given by Eqs 10.6.2 through 10.6.5 with the parameter values of Table 10.2. z_c in this case is 0.3074. The translated version of this form of the Peng-Robinson EoS is referred to as t-PR.

4.Prediction of vapor pressures for the C_1 to C_{20} n-alkanes from 0.1 mm Hg to the critical is presented in Table 10.3 and in graphical form for C_3 and toluene in Figures 10.3 and 10.4 respectively. The following observations can be made:

a.Reasonably good predictions are obtained with the t-PR EoS down to low vapor pressure values (about 10^{-4} bar), where it gives significantly better results than the original one.

b.The vdW-711 and t-vdW EoS give also very good results, but down

Table 10.2 Parameter Values in Eqs 10.6.1, 10.6.3, 10.6.4 and 10.6.6

	SRK(1)	PR(2)	t-vdW(3)	t-PR(3)
d_0	0.47979	0.379642	0.483798	0.384401
d_1	1.57600	1.485030	1.643232	1.522760
d_2	-0.19250	-0.164423	-0.288718	-0.213808
d_3	0.02500	0.016666	0.066013	0.034616
d_4	---	---	0.0	-0.001976
k_0	---	---	0.036722	-0.014471
k_1	---	---	0.063541	0.067498
k_2	---	---	-0.076221	-0.084852
k_3	---	---	0.060362	0.067298
k_4	---	---	-0.015772	-0.017366
l_0	---	---	-7.099630	-10.244700
l_1	---	---	-21.156900	-28.631200

(1) Soave (1979)

(2) Robinson and Peng (1978) for $\omega > 0.5$; for $\omega \leq 0.5$, see Table 8.3

(3) Magoulas (1990)

Table 10.3 Prediction of Vapor Pressures with the vdW-711, PR, t-vdW and t-PR EoS (from 0.1 mm Hg to the Critical)

	Aver. Abs. Error %				Max. Abs. Error %			
N_c	vdW -711	PR	t-vdW	t-PR	vdW-711	PR	t-vdW	t-PR
C_1	1.7	0.8	1.8	1.0	2.9	1.6	2.6	2.8
C_2	1.9	2.9	1.9	1.5	4.9	19.8	4.5	9.4
C_3	1.5	3.0	1.4	1.7	2.4	20.4	2.2	10.9
n-C_4	1.3	3.1	1.3	1.9	2.1	21.1	2.1	11.7
n-C_5	1.3	3.0	1.3	2.0	2.4	22.8	2.4	13.4
n-C_6	1.2	2.9	1.2	2.0	2.3	21.4	2.3	12.1
n-C_7	1.3	3.3	1.3	2.1	2.3	24.3	2.4	13.1
n-C_8	1.3	3.6	1.3	2.1	2.5	25.2	2.8	12.8
n-C_9	1.4	3.7	1.4	1.9	2.6	24.3	2.6	16.0
n-C_{10}	1.7	3.9	1.6	1.8	2.9	23.9	3.0	8.5
n-C_{11}	1.7	2.7	1.7	1.7	3.1	18.6	3.2	7.6
n-C_{12}	1.8	2.6	1.7	1.7	3.3	17.2	3.4	6.8
n-C_{13}	2.1	2.2	2.1	1.7	3.6	14.5	3.6	4.7
n-C_{14}	2.1	2.3	2.1	1.6	3.9	14.5	3.9	5.0
n-C_{15}	2.3	1.9	2.3	1.8	4.2	11.9	4.1	3.2
n-C_{16}	2.7	1.7	2.7	1.7	4.6	9.7	4.6	3.4
n-C_{17}	3.0	1.4	3.0	2.0	4.9	6.8	4.8	4.2
n-C_{18}	3.3	1.3	3.5	2.4	5.4	4.5	5.3	4.8
n-C_{19}	3.1	1.5	3.0	1.9	5.5	7.0	5.3	3.5
n-C_{20}	3.8	1.6	3.3	2.0	6.1	6.0	5.7	3.5
Overall	2.1	2.5	2.0	1.8	3.6	15.8	3.5	7.6

to much lower pressures. Thus they predict the very accurate data for propane of Goodwin and Haynes (1982) down to the triple point of $1.6 \cdot 10^{-9}$ bar with remarkable accuracy.

5. The poorer predictions at very low pressures are due - in addition to uncertainties of the experimental data in that region - to the fact that, as suggested by Table 10.1, the *m* value is not a true constant in all these EoS. This, however, is important only at low T_r values as Eq.10.4.2 indicates and Figure 10.5 demonstrates. Here the error in predicted vapor pressures for n-butane with the PR EoS is plotted versus T_r for specified errors in *m*. Notice, for example, that an error in *m* of 2% has a negligible effect on the predicted P^s value at T_r = 0.8, but a considerable one at T_r = 0.4.

6. Prediction results for saturated liquid volumes with the t-PR and t-vdW EoS are shown in Figure 10.1. Notice the significant improvement of the t-PR over the PR EoS, while for the t-vdW EoS the improvement over the vdW-711 EoS occurs only at high ω values.

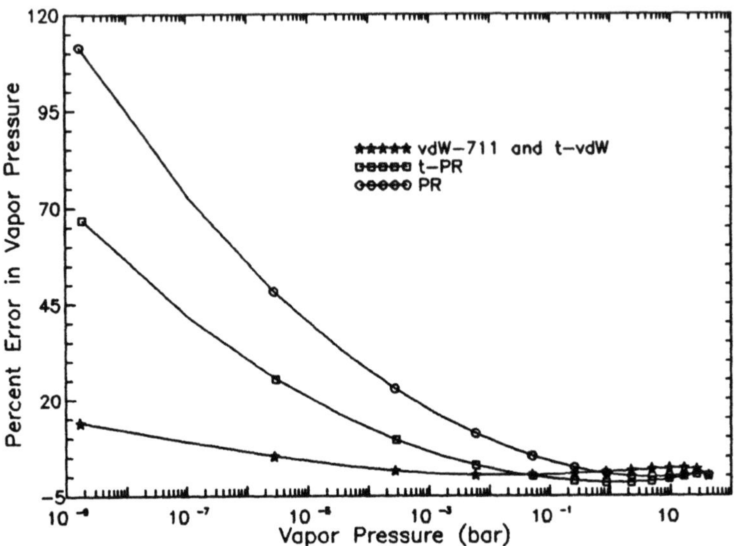

Figure 10.3 Prediction of vapor pressure for propane. (Experimental data from Goodwin and Haynes, 1982.)

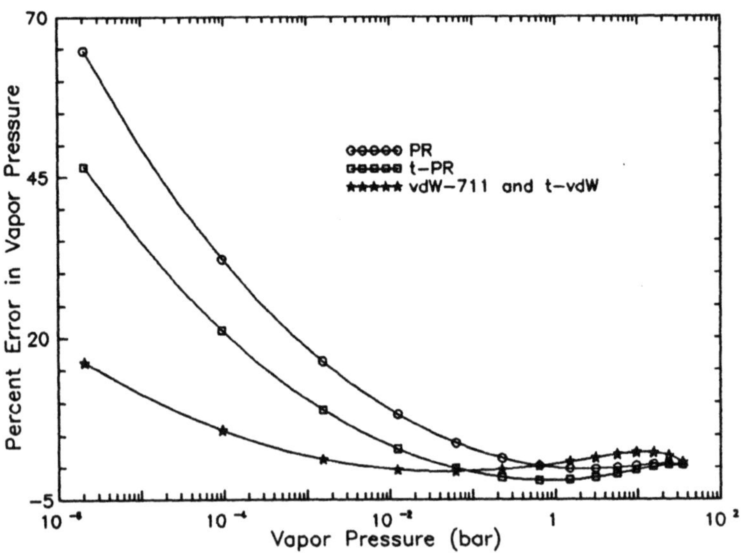

Figure 10.4 Prediction of vapor pressure for toluene. (Experimental data from Daubert and Danner.)

7. In applying cubic EoS for the prediction of vapor pressures attention should be paid to the accuracy of the critical properties and ω values. Small errors in these values can have a significant impact on the accuracy of the obtained P^s predictions, as demonstrated with the typical results shown in Figure 10.6 for the critical temperature, where the effect is most pronounced.

Notice, for example, that an uncertainty of 3 K (0.4%) in the T_c value for nC_{16} leads to a tripling of the error in P^s obtained with the SRK EoS. And such an uncertainty is not unusual for large compounds. The following, for example, T_c values for nC_{16} are given in the literature: 717 K, 720.6 K, and 724.3 K.

8. Vapor pressures and saturated liquid and vapor volumes with the SRK, PR, vdW-711, t-PR and t-vdW can be calculated with the Program EOS presented in Appendix E.

9. For polar compounds the simple dependency of alpha on T_r indicated by Eq.10.4.2 is not, as should be expected (why?), sufficient; rather a temperature dependency that involves non-generalized parameters is needed (Mathias, 1983; Stryjek and Vera, 1986; Gupte et al, 1986; Androulakis et al, 1989).

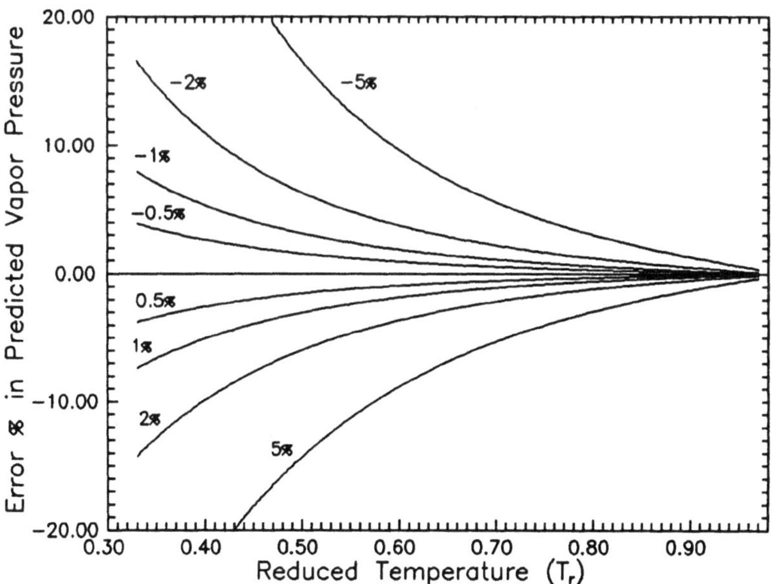

Figure 10.5 The effect of the uncertainty in the m value on the accuracy of vapor pressure predictions for n-butane with the PR EoS.

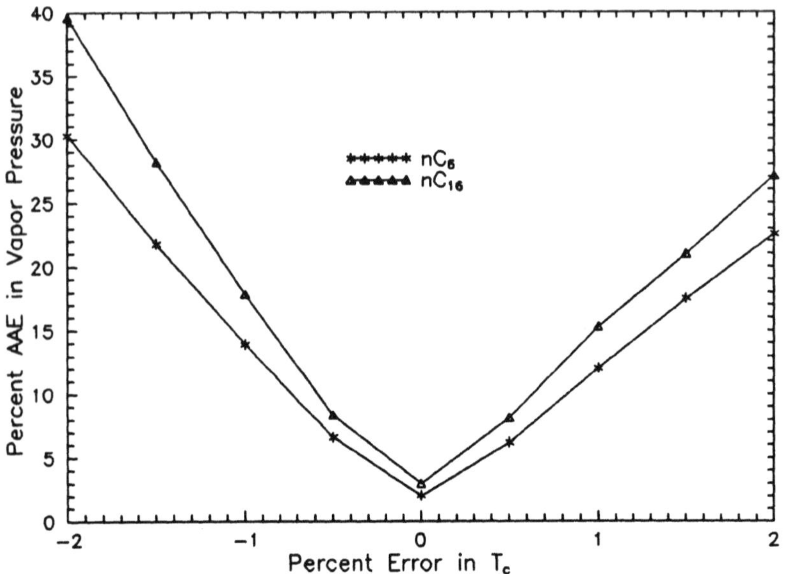

Figure 10.6 Effect of the uncertainty in T_c on the accuracy of predicted vapor pressures using the SRK EoS, holding the P_c and ω values constant.

10.7 Concluding Remarks

1.For nonpolar/weakly polar compounds very good prediction of vapor pressures is obtained with the cubic EoS discussed here. At very low P^s values, however, only the t-vdW and vdW-711 EoS seem to give reliable results, down to about 10^{-5} bar, as suggested for example by the cases of propane and toluene presented in Figures 10.3 and 10.4 respectively.

2.Use of the correct T_c, P_c, and ω values is essential for the reliable prediction of vapor pressures.

3.Both the t-PR and t-vdW EoS give excellent prediction of saturation liquid volumes.

4.Prediction of vapor and gaseous volumes, enthalpies of vaporization and enthalpy and entropy departures with cubic EoS are discussed in Sections 8.10.8, 8.14.1 and 9.8.5 respectively.

5.Application of cubic EoS in the correlation and prediction of vapor-liquid equilibria of mixtures is considered in Chapter 14.

References

Androulakis, I., Kalospiros, N., Tassios, D. P., 1989. *Fluid Phase Equilibria*, **45**, 135.

Daubert, T.E., Danner, R.P., 1985; 1986. *Data Compilation Tables of Properties of Pure Compounds*, DIPPR, AIChE, N.Y. (They provide correlations of available experimental data. Use, thus, of these correlations outside the range of the data involves extrapolation.)

Dymond, J.H., 1985. *The Second International IUPAC Workshop*, Paris.

Goodwin, R.D., Haynes, W.M., 1982. *Thermophysical Properties of Propane from 85 to 700 K at Pressures to 70 MPa*, Monograph 170, National Bureau of Standards, Washington, D.C.

Gupte, P.A., Rasmussen, P., Fredenslund, Aa., 1986. *Ind. Eng. Chem. Fundam.*, **25**, 236.

Magoulas, K., Tassios, D., 1990. *Fluid Phase Equilibria*, **56**, 119.

Mathias, P.M., 1983. *Ind. Eng. Chem. Process Des. Dev.*, **22**, 385.

Peneloux, E., Rauzy, E., Freze, R., 1982. *Fluid Phase Equilibria*, **8**,7.

Peng, D.Y., Robinson, D.B., 1976. *Ind. Eng. Chem. Fundam.*, **15**, 59.

Robinson, D.B., Peng, D.Y., 1978. *The Characterization of the Heptanes and Heavier Fractions for the GPA Peng-Robinson Programs*. GPA Res. Rep. 28.

Salerno, S., Cascella, M., May, D., Watson, P., Tassios, D., 1986. *Fluid Phase Equilibria*, **27**, 15.

Soave, G., 1972. *Chem. Eng. Sci.*, **27**, 1197.

Soave, G., 1979. *Chem. Eng. Sci.*, **34**, 225.

Stryjek, R., Vera, J.H., 1986. *Can. J. Chem. Eng.*, **64**, 323.

Van Ness, H.C., Abbott, M.M., 1982. *Classical Thermodynamics of Non Electrolyte Solutions*, McGraw-Hill, New York.

Watson, P., Cascella, M., May, D., Salerno, S., Tassios, D., 1986. *Fluid Phase Equilibria*, **27**, 35.

Problems

10.1 What requirements, with respect to pure compounds, must be met by EoS used in phase equilibria calculations?

10.2 Develop Eq.(A) of Example 10.1.

10.3 Using the original vdW EoS calculate the vapor pressure of n-octane at 370.76 K and compare it with the value given in Table 10.P.1.

10.4 Using the experimental value of the vapor pressure for nC_8 at 370.76 K of

Table 10.P.1 Vapor Pressure Data for n-Octane

T (K)	370.76	404.98	423.15	473.15
P^s (bar)	0.433	1.200	1.905	5.472

Table 10.P.1, calculate the value of a in the vdW EoS that reproduces this vapor pressure. Repeat the calculations, using the second and third vapor pressure data points given in Table 10.P.1, and in each case determine the value of $\alpha = a/a_c$. What correlation of alpha to T_r do you find best for the obtained values?

10.5 Using the correlation of alpha to T_r developed in the previous Problem, determine the vapor pressure of nC_8 at 473.15 K.

10.6.a. Show that volume translation, i.e. the replacement of V by $V' = (V + t)$ in any EoS, does not effect the calculated vapor pressure.

b. Demonstrate it numerically by redoing Problem 10.5 using an arbitrary translation.

10.7 Calculate the vapor pressure of nC_6 at 286.2 K using the vdW-711, PR, and SRK EoS and the computer Program EOS of Appendix E (expt. value: 0.117 bar).

10.8 Compare the saturated liquid volume obtained in the previous Problem with that given by the correlation of Hankinson and Thomson (Chapter 8).

10.9.a. Develop expressions for the second virial coefficient from the vdW, vdW-711, SRK, and t-PR EoS.

b. Calculate B values from these expressions for n-octane and plot them, along with the recommended values (Dymond, 1985) given in Table 10.P.2, versus T_r.

Table 10.P.2 Second Virial Coefficients (cm^3/mol) for n-Octane

T (K)	380	400	420	460	480	500	525	550	575
B	-1990	-1690	-1490	-1170	-1025	-920	-815	-725	-650

10.10 How do you explain:

a. the difference between the B values obtained from the vdW-711 EoS and the recommended ones?

b. the fact that, in spite of this difference, the EoS gives good predictions for saturated vapor volumes (except close to the critical point)?

10.11 Examine the effect of the uncertainty in the T_c, P_c and ω values on the predicted vapor pressures and saturated liquid volumes for a compound of your choice using the SRK, t-PR, and t-vdW EoS.

10.12.a. Discuss the problems associated with the sharp change of the translation with T_r close to $T_r = 1$. (See Androulakis et al.)

b. Answer the question posed in comment 9 of Section 10.6.

10.13 Develop the following recursive formula to be used in calculating vapor pressures through an equation of state (Van Ness and Abbott, 1982):

$$P_{j+1}^s = P_j^s\,[(z^v - z^l)_j - \ln(f^v/f^l)_j]/(z^v - z^l)_j$$

11
Properties of Mixtures

11.1 Introduction

Evaluation of thermophysical properties represents the second main objective of Chemical Engineering Thermodynamics. Pure compounds were considered in the previous three Chapters, a general discussion of mixtures in this one, and phase and chemical reaction equilibria in the ensuing Chapters 12 through 15.

It is important that we state clearly from the beginning the nomenclature to be used in this Chapter, to eliminate potential confusion in the representation of pure compound and mixture properties. We define:

M: Mixture molar property (V, H, etc.)
N: Number of moles in a mixture
NM: Total mixture property (NV, NS, etc.)
M_i: Pure compound molar property (V_i, H_i, etc.)

Consider now a homogeneous mixture containing N_1 moles of compound 1, N_2 of 2,... and N_k of k, where k is the number of components in the mixture, for a total of N moles; and the following question: How is the total property of this mixture, for example the total volume NV, to be determined at some specified temperature T and pressure P?

The simplest answer to this question would be provided by the equation:

$$NM = N_1M_1 + N_2M_2 + \ldots\ldots + N_kM_k = \sum N_iM_i \qquad (11.1.1)$$

with all M_i's evaluated at the same T and P.

Unfortunately, nature is not very obliging and this equation is not valid. Instead the correct expression is:

$$NM = \sum N_i\bar{M}_i$$

or, dividing by N:

$$M = \sum x_i\bar{M}_i$$

where $\bar{M}_i$ is called the *partial molar property* of component *i*. Thus, the molar volume V of a binary mixture is not given by:

$$V = x_1 V_1 + x_2 V_2, \text{ but by: } V = x_1 \bar{V}_1 + x_2 \bar{V}_2$$

In other words, a given property of compound *i* has a different value when *i* is the component of a mixture from that in its pure state, both calculated at the same temperature and pressure. This difference results from differences in the:

* inter(unlike)- and intra(like)-molecular forces, and
* molecular size and shape,

of the components of the mixture.

If such differences are not present, a hypothetical case of course, the mixture is referred to as an *ideal* one and, as we will see in Section 11.12, Eq. 11.1.1 holds except when entropy is involved, where the entropy of mixing must be accounted for (Section 11.12.2).

11.2 Objective and Approach

The main objective in this Chapter is to develop familiarity with mixture properties and the methods used for their evaluation. More specifically to:

1. Understand the concept of partial molar property, which describes the behavior of the component of a mixture, and contrast it to its behavior in the pure state.
2. Discuss the adjustments that must be made in the methods that we used in the prediction of pure compound properties:
   * the Corresponding States Principle; and
   * Equations of State,

   so that they can be used for mixtures.
3. Extend the concept of pure compound fugacity into that of the component of a mixture, and establish the methodology for determining it.
4. Develop expressions that isolate the mixing effect by comparing mixture properties to their values:
   a. in the absence of mixing, and
   b. in a hypothetical ideal solution.

We start with the definition and a brief description of partial molar properties (pmp) and then examine:

1. how mixture properties are obtained from pmp;
2. the reverse case, i.e. how pmp are evaluated from mixture properties;

3. the interdependency among the pmp of a mixture.

We proceed with a discussion of the so-called *mixing* and *combining* rules, which allow us to extend pure component and binary system information into the determination of mixture properties.

We turn then to the very important concept of the fugacity of the component of a mixture, and proceed to outline the two main approaches used in its determination:

1. equations of state; and
2. the *standard* state fugacity.

To help us appreciate the importance of fugacities, we apply them to the determination of the solubility of solids and liquids in gases. (Extensive application of fugacities will be made in the ensuing four Chapters where phase and chemical equilibria are discussed.)

We proceed with a discussion of two auxiliary mixture quantities, which help us isolate the mixing effects from:

1. pure component behavior, through the so-called *properties of mixing*;
2. ideal solution behavior, through the so-called *excess properties*.

We close with some concluding remarks.

11.3 Partial Molar Properties

11.3.1 Definition

Consider a homogeneous mixture made up of N_1, N_2, ..., N_k moles of components 1, 2, ..., k, respectively. To this mixture we add a small amount of component i, dN_i, while maintaining the temperature T, pressure P, and number of moles of the other components N_j constant. We measure the resulting change in the total property, $d(NM)$, and define the ratio $d(NM)$ over dN_i:

$$\bar{M}_i = \left(\frac{\partial(NM)}{\partial N_i}\right)_{T,P,N_j} \tag{11.3.1}$$

as the *partial molar property* $\bar{M}_i$ of component i of this mixture at the specified composition, temperature, and pressure. (The subscript j, here and throughout the Chapter, assumes all values from 1 through k, except i.)

Notice that the change must be determined at constant temperature, pressure, and number of moles of the other components. If, for example, the addition of the dN_i moles takes place at constant pressure - say the open atmosphere - we must make sure that any change in temperature, resulting from the mixing, is accommodated for by the addition or removal of the necessary amount of heat so that it remains also constant.

It follows from Eq.11.3.1 that, for a mixture of specified components, the partial molar properties are functions of its composition, temperature, and pressure:

$$\bar{M}_i = f(x_1, x_2, \ldots, T, P) \tag{11.3.2}$$

We demonstrate the physical meaning of partial molar properties with the simple case of the volume of a binary mixture in the next Example.

11.3.2 Example 11.1

Consider a mixture containing 3 moles of methanol(m) and 7 moles of water(w) at 25°C and 1 atm. To this mixture we add a small amount of water, say 0.1 mole (see Problem 11.2.a), while maintaining the temperature, pressure, and number of moles of methanol constant. The volume of the mixture increases by 1.78 cm^3.

a. What is the partial molar volume of water?

b. What is the effect of mixing on the molar volume of water?

c. What would the partial molar volume of water be equal to, if the solution were ideal?

Given: V_w (25°C, 1 atm) = 18.1 cm^3/mol.

a. From the definition:

$$\bar{V}_w = \left(\frac{\partial(NV)}{\partial(N_w)}\right)_{T,P,N_m} \approx \left(\frac{\Delta(NV)}{\Delta(N_w)}\right)_{T,P,N_m} = \frac{1.78}{0.10} = 17.8\ \frac{\text{cm}^3}{\text{mol}}$$

b. The molar volume of water in the mixture, i.e. its partial molar volume, is smaller by 1.7% than its value in the pure state at the same T and P.

c. In this case, $\bar{V}_w = V_w = 18.1$ cm^3/mol.

Comments

1. The mixing effect on volume is usually small, when the mixture is well below its critical point, but it can be very pronounced as this point is approached. Thus, Prausnitz et al (1986) report that for the saturated liquid mixture CO_2(1) - n-butane(2), at $x_1 = 0.6$ and 71°C: $\bar{V}_2 = 20$ cm^3/mol, while V_2(saturated liquid at 71°C) = 115 cm^3/mol. Actually at the critical point, which at this temperature

occurs at $x_1 = 0.713$ and $P = 81.7$ bar (Olds et al, 1949), $\bar{V}_2 = -155$ cm^3/mol.

2. The mixing effect can be also be very pronounced for other properties, especially fugacity, even well away from the mixture critical point.

11.4 Mixture Properties from Partial Molar Properties

11.4.1 The Problem

We stated in the Introduction that, unfortunately, the simple Eq.11.1.1 cannot describe mixture properties.

If, on the other hand, the fundamental equation for this mixture were available, all properties of the system could be determined as we saw in Section 9.5.1. Thus, if the fundamental equation for the Gibbs free energy of the mixture which, using the nomenclature of this chapter becomes:

$$NG = f(T, P, N_1, N_2, \ldots, N_k) \tag{9.4.13}$$

were available, then its volume - as an example - could be determined:

$$NV = \left(\frac{\partial(NG)}{\partial P}\right)_{T,N_i}$$

Unfortunately, the analytical relationship for Eq.9.4.13 is not available as we also saw in that Section.

11.4.2 The Approach

We resort therefore, as we did in Section 9.5.2, to the fundamental postulate according to which the total property NM of a homogeneous mixture - at a specified temperature T and pressure P - is a function of the amounts of its components only:

$$NM = f(N_1, N_2, \ldots, N_k) \tag{11.4.1}$$

Differentiation yields:

$$d(NM) = \left(\frac{\partial(NM)}{\partial N_1}\right)_{T,P,N_j} dN_1 + \ldots + \left(\frac{\partial(NM)}{\partial N_k}\right)_{T,P,N_j} dN_k \tag{11.4.2}$$

Notice that the derivatives are, according to the definition, Eq.11.3.1, the partial molar properties of NM. Thus Eq.11.4.2 becomes:

$$d(NM) = \sum \bar{M}_i dN_i \qquad (11.4.3)$$

Integration of Eq.11.4.3 from zero to the actual values of the number of moles, at constant temperature, pressure, and composition - a path along which all $\bar{M}_i$ values remain constant according to Eq.11.3.2 - yields:

$$NM = \sum N_i \bar{M}_i \qquad (11.4.4a)$$

or, dividing by the total number of moles N:

$$M = \sum x_i \bar{M}_i \qquad (11.4.4b)$$

i.e. the equations presented in the Introduction.

11.4.3 Comments

1.Example 11.1 and Eq.11.4.4 indicate that the partial molar property of a given component i represents the value of this property when i is surrounded by other compounds and itself, while the pure compound value when i is in the pure state.

2.The partial molar property can be thus visualized as an 'effective' value of this property in the given mixture. For example, the total volume of a mixture is not determined by using the values of the molar volumes of the pure components but rather the 'effective' values of the volumes of its components in the solution, the partial molar volumes (Eq.11.4.4).

3.If the partial molar properties for the components of a mixture are known at a given T, P, and overall composition, the corresponding property of the mixture at the same conditions can be evaluated.

11.5 Partial Molar Properties from Mixture Properties

11.5.1 The Analytical Expression

In the previous Section we developed an expression for the evaluation of a mixture property M in terms of the corresponding partial molar properties. In this Section we consider the reverse case: assuming that we have an expression for M in terms of composition, at a given pressure P and temperature T (Eq.11.4.1):

$$M = f(x_1, x_2, \ldots, x_k) \qquad (11.5.1)$$

we will develop an expression for the evaluation of the mixture partial properties.

Starting with the definition of $\bar{M}_i$, Eq.11.3.1, we obtain:

$$\bar{M}_i = \frac{\partial(NM)}{\partial N_i} = M + N\frac{\partial M}{\partial N_i} \qquad (11.5.2)$$

where the partial derivatives are taken at constant T, P, and number of moles of the other components.

To evaluate the derivative $(\partial M/\partial N_i)$ we note that in terms of independent variables, since $\Sigma x_i = 1$, Eq.11.5.1 should be written as:

$$M = f(x_1, x_2, \ldots, x_{i-1}, x_{i+1}, \ldots, x_k) \qquad (11.5.1a)$$

Using chain-rule differentiation:

$$\bar{M}_i = M + N\sum\left(\frac{\partial M}{\partial x_j}\frac{\partial x_j}{\partial N_i}\right) = M + N\sum\left(\frac{\partial M}{\partial x_j}\frac{\partial(N_j/N)}{\partial N_i}\right) = M - \sum x_j\frac{\partial M}{\partial x_j} \qquad (11.5.3)$$

where the summation is over all components except i and $(\partial M/\partial x_j)$ is taken at constant T, P, and x_l ($l \neq i$ and j).

Eq.11.5.3 is more convenient to use than Eq.11.3.1 because, in the typical case, we have an expression of M as a function of composition rather than of NM as a function of the number of moles of the mixture components.

11.5.2 Example 11.2

The molar vapor volume of a mixture of acetonitrile(1) and acetaldehyde(2) at 100°C and 1.013 bar is given by the following expression:

$$V = \frac{RT}{P} + [ay_1^2 + by_2^2 + 2cy_1y_2], \quad \frac{\text{cm}^3}{\text{mol}} \qquad (A)$$

where: a = -1935; b = -530; c = -2440, all in cm³/mol. Calculate $\bar{V}_1$ as a function of composition.

From Eq.11.5.3:

$$\bar{V}_1 = V - y_2\frac{\partial V}{\partial y_2} \qquad (B)$$

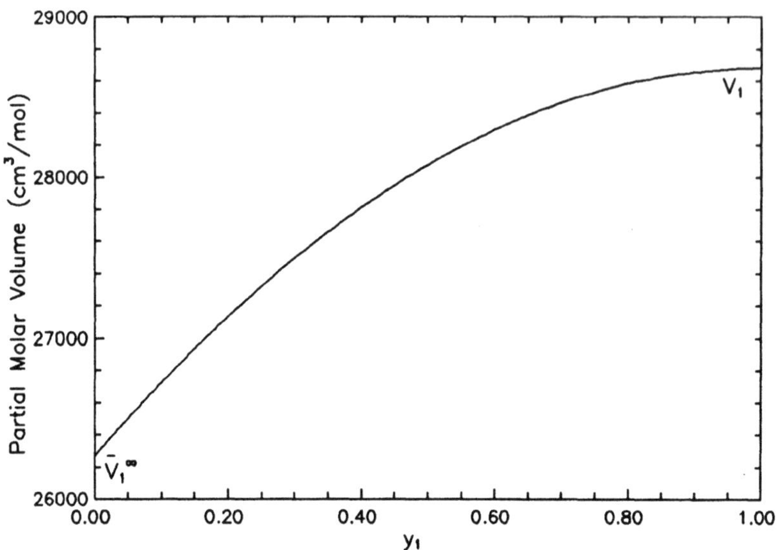

Figure 11.E.2 Variation of the partial molar volume of component (1) with y_1 for the system acetonitrile(1) - acetaldehyde(2) at 100°C and 1.013 bar.

where the partial derivative is taken at constant T and P, but not y_1:

$$\frac{\partial V}{\partial y_2} = 2ay_1(-1)+2by_2+2cy_1+2cy_2(-1) \tag{C}$$

Introduction of Eqs (A) and (C) into (B) yields, after some algebra:

$$\bar{V}_1 = \frac{RT}{P}+a(y_1^2+2y_1y_2)+y_2^2(2c-b) \ , \ \frac{\text{cm}^3}{\text{mol}}$$

The results are presented in Figure 11.E.2, where also shown are: the molar volume of pure 1, V_1, and the value of $\bar{V}_1$ at the limit $y_1 = 0$, referred to as "the partial molar volume of component 1 at infinite dilution in 2", $\bar{V}_1^{\infty}$.

Comments

1.Notice that as the composition of acetonitrile becomes smaller, its deviation from its pure compound behavior becomes more pronounced: it is only -0.3% at $y_1 = 0.8$, but -8.4% in the limit $y_1 = 0.0$.

2.We will see that this phenomenon is typical with other thermodynamic properties as well. It seems that the molecules of component i - in this case those of acetonitrile - 'feel' the molecules of the other components - in this case those of acetaldehyde - as 'hostile'. And this 'feeling' becomes stronger, as the number of the other molecules increases.

3. This 'hostility' increases also with increasing dissimilarity. Thus water mixes completely with methanol in all proportions but not with pentanol, where the dissimilarity is increased (why?).

4. For ideal solution behavior, $\bar{V}_i = V_i$, as it was the case in the methanol-water system of Example 11.1 (Part c.).

5. Notice that as y_1 approaches 1, $\bar{V}_1$ tends to V_1, i.e. ideal solution behavior is approached with respect to this component.

6. It should be emphasized that ideal solution behavior of a gas mixture does not necessarily mean ideal gas behavior. For, if this mixture were to behave as an ideal gas, then:

$$\bar{V}_1 = V_1(ideal\ gas) = \frac{RT}{P} = 30625\ \frac{\text{cm}^3}{\text{mol}}$$

7. This is why an ideal solution of gases is also referred to as an "ideal solution or mixture of real gases". Thus an ideal gas is an ideal solution, but the reverse is not necessarily the case.

11.6 Interdependency of the Partial Molar Properties

11.6.1 The Gibbs-Duhem Equation

In the general case the total mixture property NM is a function of the temperature T and pressure P in addition to the number of moles of its components N_1, N_2,..., N_k and Eq.11.4.1 becomes:

$$NM = f(T, P, N_1, \ldots, N_k) \tag{11.6.1}$$

Consequently:

$$d(NM) = \left(\frac{\partial(NM)}{\partial T}\right)_{P,N_i} dT + \left(\frac{\partial(NM)}{\partial P}\right)_{T,N_i} dP + \sum\left(\frac{\partial(NM)}{\partial N_i}\right)_{T,P,N_j} dN_i =$$

$$= N\left(\frac{\partial M}{\partial T}\right)_{P,x_i} dT + N\left(\frac{\partial M}{\partial P}\right)_{T,x_i} dP + \sum \bar{M}_i dN_i \tag{11.6.2}$$

From Eq.11.4.4a:

$$d(NM) = \sum N_i d\bar{M}_i + \sum \bar{M}_i dN_i$$

Comparison with Eq.11.6.2, followed by division by N, yields:

$$\sum x_i d\bar{M}_i - \left(\frac{\partial M}{\partial T}\right)_{P,x_i} dT - \left(\frac{\partial M}{\partial P}\right)_{T,x_i} dP = 0 \qquad (11.6.3)$$

This expression represents the *Gibbs-Duhem* equation and indicates that the intensive properties of the mixture: temperature, pressure and partial molar properties, cannot vary independently. Restricted to constant T and P, Eq.11.6.3 becomes:

$$\sum x_i d\bar{M}_i = 0 \qquad (11.6.4)$$

which is referred to as the *restricted* form of the Gibbs-Duhem equation.

We will present two applications of the restricted form of the Gibbs-Duhem equation in Examples 11.3 and 11.4, while Eq.11.6.3 will be used in Chapter 13 to evaluate the thermodynamic consistency of vapor-liquid equilibrium data.

11.6.2 Example 11.3

The following expressions are proposed for the composition dependency of the partial molar volumes of a binary mixture at a given temperature and pressure:

$$\bar{V}_1 = V_1(1 + ax_2) \qquad (A)$$

$$\bar{V}_2 = V_2(1 + bx_1) \qquad (B)$$

V_1 and V_2 are the molar volumes of pure 1 and 2 respectively, at the same temperature T and pressure P as the mixture; and a and b are constants, functions of T and P only. Is there any restriction on the values of a and b?

Introduction of Eqs (A) and (B) into the restricted form of the Gibbs-Duhem equation yields:

$$x_1(aV_1)dx_2 + x_2(bV_2)dx_1 = 0$$

Since, $dx_1 = -dx_2$:

$$(x_1 V_1 a - x_2 V_2 b)dx_2 = 0 \qquad (C)$$

In order that Eq.(C) is valid for all values of dx_2: $x_1 V_1 a - x_2 V_2 b = 0$

or:

$$a = b\frac{x_2}{x_1}\frac{V_2}{V_1}$$

The only way that a and b are constants, independent of composition, is that: $a = b = 0$.

We conclude, therefore, that the interdependency between the two partial molar volumes, required by the Gibbs-Duhem equation, restricts the values of a and b to zero. Under these conditions, Eqs (A) and (B) refer to an ideal solution.

11.6.3 Example 11.4

The partial molar volume of water(w) in its mixture with methanol(m) at 25°C and 1 atm can be approximated by:

$$\bar{V}_w = 18.1 + a x_m^2 \;;\; a = -3.2 \text{ cm}^3/\text{mol} \tag{A}$$

Develop an expression for the partial molar volume of methanol at the same conditions. Given: V_m (1 atm and 25°C) = 40.7 cm³/mol

From the restricted form of the Gibbs-Duhem equation, since P and T remain constant:

$$x_w d\bar{V}_w + x_m d\bar{V}_m = 0$$

Solution for $d\bar{V}_m$, and evaluation of $d\bar{V}_w$ from Eq.(A), yields:

$$d\bar{V}_m = \frac{-x_w}{x_m} d\bar{V}_w = \frac{-x_w}{x_m} 2 a x_m dx_m = 2 a x_w dx_w$$

Integration yields:

$$\bar{V}_m = a x_w^2 + c$$

where c is the integration constant. At $x_w = 0$, $c = V_m$, and:

$$\bar{V}_m = 40.7 + a x_w^2 \tag{B}$$

We conclude, therefore, that if the partial molar property $\bar{M}_i$ for the one component of a binary mixture is known as a function of composition at a given temperature T and pressure P from experimental measurements, the partial molar property $\bar{M}_j$ of the other component can be determined at the same temperature and pressure without the need for further experimental work. The value for the property in the pure state M_j, at the same T and P as the mixture, must also be known however.

11.7 Evaluation of Mixture Properties

11.7.1 The Problem and Approach

Eq.11.4.4 indicates that determination of mixture properties requires values of the corresponding partial molar properties. But according to Eq.11.5.3, evaluation of the partial molar properties requires knowledge of the corresponding mixture property.

Mixture properties are consequently obtained from experimental measurements or, in their absence, through estimation methods.

Experimental data are available for several binary systems, but for very few multicomponent ones that represent the common industrial case. Use of estimation techniques represents the typical methodology for mixtures, therefore, just as it did with pure compounds.

In discussing physical and thermodynamic properties of pure fluids in Chapters 8 and 9, we concluded that the main estimation techniques are:

1. Pitzer's formulation of the Corresponding States Principle (CSP), and
2. Equations of State (EoS), especially the virial and cubic ones.

Furthermore, the success of these methods in the estimation of pure compound properties suggests their application to mixtures as well.

This, in turn, leads to the following question: what adjustments must be made to these techniques, so that they can be used for mixtures?

If, for example, we are to estimate the molar volume of a mixture of nitrogen and n-butane - at a specified temperature, pressure, and composition - using the Pitzer correlation, what values will be used for T_c and P_c to determine the reduced temperature and pressure values needed for the corresponding Tables (Appendix D)?

Or, if we were to use the Redlich-Kwong - or any other - cubic EoS for the same purpose, how will the parameter values for the EoS be determined?

It should be apparent that, whatever the method used, it should rely on the information more likely to be available:

1. pure component properties; and,
2. on occasion, binary data.

Otherwise, if extensive mixture data are needed, the predictive character of the technique is lost.

Evaluation of mixture parameters from pure component and, if available, binary parameters is carried out by using the so-called *mixing* and *combining* rules.

We will discuss first the virial equation, because of its theoretical basis, and then consider the empirical approaches: Pitzer's CSP and cubic EoS.

11.7.2 The Virial Equation

a. Mixture Coefficients

Because of its theoretical basis, i.e. its development from Statistical Mechanics, the virial equation (Section 8.9) represents the only EoS where rigorous *mixing rules* for the mixture coefficients are available. Thus:

$$B = \sum\sum y_i y_j B_{ij} \tag{11.7.1}$$

B_{ii} is the second virial coefficient of pure i, B_{jj} of pure j, and B_{ij} is the second virial coefficient for the pair i-j, referred to as *cross virial coefficient* (not to be confused with the mixture second virial coefficient given by Eq.11.7.1).

Obviously $B_{ij} = B_{ji}$ and, as with the case of a pure compound, B_{ij} is a function of temperature only. The summations, finally, are over all mixture components. Thus, for a binary mixture:

$$B = y_1^2 B_{11} + y_2^2 B_{22} + 2 y_1 y_2 B_{12}$$

Similar expressions exist for the third and higher coefficients. We will restrict, however, our discussion to the virial equation truncated after the second term since higher coefficients, as we have seen, are rarely available.

Values for the cross coefficient B_{ij} are determined from experimental data, such as *PVT* ones, for the binary system i-j. But such experimental B_{ij} values are not often available and, consequently, we must resort to some estimation technique, just as we did with B for pure compounds.

Use of the Tsonopoulos and the Abbott correlations, Examples 8.5 and 8.6, for the pair i-j requires values for the:

* critical temperature, $T_{c_{ij}}$, and pressure $P_{c_{ij}}$; and
* acentric factor, ω_{ij}.

The following rules are recommended for this purpose:

$$T_{c_{ij}} = (T_{c_i} T_{c_j})^{1/2} \tag{11.7.2}$$

$$P_{c_{ij}} = \frac{z_{c_{ij}} R T_{c_{ij}}}{V_{c_{ij}}} \tag{11.7.3}$$

$$\omega_{ij} = \frac{\omega_i + \omega_j}{2} \tag{11.7.4}$$

where:

$$z_{c_{ij}} = \frac{z_{c_i} + z_{c_j}}{2} \tag{11.7.5}$$

$$V_{c_{ij}} = \left(\frac{V_{c_i}^{1/3} + V_{c_j}^{1/3}}{2}\right)^3 \tag{11.7.6}$$

A single subscript refers to pure component values and a double one to binary parameters.

Eqs 11.7.2 through 11.7.6 provide recipes for the determination of these binary parameters by combining pure component properties. They are, hence, referred to as *combining* rules. Notice that:

1. The equations are purely empirical. Their only justification is that, when used with the Tsonopoulos correlation, they lead to reasonable estimates of B_{ij} values.
2. These binary parameters have no particular physical meaning. They are just constructs, useful for the estimation of mixture properties.
3. Eq.11.7.2 expresses what is referred to as the *geometric mean* combining rule, whose origin and limitations are discussed in Section 7.6.4.
4. The performance of these combining rules becomes poorer, as we should expect, with increasing dissimilarity between species i and j. It would indeed be naive to expect that pure component properties alone would suffice for the prediction of mixture properties, considering the difference between intra- and inter- (like and unlike species) molecular forces. And this difference becomes, of course, more pronounced as the dissimilarity between the species increases.

For improved thus results, Eq.11.7.2 is written as:

$$T_{c_{ij}} = (1 - l_{ij})(T_{c_i} T_{c_j})^{1/2} \tag{11.7.7}$$

The parameter l_{ij} is referred to as the *interaction coefficient* and it is assumed to be independent of temperature and pressure. Values for l_{ij} are obtained from experimental B_{ij} values.

b. Comments on the Interaction Coefficient

1. l_{ij} can be set equal to zero, only for mixtures of compounds that are very similar.
2. For methane i with some hydrocarbon j, Tsonopoulos (1979) suggests that the following empirical expression:

$$l_{ij} = 0.0279\,(\ln Nc_j)^2 \tag{11.7.8}$$

correlates successfully the available l_{ij} values - obtained by fitting experimental data - to the number of carbon atoms in the hydrocarbon j, Nc_j.

Notice that, as expected, increasing dissimilarity - difference in the molar volumes in this case, as expressed through increasing Nc_j values - leads to larger l_{ij} values. Thus, l_{ij} is equal to 0.013 for the C_1-C_2 binary, but to 0.148 for the C_1-C_{10} one.

3. For ethane or ethylene i with some other hydrocarbon j, Tsonopoulos suggests:

$$l_{ij} = 0.0117(Nc_j - Nc_i - 1);\; Nc_j > Nc_i \tag{11.7.8a}$$

This expression can be used for other hydrocarbon pairs, with less confidence however.

4.The same reference gives l_{ij} values for several gases (CH_4, C_2H_4, CO_2, N_2, etc.) with nonpolar and some polar compounds.

5.In general l_{ij} values for nonpolar/weakly polar systems are small, typically in the range 0.0 to 0.2, and can be often estimated with reasonable accuracy. The reliability, thus, of the obtained B_{ij} values - and consequently of the calculated mixture properties - for such systems is rather high.

6.For systems containing polar components, however, l_{ij} values can be larger; and in the few cases where there is strong solvation, i.e. hydrogen bonding between the molecules of the different species, l_{ij} can assume negative values that are functions of temperature. Thus for the acetonitrile-acetaldehyde system, $l_{12}(40°C) = -0.32$; $l_{12}(100°C) = -0.23$ (see Problem 11.41).

7.Even though some guidelines for estimating l_{ij} values for polar systems are given by Tsonopoulos (1974; 1975), the reliability of the obtained B_{ij} values is rather low (unless, of course, l_{ij} for the system under consideration is available from experimental data).

8.For the estimation of B_{ij} values in the Hayden and O'Connell (1975) correlation, see the original article and also Prausnitz et al (1980).

9.When truncated after B, the virial equation should be used in the pressure range given by the following empirical expression (Prausnitz et al, 1980):

$$P \leq \frac{T}{2}\frac{\sum y_i P_{c_i}}{\sum y_i T_{c_i}} \tag{11.7.9}$$

For a pure compound, this equation reduces to Eq.8.9.5.

11.7.3 The Pitzer Corresponding States Correlation

In using the Pitzer correlation (Section 8.7.2) we need values of the mixture critical properties and acentric factor, which are calculated from empirical mixing rules. The obtained values are not, consequently, the true critical properties of the mixture. They are referred to, appropriately, as *pseudocritical* properties.

When using Pitzer's correlation with the Tables of Lee and Kesler, discussed in Chapters 8 and 9, the following mixing rules are recommended by the authors (the subscript m denotes mixture):

$$T_{c_m} = \frac{1}{8V_{c_m}}\sum\sum y_i y_j (V_{c_i}^{1/3} + V_{c_j}^{1/3})^3 (T_{c_i} T_{c_j})^{1/2} \qquad (11.7.10)$$

$$P_{c_m} = (0.2905 + 0.085\,\omega_m)\frac{RT_{c_m}}{V_{c_m}} \qquad (11.7.11)$$

$$\omega_m = \sum y_i \omega_i \qquad (11.7.12)$$

where:

$$V_{c_m} = \frac{1}{8}\sum\sum y_i y_j (V_{c_i}^{1/3} + V_{c_j}^{1/3})^3 \qquad (11.7.13)$$

$$V_{c_i} = (0.2905 + 0.085\,\omega_i)\frac{RT_{c_i}}{P_{c_i}} \qquad (11.7.14)$$

For improved results, an interaction coefficient in the product $(T_{c_i} T_{c_j})^{1/2}$ is required in Eq.11.7.10 (see Reid et al, 1987), but it is rarely used in practice.

11.7.4 Cubic Equations of State

a. Mixture Coefficients

The following general mixing rules are used with cubic (Chapters 8 and 10) equations of state:

$$a = \sum\sum x_i x_j a_{ij} \qquad (11.7.15)$$

$$b = \sum\sum x_i x_j b_{ij} \qquad (11.7.16)$$

where:

$$a_{ii} = a_{c_i}\alpha_i \qquad (11.7.17)$$

$$b_{ii} = b_i \qquad (11.7.18)$$

Typically, a geometric mean combining rule is used for the interaction parameter a_{ij}:

$$a_{ij} = (a_{ii} a_{jj})^{1/2} \qquad (11.7.19)$$

and an arithmetic mean one, for the parameter b_{ij}:

$$b_{ij} = \frac{b_{ii} + b_{jj}}{2} \tag{11.7.20}$$

In the latter case, Eq.11.7.16 becomes:

$$b = \sum x_i b_i \tag{11.7.21}$$

Finally for translated EoS, the following mixing rule is required for the translation parameter of the mixture t (Peneloux et al, 1982):

$$t = \sum x_i t_i \tag{11.7.22}$$

As with the case of B_{ij}, for improved results an interaction coefficient k_{ij} is needed in Eq.11.7.19:

$$a_{ij} = (1 - k_{ij})(a_{ii} a_{jj})^{1/2} \tag{11.7.23}$$

Notice finally the similarity of Eqs.11.7.15 and 11.7.16 with Eq.11.7.1. The first two, however, are purely empirical; the latter is developed from Statistical Mechanics.

b.Comments on the Interaction Coefficient

1.Values for the interaction coefficient are obtained from experimental data involving a mixture of compounds i and j.

2.Again as with the case of the virial equation, they are small, typically between 0.0 and 0.2 for nonpolar or weakly polar systems, and can be larger - or even negative - for polar ones.

3.The values vary from one EoS to the other.

4.For hydrocarbons that are not very different in size, zero values for the interaction coefficient can be used.

5.For systems containing CO_2 and n-paraffins, k_{ij} values for the Peng-Robinson (PR) EoS are given by Kato et al (1981).

6.For systems containing CO_2, H_2S, N_2, and CO with hydrocarbons, k_{ij} values for the Soave-Redlich-Kwong (SRK) EoS are given by Graboski and Daubert (1978).

7.For H_2-hydrocarbon systems, see Lin (1980), and Tsonopoulos and Heidman (1986).

8.For the volumetric properties of gaseous water with some nonpolar gases, see De Santis et al (1974).

We examine the prediction of the volumetric behavior of binary mixtures in the next two Examples 11.5 and 11.6; and the evaluation of interaction coefficients, in Example 11.7. We proceed, then, with some general comments on the estimation of mixture properties, followed by the discussion of fugacities in mixtures.

11.7.5 Example 11.5

Estimate the compressibility factor of a methane(1) - n-butane(2) vapor mixture at 400 psia, 100°F, and $y_1 = 0.8942$. Use the virial equation, Pitzer correlation, and SRK EoS. Experimental value: $z = 0.9135$ (Sage et al, 1940). Critical property and ω values are presented below:

	T_c (K)	P_c (bar)	z_c	V_c (cm³/mol)	ω	T_r	P_r
C_1	190.58	46.04	0.288	99.1	0.011	1.632	0.599
nC_4	425.18	37.97	0.274	255.1	0.193	0.732	0.726

a. The Virial Equation

From Eq. 11.7.9, $P < 32$ atm = 473 psia and, consequently, the virial equation truncated after B is applicable. From the Tsonopoulos correlation:

$$B_{11} = -38.4 \text{ cm}^3/\text{mol}; \quad B_{22} = -650.2 \text{ cm}^3/\text{mol}$$

For B_{12} using Eqs 11.7.2 through 11.7.8: $l_{12} = 0.054$; $T_{c_{12}} = 269.3$ K; $V_{c_{12}} = 164.2$ cm³/mol; $z_{c_{12}} = 0.281$; $\omega_{12} = 0.102$; $P_{c_{12}} = 38.1$ bar. Thus: $B_{12} = -145.7$ cm³/mol, and:

$$B = y_1^2 B_{11} + y_2^2 B_{22} + 2y_1 y_2 B_{12} = -65.6 \text{ cm}^3/\text{mol},$$

which gives: $z = 1 + BP/RT = 0.930$, in very good agreement with the experimental value (error 1.8%).

To examine the effect of l_{12}, we calculate B_{12} using $l_{12} = 0.0$, obtaining a value of $B_{12}(l_{12} = 0.0) = -164{,}6$ cm³/mol. This leads to $B = -69.1$ cm³/mol and $z = 0.926$, which is also in very good agreement with the experimental value (error: 1.4%). (The slightly better result for $l_{12} = 0.0$ may be due to the uncertainty in the experimental data.)

b. The Pitzer Correlation

From Eqs 11.7.10 through 11.7.14: $V_{c_1} = 100.3$; $V_{c_2} = 285.7$; $V_{c_m} = 116.7$, all in cm³/mol; $T_{c_m} = 224.0$ K; $\omega_m = 0.030$; $P_{c_m} = 46.7$ bar.

Therefore: $T_r = 1.389$; $P_r = 0.563$; and: $z^{(0)} = 0.932$; $z^{(1)} = 0.046$; and $z = 0.933$, which is also in very good agreement with the experimental value (error: 2.1%)

c. The SRK EoS

Using the expressions from Table 8.3:

$a_{c_{11}} = 2.331 \cdot 10^6$ bar cm⁶/mol² ; $a_{11} = 1.727 \cdot 10^6$ bar cm⁶/mol²

$b_1 = 29.82$ cm³/mol

$a_{c_{22}} = 1.407 \cdot 10^7$ bar cm^6/mol^2 ; $a_{22} = 1.742 \cdot 10^7$ bar cm^6/mol^2
$b_2 = 80.67$ cm^3/mol

From Eq.11.7.23, using $k_{12} = 0.0$, valid for this mixture and the SRK EoS:
$a_{12} = 5.485 \cdot 10^6$ bar cm^6/mol^2

Thus, from Eqs 11.7.15 and 11.7.21 respectively:
$a = 2.6136 \cdot 10^6$ bar cm^6/mol^2
$b = 35.21$ cm^3/mol

To solve for the compressibility factor, it is more convenient to write the SRK EoS in the following form:

$$z^3 - z^2 + (A - B - B^2)z - AB = 0$$

where:

$$A = \frac{aP}{(RT)^2} = 0.1078 \quad ; \quad B = \frac{bP}{RT} = 0.0375$$

Solution of the cubic equation for the largest z value gives: $z = 0.931$, which is also in very good agreement with the experimental value (error 1.9%). The results are summarized in Table 11.E.5.

Table 11.E.5 Tabulation of Results for Example 11.5

Method	z	% Error
Virial (l_{12}=0.054)	0.930	1.8
Virial (l_{12}=0.0)	0.926	1.4
Pitzer	0.933	2.1
SRK	0.931	1.9

11.7.6 Example 11.6

Estimate the compressibility factors of the methane - n-butane mixture of Example 11.5 - at the same temperature and composition, but at increased pressures - and compare them with the experimental values (Sage et al) given in Table 11.E.6. Use the SRK EoS.

Following the same approach as in Example 11.5, and using the computer Program PVT of Appendix E, we obtain the results presented in Table 11.E.6, which can be considered satisfactory up to 1500 psia (100 bar). Notice, however, that:

* both compounds are nonpolar, and
* the mixture is about 90% methane, whose reduced temperature is high (1.632).

Table 11.E.6 Prediction of z Values for a Methane - n-Butane Mixture as a Function of Pressure with the SRK EoS. $y(CH_4)$ = 0.8942; T = 100°F

P (psia)	z (exp)	z (SRK)	% Error
400	0.9135	0.931	1.9
600	0.8769	0.897	2.3
800	0.8444	0.866	2.5
1000	0.8154	0.837	2.6
1500	0.7587	0.781	2.9
2000	0.7274	0.759	4.3
2500	0.7212	0.768	6.5
3000	0.7470	0.796	6.6

11.7.7 Example 11.7

Using the saturated vapor compressibility factor data for the system n-butane - carbon dioxide (Olds et al) presented in Table 11.E.7:
a.evaluate the performance of the SRK, PR, and vdW-711 EoS using a zero value for the interaction coefficients;
b.determine the interaction coefficient values for the three EoS.

a.Compressibility factor values were calculated using the Program PVT of Appendix E and are presented in Table 11.E.7. The larger errors at 618 psia should be attributed, in addition to the saturation condition and the dissimilarity of the components of the mixture, to the higher pressure.

b.Using the same Program, the interaction coefficient values presented in Table

Table 11.E.7 Compressibility Factors and Interaction Coefficients for n-Butane - Carbon Dioxide

				% Error (k_{12}=0.0)			Inter. Coefficient		
z	T(°F)	P(psia)	$y(nC_4)$	SRK	PR	vdW-711	SRK	PR	vdW-711
0.822	100	372	0.1694	-1.2	-3.0	-2.9	0.077	0.188	0.195
0.495	280	618	0.8273	-7.3	-11.6	-15.5	0.081	0.144	0.213

11.E.7 were obtained by trial-and-error. (They could of course be obtained directly by regression.) Notice that the obtained values vary between the two sets of conditions.

Two reasons account for this. First, k_{ij} is not a constant, strictly speaking. Second, and the most important in this case, the obtained values reflect the uncertainty in the experimental data. In practice, therefore, k_{ij} values are obtained through regression of a set of data, typically vapor-liquid equilibrium ones.

11.7.8 Comments on the Estimation of Mixture Properties

1.The predictive techniques for pure compounds can be applied to mixtures as well. The quality of the obtained results, however, depends on the need for, and the availability of, interaction coefficient values.

2.For the volumetric behavior of vapor and gaseous mixtures of hydrocarbons up to moderate pressures, the cubic equations of state discussed in Chapter 10 give reasonable results even with $k_{ij} = 0$. At very high pressures, the Pitzer correlation (with the Lee-Kesler Tables) should be used.

3.In the presence of gases such as CO_2, CO, H_2S, N_2, etc. in the hydrocarbon mixture, however, interaction coefficients should be used for accurate results, especially at high pressures. They are also needed, but to a lesser degree, in the presence of CH_4.

4.For the volumetric behavior of polar - and, of course, nonpolar - mixtures at low pressures, the virial equation gives reasonable accuracy, provided that l_{ij} values are available.

5.For polar systems at high pressures the cubic equations of state modified for polar compounds (Section 14.7.3), and for still higher ones the Pitzer-Lee-Kesler approach, give the most reliable results. The accuracy is poorer, however, than that of the previous cases.

6.For volumes of nonpolar/weakly polar saturated liquid mixtures, the translated EoS give very good results as we will see in Chapter 14. The correlation of Hankinson and Thomson, Chapter 8, gives also good results.

7.For enthalpy and entropy departure values of nonpolar/weakly polar mixtures, the cubic EoS discussed here give reasonable accuracy up to moderate pressures. At higher pressures, the Pitzer-Lee-Kesler approach should be used.

8.For more details on the estimation of the volumetric behavior and of enthalpy and entropy departure values for gas and liquid mixtures, see Reid et al.

11.8 Fugacities in Mixtures

Fugacities of the components of a mixture are extremely important in Chemical Engineering applications for they are essential in chemical and in phase equilibrium calculations. We will start our discussion of this subject by presenting in this Section:

1. the definition of the fugacity of component i in a mixture and how it can be evaluated in principle;
2. the relationship between the fugacity of a mixture and those of its constituents, which will demonstrate that the component fugacities are not partial molar properties of the mixture fugacity; and
3. the relationship between the fugacity of a component i in a mixture and that of pure i, for the special case of an ideal solution.

We proceed then with a presentation - in Sections 11.9 and 11.10 - of the two approaches used in the evaluation of fugacities of mixture components:

a. equations of state, and
b. the *standard state* fugacity,

and apply them - in Section 11.11 - in the determination of the solubility of solids and liquids in gases. Extensive application of fugacities is made, of course, in the Chapters where phase and chemical equilibria are discussed (Chapters 12, 13, 14 and 15).

11.8.1 Fugacity of the Components of a Mixture

The fugacity of the component i, $\hat{f}_i$, in a homogeneous mixture is defined in a fashion similar to that of a pure compound:

$$d\mu_i = d\overline{G}_i = RTd\ln\hat{f}_i = \overline{V}_i dP, \; T=const. \qquad (11.8.1)$$

with the proviso that:

$$\lim_{P\to 0} \frac{\hat{f}_i}{x_i P} = 1 \qquad (11.8.1a)$$

For the case of an ideal gas Eq.11.8.1 yields:

$$RTd\ln(x_i P) = \overline{V}_i(id)dP = \frac{RT}{P}dP$$

Subtraction from Eq.11.8.1 yields after rearrangement and integration:

$$\ln\hat{\phi}_i = \ln\frac{\hat{f}_i}{x_i P} = \frac{1}{RT}\int_0^P \left(\bar{V}_i - \frac{RT}{P}\right)dP \qquad (11.8.2)$$

where $\hat{\phi}_i$, is the *fugacity coefficient* of component i of the mixture.

Determination, hence, of the fugacity of the component i of a mixture is carried out - as in the case of a pure compound - through its fugacity coefficient. Evaluation of the latter, however, requires partial molar volume data for component i in this mixture - from zero pressure to that of the system - all at the temperature and composition of the system. Such information is of course very rare, and fugacities are typically calculated using one of the following two approaches:

1. through equations of state, and
2. through the concept of standard state fugacity.

The first approach is discussed in Section 11.9, and can be successfully applied to vapor mixtures; and to liquid mixtures of nonpolar or weakly polar compounds (i.e. hydrocarbons, CO, CO_2, H_2S, H_2, etc.).

For polar liquid mixtures, the available equations of state do not provide yet reliable results, especially in multicomponent cases. Fugacities in such mixtures, therefore, are calculated with the second approach, discussed in Section 11.10.

For a physical interpretation of the fugacity of the component i of a mixture, we will demonstrate in the next Chapter that when two phases - for example liquid (l) and vapor (v) - are in equilibrium with each other at a given temperature and pressure, then:

$$\hat{f}_i^l = \hat{f}_i^v \qquad (11.8.3)$$

We can interpret, therefore, fugacity in this case as the escaping tendency of component i in each phase, just as we did for pure compounds.

11.8.2 Relationship Between f and $\hat{f}_i$

Following Eq.11.8.1 the fugacity of a mixture, f, is related to its chemical potential, μ, through the expression:

$$d\mu = RTd\ln f, \; T = const. \qquad (11.8.4)$$

Integration at constant T and composition, from a low pressure P^*, where the mixture behaves as an ideal gas, to pressure P, yields:

$$\mu[T,P,x_1,x_2,...] - \mu^*[T,P^*,x_1,x_2,...] = RT\ln(\frac{f}{P^*}) \qquad (11.8.5)$$

Integration of Eq.11.8.1, under the same conditions yields:

$$\mu_i[T,P,x_1,x_2,...] - \mu_i^*[T,P^*,x_1,x_2,...] = RT\ln\left(\frac{\hat{f}_i}{x_i P^*}\right) \tag{11.8.6}$$

Multiplication by x_i of both sides of Eq.11.8.6 yields:

$$x_i\mu_i - x_i\mu_i^* = x_i RT\ln\left(\frac{\hat{f}_i}{x_i P^*}\right) \tag{11.8.7}$$

where the terms in the brackets are dropped for simplicity.

Summation over all the components of the mixture leads to:

$$\sum[x_i\mu_i - x_i\mu_i^*] = \mu - \mu^* = RT\sum x_i\left[\ln\left(\frac{\hat{f}_i}{x_i}\right) - \ln P^*\right] \tag{11.8.8}$$

Comparison with Eq.11.8.5 gives:

$$\ln f = \sum x_i \ln\left(\frac{\hat{f}_i}{x_i}\right) \tag{11.8.9}$$

Subtracting $\ln P$ from both sides yields:

$$\ln\phi = \sum x_i \ln\hat{\phi}_i \tag{11.8.10}$$

Eqs 11.8.9 and 11.8.10 indicate that $\ln(\hat{f}_i/x_i)$ and $\ln\hat{\phi}_i$ are partial molar properties; $\hat{f}_i$ and $\hat{\phi}_i$ are not. This is why we use the symbol (^) and not (‾) to differentiate the fugacity and fugacity coefficient of a mixture component i from that of pure i.

11.8.3 The Case of an Ideal Solution

Subtraction of Eq.9.11.4 from Eq.11.8.2 yields:

$$\ln\frac{\hat{f}_i}{x_i f_i} = \frac{1}{RT}\int_0^P (\bar{V}_i - V_i)dP \tag{11.8.11}$$

where $\bar{V}_i$ is the partial molar volume of i, and V_i the molar volume of pure i, both at the same P and T. In the case of an ideal solution, $\bar{V}_i = V_i$, and consequently:

$$\hat{f}_i = x_i f_i \tag{11.8.12}$$

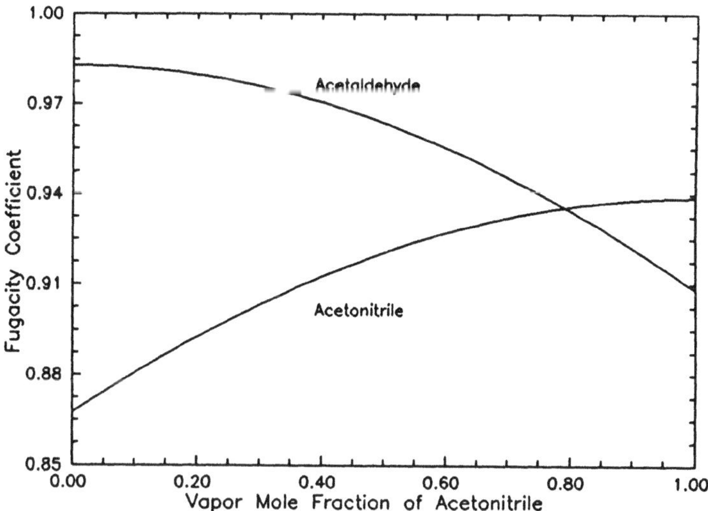

Figure 11.1 Fugacity coefficients for the system acetonitrile(1) - acetaldehyde(2) at 1 bar and 100°C with the virial equation and experimental B_{ii} and B_{ij}.

This equation is also known as the Lewis-Randall rule. It can be used for the determination of the fugacity of a mixture component through its fugacity in the pure state - at the same P and T as the mixture and in the same phase.

Eq.11.8.12 is applicable to ideal mixtures only, i.e. mixtures of similar compounds. It does provide, however, a good approximation for $\hat{f}_i$ in nonideal systems at high mole fractions of component i, where - as we saw in Example 11.2 - ideal solution behavior with respect to component i is approached. The accuracy declines, of course, with decreasing mole fraction of component i; and with increasing nonideality of the solution.

11.8.4 Comments

1.At low pressures, say below 5 bar, vapor phase fugacity coefficient values are not very different from unity as Figure 11.1 demonstrates.

2.Systems containing organic acids, where dimerization due to hydrogen bonding occurs, demonstrate large deviations from ideality, even at subatmospheric pressures. Thus, Prausnitz et al (1980) report for the system acetaldehyde-acetic acid at 90°C and 0.25 atm fugacity coefficient values in the range 0.4 to 1.5.

3.At high pressures, fugacity coefficients deviate substantially from unity as demonstrated in Figures 11.2 and 11.3.

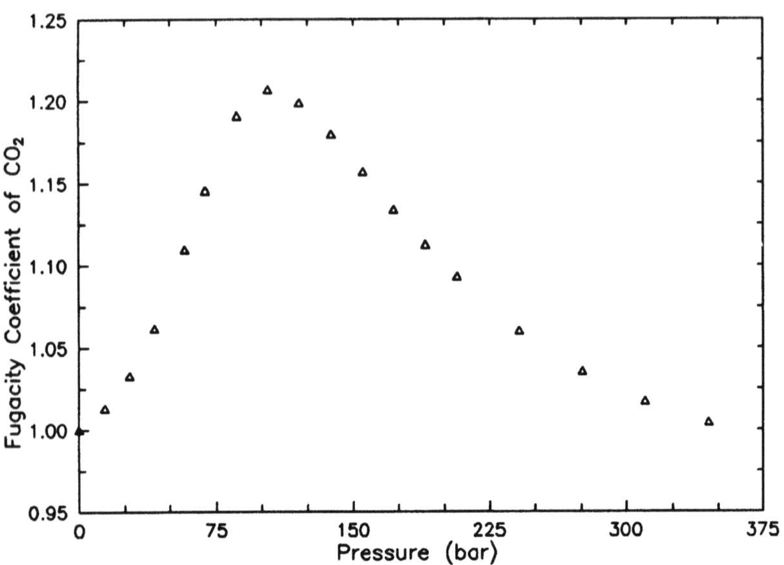

Figure 11.2 Fugacity coefficient of CO_2 (1) in its mixture with nC_4 at 478 K and y_1 = 0.17. (Calculated from the volumetric data of Olds et al.)

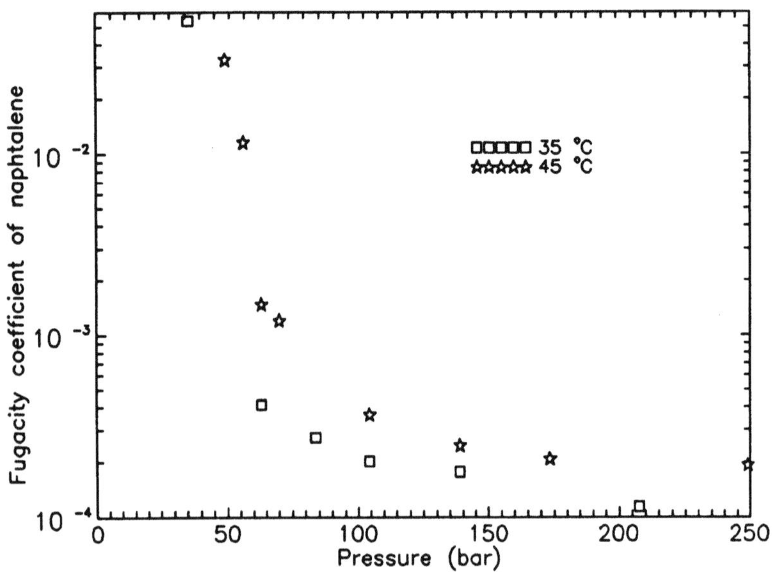

Figure 11.3 Fugacity coefficients of napthalene in ethane at saturation. (Calculated from the solubility data of Johnston et al, 1982.)

11.9 Fugacities with Equations of State

To express the fugacity of component *i* of a mixture through an EoS, we note that - in the typical case - the EoS is in the form:

$$P = f(V,\ T,\ comp.) \tag{11.9.1}$$

rather than in the form:

$$V = f(P,\ T,\ comp.) \tag{11.9.2}$$

that is needed to evaluate $\bar{V}_i$ in Eq.11.8.2. For this reason, the fugacity coefficient is determined using the following thermodynamic transformation of Eq.11.8.2, where the independent variable is V rather than P (Smith and Van Ness, 1987):

$$\ln\hat{\phi}_i = z - 1 - \ln z - \int_{\infty}^{V} \left[\left(\frac{\partial Nz}{\partial N_i} \right)_{T,V,N_j} - 1 \right] \frac{dV}{V} \tag{11.9.3}$$

which can be conveniently used with equations of state having the form of Eq.11.9.1.

The most often used equations of state for the evaluation of fugacity coefficients are the virial and cubic ones.

11.9.1 Fugacities With the Virial Equation

We restrict our discussion to the virial equation truncated after the second term, because third or higher virial coefficients are rarely available. Use of the volume series with Eq.11.9.3 yields:

$$\ln\hat{\phi}_i = \frac{2}{V}\sum y_j B_{ij} - \ln z \tag{11.9.4}$$

while use of the pressure series directly into Eq.11.8.2 yields:

$$\ln\hat{\phi}_i = [2\sum y_j B_{ij} - B]\frac{P}{RT} \tag{11.9.5}$$

V, z, and B are the molar volume, compressibility factor, and second virial coefficient of the mixture respectively, B_{ij} is the second virial coefficient of the pair *i-j*, and the summation is over all the components of the mixture. Thus, for a binary mixture, Eq.11.9.5 becomes:

$$\ln\hat{\phi}_1 = \frac{P}{RT}(B_{11} + y_2^2\delta_{12}) \tag{11.9.5a}$$

$$\ln\hat{\phi}_2 = \frac{P}{RT}(B_{22} + y_1^2\delta_{12}) \tag{11.9.5b}$$

where: $\delta_{12} = 2B_{12} - B_{11} - B_{22}$

No comparison between Eqs 11.9.4 and 11.9.5 is available but we saw in Chapter 8 that, for pure compounds, Eq.11.9.5 performs at least as well as Eq.11.9.4. In addition, it is more convenient to use since it does not require the calculation of the molar volume. It is, therefore, recommended for mixtures as well.

The virial equation is, of course, applicable to the vapor phase only and, when truncated after B, should not be used outside the pressure range of Eq.11.7.9.

Finally, because of its rigorous mixing rules, and the success of the empirical correlations for the estimation of B_{ii} and B_{ij}, the virial equation finds extensive use in the estimation of vapor fugacities for low pressure vapor-liquid equilibrium calculations, especially in systems containing polar components. The liquid phase fugacities, in such cases, are calculated using the standard state fugacity approach that will be discussed in Section 11.10.

11.9.2 Fugacities With Cubic Equations of State

Fugacities of nonpolar/weakly polar mixtures, for both the liquid and vapor phases, are typically calculated with cubic EoS. Use of the SRK, PR, and vdW-711 EoS with Eq.11.9.3 gives:

SRK:

$$\ln\hat{\phi}_i = \frac{B_i}{B}(z-1) - \ln(z-B) - \frac{A}{B}\left(\frac{2\sum x_j A_{ij}}{A} - \frac{B_i}{B}\right)\ln\left(1+\frac{B}{z}\right) \tag{11.9.6}$$

PR:

$$\ln\hat{\phi}_i = \frac{B_i}{B}(z-1) - \ln(z-B) - \frac{A}{2B\sqrt{2}}\left(\frac{2\sum x_j A_{ij}}{A} - \frac{B_i}{B}\right)\ln\left(\frac{z+2.414B}{z-0.414B}\right) \tag{11.9.7}$$

vdW-711:

$$\ln\hat{\phi}_i = -\ln(z+t'-B) + \frac{B_i - t'_i}{z+t'-B} - \frac{2\sum A_{ij}x_j}{z+t'} + \frac{At'_i}{(z+t')^2} \tag{11.9.8}$$

where: $$A = \frac{aP}{(RT)^2} \; ; \; A_{ij} = \frac{a_{ij}P}{(RT)^2}$$

$$B_i = \frac{b_i P}{RT} \; ; \; B = \frac{bP}{RT} \; ; \; t' = \frac{tP}{RT}$$

a: from Eq.11.7.15, with a_{ij} from Eq.11.7.23;
b: from Eq.11.7.21; t: from Eq.11.7.22
For liquid phase fugacities, the liquid phase compressibility factor and compositions are used; for vapor phase ones, accordingly.

Comments

1.Lin and Daubert (1980) did an extensive study on the evaluation of fugacities for hydrocarbon mixtures with the SRK and PR EoS. Their findings can be summarized as follows:

a.At low to moderate pressures, very good results are obtained with both EoS and the use of interaction coefficients has a negligible effect.

b.At higher pressures, interaction coefficients are needed and their impact exceeds 10%.

c.For gas (H_2, CO_2, CO, N_2, H_2S, etc.)-hydrocarbon mixtures, interaction coefficients are needed, especially for high pressures and for saturation conditions.

2.Because the calculation of fugacities requires the evaluation of derivatives with respect to composition (Eq.11.9.3), the obtained results are more sensitive to the accuracy of the k_{ij} - and l_{ij} for the virial - values, than are total mixture properties, such as volume. (See Problem 11.51).

3.We will return to interaction coefficients when we discuss high pressure vapor-liquid equilibrium, where they play a key role since they are involved in the calculation of both, vapor and liquid phase fugacities.

Evaluation of fugacities with EoS is presented in the next Example.

11.9.3 Example 11.8

Estimate the fugacity coefficients of methane(1) and n-butane(2) in a saturated vapor mixture at: T = 311.1 K; P = 27.60 bar; and y_1 = 0.824, using the virial and SRK equations of state. Tabulate your results and comment on their accuracy.

a. The Virial Equation

The virial equation, truncated after B, is applicable here because Eq. 11.7.9 is satisfied. Use of Eqs 11.9.5a and 11.9.5b and information from Example 11.5: $B_{11} = -38.4$; $B_{22} = -650.2$; $B_{12} = -145.7$, and $\delta_{12} = 397.2$, all in cm^3/mol, yields:

$$\hat{\phi}_1 = 0.973 \ ; \ \hat{\phi}_2 = 0.666$$

Use of the $B_{12} = -164.6$, $\delta_{12} = 359.4$, cm^3/mol, that correspond to $l_{12} = 0.0$, gives:

$$\hat{\phi}_1 = 0.971 \ ; \ \hat{\phi}_2 = 0.648$$

b. The SRK EoS

Step 1. Calculate z needed in Eq. 11.9.6 from:

$$z^3 - z^2 + z\,(A - B - B^2) - AB = 0$$

Using the a_{ii} and b_i values from Example 11.5: $z = 0.904$

Step 2. Calculate $\hat{\phi}_1$ and $\hat{\phi}_2$

From Eq. 11.9.6: $\hat{\phi}_1 = 0.975$; $\hat{\phi}_2 = 0.655$

c. Tabulation of Results. See Table 11.E.8.

Table 11.E.8 Results of Example 11.8

Method	$\hat{\phi}_1$	$\hat{\phi}_2$
Virial, l_{12}= 0.054	0.973	0.666
Virial, l_{12}= 0.0	0.971	0.648
SRK	0.975	0.655

The values obtained from the virial equation using l_{12}= 0.054 should be close to the correct values. Both equations, thus, give reasonable accuracy for component 1 even with zero values for the interaction coefficients; but not so for component 2. In general nonzero k_{ij} values are needed for C_1-alkane mixtures due to size difference. Their importance becomes more pronounced at small values of the mole fraction of the heavier component, butane in this case ($y_2 = 0.176$).

We turn next to the second method of expressing mixture component fugacities, which utilizes the standard state fugacity approach.

11.10 Fugacities with the Standard State Method

11.10.1 The Problem and Approach

For polar systems the intermolecular forces involved in the typical case lead to nonideal mixture behavior that cannot be successfully described by the available equations of state and commonly used mixing rules, especially in the liquid phase. We define, therefore, the liquid phase fugacity of component i in such a mixture with respect to its value at some reference state, called the *standard state* fugacity $f_i°$:

$$\hat{f}_i = x_i \gamma_i f_i^\circ \tag{11.10.1}$$

where γ_i is called the *activity coefficient* (gamma).

Notice that, for a given value of x_i, the value of $\hat{f}_i$ is specified by Eq.11.8.2. The value, hence, of the activity coefficient depends on the choice of the standard state, whose complete description requires specification of the:

* temperature,
* pressure, and
* mole fraction of component i.

The temperature and pressure of the system are generally accepted, and the choice of composition depends on whether compound i is *condensable* or *noncondensable*. (A compound is considered condensable if its critical temperature is higher than the temperature of the system; and noncondensable, if it is lower.)

In addition - for convenience purposes - we want the activity coefficient to be equal to one at the standard state. The choice, thus, of the composition where γ_i = 1.0 is often referred to as the *normalization* of the activity coefficient.

It is apparent, finally, that the approach denoted by Eq.11.10.1 applies also to mixtures that are not very nonideal, such as hydrocarbon ones.

11.10.2 The Standard State Fugacity for a Condensable Component

For a condensable component, the following normalization is used:

$$\gamma_i \rightarrow 1.0 \quad \text{as} \quad x_i \rightarrow 1.0 \tag{11.10.2}$$

In this case, the standard state fugacity represents the fugacity of pure liquid i at the temperature T and pressure P of the mixture; and comparison of Eqs 11.10.1 and 11.8.12 suggests that the activity coefficient can be viewed as the 'correction' factor to the Lewis-Randall rule.

If the components of a system are subcritical, this type of normalization applies to all of them. It is referred to, appropriately, as following the *symmetric convention*.

The standard state fugacity is evaluated through Eq.9.11.14. Introduction into Eq.11.10.1 yields the expression for the fugacity of the component i of a liquid mixture:

$$\hat{f}_i = x_i \gamma_i P_i^s \phi_i^s \exp\left(\frac{V_i(P-P_i^s)}{RT}\right) \tag{11.10.3}$$

where:

P_i^s: vapor pressure of pure i at T

ϕ_i^s: fugacity coefficient of the saturated vapor at T

V_i: average value of the liquid molar volume at T and the pressure range P_i^s to P.

The exponential term in Eq.11.10.3 is the Poynting effect (Pe) and if the temperature of the system is close to the critical one of pure i, it must be replaced by the expression of Eq.9.11.13.

11.10.3 Example 11.9

Hermsen and Prausnitz (1963) report that at 25°C and 294.11 mm Hg a liquid mixture of benzene(1) and cyclopentane(2) with $x_1 = 0.1417$ is in equilibrium with a vapor one where $y_1 = 0.0655$. From the information given below:

Compound	P^s (mm Hg)	V_i	B_{ii}	B_{12}	δ_{12}
		cm³/mol			
1	94.95	89.39	-1314	-1176	16
2	317.66	94.71	-1054		

determine:

a. the standard state fugacity of benzene;

b. the fugacity of benzene in the vapor phase;

c. the same, but for the liquid phase;

d. the activity coefficient of benzene;

e. the fugacity of benzene in the vapor and liquid phases assuming ideal mixture, i.e. using the Lewis-Randall rule.

a. From Eqs 9.11.14, 9.11.10, and 9.11.13: $f_1^\circ = \phi_1^s P_1^s (Pe)_1$

$$\phi_1^s = \exp\left(\frac{B_{11}P_1^s}{RT}\right) = 0.993 \quad ; \quad (Pe)_1 = \exp\left(\frac{V_1(P-P_1^s)}{RT}\right) = 1$$

and $f_1^\circ = 94.30$ mm Hg.

b. From Eq.11.9.5a: $\hat{\phi}_1 = 0.980$ and $\hat{f}_1 = y_1 \hat{\phi}_1 P = 18.88$ mm Hg.

c. We cannot use Eq. 11.10.3 to determine the fugacity of benzene, because its activity coefficient is not known. Since, however, the two phases are at equilibrium, then according to Eq. 11.8.3:

$$\hat{f}_1^{\,l} = \hat{f}_1^{\,v} = 18.88 \text{ mm Hg}$$

d. From Eq. 11.10.1: $\gamma_i = \dfrac{\hat{f}_i}{x_i f_i^{\circ}} = 1.413$

e. According to the Lewis-Randall rule:

* Vapor Phase:

$$\hat{f}_1 = y_1 f_1(\text{pure vapor at the } T \text{ and } P \text{ of the solution}) = y_1 \phi_1 P$$

From Eq. 9.11.10: $\phi_1 = 0.979$; and $f_1 = 287.9$ mm Hg. Hence:

$$\hat{f}_1(\text{LR}, v) = 18.86 \text{ mm Hg}$$

* Liquid Phase:

$$\hat{f}_1 = x_1 f_1(\text{pure liquid at } T \text{ and } P \text{ of solution})$$

f_1 is given in Part(a), and $\hat{f}_1(\text{LR}, l) = 13.36$ mm Hg.

<u>Comments</u>

1. The standard state fugacity of benzene is practically the same with its vapor pressure (why?).

2. Notice that for the vapor phase fugacity:

$$\hat{f}_1(real) = y_1 \hat{\phi}_1 P; \quad \hat{f}_1(LR) = y_1 \phi_1 P$$

The difference is, thus, caused by that between $\hat{\phi}_1$ and ϕ_1; and, since both are evaluated at the same T and P, the difference reflects the nonideal solution behavior of the *vapor* phase. Because of the very low pressure in this case, however, the molecules are too far apart to feel the effect of the change in their molecular environment and the difference is negligible.

3. For the liquid phase fugacity, on the other hand, we have:

$$\hat{f}_1(real) = x_1 \gamma_1 P_1^s \phi_1^s (Pe)_1 \quad ; \quad \hat{f}_1(LR) = x_1 P_1^s \phi_1^s (Pe)_1$$

and the difference is the activity coefficient, which reflects the nonideal solution behavior of the *liquid* phase. In this case, the molecules are much closer, and the effect is more pronounced. This is demonstrated by the value of $\gamma_1 = 1.413$, even though the two components are fairly similar.

11.10.4 The Standard State Fugacity for a Noncondensable Component

Consider a liquid mixture of ethanol and methane at room temperature.

Use of the symmetric convention to express the fugacities of the two components would require, according to Eq.11.10.3, values for the vapor pressure of pure ethanol and of pure methane at room temperature.

While ethanol (T_c=513.9 K) can exist as a liquid at this temperature, methane (T_c=190.6 K) cannot and, consequently, its vapor pressure cannot be determined.

It becomes apparent that some other type of normalization is required for a supercritical (noncondensable) component and it is given by:

$$\gamma_i^* \rightarrow 1.0 \quad \text{as} \quad x_i \rightarrow 0.0 \tag{11.10.4}$$

where the asterisk depicts the different type of normalization involved in this case. Hence from Eq.11.10.1:

$$\lim_{x_i \to 0} \frac{\hat{f}_i}{x_i} = f_i^\circ = H_i \tag{11.10.5}$$

and:

$$\hat{f}_i = x_i \gamma_i^* H_i \tag{11.10.6}$$

H_i is called the Henry's constant, evaluated at the temperature and pressure of the system, and represents the standard state fugacity for this type of normalization of the activity coefficient; and γ_i^*, is the 'correction' factor to Henry's law:

$$\hat{f}_i = x_i H_i \tag{11.10.7}$$

Notice that while determination of f_i° for a condensable component requires knowledge of pure compound properties only, evaluation of H_i requires low composition measurements, so that the limit in Eq.11.10.5 can be determined.

The Henry's constant is - as any other fugacity - a function of temperature and pressure, and its dependency on pressure becomes significant at elevated pressures, especially as the critical point is approached. Details and techniques for estimating this dependency for systems that do not deviate much from ideality, are given by Prausnitz et al (1986).

Finally, it is apparent that for liquid mixtures of condensable and noncondensable components two types of normalization must be used. This approach to normalization is referred to, therefore, as following the *unsymmetric convention*.

Such a normalization is shown in Figure 11.4. Notice that, in the typical case, the activity coefficient γ_i (for the condensable component) is larger than one, while the activity coefficient γ_i^* (for the noncondensable one) is less than one.

The unsymmetric normalization is used in the following Example.

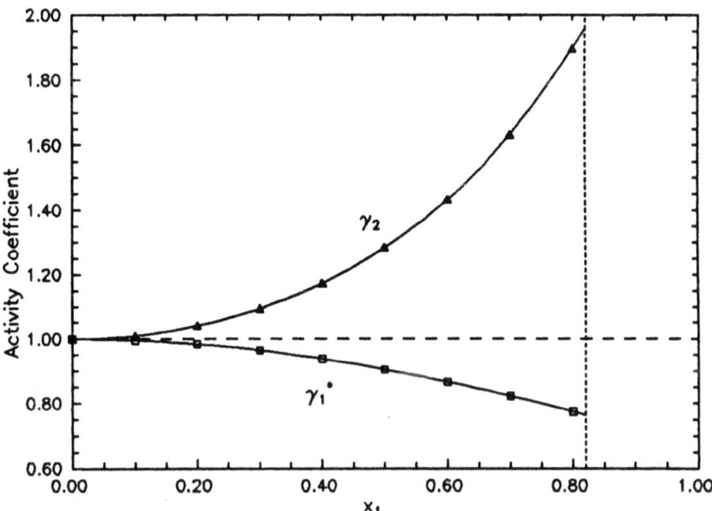

Figure 11.4 Typical variation of γ and γ^* with composition for a binary system.

11.10.5 Example 11.10

The solubility of CO_2(1) in benzene(2) at 293.2 K and a partial pressure of one bar is: $x_1 = 9.5 \cdot 10^{-4}$.

a. What is meant by 'partial pressure' and how is it related to fugacity?

b. Determine the Henry's constant;

c. What is the solubility of CO_2 at the same temperature and a partial pressure of 2 bar?

Assume, for simplicity and the low pressures involved, ideal vapor behavior.

a. The partial pressure is defined by: $\bar{P}_i = y_i P$. It represents, thus, the part of the total pressure that is due to component i. Notice that since: $\hat{f}_i = y_i \hat{\phi}_i P$, $\bar{P}_i$ is related to its vapor phase fugacity through: $\bar{P}_i = \hat{f}_i / \hat{\phi}_i$

b. From the equality of fugacities of CO_2 in the vapor and liquid phases:

$$y_1 \hat{\phi}_1 P = x_1 \gamma_1^* H_1$$

From the ideal vapor assumption: $\hat{\phi}_1 = 1.0$; and, since in this case x_1 is close enough to zero, we can safely assume $\gamma_1^* = 1.0$. Hence: $H_1 = 1/x_1 = 1053$ bar.

c. At $\bar{P}_1 = 2$ bar: $x_1 = 2/H_1 = 0.0019$, which is again close enough to zero for $\gamma_1^* = 1.0$.

Note: At low pressures, the solubility of practically all gases in a variety of liquids is very small and Henry's law, i.e. $\gamma_i^* = 1.0$, can be safely assumed. This is not the case, however, at high pressures as we will see when considering high pressure vapor-liquid equilibrium in Chapter 14.

11.11 An Application of Fugacities: Solubility of Solids and Liquids in Gases

11.11.1 Solids

Consider the equilibrium between a solid, component 2, and some gas, component 1, at some temperature T and pressure P. Since the solubility of the gas in the solid is negligible, the equality of fugacities for component 2 in the two phases yields:

$$y_2 \hat{\phi}_2 P = P_2^s \phi_2^s (Pe)_2 \tag{11.11.1}$$

where:

$$(Pe)_2 = \exp\left(\frac{V_2(P - P_2^s)}{RT}\right)$$

P_2^s, V_2 are the solid vapor pressure and molar volume respectively at T. The solubility of the solid in the vapor phase is, therefore:

$$y_2 = \frac{P_2^s}{P} E \tag{11.11.2}$$

where:

$$E = \frac{\phi_2^s (Pe)_2}{\hat{\phi}_2} \tag{11.11.3}$$

At very low pressures, the two fugacity coefficients and the Poynting effect are equal to unity, and:

$$y_2 = P_2^s / P$$

is referred to as the *ideal gas law* solubility of the solid in the gas. Eq. 11.11.2 becomes, therefore:

$$y_2 = y_2(\textit{ideal gas law})\, E$$

As the pressure increases, however, E becomes progressively more important. Actually, E is always larger than one, leading to increased -

over the ideal gas law - solid solubility. It is referred to, appropriately, as the *enhancement* factor.

11.11.2 Comments

1.Evaluation of y_2 involves a trial-and-error calculation.

2.The enhancement factor can assume very large values at high pressures as demonstrated with the case of fluorene in ethylene (Figure 11.5).

3.Notice that at a given pressure the solubility enhancement is more pronounced at temperatures just above the critical of the gaseous solvent (T_c of ethylene equals 282.4 K).

4.The enhancement factor contains three terms: ϕ_2^s, $(Pe)_2$, and $\hat{\phi}_2$. The first is practically equal to one, since the solid vapor pressure is low. The second represents the effect of pressure on the fugacity of the solid, and it increases with increasing pressure. It rarely, however, assumes values larger than 2 to 3. The third, which reflects mainly the difference in the 1-1, 2-2, and 1-2 intermolecular forces, is thus the most important one for it can assume values much smaller than 1 (Figure 11.3).

5.These large solubilities of solids in supercritical gases represent the basis for the increasingly popular field of *Supercritical Extraction*. Consider, for example, a solid mixture containing a valuable component A. If a gas can be found that through strong interaction with A - but not with the other components of the mixture - can act as a good 'gas' solvent, then recovery of A can be effected. For once the resulting, from the treatment of the solid, gas mixture is decompressed, E becomes equal to 1 and the solid precipitates.

6.For best results, temperatures just over the gas T_c and pressures substantially larger than its P_c should be used as Figure 11.5 suggests. Under these conditions *liquid-like solubilities are approached combined with the large gas diffusivities*.

7.The main advantage of supercritical extraction, over the conventional one or distillation, is that solvent free products - of essential importance in the food and pharmaceutical industries - can be recovered. Substantial energy savings are also possible.

8.For a review of the status of industrial applications of supercritical extraction, see Basta (1985); McHugh and Krukonis (1986); and Johnston and Penninger (1989).

9.The thermodynamic treatment of supercritical extraction is rather difficult for three main reasons:

a.the system is highly asymmetric, both in terms of molecular size and

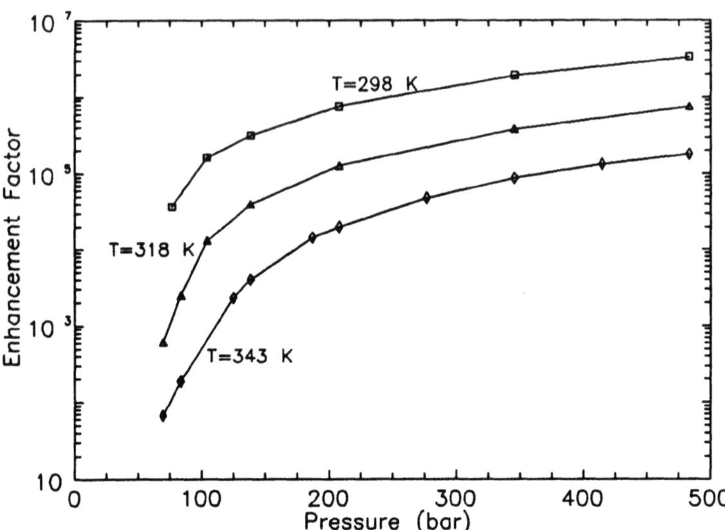

Figure 11.5 Variation of the enhancement factor with pressure as a function of temperature for the system fluorene-ethylene(from the solubility data of Johnston et al, 1982). Lines for identification only.

of intermolecular forces, as demonstrated by the difference in the critical properties of the gas and the solute;

b.the system is close to the gas critical temperature; and

c.pure component property values : T_c, P_c, ω, and vapor pressures are often unavailable, or unreliable, for the solutes.

10.Quantitative calculations are carried out through equations of state [see, for example: Prausnitz, et al (1986); and Johnston et al, 1982]; or empirical approaches (Ziger and Eckert, 1983). For a review of the subject, see Williams (1981); Starling et al (1984); and especially McHugh and Krukonis; and Johnston and Penninger.

We discuss, next, the calculation of solubilities of liquids in gases.

Superficial(!) Extraction

"Theron Brayman of Orrville, Ohio, passed along a report of a new process that is said to remove 95% of the cholesterol from various dairy products without effecting their appearance, consistency, or taste. The process is reported to use pure carbon dioxide and to be similar to a technique known as superficial(!) fluid extraction." (Chem. & Eng. News, Dec. 22, 1986, p.44.)

11.11.3 Liquids

The approach for determining the solubility of liquids in gases is similar to that of solids, except for the complication introduced by the fact that, here, the gas is soluble in the liquid.

At high pressures, the solubility of the gas in the liquid phase can be substantial and the subject will be discussed under high pressure vapor-liquid equilibrium. Here we will concentrate in the simpler case of low pressures, where the gas solubility is typically low enough for Henry's law to apply. Component 1 is again the gas, and 2 the liquid.

From the equality of the fugacities of component 2 in the two phases:

$$y_2 = \frac{x_2 P_2^s \phi_2^s (Pe)_2}{\hat{\phi}_2 P} \tag{11.11.4}$$

where γ_2, due to the small value of x_1, is assumed to be equal to one. Eq. 11.11.4 is similar to Eq.11.11.2 except that, here, x_2 is not equal to 1.0.

To calculate $x_2 = 1 - x_1$, notice that for the gas:

$$x_1 = \frac{P\hat{\phi}_1 y_1}{H_1}$$

Evaluation of y_2, thus, requires again a trial-and-error calculation.

Comments

1.At low pressures, $\hat{\phi}_2$ does not deviate much from unity and the solubility of the liquid does not depend on the nature of the gas.

2.At increased pressures, however, very small values of $\hat{\phi}_2$ are realized that lead to large solubilities, just as it was the case with solids, and the nature of the gas becomes very important.

3.Thus, Prausnitz et al (1986) report that the solubility of n-decane in carbon dioxide, at 350 K and 100 atm, is about:

* ten times larger than that in nitrogen; and,
* forty times larger than that in hydrogen.

4.The solubility of liquids (and solids) in gases must be taken into account in gas-cooling and gas-transportation processes, especially of natural gas, to avoid plugging by precipitating liquids (and solids) due to drop in temperature and pressure.

5.The concept of supercritical extraction can be of course applied to mixtures of high-boiling liquids as well.

11.12 Property Changes of Mixing

11.12.1 Definition

Consider the mixing of x_1 moles of compound 1, x_2 moles of 2, ..., and x_k moles of compound k, for a total of one mole. Pure compounds and mixture are at the same temperature T and pressure P so that the change, from pure compounds to mixture, reflects only mixing effects.

The *property change due to mixing* - or just the *property of mixing*, as it is referred to - is defined by:

$$\Delta M = M - \sum x_i M_i \tag{11.12.1}$$

and represents the measure of this change. ΔH, for example, is the amount of heat that must be added or removed so that the mixture attains the same temperature as that of the pure compounds, both at the same pressure.

Replacement of M in Eq.11.12.1 through Eq.11.4.4b yields:

$$\Delta M = \sum x_i (\bar{M}_i - M_i) \tag{11.12.2}$$

or, $$\Delta M = \sum x_i (\Delta \bar{M}_i) \tag{11.12.3}$$

where: $$\Delta \bar{M}_i = \bar{M}_i - M_i \tag{11.12.4}$$

represents the effect of mixing on component i.

Rearrangement of Eq.11.12.1 yields:

$$M = \sum x_i M_i + \Delta M \tag{11.12.1a}$$

indicating that knowledge of ΔM - along with the corresponding properties of the pure compounds, all at the same T and P - suffices for the determination of the mixture property M. For example, knowledge of ΔH and of the pure compound enthalpies, all at the same P and T, determines the enthalpy of the mixture.

We consider next the property changes associated with the formation of the simplest possible mixture, an ideal one.

11.12.2 The Case of an Ideal Solution

An ideal solution (also referred to as an ideal mixture) meets the following requirements:

1. the forces between unlike and like molecules are the same, and
2. the molecular sizes and shapes are also the same.

Using (*id*) to designate such behavior, we proceed to develop expressions for $\Delta M(id)$, where $M = V, H, G$ and S. From Eq.11.12.1:

$$\Delta V(id) = V(id) - \sum x_i V_i = \sum x_i [\bar{V}_i(id) - V_i] \quad (11.12.5)$$

$$\Delta H(id) = H(id) - \sum x_i H_i = \sum x_i [\bar{H}_i(id) - H_i] \quad (11.12.6)$$

$$\Delta G(id) = G(id) - \sum x_i G_i = \sum x_i [\bar{G}_i(id) - G_i] \quad (11.12.7)$$

$$\Delta S(id) = S(id) - \sum x_i S_i = \sum x_i [\bar{S}_i(id) - S_i] \quad (11.12.8)$$

In the first two cases, the ideal solution assumption dictates that:

$$\bar{V}_i(id) = V_i$$

$$\bar{H}_i(id) = H_i$$

and, consequently:

$$\Delta V(id) = 0 \quad (11.12.5a)$$

$$\Delta H(id) = 0 \quad (11.12.6a)$$

For the case of $\Delta G(id)$ and $\Delta S(id)$, however, the answer is not so apparent. Rather, to develop the appropriate expressions, we integrate Eq.11.8.1 from the fugacity of pure i, f_i, to that of i in an ideal solution at the same T and P which, according to Eq.11.8.12, is equal to $(x_i f_i)$:

$$\bar{G}_i(id) - G_i = RT\ln\frac{x_i f_i}{f_i} = RT\ln x_i \quad (11.12.9)$$

Hence:

$$\Delta G(id) = RT\sum x_i \ln x_i \quad (11.12.7a)$$

Finally, since $G = H - TS$, $\Delta G = \Delta H - T\Delta S$:

$$\Delta S(id) = -R\sum x_i \ln x_i \quad (11.12.8a)$$

Comments

1. An ideal mixture is, of course, a hypothetical one. Since, however, the second requirement for such behavior becomes important only at large differences in size, ideal mixture behavior is approximated by molecules of similar chemical structure.

2. Thus a mixture of pentane and hexane, or of ethanol and propanol, would approximate an ideal mixture, especially in the gaseous phase, where the larger distances among molecules tend to reduce the effect of the difference in inter- and intra-molecular forces.

3. According also to Section 11.8.3, the larger the composition of a component of a mixture, the closer its behavior approaches that in an ideal mixture.

4. Notice that $\Delta M(id) = 0$ for volume, internal energy and enthalpy; but it is different from zero, as it should be expected, for entropy and the associated properties, the Gibbs and Helmholtz free energies.

5. It is apparent, finally, that since from Eq. 11.12.1a:

$$M(id) = \sum x_i M_i + \Delta M(id)$$

the properties of an ideal mixture are known provided that the corresponding values of the components in their pure state, at the same P and T as the mixture, are known. $M(id)$ represents, therefore, a convenient reference for comparing real mixture behavior to, as discussed next.

11.13 Excess Properties

11.13.1 Definition

The difference between the value of the property M of a mixture at a given pressure, temperature and composition, and its value $M(id)$ if the mixture were to behave as an ideal one at the same conditions, represents the *excess* property M^E:

$$M^E = M - M(id) \tag{11.13.1}$$

Similarly, for the component i of a mixture:

$$\bar{M}_i^E = \bar{M}_i - \bar{M}_i(id) \tag{11.13.2}$$

and it is easily seen that:

$$M^E = \sum x_i \bar{M}_i^E = \sum x_i [\bar{M}_i - \bar{M}_i(id)] \tag{11.13.3}$$

From the theoretical point of view, excess properties provide a quantitative measure of the effect of the differences:

* between inter- and intra-molecular forces; and
* in the size and shape of the molecules;

on the mixture behavior.

From the practical point of view, the most important excess property is that of the Gibbs free energy, because of its relationship to the activity coefficients of the components of the mixture, as it will be demonstrated next.

11.13.2 The Excess Gibbs Free Energy and the Activity Coefficient

Integration of Eq.11.8.1:

$$d\bar{G}_i = RTd\ln\hat{f}_i$$

between the ideal and real solution states at constant temperature, pressure, and composition, yields:

$$\bar{G}_i^E = \bar{G}_i - \bar{G}_i(id) = RT\ln\left(\frac{x_i\gamma_i f_i^\circ}{x_i f_i^\circ}\right) = RT\ln\gamma_i \qquad (11.13.4)$$

and, according to Eq.11.13.3:

$$G^E = \sum x_i\bar{G}_i^E = RT\sum x_i\ln\gamma_i \qquad (11.13.5)$$

which indicate that $RT\ln\gamma_i$ is a partial molar property of the excess Gibbs free energy. It follows then, that:

$$\ln\gamma_i = \left(\frac{\partial(NG^E/RT)}{\partial N_i}\right)_{T,P,N_j} \qquad (11.13.6)$$

Comments

1.Notice that, since for an ideal mixture the excess Gibbs free energy is equal to zero, all activity coefficients in such a mixture are equal to one.

2.Eq.11.13.6 provides the basis for the development of analytical expressions for the activity coefficient, such as those of van Laar, Margules, Wilson, etc. To this purpose, qualitative and quantitative arguments about the deviation of a given solution from an ideal one at the same conditions, are used to develop analytical expressions for G^E as functions of the composition, temperature, and pressure. Appropriate differentiation, using Eq.11.13.6, yields then the analytical expression for the activity coefficient of each solution component. We will discuss this approach in Chapter 13.

11.14 Concluding Remarks

1.Differences in:
a.inter- and intra-molecular forces, and
b.the size and shape of the molecules,
lead to nonideal mixture behavior.

2.As a result, mixture properties cannot be determined from pure compound properties alone.

3.Experimental data are available for a variety of binary mixtures, but rarely for multicomponent ones that are needed in the typical industrial application.

4.We resort therefore, as with pure compounds, to predictive techniques, mainly the corresponding states principle and equations of state.

5.Vapor phase mixture properties at low pressures are calculated with the virial equation. Interaction coefficients are useful for nonpolar systems, important for polar ones, and essential where solvation occurs.

6.For vapor and gaseous nonpolar/weaklypolar systems at higher pressures cubic EoS are used. Reasonable accuracy is obtained for hydrocarbon systems without interaction coefficients, which are essential however in the presence of gases such as CO, CO_2, H_2, H_2S, etc., and to a lesser degree of CH_4.

7.No reliable method exists for vapor and gaseous polar systems at high pressures; best estimates are obtained with cubic EoS, modified for polar compounds (Section 14.7.3), and for very high pressures with the Pitzer-Lee-Kesler approach.

8.Component fugacities in nonpolar/weakly polar liquid mixtures are calculated with EoS, typically cubic ones, and use of k_{ij} is essential for accurate results, except for mixtures of similar hydrocarbons, such as of alkanes (that do not contain CH_4, however).

9.For polar liquid mixtures, the standard state fugacity approach is used for this purpose. For a subcritical component, it is expressed through its vapor pressure; for a supercritical one, through its Henry's constant in the solution, requiring thus mixture data (see also Section 14.7.3).

10.In gas-solid and gas-liquid systems, large deviations from ideal solution behavior in the vapor phase at high pressures can lead to high solubilities of solids and of liquids in compressed gases.

11.We have just touched on the subject of phase equilibria which, along with chemical reaction equilibria, are of paramount importance in our profession representing, as we have seen, the third main objective of

Chemical Engineering Thermodynamics. Our discussion of phase equilibria in mixtures will continue in Chapters 13 and 14, while chemical reactions in them will be considered in Chapter 15. The thermodynamic basis for such equilibria will be developed in the next Chapter.

References

Basta, N., 1985. *Chem. Eng.*, Feb. 4, 14.

De Santis, R., Breedveld, G.J.F., Prausnitz, J.M., 1974. *Ind. Eng. Chem. Process Des. Dev.*, **13**, 574.

Graboski, M.S., Daubert, T.E., 1978. *Ind. Eng. Chem. Process Des. Dev.*, **17**, 448.

Hayden, J.G., O'Connell, J.P., 1975. *Ind. Eng. Chem. Process Des. Dev.*, **14**, 221.

Hermsen, R.W., Prausnitz, J.M., 1963. *Chem. Eng. Sci.*, **18**, 485.

Johnston, K.P., Penninger, J.M.L., Editors, 1989. *Supercritical Fluid Science and Technology*, ACS Symposium Series No 406, ACS, Washington, D.C.

Johnston, K.P., Ziger, D.H., Eckert, C.A., 1982. *Ind. Eng. Chem. Fundam.*, **21**, 191.

Kato, K., Nagahama, K., Hirata, M., 1981. *Fluid Phase Equilibria*, **7**, 219.

Lin, H., 1980. *Ind. Eng. Chem. Process Des. Dev.*, **19**, 501.

Lin, C., Daubert, T.E., 1980. *Ind. Eng. Chem. Process Des. Dev.*, **19**, 51.

McHugh, M.A., Krukonis, V.J., 1986. *Supercritical Fluid Extraction*, Butterworths, Boston.

Olds, R.H., Reamer, H.H., Sage, B.H., Lacey, W.N., 1949. *Ind. Eng. Chem.*, **41**, 475.

Peneloux, A., Rauzy, E., Freze, R., 1982. *Fluid Phase Equilibria*, **8**, 7.

Prausnitz, J.M., Lichtenthaler, R.N., de Azevedo, E.G., 1986. *Molecular Thermodynamics of Fluid-Phase Equilibria*, Prentice-Hall, Englewood Cliffs, N.J.

Prausnitz, J.M., Anderson, T.F., Grens, E.A., Eckert, C.A., Hsieh, R., O'Connell, J.P., 1980. *Computer Calculations for Multicomponent Vapor-Liquid and Liquid-Liquid Equilibria*, Prentice-Hall, Englewood Cliffs, N.J.

Prausnitz, J.M., Carter, W.B., 1960. *AIChE J.*, **6**, 611.

Reamer, H.H., Sage, B.H., Lacey, W.N., 1951. *Ind. Eng. Chem.*, **43**, 976.

Reid, R.C., Prausnitz, J.M., Poling, B.E., 1987. *The Properties of Gases and Liquids*, McGraw-Hill, New York.

Sandler, S., 1977. *Chemical Engineering Thermodynamics*, Wiley, New York.

Sage, B.H., Budenholzer, R.A., Lacey, W.N., 1940. *Ind. Eng. Chem.*, **32**, 1262.

Smith, J.M., Van Ness, H.C., 1987. *Introduction to Chemical Engineering Thermodynamics*, McGraw-Hill, New York.

Starling, K.E., Khan, M.A., Watanasiri, S., 1984. *AIChE Meeting*, San Francisco.
Tsonopoulos, C., 1974;1975. *AIChE J.*, **20**, 263; **21**, 827.
Tsonopoulos, C., 1979. In *Equations of State in Engineering and Research*, Chao, K.C., Robinson, R.L., Editors. Advances in Chemistry Series, Vol. **182**, Am. Chem. Soc., Washington, D.C.
Tsonopoulos, C., Heidman, J.L., 1986. *Fluid Phase Equilibria*, **29**, 391.
Williams, D.F., 1981. *Chem. Eng. Sci.*, **36**, 1769.
Ziger, D.H., Eckert, C.A., 1983. *Ind. Eng. Chem. Process Des. Dev.*, **22**, 582.

Problems

11.1 What is responsible for the difference in values between partial molar and pure component properties?

11.2 With reference to Example 11.1:

a. why is a only small amount of water added?

b. can you determine the partial molar volume of methanol from the given data?

11.3 Under what conditions will a mixture behave as an ideal one?

11.4 What factors effect the values of the partial molar volume of a component of a binary mixture?

11.5 The enthalpy of a binary liquid mixture at a given temperature and pressure is given by:

$$H = ax_1 + bx_2 + cx_1x_2$$

where:

a = 15000; b = 20000; and c = -2000; all in J/mol.

a. Calculate the enthalpies of pure 1 and 2;

b. Plot the partial molar enthalpies of 1 and 2 versus x_1;

c. Compare $\bar{H}_1$ at $x_1 = 0.2$ and at $x_1 = 0.8$, to H_1;

d. Discuss the effect of composition on $\bar{H}_1$ and $\bar{H}_2$.

11.6 With reference to the Gibbs-Duhem equation:

a. give two applications;

b. develop a specific example where you apply it.

11.7 The partial molar enthalpy of component 1 in a binary mixture, at a specified temperature T and pressure P, is given by:

$$\bar{H}_1 = H_1 + ax_2^2 + bx_2^3$$

where: H_1 = 10000 J/mol, is the enthalpy of pure 1 at T and P, and a = -500 and b = -1000, both in J/mol, are functions of T and P only. Develop the expression for $\bar{H}_2$ as a function of x_1 (any problems?).

11.8 With reference to the mixing rules:

a. what is their purpose?

b. what is the combining rule?

c. what is the purpose of the interaction coefficients?

11.9 Smith and Van Ness (1987) suggest that the second virial coefficients of Prausnitz and Carter (1960) for the system acetonitrile(1) - acetaldehyde(2) can be represented approximately as follows in the range 325 K to 375 K:

$$B_{11} = -8.55\left(\frac{1000}{T}\right)^{5.50} \quad ; \quad B_{22} = -21.5\left(\frac{1000}{T}\right)^{3.25} \quad ; \quad B_{12} = -1.74\left(\frac{1000}{T}\right)^{7.35}$$

where: T in K, and B in cm³/mol.

a. Calculate the molar volume of an equimolar mixture at 370 K and 1.5 bar.

b. Give your justified estimate of the accuracy of the obtained value.

11.10.a. In Problem 11.9, you used experimental values for the second virial coefficients. Explain if use of the term 'experimental' is justified.

b. Repeat the Problem, using values obtained through the empirical correlation of your choice.

11.11 Repeat Problem 11.9, using the PR equation of state.

11.12 Tabulate your results on the acetonitrile-acetaldehyde system (Problems 11.9 through 11.11), compare them to the correct value, and explain the observed differences.

11.13 With reference to Problem 11.9:

a. plot the partial molar volume of component 2 versus y_2, and comment on the effect of y_2;

b. repeat the plot, assuming the mixture to behave as: (1) an ideal mixture of real gases, and (2) an ideal gas, and discuss the accuracy of the obtained values.

11.14.a. What is the minimum information needed to estimate the volume of a nonpolar vapor or gaseous mixture?

b. The same for a polar one.

c. What method would you use in each case?

d. What accuracy would you expect?

Consider both, low and high pressures, and complete the following Table:

System	Pressure	Proposed method	Information needed	Expected accuracy
Nonpolar	Low			
Nonpolar	High			
Polar	Low			
Polar	High			

11.15.a. With reference to the interaction coefficient l_{12} in the Tsonopoulos correlation for B_{12}:

1.what is l_{12} a function of ?
2.can it be estimated from pure component properties? If yes, when?
3.discuss the type of experimental measurement that you would need to evaluate it.

b.The same, but for k_{12} (SRK).

11.16 For an equimolar vapor mixture of methane and hydrogen sulphide at 160 °F and 1000 psia:

a.what method would you use to estimate the molar volume V of this mixture?

b.present the reasons for your decision;

c.calculate V using this method;

d.could you get a better answer?

Experimental value: 5.52 ft^3/lb-mol (Reamer et al, 1951)

11.17 A gas mixture consisting of 80% (mole) methane and the rest propane exerts a pressure of 6 MPa when stored in a cylinder at 280 K. What is the maximum temperature that the cylinder may be exposed to if the pressure is not to exceed 12 MPa?

11.18 A gas mixture consisting of 75% (mole) methane and the rest propane enters a pipeline of constant cross-section at 310 K and 70 bar with a linear velocity of 10 m/s, and leaves at 280 K and 20 bar. If the flow is steady, what is the linear velocity of the gas leaving the pipe?

11.19 One thousand moles of an equimolar mixture of nitrogen and carbon dioxide at 1 bar and 325 K are compressed isothermally to 200 bar. What is the corresponding enthalpy and entropy changes?

11.20.a. Show that if the volumetric behavior of a gas mixture is described by:

$$z = 1 + \frac{BP}{RT} \quad \text{then :} \quad \ln\phi = \frac{BP}{RT}$$

b. Show that if this mixture is a binary one, then:

$$\ln\hat{\phi}_i = \frac{P}{RT}(B_{ii} + y_j^2\delta_{ij})$$

11.21 Why do we use $\hat{f}_i$ and not $\bar{f}_i$?

11.22 Starting with: $RTd\ln\hat{f}_i = \bar{V}_i dP$, develop the Lewis-Randall rule.

11.23 Express the partial pressure of a component of a vapor mixture in terms of its fugacity.

11.24 For the acetonitrile(1) - acetaldehyde(2) system at: $y_1 = 0.1$, $T = 373$ K and P=1 bar, calculate the fugacity of component 1 using:

a.the virial equation with the B_{ii} and B_{ij} values given in Problem 11.9;

b.the same equation, with B_{ii} and B_{ij} estimated with the method of your choice;

c.the PR EoS.

Tabulate your results, and explain the obtained % errors.

11.25 The fugacity coefficient of n-butane(1) in a vapor mixture with nitrogen(2) at $y_1 = 0.5024$, 370°C, and 100 bar is needed.

a. What method would you recommend?

b. What accuracy would you expect?

c. Calculate the fugacity coefficient with the recommended method.

d. What information is needed to determine this fugacity coefficient experimentally?

11.26 If component 1 of a binary equimolar vapor mixture is nonpolar, complete the following Table:

Component 2	Pressure	Range of $\hat{\phi}_i$ values	
		Comp. 1	Comp. 2
Nonpolar	Low		
Nonpolar	High		
Polar	Low		
H-bonded	Low		

11.27.a. Calculate the fugacity of steam at 300°C and 10 bar.

b. Using this answer, estimate the fugacity of steam(1) in a mixture with n-butane(2) at the same conditions, and at two mole fractions, $y_1 = 0.2$ and $y_1 = 0.8$, using the Lewis-Randall rule.

c. Which of the two values is more accurate? Explain.

d. How could you get a better estimate?

11.28 In Figure 11.4, the curves for the two activity coefficients are terminated before $x_1 = 1.0$. Why?

11.29 At 346 K and 1 atm, a liquid mixture of acetone(1) and benzene(2) with $x_1 = 0.20$ is in equilibrium with a vapor phase where $y_1 = 0.40$. Assuming ideal gas behavior for the vapor phase, calculate:

a. the fugacity of acetone in the liquid phase;

b. the standard state fugacity for acetone at these conditions;

c. the activity coefficient values.

Estimate the error involved in each case as a result of the ideal gas assumption.

11.30 What experimental measurements would you carry out to determine the Henry's constant of methane in ethanol at 25°C and 10 bar?

11.31 The solubility of methane in n-heptane at 220°F and a partial pressure of 1 atm is: $x = 2.83 \cdot 10^{-4}$.

a. What is the solubility of methane at the same temperature and a partial pressure of 5 atm?

b. State any approximations you make and estimate their impact on your answer.

11.32 Do Problem 11.61, part c.

11.33 The fugacity of component 1 in a binary liquid mixture at some temperature T and pressure P is given by:

$$\hat{f}_1 = x_1 + 4x_1^2 - 3x_1^3$$

a. What error would result from evaluating $\hat{f}_1$ by using the Lewis-Randall rule at $x_1 = 0.5$ and the same T and P?

b. Calculate the Henry's constant of component 1 in this mixture at the given T and P.

11.34 With reference to supercritical extraction:

a. what is it?

b. what are its advantages over other separation techniques?

c. what operating conditions are desirable?

d. what are the difficulties in its thermodynamic modelling?

e. give a possible application of your choice.

11.35 The vapor pressure of solid n-dodecane at -10°C is 0.6 Pa. Calculate its solubility in methane at -10°C and 100 bar.

11.36 For an ideal mixture:

a. which property changes of mixing are equal to zero and which are not?

b. show that all activity coefficients are equal to one.

11.37 Discuss the importance of excess properties.

11.38 Redlich and Kister have proposed the following polynomial expression to represent G^E of a binary mixture:

$$G^E = x_1x_2[A+B(x_1-x_2)+C(x_1-x_2)^2+\ldots]$$

For moderately nonideal systems, the two first terms are usually sufficient. Show that for such systems:

$$RT\ln\gamma_1 = (A+3B)x_2^2-4Bx_2^3$$

$$RT\ln\gamma_2 = (A-3B)x_1^2+4Bx_1^3$$

11.39 A binary solution has the following properties:

a. its entropy is the same as that of a hypothetical ideal solution of the same composition, and

b. its enthalpy - per mole of mixture - exceeds that of the corresponding ideal solution by an amount (bx_1x_2), where b is a function of pressure only. Show that the activity coefficients are given by:

$$RT\ln\gamma_1 = bx_2^2$$

$$RT\ln\gamma_2 = bx_1^2$$

11.40 What advantages, if any, does the virial equation have over the other methods in the estimation of mixture properties?

11.41 Explain positive and negative values for l_{ij} in Eq.11.7.7 in terms of intermolecular forces.

11.42 With reference to this Chapter:

a. what is the main objective?

b. describe briefly three important items that you have learned including an application in each case.

11.51 Demonstrate, with whatever method you want, that k_{12} (or l_{12}) is more important for the accurate determination of the partial molar volume of a component of a mixture, than for that of the mixture volume.

11.52 For the system acetonitrile-acetaldehyde:

a. using the B_{12} data of Prausnitz and Carter, Problem 11.9, calculate k_{12} in the Tsonopoulos correlation as a function of the temperature T.

b. how do you explain this variation with T?

11.53 Dave and George want to estimate the molar volume of a methane - n-butane mixture at: $y_1 = 0.8$, 150°C, and 100 atm.

George: Let us use the SRK EoS;

Dave: What about k_{12}?

George: They are both alkanes, we can use $k_{12} = 0.0$;

Dave: Yes, but with methane present we would get better accuracy if we could find a value for k_{12}. Perhaps we can use the virial equation, where we have l_{12} from the Tsonopoulos correlation;

George: Ya, but the pressure is too high;

Dave: Well, even if we cannot use the virial, there must be a way of using this l_{12} value.

a. What advice would you give to them?

b. Calculate the molar volume following your advice;

c. What accuracy do you expect?

Hint: Develop first an expression for the second virial coefficient through the SRK EoS.

11.54.a. Show that use of the SRK EoS for the entropy departure of a binary system gives:

$$S^\circ - S = -R\ln\left(\frac{V-b}{V^\circ}\right) - \frac{1}{b}\ln\left(\frac{V+b}{V}\right)Q$$

where:

$$Q = \left[y_1^2 + y_1y_2(1-k_{12})\frac{a_{22}}{(a_{11}a_{22})^{1/2}}\right]\frac{da_{11}}{dT} + \left[y_2^2 + y_1y_2(1-k_{12})\frac{a_{11}}{(a_{11}a_{22})^{1/2}}\right]\frac{da_{22}}{dT}$$

b. Do Problem 11.19, using the SRK EoS with $k_{12} = 0.0$.

c. Compare the answer with the one you obtained in Problem 11.19 and comment on their accuracy.

11.55 Show that:

$$\left[\frac{\partial \ln H_{i,s}}{\partial P}\right]_T = \frac{\bar{V}_{i,s}^{\infty}}{RT}$$

where $H_{i,s}$ and $\bar{V}_{i,s}^{\infty}$ are the Henry's constant and infinite dilution partial molar volume respectively of component i in solvent s.

11.56 In Example 11.10 we neglected the effect of pressure on the Henry's constant. What error does this lead to?

11.57 The solubility of methane in water, at 25°C and a partial pressure of 1 atm, is $x = 0.00025$.

a. Calculate the solubility of methane at a total pressure of 5 atm making any simplifying assumptions you want.

b. Comment on the impact of these assumptions on the accuracy of your final answer.

c. Is there a better method? If yes, which one?

11.58 Consider the solubility of a solid(2) in a gas(1) as a function of pressure P at some temperature T. Assuming that the vapor phase nonideality is described by the virial equation truncated after B, show that there is a minimum solubility of 2 in 1 given by:

$$y_2(minimum) = \frac{-cB_{12}P_2^s}{RT}$$

$$P(at\ minimum) = \frac{-RT}{B_{11}+2B_{12}}$$

where c is a constant depending only on the units used. Neglect ϕ_2^s and Poynting effect and discuss any exceptions.

11.59 At 30°C the solubilities of ethane and acetylene in n-octane are about the same. However, at the same temperature, the solubility of acetylene in dimethyl formamide is very much higher than that of ethane. How do you explain this in terms of intermolecular forces?

11.60.a. Show that for a binary system: $\bar{M}_1 = M + x_2(dM/dx_1)$

b. Show that for the same case, the intercepts of a tangent line at point x_1 of a M vs. x_1 curve represent the two partial molar values at x_1.

11.61 The following heat of mixing data are reported for a mixture of two hydrocarbons at 40°C:

x_1	ΔH (kJ/mol)
0.083	0.250
0.176	0.488
0.268	0.670
0.353	0.790
0.428	0.863
0.720	0.775
0.780	0.669
0.850	0.510
0.900	0.362

a. Calculate $\bar{H}_1$ and $\bar{H}_2$ at $x_1 = 0.25$ using the approach of the previous Problem.

b. How would you get a more accurate answer? Be specific.

c. Repeat part (a) using the following expression developed from the data:

$$\Delta H = x_1 x_2 [866.47 + 110.13(x_1 - x_2) + 30.07(x_1 - x_2)^2]$$

d. Compare the results of parts (a) and (c), and explain any differences.

11.62 How do you explain in terms of molecular behavior the significant differences in solubility of n-decane in carbon dioxide, nitrogen and hydrogen mentioned in the Comments of Section 11.11.3?

11.63 For the system ethylene(1)-naphthalene(2) plot the solubility of (2) in (1) at 285 K versus pressure assuming that the vapor phase is described by the:

a. ideal gas law;

b. virial equation truncated after B using your best estimate for l_{12};

c. Peng-Robinson EoS with $k_{12} = 0.02$ (Prausnitz et al, 1986, p.178).

Discuss the accuracy of the three approaches as a function of pressure.

Given: $\log P_2^s(\text{Pa}) = 13.575 - 3729.3/T$, T in K.

11.64 Correlate the hexamethylbenzene - ethane data of Johnston et al at 25°C, 45°C, 70°C with the PR EoS.

11.65 Discuss the standard state for an electrolyte in water (see Sandler, 1977).

12
Equilibrium and Stability

12.1 Introduction

We are now ready, having concluded the general discussion of the thermophysical properties of mixtures, to consider the third objective of Chemical Engineering Thermodynamics: Phase and Chemical Reaction Equilibrium.

These two areas are of paramount importance in practice: design of distillation and extraction columns, gas absorbers, evaporators, reactors, etc. is based on the analytical description of phase or chemical reaction equilibrium and, sometimes, of both. And one rarely finds a chemical plant or refinery where several of these units are not present.

But what is the reason for the existence of two phases in equilibrium with each other? Why do two species react to form an equilibrium mixture of products and reactants? What is the force that drives these systems into equilibrium?

From the more practical point of view, what is the basis for developing the analytical description of these equilibrium states? In previous Chapters, for example, we accepted that when two phases are in equilibrium at a given temperature and pressure, the fugacity of any component in the one phase is equal to that in the other phase. [We used this equality to arrive at a physical interpretation of fugacity in Chapters 9 and 11; to determine vapor pressures of pure fluids in Chapter 10 using equations of state; and to calculate gas solubilities in liquids, or solid and liquid solubilities in compressed gases, in Chapter 11.] Where does this equality of fugacities come from?

We will answer these questions and develop the basic equations required for the analytical description of phase and of chemical reaction equilibrium in this Chapter and proceed to apply them in these areas in the ensuing Chapters 13, 14 and 15.

12.2 Objective and Approach

The main objective in this Chapter is to develop an understanding of:

1. The *criteria* that must be met so that a given system attains a state of thermodynamic equilibrium under certain constraints;
2. How these criteria are used to develop the framework of thermodynamic equations that will be used, in the ensuing Chapters, for the determination of numerical answers, such as compositions in phase equilibrium; or, equilibrium conversions in reacting systems; and
3. The criteria that must be met so that the attained equilibrium is a *stable* one, and the consequences if they are not met.

We start by establishing the *criteria* that must be met so that a closed system attains equilibrium under certain constraints, and proceed to demonstrate how these rather abstract criteria are used as the basis for developing a framework of equations for obtaining numerical answers. To this purpose, we will consider systems involving:

a. two phases in equilibrium;
b. one chemical reaction occuring in a single phase; and
c. the more general case of multiphase/multireaction equilibrium;

concentrating on the important, from the practical point of view, constraint of constant temperature and pressure.

Using, next, the number of the equations that must be satisfied by a given equilibrium system in combination with the number of system variables, we establish the *degrees of freedom* for the system: the number of independent variables, whose specification is necessary for the complete determination of the system.

We proceed then to identify the criteria that must be met so that the attained equilibrium is a *stable* one, and examine, briefly, the phase transitions resulting when these *stability* criteria are not met for a pure and a two-component fluid. We close with some concluding remarks.

12.3 The Thermodynamic Criteria of Equilibrium

We know from experience that a given system under a set of constraints will evolve with time into a state where no more macroscopic changes occur. We refer to this state as an *equilibrium* state. Thus, if a reactive system reaches equilibrium in a closed isolated from its environment vessel, the temperature, pressure, and composition do not vary any more with time. There are no driving forces to effect further change.

We examine, next, the thermodynamic criteria for the equilibrium state of a system as a function of the constraints imposed on it.

12.3.1 A Closed System of Constant Internal Energy and Volume

The rate of entropy change, dS/dt, for a closed system is given by:

$$\frac{dS}{dt} = \frac{dS_r}{dt} + \frac{dS_g}{dt} \tag{12.3.1}$$

The first term, dS_r/dt, represents the variation of entropy with time due to reversible changes in the system; and the second, dS_g/dt, the rate of entropy generation due to irreversible changes in it.

To determine dS_r/dt, let us assume that the system is undergoing a reversible change and apply the first law to it:

$$\frac{dU}{dt} = \frac{TdS_r}{dt} - \frac{PdV}{dt} \tag{12.3.2}$$

If now the internal energy U and volume V are constant, $dS_r/dt = 0$, and Eq.12.3.1 becomes:

$$\frac{dS}{dt} = \frac{dS_g}{dt} \tag{12.3.3}$$

Since dS_g is always positive, the entropy will continue to increase until the system attains equilibrium, where it reaches its maximum value:

$$S = maximum \tag{12.3.4}$$

We conclude, therefore, that if a system is allowed to reach equilibrium - at constant mass, internal energy, and volume - its entropy assumes the maximum possible value under these constraints. Consequently, $(dS/dt) = 0$; or just:

$$dS = 0 \tag{12.3.5}$$

Entropy, in this case, represents the potential that drives the system into the equilibrium state.

Notice that Eq.12.3.5 is also satisfied if the entropy reaches a minimum value and, consequently, the second equilibrium requirement is $(d^2S/dt^2) < 0$; or:

$$d^2S < 0 \tag{12.3.6}$$

We will explore the consequences of this inequality in Section 12.8 - when we discuss the concept of system stability - while the implications of Eq.12.3.5 are considered in the next Example.

Nature's Favorable States

Nature, wrote Planck, seems to 'favor' certain states. The irreversible increase in entropy dS_g/dt describes a system's approach to a state which 'attracts' it, which the system prefers and from which it will not move on its own 'free will'. 'From this point of view, Nature does not permit processes whose final states she finds less attractive than their initial states. Reversible processes are limiting cases. In them, Nature has an equal propensity for initial and final states; this is why the passage between them can be made in both directions.' (Prigogine and Stengers, 1984, p.121)

12.3.2 Example 12.1

A given amount of a mixture of k nonreacting compounds is enclosed into a well insulated flask and is allowed to reach equilibrium, whereby two phases are present: a vapor and a liquid one.

Show that the temperature, pressure, and chemical potential in the vapor phase, $T(v)$, $P(v)$, and $\mu_i(v)$, are the same with those in the liquid phase, $T(l)$, $P(l)$, and $\mu_i(l)$, where i stands for any of the mixture components.

From Eq.12.3.5:

$$dS(l) + dS(v) = 0 \tag{A}$$

where (l) and (v) denote the liquid and vapor phase respectively.

From Eq.9.4.5 applied to the two phases:

$$dS(l) = \frac{dU(l) + P(l)dV(l) - \Sigma\mu_i(l)dN_i(l)}{T(l)} \tag{B}$$

$$dS(v) = \frac{dU(v) + P(v)dV(v) - \Sigma\mu_i(v)dN_i(v)}{T(v)} \tag{C}$$

Furthermore, since U, V, and N_i, are constant:

$$dU(l) = -dU(v) \tag{D}$$

$$dV(l) = -dV(v) \tag{E}$$

$$dN_i(l) = -dN_i(v) \tag{F}$$

Introduction of equations: (B), (C), (D), (E), and (F) into (A) yields after rearrangement:

$$\left(\frac{1}{T(l)}-\frac{1}{T(v)}\right)dU(l)+\left(\frac{P(l)}{T(l)}-\frac{P(v)}{T(v)}\right)dV(l)-\sum\left(\frac{\mu_i(l)}{T(l)}-\frac{\mu_i(v)}{T(v)}\right)dN_i(l) = 0$$

This equation must be valid for all changes of the independent variables, $U(l)$, $V(l)$, and $N_i(l)$, and, consequently, each bracketed term must be equal to zero. Hence:

$$T(l) = T(v);$$
$$P(l) = P(v);$$
$$\mu_i(l) = \mu_i(v),\ i = 1, k$$

Thus, when a closed system reaches equilibrium, constrained to constant internal energy and volume:

a. the temperature and pressure must be the same in both phases, i.e. no heat transfer across, or movement of, the interface can occur, and

b. the chemical potential of all species must be also the same in both phases, i.e. no mass transfer across the interface can take place.

12.3.3 A Closed System of Constant Temperature and Pressure

Let us now consider a closed system approaching equilibrium - this time at constant temperature and pressure - and develop the corresponding equilibrium criteria. For a reversible change in such a system, from the first law (Eq. 12.3.2):

$$\frac{dS_r}{dt} = \frac{1}{T}\frac{dU}{dt}+\frac{P}{T}\frac{dV}{dt} \tag{12.3.2a}$$

Introduction into Eq. 12.3.1 yields:

$$\frac{dS}{dt} = \frac{1}{T}\frac{dU}{dt}+\frac{P}{T}\frac{dV}{dt}+\frac{dS_g}{dt} \tag{12.3.7}$$

Since $G = U + PV - TS$ then, at constant T and P:

$$dG = dU + PdV - TdS$$

and Eq. 12.3.7 becomes: $$\frac{dG}{dt} = T\frac{-dS_g}{dt}$$

which indicates that, as the system moves towards equilibrium, its Gibbs free energy decreases continuously until it reaches its minimum value, where no more change is possible, i.e. where equilibrium is attained. Thus, in this case, *the Gibbs free energy represents the potential that drives the system towards equilibrium.*

The conditions, therefore, for equilibrium of a closed system at constant temperature and pressure are:

$$G = minimum \tag{12.3.8}$$

$$dG = 0 \tag{12.3.9}$$

$$d^2G > 0 \tag{12.3.10}$$

We apply Eqs 12.3.9 and 12.3.10 to a simple reacting system in the next Example 12.2. We proceed then to develop in Sections 12.4, 12.5, and 12.6, using the same equation, the basis for calculations in the very important for practical applications cases:

* phase equilibrium;
* chemical reaction equilibrium; and
* combination of them;

all occuring at constant mass, temperature, and pressure.

The consequences of inequality (12.3.10) are considered in Section 12.10.

12.3.4 Example 12.2

N_2O_4 decomposes according to the reaction: $N_2O_4 \rightarrow 2NO_2$. One mole of N_2O_4 is placed into a piston-and-cylinder arrangement, which is kept at a constant temperature of 25°C and a pressure of 1 atm.

Determine the composition of the mixture at equilibrium by finding the amount of N_2O_4 consumed when the Gibbs free energy of the system reaches its minimum value. It is given that at 298 K and 1 atm: $G(N_2O_4) = 97.90$; $G(NO_2) = 52.00$, both in kJ/mol.

Let: N_2O_4, component 1; NO_2, component 2; and e, the number of moles of N_2O_4 consumed up to a given moment. At this point:

Moles	Mole Fractions
$N_1 = 1 - e$	$y_1 = (1 - e)/N$
$N_2 = 2e$	$y_2 = 2e/N$
Total: $N = 1 + e$	

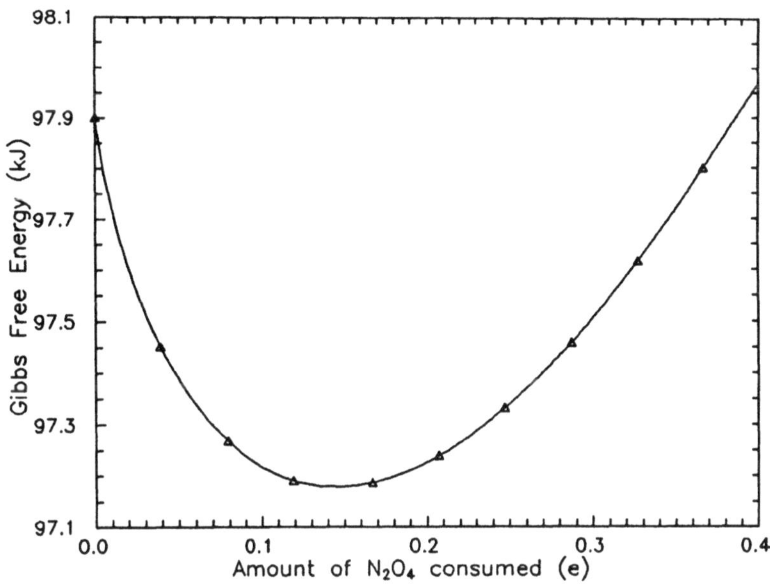

Figure 12.E.2 Determination of the equilibrium conversion of the reaction $N_2O_4 \rightarrow 2NO_2$ by minimizing the Gibbs free energy of the reacting mixture.

The total Gibbs free energy of the mixture is given, according to Eq.11.12.1a, by:

$$G = N_1 G_1 + N_2 G_2 + \Delta G \tag{A}$$

where: $G_1 = 97.90$ and $G_2 = 52.00$, both in kJ/mol.

Since at these conditions the mixture can be considered to behave as an ideal gas, from Eq.11.12.7a:

$$\Delta G = NRT(y_1 \ln y_1 + y_2 \ln y_2) \tag{B}$$

Introduction of Eq.(B) into Eq.(A) yields an expression for the total Gibbs free energy as a function of e:

$$G = N_1 G_1 + N_2 G_2 + NRT(y_1 \ln y_1 + y_2 \ln y_2) = f(e) \tag{C}$$

The number of moles of N_2O_4 consumed at equilibrium can be now determined by finding the value of e for which the total Gibbs free energy assumes its minimum value.

This can be done either graphically or analytically, the former giving a better picture of the system's progress towards equilibrium, as shown in Figure 12.E.2, and a value of $e = 0.144$ is obtained. Hence: $y_1 = 0.748$ and $y_2 = 0.252$.

12.4 The Basis for Phase Equilibrium Calculations at Constant Temperature and Pressure

12.4.1 The Equality of Chemical Potentials

Consider a closed system containing k nonreacting components and two phases I and II, that reaches equilibrium at constant temperature T and pressure P. We will demonstrate next that the requirement that the total Gibbs free energy of the system is minimized at equilibrium, leads to the equality of chemical potentials in the two phases for any component i.

From Eq.12.3.9:

$$dG = d[G(I) + G(II)] = 0 \tag{12.3.9}$$

From Eq.9.4.14 at constant T and P:

$$dG(I) = \Sigma \mu_i(I) dN_i(I) \tag{12.4.1a}$$

$$dG(II) = \Sigma \mu_i(II) dN_i(II) \tag{12.4.1b}$$

Since the system is closed:

$$dN_i(I) = -dN_i(II) \tag{12.4.1c}$$

Introduction of Eqs 12.4.1a, 12.4.1b, and 12.4.1c into Eq.12.3.9 yields:

$$\sum [\mu_i(I) - \mu_i(II)] dN_i(I) = 0 \tag{12.4.2}$$

In order that Eq.12.4.2 be valid for all values of the independent variables: $N_1(I)$, $N_2(I)$, ..., $N_k(I)$, each bracketed term must be equal to zero. Hence:

$$\mu_i(I) = \mu_i(II) \tag{12.4.3}$$

We conclude, therefore, that the chemical potential of component i has the same value in both phases that are in equilibrium at constant temperature and pressure. (Should this be expected from the discussion of Section 12.3.2?)

This is demonstrated numerically in the next Example, while in Example 12.4 we use this equality to develop the Clapeyron equation; and in Section 12.4.4 to conclude that the fugacities of component i in the two phases are also equal to each other.

12.4.2 Example 12.3

Show that the vaporization of saturated water at 1 atm to saturated steam at the same pressure, leads to no change of the chemical potential: the chemical potential in the two equilibrium phases is the same.

From the definition: $\mu = G = H - TS$, and using the Steam Tables:

$$\mu(v) = 18\,[2676.0 - (373.15)(7.3554)] = -1236.0\,\text{J/mol}$$

$$\mu(l) = 18\,[419.1 - (373.15)(1.3069)] = -1234.3\,\text{J/mol}$$

and the two values are equal to each other within the accuracy of the Tables.

Notice, thus, that while vaporization of saturated water into saturated steam leads into an enthalpy change of 40.6 kJ/mol, the corresponding Gibbs free energy change is equal to zero.

12.4.3 Example 12.4. The Clapeyron Equation

Starting with Eq. 12.4.3, show that for a pure fluid:

$$\frac{dP^s}{dT} = \frac{\Delta H_v}{T[V(v) - V(l)]}$$

where: P^s is the vapor pressure, ΔH_v is the molar enthalpy of vaporization, V is the molar volume, and (v) and (l) refer to saturated vapor and liquid respectively, all at the temperature T.

From Eq. 12.4.3 applied to a pure liquid in equilibrium with its vapor at temperature T:

$$d\mu(l) = d\mu(v)$$

and from Eq. 9.9.2:

$$V(l)dP^s - S(l)dT = V(v)dP^s - S(v)dT$$

Thus:

$$\frac{dP^s}{dT} = \frac{S(v) - S(l)}{V(v) - V(l)} = \frac{H(v) - H(l)}{T[V(v) - V(l)]} = \frac{\Delta H_v}{T[V(v) - V(l)]}$$

This equation was developed with a different approach, as we saw in Chapter 4, by Clapeyron before the chemical potential was introduced by Gibbs. It is also known as the Clausius-Clapeyron equation, and is often used for the evaluation of the enthalpy of vaporization from vapor pressure and saturated volume data (Section 8.14).

12.4.4 The Equality of Fugacities

Starting with the relationship between chemical potential and fugacity:

$$d\mu_i = RTd\ln\hat{f}_i \quad [T = const.] \tag{11.8.1}$$

and integrating - at constant T and P - from [$\mu_i(I)$ and $\hat{f}_i(I)$] to [$\mu_i(II)$ and $\hat{f}_i(II)$], where (I) and (II) refer to two phases at equilibrium, we obtain:

$$\mu_i(II) - \mu_i(I) = RT\ln\frac{\hat{f}_i(II)}{\hat{f}_i(I)}$$

which is equal to zero according to Eq.12.4.3. Thus:

$$\hat{f}_i(II) = \hat{f}_i(I) \tag{12.4.4}$$

We conclude, therefore, that when two phases reach equilibrium - at constant T and P - the fugacity of any component i has the same value in both phases.

12.4.5 Comments

1.Eq.12.4.4 represents the famous *equality of fugacities* that we already used in Chapters 9 and 11 to provide a physical interpretation of fugacity.

2.Use of the equality of these abstract fugacities in the determination of numerical answers in phase equilibrium problems was made:

* in Chapter 10, in evaluating vapor pressures of pure fluids through cubic equations of state; and, briefly,
* in Chapter 11, in evaluating solubilities of gases in liquids, and of solids and liquids in compressed gases.

3.Eq.12.4.4 represents, thus, the starting point for phase equilibrium calculations as we will see in Chapters 13 and 14.

4.Eq.12.4.3 was first developed by Gibbs, who provided thus the solution of the phase equilibrium problem in the abstract level (see Chapter 4); its translation however into the real world is a very difficult process as we will see in Chapters 13 and 14.

We consider next the implications of Eq.12.3.9 in the case of a reacting system reaching equilibrium at constant T and P in a single phase.

12.5 The Basis for Chemical Reaction Equilibrium Calculations at Constant Temperature and Pressure

12.5.1 The Problem and Approach

Consider a homogeneous mixture of compounds A, B, C, and D, that is made initially of N_A^o, N_B^o, N_C^o, and N_D^o moles of each. The mixture reacts according to:

$$aA + bB \rightleftharpoons cC + dD \tag{12.5.1}$$

and is allowed to reach equilibrium at constant T and P.

We will demonstrate next that the minimization of the system's Gibbs free energy at equilibrium leads, in this case, to the *equilibrium constant K. This constant represents the basis for numerical calculations of this type of equilibrium*, such as in determining the number of moles of each of the four components of the aforementioned mixture when equilibrium is reached. To this purpose, we will use the following approach:

a.Define the *stoichiometric coefficients* and *numbers* for the reaction;
b.Introduce the *progress of the reaction variable*;
c.Express the partial molar Gibbs free energy of the components of the mixture through the concept of standard state; and
d.Develop the expression for the equilibrium constant K.

We will then see in Example 12.7, and in detail in Chapter 15, how K is used to carry out calculations for this type of equilibrium.

12.5.2 Stoichiometric Coefficients and Numbers

We define a, b, c, and d in the reaction described by Eq.12.5.1 as the *stoichiometric coefficients*. It is, however, more convenient - especially for computer calculations - to write the reaction in the following algebraic form for the general case of k components:

$$\nu_1 A_1 + \nu_2 A_2 + \dots + \nu_k A_k = 0 \tag{12.5.2}$$

where: ν_1, ν_2, ..., ν_k, are referred to as the *stoichiometric numbers*; they are negative for reactants and positive for products. Hence Eq.12.5.1 becomes:

$$\nu_1 A_1 + \nu_2 A_2 + \nu_3 A_3 + \nu_4 A_4 = 0 \tag{12.5.3}$$

where: $A_1 = A;\ A_2 = B;\ A_3 = C;\ A_4 = D;$ and

$$\nu_1 = -a;\ \nu_2 = -b;\ \nu_3 = c;\ \nu_4 = d.$$

12.5.3 The Progress of the Reaction Variable

Assume that at some point in the path of the reacting system towards equilibrium a small amount of compound j, dN_j, is consumed (or produced). It is easily seen that the corresponding amounts of the other compounds are then given by:

$$\begin{aligned} dN_1 &= \frac{\nu_1}{\nu_j} dN_j \\ dN_2 &= \frac{\nu_2}{\nu_j} dN_j \\ &\cdots\cdots\cdots \\ dN_k &= \frac{\nu_k}{\nu_j} dN_j \end{aligned} \tag{12.5.4}$$

Simple rearrangement yields:

$$\frac{dN_1}{\nu_1} = \frac{dN_2}{\nu_2} = \dots = \frac{dN_k}{\nu_k} = \frac{dN_j}{\nu_j} = de \tag{12.5.5}$$

where e has units of moles.

Notice that if the initial amounts of the components of the reacting system (N_1°, N_2°, ..., N_k°) are specified, and the value of e at some point of the path towards equilibrium is known, their amounts at that point can be evaluated by integrating Eq. 12.5.5:

$$\begin{aligned} N_1 &= N_1^\circ + \nu_1 e \\ N_2 &= N_2^\circ + \nu_2 e \\ &\cdots\cdots\cdots \\ N_k &= N_k^\circ + \nu_k e \end{aligned} \tag{12.5.6}$$

Hence, the value of e at a given point - of the progress of the reaction towards equilibrium - suffices to determine the number of moles of all components, i.e. establishes the progress of the reaction up to this point. It is referred to, appropriately, as the *progress of the reaction variable*, or just the *progress variable*.

The total Gibbs free energy of the mixture reacting according to Eq. 12.5.2 is given by Eq.11.4.4a:

$$G = \sum N_i \bar{G}_i$$

and at equilibrium, reached at constant P and T, from Eq.12.3.9:

$$dG = \sum \bar{G}_i dN_i + \sum N_i d\bar{G}_i = 0$$

Notice that:
a. the second summation is equal to zero, according to the Gibbs-Duhem equation, and
b. $dN_i = v_i\, de$, according to Eq.12.5.5.
Thus:

$$\sum \bar{G}_i\, v_i\, de = 0 \quad \text{or} \quad \sum v_i \bar{G}_i = 0 \qquad (12.5.7)$$

We will consider next the evaluation of $\bar{G}_i$, and then proceed to develop the expression for the equilibrium constant starting with Eq.12.5.7.

To become familiar, however, with the progress variable e consider first the following two Examples.

12.5.4 Example 12.5

Consider the reaction: $2A \rightarrow B$, taking place at a given temperature and pressure. Starting with one mole of A, it is found that 0.4 moles of A are converted when equilibrium is reached. Evaluate e:
a. at the beginning of the reaction, $e(0)$;
b. when 0.2 moles of A are converted, $e(0.2)$; and
c. at equilibrium, $e(e)$.

According to Eq.12.5.6: $N_A = N_A^{\circ} - 2e$; $N_B = N_B^{\circ} + e$. Thus:

a. $e(0) = 0$

b. $0.8 = 1 + (-2)e; \rightarrow e(0.2) = 0.1$

c. $0.6 = 1 + (-2)e \rightarrow e(e) = 0.2$

12.5.5 Example 12.6

Let us now consider the reverse case of Example 12.5: Given the initial amounts of the reactants: $N_A^{\circ} = 1.0$ and $N_B^{\circ} = 0.0$ moles, determine their amounts when: $e = 0.0; 0.1; 0.2$.

From Eq.12.5.6 again:

$$N_A = N_A{}^\circ - 2e$$
$$N_B = N_B{}^\circ + e$$

Hence:

e	N_A	N_B
0.0	1.0	0.0
0.1	0.8	0.1
0.2	0.6	0.2

Comments

1. We conclude, therefore, that:

a. If the number of moles of one of the components of the reacting system at a given point in the progress of the reaction is known, and its initial amount is specified, e can be determined; and

b. If e is known, the amounts of all components of the reacting system present can be determined from their initial amounts.

2. From the practical point of view the second option is the most interesting, especially if e can be determined without resorting to experimental measurements; we will see in Example 12.7, and in detail in Chapter 15, that this is accomplished through the equilibrium constant K.

But, first, we consider the expression for the partial molar Gibbs free energy of the reaction components through which K will be determined.

12.5.6 Evaluation of $\bar{G}_i$: The Standard State Fugacity and Gibbs Free Energy

Integration of Eq.11.8.1:

$$d\bar{G}_i = RTd\ln\hat{f}_i$$

at constant temperature T from some reference state - designated by the superscript ° - to the final state, yields:

$$\bar{G}_i = G_i{}^\circ + RT\ln\frac{\hat{f}_i}{f_i^\circ} \tag{12.5.8}$$

In other words, $\bar{G}_i$ can be evaluated from the fugacity of component i in the mixture, only relative to its value at some reference state, G_i°, and the value of the fugacity in that state, $f_i{}^\circ$.

Establishment of the reference state - referred to as *standard state* -

requires, as discussed in Section 11.10, specification of the:

* mole fraction of i,
* temperature, and
* pressure.

The following standard state has been accepted as the most convenient one in the typical case:

* mole fraction: pure i at the same phase as the reacting system;
* temperature: that at which the reaction takes place; and
* pressure: 1 atm (or 1 bar).

In addition, for gas phase reactions, we specify that f_i° = 1 atm (or 1 bar), i.e. ideal gas behavior in the standard state.

12.5.7 The Expression for the Equilibrium Constant K

We return now to Eq. 12.5.7, obtained from the minimization of the Gibbs free energy of the reacting system at constant T and P, and introduce $\bar{G}_i$ from Eq. 12.5.8:

$$\sum \nu_i \left(G_i^\circ + RT \ln \frac{\hat{f}_i}{f_i^\circ} \right) = 0 \tag{12.5.9}$$

Rearrangement gives:

$$\ln K = \frac{-\Delta G^\circ}{RT} \tag{12.5.10}$$

where:

$$K = \prod \left(\frac{\hat{f}_i}{f_i^\circ} \right)^{\nu_i} = \prod \hat{a}_i^{\nu_i} \tag{12.5.11}$$

$$\Delta G^\circ = \sum \nu_i G_i^\circ \tag{12.5.12}$$

and the product and summation include all components of the reacting mixture.

K is called the *equilibrium constant of the reaction*, ΔG° the *standard Gibbs free energy of the reaction*, and $\hat{a}_i$ is the *activity* of component i in the mixture, defined by:

$$\hat{a}_i = \frac{\hat{f}_i}{f_i^\circ} \tag{12.5.13}$$

Notice that, since all G_i^o values are determined at the temperature of the reaction but at a pressure always equal to 1 atm (or 1 bar), *K is a function of temperature only.*

12.5.8 Example 12.7

Repeat Example 12.2 using the Gibbs free energy data given there.

From Eq. 12.5.10 and, since the G values are given at the standard state - temperature of the reaction and pressure of 1 atm:

$$\ln K = -\Delta G^o/RT = -[(-1)(97.90) + (2)(52.00)]/RT = -2.463$$

$$K = 0.085$$

From the definition of K, since $f_1^o = f_2^o = 1$ atm:

$$K = \frac{\hat{f}_2^2}{\hat{f}_1} \tag{A}$$

From Chapter 11 assuming ideal gas behavior: $\hat{f}_i = y_i P$. Therefore:

$$K = \frac{y_2^2}{y_1} \tag{B}$$

To determine the number of moles of component 2 at equilibrium we note that:

$$N_1 = 1 - e \quad N_2 = 2e \quad N = 1 + e$$

Thus,

$$y_1 = \frac{1-e}{1+e}; \quad y_2 = \frac{2e}{1+e}$$

and:

$$K = \frac{[2e/(1+e)]^2}{(1-e)/(1+e)} = \frac{4e^2}{1-e^2} = 0.085$$

Solving: $e = 0.144$, which is the same - as expected, of course - with that of Example 12.2, where e was obtained by directly minimizing the Gibbs free energy of the mixture.

12.5.9 Comments

1. Example 12.7 demonstrates that knowledge of K leads to the evaluation of the equilibrium conversion.

2. On the other hand, evaluation of K requires knowledge of the free energies of the components of the reacting systems at the temperature of the reaction. Both items will be explored in Chapter 15.

Undergraduate students can now move on to Section 12.7, accepting the following statements - that will be developed in the next Section - and pertain to a closed system that reaches equilibrium at constant temperature and pressure:

a. If more than two phases are present, they are all in equilibrium with each other;

b. In a single phase system with multiple reactions taking place, all reactions are in equilibrium; and

c. In a multiphase/multireaction system, all phases are in equilibrium with each other; and all reactions are in equilibrium in each phase.

12.6 The Basis for Multiphase/Multireaction Equilibrium Calculations at Constant Temperature and Pressure

Multiphase equilibrium, mainly liquid-liquid-vapor, is encountered occasionally in the chemical and petroleum industries. Multiple reactions, on the other hand, represent the typical case; and, in a few rare cases, they occur in more than one phases.

In the interest of simplicity, we will consider in this Section the following cases:

a. three-phase equilibrium;

b. multiple reactions in a single phase; and

c. multiple reactions in a two-phase system.

12.6.1 Three-Phase Equilibrium

Consider a closed system containing N_1, ..., N_k moles of k nonreacting components, distributed among three phases: I, II, and III, that reaches equilibrium at constant temperature T and pressure P. The independent variables will be the number of moles in:

Phase I: $N_1(I)$, $N_2(I)$, ..., $N_k(I)$, and

Phase II: $N_1(II)$, $N_2(II)$, ..., $N_k(II)$,

while the number of moles in the third phase is given by the material balances:

$$N_i(III) = N_i - N_i(I) - N_i(II), \; i = 1, k \tag{12.6.1}$$

The total Gibbs free energy of the system is:

$$G = \sum N_i(I)\mu_i(I) + \sum N_i(II)\mu_i(II) + \sum N_i(III)\mu_i(III) \qquad (12.6.2)$$

which assumes its minimum value when equilibrium is reached at constant T and P.

Differentiation of Eq. 12.6.2 and introduction into Eq. 12.3.9 yields:

$$\sum \mu_i(I)dN_i(I) + \sum \mu_i(II)dN_i(II) + \sum \mu_i(III)dN_i(III) = 0 \qquad (12.6.3)$$

as each of: $\Sigma N_i(I)\, d\mu_i(I)$, $\Sigma N_i(II)\, d\mu_i(II)$, and $\Sigma N_i(III)\, d\mu_i(III)$, equals zero by the Gibbs-Duhem equation. In addition, Eq. 12.6.1 gives:

$$dN_i(III) = -[dN_i(I) + dN_i(II)], \; i = 1, k \qquad (12.6.1a)$$

Introduction of Eq. 12.6.1a into Eq. 12.6.3 yields, after rearrangement:

$$\sum [\mu_i(I) - \mu_i(III)]dN_i(I) + \sum [\mu_i(II) - \mu_i(III)]dN_i(II) = 0 \qquad (12.6.4)$$

In order, now, that Eq. 12.6.4 be satisfied for arbitrary changes in all independent variables:

$$\mu_i(I) = \mu_i(III)$$
$$\mu_i(II) = \mu_i(III)$$

for all values of i from 1 to k. Thus:

$$\mu_i(I) = \mu_i(II) = \mu_i(III) \qquad (12.6.5)$$

and, it follows from Section 12.4.4 that:

$$\hat{f}_i(I) = \hat{f}_i(II) = \hat{f}_i(III) \qquad (12.6.6)$$

We conclude that when a three-phase closed system reaches equilibrium at constant T and P, the chemical potential and fugacity of component i must be the same in all three phases.

12.6.2 Multiple Reactions in a Single-Phase System

Consider a homogeneous closed system made initially of N_1^o, ..., N_k^o moles of k components, where r chemical reactions take place at constant T and P:

$$\nu_{1,1}A_1 + \nu_{2,1}A_2 + ... + \nu_{k,1}A_k = 0$$

$$.........................$$

$$\nu_{1,r}A_1 + \nu_{2,r}A_2 + ... + \nu_{k,r}A_k = 0$$

The total Gibbs free energy of the system at equilibrium is:

$$G = \sum N_i \bar{G}_i \tag{12.6.7}$$

where: $$N_i = N_i^\circ + \nu_{i,1} e_1 + ... + \nu_{i,r} e_r = N_i^\circ + \sum \nu_{i,j} e_j \tag{12.6.8}$$

and: $i = 1, k$; $j = 1, r$; and e_j is the equilibrium value of the progress of the jth reaction variable.

Introduction of Eq.12.6.7 into Eq.12.3.9, combined with the Gibbs-Duhem equation, yields:

$$\sum \bar{G}_i dN_i = 0$$

and, since from Eq.12.6.8, $dN_i = \sum \nu_{i,j}\, de_j$

$$\sum \bar{G}_i \nu_{i,1} de_1 + ... + \sum \bar{G}_i \nu_{i,r} de_r = 0 \tag{12.6.9}$$

In order that Eq.12.6.9 be satisfied for arbitrary changes in all independent variables, e_1, e_2, ..., e_r:

$$\sum \bar{G}_i \nu_{i,j} = 0, \quad j = 1, r \tag{12.6.10}$$

Thus when a single-phase system, where multiple reactions take place, reaches equilibrium, so does each of the reactions involved.

12.6.3 Multiple Reactions in a Two-Phase System

Consider a closed system of two phases, I and II, made initially of N_1°, ..., N_k° moles of k components involved in r reactions, which reaches equilibrium at constant T and P.

The total Gibbs free energy is given by:

$$G = \sum N_i(I)\bar{G}_i(I) + \sum N_i(II)\bar{G}_i(II)$$

At equilibrium (Eq.12.3.9) use of the Gibbs-Duhem equation gives:

$$\sum \bar{G}_i(I) dN_i(I) + \sum \bar{G}_i(II) dN_i(II) = 0 \tag{12.6.11}$$

The total number of moles of component i present at equilibrium is:

$$N_i = N_i(I) + N_i(II) = N_i^\circ + \sum \nu_{i,j} e_j(I) + \sum \nu_{i,j} e_j(II) \tag{12.6.12}$$

and consequently:

$$dN_i(II) = \sum \nu_{i,j} de_j(I) + \sum \nu_{i,j} de_j(II) - dN_i(I) \tag{12.6.12a}$$

Introduction of $dN_i(II)$ from Eq.12.6.12a into Eq.12.6.11 yields:

$$\begin{aligned}&\sum \bar{G}_i(I)dN_i(I)+\sum \bar{G}_i(II)[\sum \nu_{i,j}de_j(I)+\\&\sum \nu_{i,j}de_j(II)-dN_i(I)] = \sum [\bar{G}_i(I)-\bar{G}_i(II)]dN_i(I)+\\&\quad\sum \bar{G}_i(II)\sum \nu_{i,j}de_j(I)+\sum \bar{G}_i(II)\sum \nu_{i,j}de_j(II) = 0\end{aligned} \tag{12.6.13}$$

In order that Eq.12.6.13 hold for arbitrary changes in all independent variables: $N_i(I)$, $i = 1, k$; $e_j(I)$ and $e_j(II)$, $j = 1, r$:

$$\bar{G}_i(I) = \bar{G}_i(II), \quad i = 1, k$$

$$\sum \nu_{i,j}\bar{G}_i(II) = 0, \quad j = 1, r$$

$$\sum \nu_{i,j}\bar{G}_i(I) = 0, \quad j = 1, r$$

In other words, in order that the two-phase system be in equilibrium at constant T and P:

* the two phases must be in equilibrium with each other; and
* all reactions must be in equilibrium in each phase.

12.6.4 Comments

1. Development of the equations of this Section can also be carried out by using Lagrangian multipliers (see, among others, Sandler, 1977).

2. Extension of the findings of Section 12.6.1 to more than three phases, and of 12.6.3 to more than two phases is straightforward (see, for example, Problem 12.51). They are rarely, however, of practical importance.

12.7 The Complete Specification of an Equilibrium System

Once we have established the equations that must be satisfied in order that a closed system be in equilibrium at some temperature T and pressure P, we consider the following question: *What is the number of variables that must be specified, so that the equilibrium state of the system can be determined?*

We consider first a nonreacting system, and proceed to determine the number of variables needed for the determination of the *intensive* and *extensive* state of the system, which will lead to the *Phase Rule* and *Duhem's theorem* respectively. The case of a reacting system will be then considered.

12.7.1 The Phase Rule

Consider a closed system, containing k nonreacting components distributed over ϕ phases, at equilibrium at some temperature T and pressure P.

The number of variables for determining the intensive state of the system are:

* for a given phase, $(k - 1)$ compositions - since $\sum x_i = 1$, for each phase - or $\phi(k - 1)$ for all phases; and

* T and P.

Hence, the total number of variables is: $\phi(k - 1) + 2$.

To establish the number of equations we note that, since the fugacity of any component i in a given phase must equal its value in the other phases, we have $(\phi - 1)$ equations for each component, or $k(\phi - 1)$ for all of them.

The degrees of freedom, F, i.e. the number of variables that must be specified for the complete determination of the system, is then given by:

F = total number of variables - number of equations, or

$$F = \phi(k-1)+2-k(\phi-1) = k+2-\phi \tag{12.7.1}$$

Eq.12.7.1 represents the *Phase Rule* and, as its development indicates, it deals with the *intensive* properties of the system (T, P, composition, and partial molar properties).

In the case of binary vapor-liquid equilibrium for example, Eq.12.7.1 indicates that knowledge of two intensive variables, say T and x_1, suffices to determine P and y_1, but not the extensive properties, such as the amounts of liquid and vapor. The condition for their determination is provided by Duhem's Theorem.

Finally, in certain cases special constraints can be imposed on the system beyond the $k(\phi - 1)$ equations resulting from the equality of fugacities. A more general expression, therefore, of the phase rule for a nonreacting system is:

$$F = k+2-\phi-SC \tag{12.7.2}$$

where SC is the number of equations resulting from the special constraints.

If we return, for example, to the aforementioned binary vapor-liquid equilibrium system and impose the requirement that an azeotrope ($y_1 = x_1$) is present, then $SC = 1$ and $F = 1$. We can no longer specify both T and x_1, but only one of them.

We proceed now with the development of Duhem's theorem.

12.7.2 The Duhem Theorem

Consider a nonreacting system containing specified amounts of its k components that reaches equilibrium at some temperature T and pressure P and let ϕ be the number of phases present. What information is needed for the complete specification of this system in terms of both *intensive and extensive* variables? In this case, thus, the complete determination of its equilibrium state, requires - in addition to T, P, and the composition of each phase - the *amounts* of each phase. Hence:

Variables:
* T and P;
* $(k - 1)\phi$ compositions;
* ϕ amounts;

total: $2 + (k - 1)\phi + \phi$.

Equations:
* $(\phi - 1)k$ equalities of the component fugacities;
* k material balances, one for each component;

total: $(\phi - 1)k + k$.

The difference between variables and equations, equal to 2, suggests that specification of two variables suffices for the complete determination of the equilibrium state of a multiphase system, provided that the initial amounts of its components are known. This is referred to as the *Duhem Theorem*. These two variables can be intensive or extensive. Keep in mind, however, that the number of *independent* intensive variables is determined by the phase rule (see next Example).

12.7.3 Example 12.8

A cylinder-and-piston assembly, containing one kg of steam at $P = 20$ kPa, is placed into a constant temperature (348.15 K) bath and, after some time has elapsed, thermal equilibrium is attained. We start increasing the pressure by pushing the piston slowly, to ensure that the system is always in equilibrium with the bath. Is the system completely specified at all pressures?

At any value of the pressure, up to its dew point value of 38.55 kPa, all the properties of the steam are known. For example, when the pressure is 30 kPa, $H = 2636.8$ kJ.

If we push the piston farther, however, we notice that as long as the two phases exist at equilibrium:

a. the pressure does not change, i.e. it becomes a dependent variable; and

b. knowledge of pressure and temperature does not suffice anymore for the determination of the extensive variables of the system for, as the phase rule dictates, pressure and temperature are not independent variables once the second phase appears. Thus, for example, the enthalpy of the mixture cannot be evaluated.

To accomplish this, an additional property - say, the total volume of the mixture - must be specified. This is because, unlike pressure, the total volume can change with the movement of the piston, i.e. volume and temperature are independent variables.

Hence to completely specify the system two properties, that can vary independently at equilibrium, must be specified.

12.7.4 A Multiphase, Multireaction System

Consider a closed system, containing k species involved in r independent reactions and distributed among ϕ phases, which is in equilibrium at some temperature T and pressure P.

The total number of intensive variables is again: $(k - 1)\phi + 2$, but the number of equations becomes: $k(\phi - 1) + r$, i.e. it is increased by the number of reactions r, since for each reaction Eq.12.6.10 must be satisfied. The degrees of freedom are, therefore:

$$F = k - r + 2 - \phi \tag{12.7.3}$$

Identification of the independent reactions that lead to an equilibrium mixture is not always apparent, and it will be discussed in Chapter 15. In the presence of special constraints, Eq.12.7.3 becomes:

$$F = k - r + 2 - \phi - SC \tag{12.7.4}$$

Finally, Duhem's theorem - as it will be demonstrated in Chapter 15 - applies to reacting systems as well.

At this point, undergraduate students can move on to Section 12.10.

12.8 Stability in Thermodynamic Systems

We emphasized in Section 12.3 that the equilibrium requirement for a closed system at constant internal energy U and volume V: $dS = 0$, could be met even if the total entropy S reached a minimum value and, consequently, the requirement:

$$d^2S < 0 \tag{12.3.6}$$

should be also satisfied.

A similar observation was made for the Gibbs free energy in the case of constant temperature T and pressure P where:

$$d^2G > 0 \tag{12.3.10}$$

We will demonstrate next that if inequality 12.3.6 is satisfied, the equilibrium state is a *stable* one. (The term 'stable' indicates that if some perturbation were to occur in the system, the latter would accommodate it and return to its original equilibrium state, consistent with the constraints of constant volume and internal energy.) This will lead to two *stability* criteria: the first involving the constant volume heat capacity; the second the derivative of pressure with respect to volume at constant temperature.

We proceed then in Section 12.9 to apply the second of these criteria to the description of pure fluid behavior by an equation of state; and in Section 12.10, to examine the implications of inequality 12.3.10 not being satisfied by discussing liquid-liquid equilibrium in binary systems.

12.8.1 Stability Criteria

To explore the implications of inequality 12.3.6, consider a single-component closed system of N moles in equilibrium at constant U and V, which is divided into two imaginary parts, A and B.

We assume, next, that a small perturbation occurs in part A, for example a small change in its temperature, and proceed to identify the criteria that must be met by the system so that it returns to its equilibrium state; they are referred to as the *stability* criteria.

To this purpose we develop, through a Taylor series expansion, an expression for d^2S that corresponds to the changes in part A: $dU(A)$, $dV(A)$, and $dN(A)$; and in part B: $dU(B)$, $dV(B)$, and $dN(B)$, subject to the constraints that are consistent with a closed system at constant U and V:

$$dU(A) + dU(B) = 0$$

$$dV(A) + dV(B) = 0$$

$$dN(A) + dN(B) = 0$$

and examine the requirements so that d^2S be positive.

The algebra involved is somewhat lengthy (see, for example Sandler; Callen, 1960, or any standard text on Advanced Thermodynamics) and leads to the following requirements:

$$C_V = \frac{1}{N}\left(\frac{\partial U}{\partial T}\right)_V > 0 \qquad (12.8.1)$$

$$\left(\frac{\partial P}{\partial V}\right)_T < 0 \qquad (12.8.2)$$

which represent the *stability* criteria. Thus, in a stable equilibrium system:

a.the heat capacity must be positive, that is an increase in temperature should lead to an increase in the internal energy; and

b.an increase in pressure at constant temperature should lead to a decrease in volume.

12.8.2 Comments

1.We have considered so far the stability of a closed, homogeneous, single-component system at constant U and V. This is referred to as *intrinsic* stability.

2.In the case of a closed single-component system, containing two phases in equilibrium at constant U and V, it can be shown following a similar approach (Sandler) that the system stability - referred to as *mutual* stability - requires that each phase is intrinsically stable.

3.For the stability criteria of a closed two-phase nonreacting multi-component system, in equilibrium at constant U and V, see Modell and Reid (1974).

4.We have commented in previous Chapters that thermodynamics alone does not provide information on heat capacities or equations of state. Rather, they are obtained from statistical mechanics and/or experimental data. Thermodynamics, however, does set restrictions on such results in terms of inequalities 12.8.1 and 12.8.2. (Sandler suggests that they "may be thought of as part of the philosophical content of thermodynamics".)

5.The well-known Le Chatelier principle: "The spontaneous reaction of a stable system induced by a displacement from equilibrium, is in the direction of restoring the system to equilibrium", can be viewed as a verbal statement of the two stability criteria.

6.Assume, for example, that due to some disturbance in the system of Section 12.8.1 the temperature in part A, $T(A)$, becomes higher than that in part B, $T(B)$. The spontaneous process is, of course, that energy will flow from part A to B.

Now, according to the first stability criterion the temperature in A will decrease and that in B will increase until they become equal. This is exactly what the Le Chatelier principle calls for.

7.Assume, now, that $P(A)$ becomes higher than $P(B)$ due to some disturbance. The spontaneous reaction for the system will be for $P(A)$ to push the imaginary wall - that separates A from B - in such a way as to increase the volume of A and decrease that of B.

According, now, to the second stability criterion this will lead to a decrease in $P(A)$ and an increase in $P(B)$, until the two pressures become equal and the system is restored in its original equilibrium state. This, again, is what the Le Chatelier principle calls for.

8.Because of our interest in PVT behavior and its representation through equations of state - especially cubic ones - we apply the second of the stability criteria to the van der Waals equation of state in the following Example, and then proceed to examine what happens when it is not met.

A student's definition of the Le Chatelier principle:

If to any system in equilibrium is added an upset, the system will tend to shift to offset the upset. (C&EN, July 31, 1989, p.76.)

12.8.3 Example 12.9

Examine if the van der Waals (vdW) equation of state (EoS):

$$P = \frac{RT}{V-b} - \frac{a}{V^2}$$

satisfies the second stability criterion. To this purpose, plot P versus V for one mole of propane at two temperatures: $T_1 = 325\text{ K} < T_c$; and $T_2 = 450\text{ K} > T_c$, and examine the slopes of the obtained curves.

From Section 8.10.2: $\quad a = \frac{27}{64}\frac{(RT_c)^2}{P_c} \quad ; \quad b = \frac{RT_c}{8P_c}$

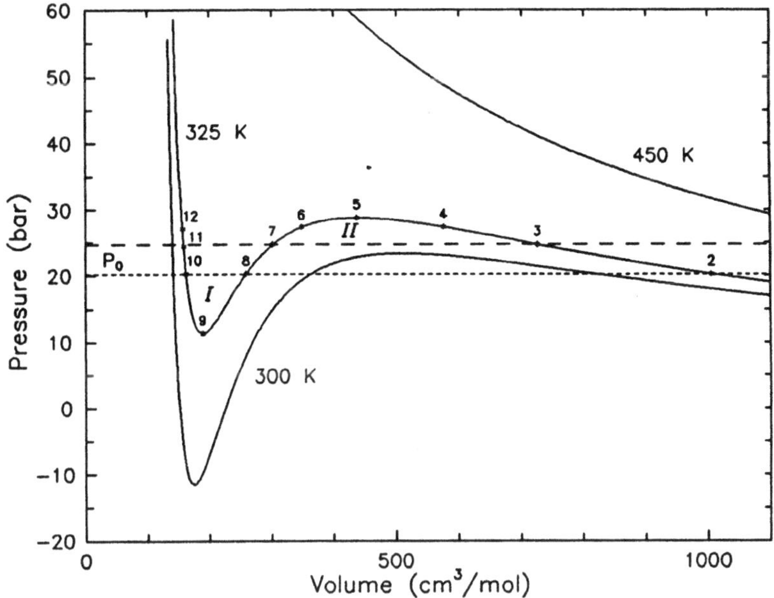

Figure 12.E.9 Variation of pressure with volume for propane with the vdW EoS.

Using: T_c = 369.82 K and P_c = 42.49 bar, we obtain:

$$a = 9.386 \cdot 10^6 \text{ bar cm}^6/\text{mol}^2; \quad b = 90.45 \text{ cm}^3/\text{mol}$$

Plots of P versus V for T_1 = 325 K and T_2 = 450 K are presented in Figure 12.E.9.

Notice that the second stability criterion is satisfied throughout the T_2 = 450 K > T_c isotherm; the same is true for the T_1 = 325 K < T_c isotherm, *except* for the section (5-6-7-8-9).

This behavior is not particular to the vdW EoS. It is exhibited by all types of equations of state, and the implications are examined next.

12.9 Phase Transitions

Let us now explore the behavior of a fluid whose vdW's, or any other EoS's, isotherm has the shape of that for propane at 325 K, Figure 12.E.9, i.e. it does not satisfy the second stability criterion throughout.

12.9.1 The Unstable Region

Consider first point 8. Since this point does not meet the second stability criterion (Ineq.12.8.2), it cannot represent a stable homogeneous phase. The system must move to points 2 or 10. (Or, it could break down into two phases represented by points 2 and 10. If these two phases, however, are to be in equilibrium, they must meet the criterion that was developed in Section 12.4: the chemical potential of the fluid must be the same in both phases. For this to occur, according to Problem 9.57, the two areas *I* and *II* - below and above the constant pressure line - must be equal. This pressure - not equal to P_0 of course - represents the vapor pressure of the fluid at the specified temperature according to the EoS used.) This applies to all states described by the line segment from point 5 to point 9 and they are referred to as *unstable*.

12.9.2 The Metastable Region

Let us now consider points 2 and 10 that satisfy the stability criterion. In addition, since areas *I* and *II* are not equal, they do not represent two phases at equilibrium. Which of the two states will the fluid find itself in at $T = 325$ K and $P = P_0$?

The only way to answer this question is to determine in which of the two states the fluid assumes the lowest Gibbs free energy. We proceed, therefore, to calculate the molar Gibbs free energy of propane as a function of pressure, using the vdW EoS.

At a constant temperature T from Eq.9.11.1:

$$d\mu = RTd\ln f = VdP$$

and:

$$\mu - \mu' = \int_{P'}^{P} VdP \tag{12.9.1}$$

We can thus calculate μ relative to its value at some reference state only choosing, arbitrarily, $P' = 1.013$ bar; $\mu' = 0.0$. (This does not present any problems, however, since we are interested in comparing μ values, not in absolute ones.) To facilitate the integration in Eq.12.9.1, we use:

$$VdP = d(PV) - PdV$$

Thus:

$$\mu - \mu' = PV - P'V' - RT\ln\frac{V-b}{V'-b} - a\left(\frac{1}{V} - \frac{1}{V'}\right)$$

The results are plotted in Figure 12.1, where the numbers correspond to those in Figure 12.E.9.

Consider, first, point 1(V = 2500 cm^3/mol; P = 9.7 bar; μ = 5872 J/mol). Figure 12.E.9 indicates that - at this pressure and temperature - propane can exist in one state only. So does Figure 12.1. (The point is outside the scale in both Figures.)

Increasing the pressure to point 2, we notice that at this pressure - according to Figure 12.E.9 - propane could also be in point 8, or 10. Point 8 has already been rejected, and we arrive at the previously stated question: will propane stay in point 2, or jump over to point 10? The answer is given by Figure 12.1: $\mu(2) < \mu(10)$ and, consequently, propane will stay in point 2.

The same question arises when the pressure reaches point 3, between it and point 11. According to Figure 12.1 both states have the same Gibbs free energy and, consequently, propane can be in either phase. The two phases, thus, exist in equilibrium with each other at this point; and the pressure of 24.7 bar, represents the vapor pressure of propane at 325 K as predicted by the van der Waals EoS. Notice also that the two areas in Figure 12.E.9, above and below this pressure, are equal as required by the equality of the chemical potentials.

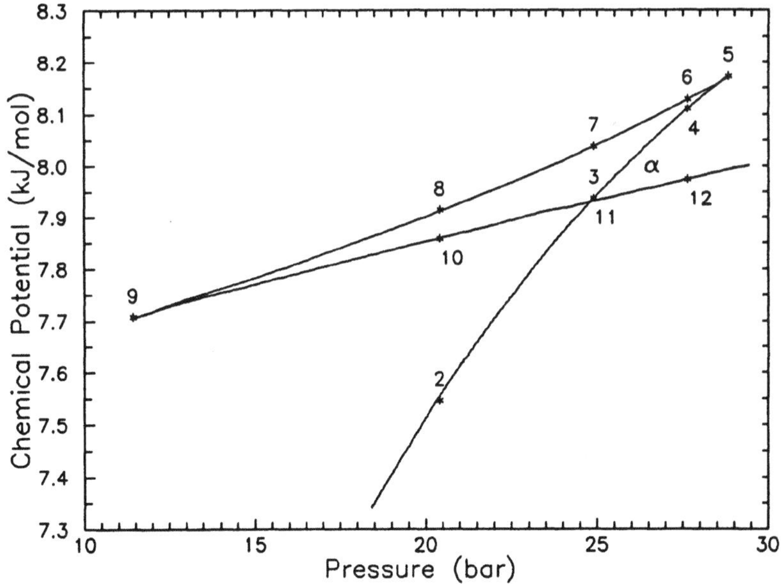

Figure 12.1 Variation of the chemical potential of propane with pressure at 325 K calculated with the vdW EoS. Reference: $\mu = 0$ at $P = 1.013$ bar.

We proceed next to point 4 and the associated ones, 6 and 12. Point 6 is rejected again, but now $\mu(12) < \mu(4)$, and point 12 represents the stable state for propane (liquid).

Let us now consider the question: Since point 4 does not violate the stability criterion, can propane exist in this state? The answer is yes. If the compression is done very slowly, without any vibrations, the system can find itself in point 4. The slightest tapping, however, will shift the system to point 12.

The states described by the segment line from 3 to 5 are referred to as *metastable*. And, using the same line of argument, we conclude that the states between 9 and 11 are also metastable. Points 5 and 9 establish the *limits of stability*.

12.9.3 Comments

1. It is apparent that the stable region of propane at a subcritical temperature, as described by the vdW EoS, is represented by:

* the curve 1-2-3 (vapor);
* the horizontal line 3-11 (vapor + liquid); and
* the curve 11-12 (liquid)

of Figure 12.E.9. For the actual behavior of a pure fluid see Figure 8.1.

2. In the two-phase region 3-11, the vapor and liquid molar volumes remain constant and their relative amounts, at some point between 3 and 11, are given by the well known inverse lever rule.

3. As the temperature increases, the angle (a) in Figure 12.1 becomes smaller as shown in Figure 12.2. And so does the two-phase region, as shown in Figure 8.1.

4. The point where the angle (a) becomes zero and the two-phase region vanishes, represents the *critical* point; and the temperature and pressure there are the critical ones (T_c and P_c).

5. Since the liquid and vapor molar volumes become identical at the critical point, the two phases become indistinguishable. This is observed experimentally, as the meniscus separating the two phases disappears.

6. The phase transitions discussed so far are characterized by the fact that, except at the critical point, certain properties become discontinuous. Thus, the volume, enthalpy, and entropy, undergo a drastic change between the liquid and vapor coexisting phases. And the heat capacity at constant pressure, $C_P = (\partial H/\partial T)_P$ becomes infinite at a phase transition: dH is finite but $dT = 0$.

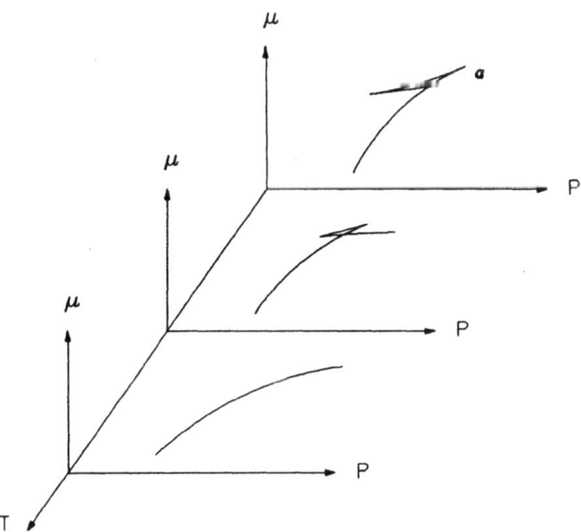

Figure 12.2 Variation of the chemical potential with pressure as a function of temperature.

7.On the other hand, as suggested by Eq.12.4.3, the chemical potential is continuous in the phase transition region. But, its first derivatives:

$$\left(\frac{\partial \mu}{\partial T}\right)_P = -S; \quad \left(\frac{\partial \mu}{\partial P}\right)_T = V$$

are discontinuous, which can be seen from Figure 12.1 for the second one. These types of phase transitions are referred to as *first-order*.

8.Higher-order phase transitions follow from the above observations about the chemical potential. Thus, for second-order phase transitions at constant T and P, both the chemical potential and its first derivatives are continuous but not the second ones, and so on. Only second-order phase transitions have been observed experimentally. For details, see Callen (1960).

9.Metastable states are often encountered in heating or cooling of liquids at constant pressure. Water, for example, can be heated above 100°C at P = 1 atm without boiling. The presence, however, of dust particles - or a slight tapping - can induce violent boiling. (To avoid such problems 'boiling chips are added in laboratory experiments). Liquids in such states are referred to as *superheated*.

10. Vapor in point 4, i.e. at a pressure higher than its vapor pressure at this temperature, is referred to as *subcooled*. So is called a liquid that is cooled below its freezing point at constant pressure.

11. How does a fluid in a metastable state move into a stable one in the absence of any external disturbance? We will see in the discussion of statistical mechanics that the microstate of a system is not static; the system undergoes rapid fluctuations continuously, through which it explores and discovers states of the minimum value for the appropriate potential, the Gibbs free energy in this case.

12. At low temperatures, the van der Waals equation of state - but others as well - predict that the limit of stability of the liquid phase occurs at negative pressures, as shown in Figure 12.E.9 for propane at 300 K. The liquid, in other words, can support a tensible force.

Such metastable behavior has been observed with water in capillary tubes and is thought to be important in the vascular system of plants (Sandler). For a review of water under tension, especially in biological systems, see Scholander (1972).

12.10 Liquid-Liquid Equilibrium in Binary Systems

12.10.1 The Mixing of Two Pure Liquids

We examine next the stability of a binary system that reaches equilibrium at constant temperature and pressure. To this purpose consider two liquid compounds 1 and 2 that are mixed at a specified temperature T and some low pressure P, and examine the variation of its Gibbs free energy with composition.

Assuming, for simplicity, that the molar excess Gibbs free energy is given by the two-suffix Margules equation:

$$G^E = Ax_1x_2 \tag{12.10.1}$$

where A is a function of temperature only, then the molar Gibbs free energy of the mixture is (Eqs 11.13.1, 11.12.1a and 11.12.7a):

$$G = x_1 G_1 + x_2 G_2 + RT(x_1 \ln x_1 + x_2 \ln x_2) + Ax_1x_2 \tag{12.10.2}$$

where G_1 and G_2 are the molar Gibbs free energies of pure 1 and 2 respectively at the same T and P as the system.

Figure 12.3 presents a plot of G versus x_1 at 300 K using, arbitrarily, $G_1 = 0$ and $G_2 = 1$ (both in kJ/mol) for three values of A: 0, 4 and 7 kJ/mol.

Consider now a mixture of a specified composition x_0. We notice that while for $A = 0$ (ideal solution) and $A = 4$ the corresponding values of G are the minimum ones for this composition as required by Eq.12.3.8, this is not the case for $A = 7$. Here the system can assume a lower value (G_α) by splitting into two phases - in equilibrium with each other - with compositions given by the points where the common tangent line touches the G curve: x_1' and x_1''; and this splitting will occur for all compositions in the range $x_1' < x_1 < x_1''$ but not outside it.

In terms of molecular behavior, the very nonideal behavior of the mixture - manifested by the large value of A - indicates that the molecules of the two species do not feel very 'comfortable' with each other, thus do not mix in all proportions.

12.10.2 Evaluation of the Miscibility Limits

To determine the composition of the two phases in equilibrium, i.e. the miscibility limits, we do not have to resort to the graphical method sug-

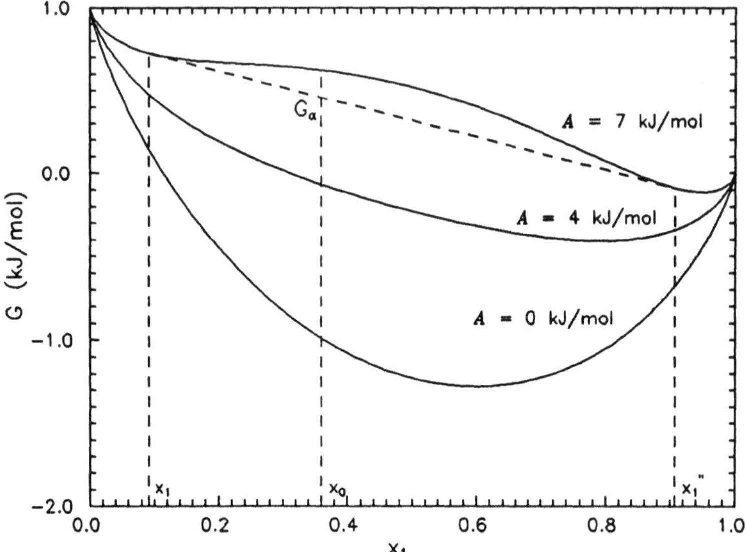

Figure 12.3 The Gibbs free energy of a binary liquid mixture for various values of A (Eq. 12.10.2).

gested by Figure 12.3. Rather we take advantage of the fact that, since the two phases are in equilibrium:

$$\hat{f}_i{}' = \hat{f}_i{}'', \quad i = 1,\, 2 \qquad (12.10.3)$$

where the single and double prime refer to the two phases. Or, since the standard state fugacity of each component is the same in both phases:

$$\gamma_i{}'x_i{}' = \gamma_i{}''x_i{}'', \quad i = 1,2 \qquad (12.10.4)$$

From Eqs 12.10.1 and 11.13.6:

$$\ln\gamma_i = \frac{A}{RT}(1-x_i)^2, \quad i = 1,\, 2 \qquad (12.10.5)$$

which leads to a system of two equations and two unknowns:

$$x_1{}'\exp\left(\frac{A}{RT}(1-x_1{}')^2\right) = x_1{}''\exp\left(\frac{A}{RT}(1-x_1{}'')^2\right) \qquad (12.10.6a)$$

$$(1-x_1{}')\exp\left(\frac{A}{RT}(x_1{}')^2\right) = (1-x_1{}'')\exp\left(\frac{A}{RT}(x_1{}'')^2\right) \qquad (12.10.6b)$$

Solution of this system by trial-and-error gives:

$$x_1{}' = 0.092 \quad ; \quad x_1{}'' = 0.908$$

which is also suggested by a visual observation of Figure 12.3.

12.10.3 The Metastable and Unstable Regions

We have considered so far the implications of the first requirement for a closed system that reaches equilibrium at a specified T and P: its Gibbs free energy must assume its minimum value (Eqs 12.3.8 and 12.3.9). We examine next the implications of the second requirement (Ineq. 12.3.10):

$$d^2G > 0$$

which dictates that the equilibrium is a stable one. More specifically we will examine the second composition derivative of the molar Gibbs free energy: $(\partial^2G/\partial x_1{}^2)_{T,P}$ at various values of x_1. From Eq. 12.10.2:

$$\left(\frac{\partial^2 G}{\partial x_1^2}\right)_{T,P} = \frac{RT}{x_1 x_2} - 2A \qquad (12.10.7)$$

At the specified temperature of 300 K, $(\partial^2 G/\partial x_1^2)_{T,P}$ is larger than zero for all values of x_1, when $A = 0$ or 4; but not when $A = 7$. In this case, the G versus x_1 curve in Figure 12.3 goes through two inflection points where:

$$\left(\frac{\partial^2 G}{\partial x_1^2}\right)_{T,P} = \frac{RT}{x_1 x_2} - 2A = 0 \tag{12.10.8}$$

The requirement $(\partial^2 G/\partial x_1^2)_{T,P} > 0$ is thus satisfied for $x_1 < 0.232$ and $x_1 > 0.768$, found by solving Eq.12.10.8, but not for $0.232 \le x_1 \le 0.768$.

If we combine our findings with those of the previous Section we conclude that for $A = 7$:

complete miscible region: $x_1 \le 0.092$
$x_1 \ge 0.908$

metastable region: $0.092 < x_1 < 0.232$
$0.768 < x_1 < 0.908$

unstable region: $0.232 \le x_1 \le 0.768$

The values of $x_1 = 0.092$ and 0.908 establish the *miscibility limits*; and the values of $x_1 = 0.232$ and 0.768, the *stability limits*. (The latter points are also referred to as *incipient instability* points).

12.10.4 The Effect of Temperature on Mutual Solubility

We have considered so far the effect of nonideality on the mixing of the two liquids, at a specified temperature of 300 K, by varying the value of A in Eq.12.10.1. We examine next the effect that temperature has on the mixing of the two liquids. To this purpose we keep, for simplicity, A independent of temperature and equal to 4 kJ/mol, a value that leads to complete miscibility at 300 K.

From Eq.12.10.7 it follows that $(\partial^2 G/\partial x_1^2)_{T,P} > 0$, and the single phase is the equilibrium phase if:

$$T > \frac{2Ax_1x_2}{R} \tag{12.10.9}$$

while $(\partial^2 G/\partial x_1^2)_{T,P} < 0$ and phase separation occurs if:

$$T < \frac{2Ax_1x_2}{R} \tag{12.10.10}$$

The limit of stability occurs at:

$$T = \frac{2Ax_1x_2}{R} \tag{12.10.11}$$

At this temperature the two liquid phases merge into one, i.e. their properties become identical, and it is referred to as the *critical solution (or consolute) temperature.*

The maximum temperature at which phase separation is possible independent of composition, occurs at $x_1 = x_2 = 0.5$:

$$T = \frac{A}{2R} \tag{12.10.12}$$

which is equal to 240.5 K in this case of $A = 4$ kJ/mol. It is referred to as the *upper* critical solution temperature. (Notice that in this case where G^E is given by Eq.12.10.1, no lower critical solution temperature is possible.)

We conclude that at temperatures higher than 240.5 K a single liquid is the equilibrium state; at lower ones, two liquid phases at equilibrium are present. This is demonstrated in the following Example.

12.10.5 Example 12.10

For the mixture of compounds 1 and 2, A in Eq.12.10.1 is equal to 4 kJ/mol. Assuming that A is temperature independent, determine:

a. the liquid-liquid coexistence curve; and

b. the *spinodal* curve (the locus of the limits of stability);

as a function of temperature.

a. The upper critical solution temperature for this mixture is given by Eq. 12.10.12:

$$T = A/2R = 240.5\text{ K}$$

To establish the liquid-liquid coexistence curve at $T < 240.5$ K we resort to Eqs 12.10.6a and b. Furthermore, because of the symmetry involved here due to the use of Eq.12.10.1:

$$x_1{}' = 1 - x_1{}'' \tag{A}$$

Eq.12.10.6a becomes thus:

$$x_1{}'\exp\left(\frac{A}{RT}(1-x_1{}')^2\right) = (1-x_1{}')\exp\left(\frac{A}{RT}(x_1{}')^2\right) \tag{B}$$

Solution of Eq.(B) at temperatures less than 240.5 K gives the value of $x_1{}'$; $x_1{}''$ is then obtained from Eq.(A). Thus at $T = 200$ K: $x_1{}' = 0.169$; $x_1{}'' = 0.831$.

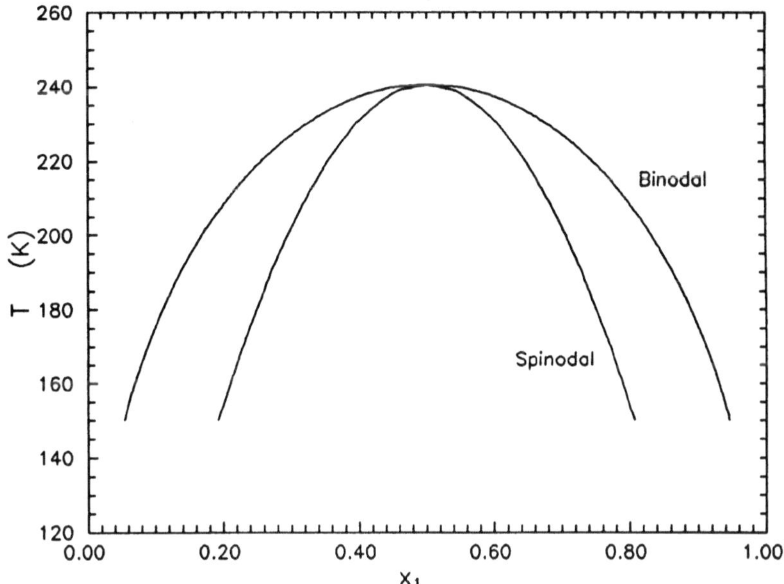

Figure 12.E.10 The binodal and spinodal curves for the system of Example 12.10.

The liquid-liquid coexistence curve - known also as the *binodal* curve - is obtained by repeating this procedure at different temperatures and it is shown in Figure 12.E.10.

b. To establish the spinodal curve we must determine at each temperature (less than 240.5 K) the limits of stability, i.e. the points where Eq. 12.10.8 is satisfied. Thus at $T = 200$ K, $x_1 = 0.295$ and 0.705. The results at different temperatures establish the spinodal curve shown also in Figure 12.E.10.

12.10.6 Comments

1. Equations 12.10.5 through 12.10.12 are based on the use of the two-suffix (one-parameter) Margules expression for the excess Gibbs free energy (Eq. 12.10.1). It is a very simple model which, even with a temperature dependent A, is applicable to a very limited number of systems where G^E is a symmetric function of x_1, mainly hydrocarbons of similar size.

2.For polar systems, and especially where hydrogen bonding is present, the dependency of G^E on x_1 is typically asymmetric and often a strong function of temperature. More realistic models for G^E, such as those discussed in the next Chapter, are needed to this purpose.

3.In such cases, in addition to the upper critical solution temperature of Section 12.10.4, a lower or both a lower and an upper one, are possible. [See Problem 12.58 and, especially, Wisniak (1983)].

4.The range between the binodal and spinodal curves represents states of supersaturation with respect to the dilute component.

5.We have considered in this Section the mixing of two liquids at a low pressure, where its effect is negligible. At higher pressures, however, the effect can be very pronounced. Actually at high pressures, and temperatures above the critical ones of the two components, 'gas-gas' equilibrium is possible. For a discussion on the effect of pressure on mutual miscibility, see Prausnitz et al (1986).

12.11 Concluding Remarks

1.An application of the first and second laws to a closed system, that undergoes change with time under certain constraints, leads to the conclusion that some potential assumes its maximum or minimum value when the system reaches equilibrium; this represents the thermodynamic requirement of equilibrium under these constraints.

2.If the constraints are constant volume V and internal energy U, the potential is the system entropy S, which assumes its maximum value at equilibrium; if they are constant temperature T and pressure P, the potential is the Gibbs free energy G of the system which assumes its minimum value at equilibrium.

3.To satisfy these requirements, the following criteria must be met:

$$dS = 0;\ d^2S < 0\ (V \text{ and } U \text{ constant})$$
$$dG = 0;\ d^2G > 0\ (T \text{ and } P \text{ constant})$$

4.These criteria lead in turn to some useful, from the practical point of view, thermodynamic relationships that must be satisfied when a system reaches a state of stable equilibrium under the given constraints.

5.Use thus of the first criterion for equilibrium reached at constant T and P, leads to a framework of equations for obtaining numerical answers in the case of phase and/or chemical equilibrium under these constraints.

6. Use of the equations that must be satisfied by a system in equilibrium combined with the number of system variables, leads to the phase rule and Duhem's theorem that provide for the complete specification of the state of the system.

7. Use of the second criterion for equilibrium reached at constant V and U; or at constant T and P; leads to the requirements so that this equilibrium is a stable one (stability requirements).

References

Callen, H.B., 1960. *Thermodynamics*, Wiley, New York.

Haynes, W.M., Goodwin, R.D., 1982. *Thermophysical Properties of Normal Butane from 85 to 700 K at Pressures to 70 MPa*, NBS, Monograph 169, Boulder, Co.

Modell, M., Reid, R.C., 1974. *Thermodynamics and its Applications*, Prentice-Hall, Englewood Cliffs, N.J.

Prausnitz, J.M., Lichtenthaler, R.N., de Azevedo, E.G., 1986. *Molecular Thermodynamics of Fluid-Phase Equilibria*, Prentice-Hall, Englewood Cliffs, N.J.

Prigogine, I., Stengers, I., 1984. *Order out of Chaos*, Bantam Books, N.Y.

Sandler, S.I., 1977. *Chemical and Engineering Thermodynamics*, Wiley, N.Y.

Scholander, P.F., 1972. *American Scientist*, **60**, 584.

Wisniak, J., 1983. *Chem. Eng. Sci.*, **38**, 969.

Problems

12.1 Trace Eq.(A), Example 12.1, back to the first and second laws.

12.2 Show that if a closed system is allowed to reach equilibrium at constant temperature and volume, its Helmholtz free energy is minimized.

12.3 For the isomerization reaction:

$$CH_3-CH_2-CH_2-CH_3 \rightleftharpoons \begin{array}{c} CH_3-CH-CH_3 \\ | \\ CH_3 \end{array}$$

it is given that at 298 K and 1 atm:

$$G(\mathrm{nC_4}) = -4100 \text{ cal/mol}$$
$$G(\mathrm{iC_4}) = -4990 \text{ cal/mol}$$

Starting with one mole of nC_4, calculate the number of its moles at equilibrium at 298 K and 1 atm by minimizing the Gibbs free energy.

12.4.a. Show that for the reaction $A \rightarrow B$, at equilibrium at a given T and P:

$$\bar{G}_A = \bar{G}_B$$

b. Use this to do Problem 12.3.

12.5 Do Problem 12.3 using the equilibrium constant K.

12.6 Consider the reaction: $A + B \rightarrow C$, taking place in the gaseous phase at $T = T_0$ and $P = 1$ atm. Starting with 2 moles of A and 1.5 of B, 0.5 moles of B are found at equilibrium. What is the equilibrium composition when starting with 2 moles each of A and B at the same T and P ?

12.7 Two moles of benzene and one mole of toluene are placed in a container and after equilibrium is reached at 90°C and 1 atm, two phases are present.

a. Can, according to the phase rule, the amounts of liquid(L) and vapor(V) phases be determined? How about the compositions of the two phases?

b. Same as (a), but according to the Duhem theorem.

c. If your answer to part (a) or (b) is yes, calculate L and V.

Make any assumptions you need, but justify them.

12.8 Consider a vapor-liquid equilibrium mixture containing three components. According to the phase rule: $F = 3 + 2 - 2 = 3$, but according to Duhem's theorem, specification of two of these variables suffices to completely describe this system. Is there a paradox here?

12.9 A mixture of water and ethanol reaches equilibrium with its vapor at a given temperature T and pressure P. According to the phase rule the intensive state of the system is specified. Explain how this applies to the partial molar volume of water in this case.

12.10 With reference to the 'fundamental postulate':

a. where did you use it?

b. do you see any relationship with the Duhem theorem?

12.11 Discuss the three more important, in your opinion, items that you have learned in this Chapter and give an example where each one could be used in practice.

12.51 Show that for a five-phase closed system of nonreacting components in equilibrium at constant T and P:

$$\hat{f}_i(I) = \hat{f}_i(II) = \hat{f}_i(III) = \hat{f}_i(IV) = \hat{f}_i(V)$$

where i is any component of the mixture and I through V are the phase numbers.

12.52 Consider the gas-phase reaction $2A \rightarrow B$ in equilibrium at constant pressure and temperature.

a. What would happen to the amount of B if the pressure were increased at constant T?

b. The same, using the Le Chatelier principle.

12.53 Determine the vapor pressure of propane at 300 K using the SRK EoS and the approach shown in Figure 12.1 and compare it to the experimental value.

12.54 Plot the enthalpy, entropy, and chemical potential of water as a function of temperature at some pressure P using the Steam Tables, and identify the discontinuities occuring at the phase transition. Neglect the effect of pressure on the enthalpy and entropy of the liquid.

12.55 Show that for any fluid: $C_P \geq \cdot C_V$. When does the equality sign hold?

12.56.a. Determine the stability limits of n-butane at 300 K with the vdW EoS.

b. Develop a P-V diagram for n-butane including the phase envelope and spinodal curve, one subcritical and one supercritical temperatures, using the vdW and t-vdW EoS. Compare your results with the experimental data presented below (Haynes and Goodwin, 1982).

T, K	P, MPa	$\rho(v)^*$	$\rho(l)$
260	0.061	1.69	614.46
300	0.258	6.52	570.49
340	0.754	18.56	519.47
380	0.938	20.00	
	1.302	30.40	
	1.598	40.00	
	1.739	45.78	453.83
	4.253		470.00
	11.721		500.00
420	3.483	133.47	328.06
430	1.103	20.00	
	2.649	60.00	
	3.493	100.00	
	3.729	120.00	
	3.880	140.00	
	3.974	160.00	
	4.030	180.00	
	4.067	200.00	
	4.321	300.00	
	7.098	400.00	

* ρ: density, kg/m^3

12.57 If in Section 12.10.1 a plot of ΔG versus x_1 were used, would the same miscibility limits with those of Figure 12.3 be obtained?

12.58 Show that if A in Eq. 12.10.1 is a function of temperature, the prediction of an upper, lower, or both critical solution temperatures is possible.

12.59 At 302 K and 1 atm a liquid mixture of n-hexane (1) and methanol (2) splits into two equilibrium phases with $x_1' = 0.25$ and $x_1'' = 0.68$.

a. Give your best estimate of the upper critical solution temperature.

b. Compare it to the experimental value of 306.5 K and discuss the reasons for the discrepancy between the two values.

Hint: Due to the asymmetry involved, use: $G^E = x_1 x_2 [A + B(x_1 - x_2)]$, known as the two-parameter Redlich-Kister equation.

12.60 If the critical solution temperature of a binary system is 350 K and the liquid phase nonideality is described by Eq. 12.10.1, develop its *P-x-y* diagram at 300 K, using: $P_1^s = 3$ bar and $P_2^s = 1$ bar.

12.61 In Section 12.7.4 the statement is made that the number of equations increases by r. Why not by ϕr, since the r reactions are at equilibrium in each phase?

13 Low Pressure Vapor-Liquid Equilibrium

13.1 Introduction

Separations using, Distillation, Extraction, Gas Absorption, etc., represent the main unit operations in the Chemical and Petroleum Industries, often reaching 50% of the total plant cost.

Essential for the successful design and operation of such units is the relationship among the compositions of the phases present:
* vapor-liquid
* liquid-liquid
* liquid-liquid-vapor
* gas-liquid
* gas-solid,

at the operating temperature and pressure.

Consider, for example, the separation of a mixture of methanol and water to recover methanol of 99% purity using the packed tower discussed in Example 1.7. The relationship between the composition of the vapor bubbles and that of the liquid surrounding them at the operating conditions of pressure and temperature - from the bottom of the tower to the top - will determine the height and reflux ratio necessary to achieve the desired separation.

This relationship represents one case of the general subject of *Phase Equilibrium* which becomes, thus, one of the most important areas of Chemical Engineering Thermodynamics applications.

We will concentrate on *Vapor-Liquid Equilibrium* (*VLE*) because of its wide applicability. In addition, the methodology involved and discussed here is typical to all types of phase equilibria. Finally, for convenience reasons that will become apparent in Section 13.4, we consider in this Chapter *Low Pressure VLE*; and *High Pressure VLE*, in the next one.

13.2 Objective and Approach

Our main objective in this Chapter, and the next, is to develop a qualitative and quantitative understanding of the factors that effect the relationship between: the compositions of the liquid and vapor phases, that are in equilibrium with each other, and the corresponding temperature and pressure; and, since we will soon see that the basis of this relationship - at the low pressures considered in this Chapter - is the *activity coefficient*, we will concentrate on the latter's:

1. Evaluation from experimental measurements;
2. Dependency on:
   a. composition,
   b. molecular nature of the mixture components, and
   c. temperature and pressure;
3. Prediction from limited data;
4. Estimation, when experimental data are not available; and
5. Use in carrying out bubble and dew point calculations that provide the numerical answers for the relationship between the two phases.

We will begin by defining the *VLE* problem, and proceed to discuss - using the equality of component fugacities in the two equilibrium phases, developed in Section 12.4, as a starting point - the methodology used in obtaining quantitative results from these abstract quantities.

This discussion will suggest two broad approaches: one for low pressures, considered in this Chapter, and an other for high pressures, considered in the next one.

We start the presentation of low pressure *VLE* by demonstrating that the main property involved here is the activity coefficient (gamma), the *measure* of the liquid phase nonideality.

Next, we consider the simple case of an ideal solution, i.e. one where all activity coefficients are equal to unity (Raoult's Law), and use it to develop familiarity with some typical *VLE* calculations.

We return then to the activity coefficient and examine its dependency on:

* composition,
* the molecular characteristics of the mixture components,
* temperature, and
* pressure,

followed by the methods used in evaluating the thermodynamic consistency of experimental data.

Analytical expressions for the activity coefficient as a function of composition and temperature will then be developed, so that extrapolation from a limited number of experimental measurements can be achieved. We will see, for example, that a single measurement in a binary *VLE* system is sufficient to provide a reasonably accurate description of the whole composition range; or, that binary *VLE* data alone can be successfully used for the prediction of multicomponent *VLE* behavior.

We will also use these expressions to demonstrate how measurements involving only pressure, temperature, and liquid phase composition - a common experimental technique - can be used to evaluate vapor phase compositions.

We return then to the general vapor-liquid equilibrium problem and discuss how the activity coefficient is used in solving it. We will see that this is accomplished through the so-called *bubble* and *dew* point calculations.

Our presentation, to that point, will pertain to systems where sufficient experimental data is available. We proceed, therefore, to discuss how *VLE* behavior can be estimated when no experimental data are available, a frequent case in industrial applications.

We will close the Chapter with some concluding remarks.

13.3 The Vapor-Liquid Equilibrium Problem

Consider a system of k components and two phases, one liquid and one vapor, being in equilibrium with each other at some temperature T and pressure P (Figure 13.1), and the following three questions:

1. What is the total number of variables?
2. How many degrees of freedom (F) are there, i.e. how many of these variables must be specified so that the intensive state of this system is determined?
3. How are the remaining ones to be determined?

To this purpose, we notice that:

1. The number of variables is:
   * liquid phase compositions: $x_1, x_2, \ldots, x_{k-1} = k - 1$
   * vapor phase compositions: $y_1, y_2, \ldots, y_{k-1} = k - 1$
   * pressure (P) and temperature (T) $= 2$
   * total $= 2k$

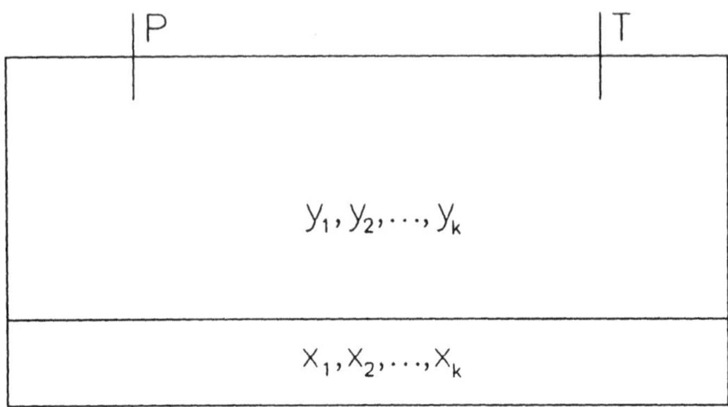

Figure 13.1 A vapor-liquid equilibrium system.

2.According to the Phase Rule (Section 12.7.1): $F = k + 2 - 2 = k$ i.e. specification of k of these $2k$ variables suffices for the complete description of the intensive state of the system.

The remaining k variables would thus have to be determined experimentally. This approach is time-consuming and costly, however, and *our objective is the evaluation of these k variables using the minimum amount of experimental information possible*. This represents the general statement of the vapor-liquid equilibrium problem. The typical cases encountered in practice, are presented in Table 13.1.

13.4 Methodology

We have seen in Chapter 12 that when a system containing two phases reaches equilibrium - at constant temperature and pressure - the total Gibbs free energy assumes its minimum value; and that this, in turn, leads to the equality of fugacities for any component i in the two phases (Section 12.4.4). In the case of vapor-liquid equilibrium, thus:

$$\hat{f}_i^v = \hat{f}_i^l \qquad (13.4.1)$$

where the superscripts (v) and (l) refer to vapor and liquid phase respectively.

We examine next the methodology used to obtain numerical answers from the equality of these two abstract quantities. To this purpose, we

Table 13.1 Typical *VLE* Problems [(1)]

Specified Variables	Variables to be determined	Method
x's and T	y's and P	Bubble Point P
x's and P	y's and T	Bubble Point T
y's and T	x's and P	Dew Point P
y's and P	x's and T	Dew Point T

[(1)] An other type of *VLE* calculation, the *flash* calculation, will be discussed in Chapter 14.

consider - for simplicity - a binary system where, according to Section 13.3, two variables must be specified in order that the intensive state of the system be completely determined. Let us, then, assume that T and x_1 are specified, and examine how P and y_1 are evaluated.

The two equations to be used for this purpose are:

$$\hat{f}_1^v = \hat{f}_1^l$$
$$\hat{f}_2^v = \hat{f}_2^l$$

Following Section 11.8:

$$\ln\hat{\phi}_i^v = \ln\frac{\hat{f}_i^v}{y_i P} = \frac{1}{RT}\int_0^P\left(\bar{V}_i^v - \frac{RT}{P}\right)dP$$

$$\ln\hat{\phi}_i^l = \ln\frac{\hat{f}_i^l}{x_i P} = \frac{1}{RT}\int_0^P\left(\bar{V}_i^l - \frac{RT}{P}\right)dP$$

If an equation of state that can describe the volumetric behavior of both phases is available, then $\hat{f}_1^v$ and $\hat{f}_2^v$ can be expressed in terms of P, T and y_1; and $\hat{f}_1^l$ and $\hat{f}_2^l$ in terms of P, T and x_1. We have, thus, a system of two equations that can be solved for the two unknowns: y_1 and P. The same methodology would be used for a multicomponent system.

In practice vapor-liquid equilibrium calculations using an equation of state, that is applicable to both phases, can be carried out with sufficient accuracy only for systems that are up to moderately nonideal, typically hydrocarbon mixtures alone or with such gases as: H_2S, CO, CO_2, etc. Furthermore, since separation of such systems by distillation - because of their typically low boiling point temperatures at atmospheric pressure - is carried out at high pressures, this methodology represents what is traditionally referred to as the *High Pressure* or *Equation of State* approach to vapor-liquid equilibrium calculations.

For nonideal systems, however, this approach does not provide yet sufficient accuracy, especially for multicomponent cases, because of limitations in the available equations of state. Such systems, in addition, have typically higher boiling points at atmospheric pressure and their separation by distillation is carried out at low pressures. These systems are described using the following methodology:

1.The vapor phase fugacity is expressed through:

$$\hat{f}_i^V = y_i \hat{\phi}_i P \tag{11.8.2}$$

and the vapor phase fugacity coefficient $\hat{\phi}_i$(phi) is evaluated with sufficient accuracy, as we have seen in Section 11.9, through an equation of state.

2.The liquid phase fugacity is expressed through the *standard* state fugacity (f_i^o) approach (Section 11.10):

$$\hat{f}_i^l = x_i \gamma_i f_i^\circ = x_i \gamma_i P_i^s \phi_i^s \exp\left(\frac{V_i(P-P_i^s)}{RT}\right) \tag{11.10.3}$$

where:

* γ_i is the activity coefficient (gamma);
* P_i^s is the vapor pressure of pure i at the system temperature T;
* ϕ_i^s is the fugacity coefficient of saturated pure liquid i at T, that can be again evaluated from an equation of state;
* V_i is the average molar volume of pure liquid i at T from P_i^s to the system pressure P; and
* the exponential term expresses the Poynting effect $(Pe)_i$.

This methodology represents what is traditionally referred to as the *Low Pressure* or *Gamma-Phi* approach to vapor-liquid equilibrium calculations, and it is discussed in this Chapter; and the high pressure one, in the next.

13.5 Activity Coefficients: Determination from Vapor-Liquid Equilibrium Measurements

To this purpose, we present first a typical method for such measurements, and then discuss how the activity coefficients are evaluated from them.

13.5.1 Experimental Measurements

Let us consider the following experiment: 25 ml of a mixture of ethanol and water is charged into a small (100 ml) flask, which is connected to a mercury manometer and a vacuum source, and is placed into a constant temperature bath.

In the typical case, liquids contain dissolved gases - mostly air - which would effect the equilibrium conditions and must be removed. To this purpose the bath temperature is lowered until the mixture solidifies and the vacuum is activated. The dissolved gases are removed, while only a small amount of the solvent mixture is lost due to its very low vapor pressure at this temperature.

The flask is then disconnected from the vacuum source and the bath temperature is raised to, say, 60°C. We notice that the pressure in the flask is increasing with time, even after the bath has reached the temperature of 60°C. As a matter of fact, the increase continues for several hours, probably over 20, until no discernible variation in pressure can be observed.

At this point we say that the liquid and vapor phases of the ethanol-water system are at equilibrium. (In practice, a magnetic stirrer is placed in the flask which helps the attainment of equilibrium, typically in one to two hours. It should be realized, of course, that 'true' equilibrium is reached only after an infinite amount of time.)

Let us now assume that, for all practical purposes, equilibrium has been reached and proceed to determine the compositions of the two phases by some analytical technique, such as chromatography.

We know, thus, the following quantities for the equilibrium system:

* pressure: P;
* temperature: T;
* composition of the liquid phase: x_1 and x_2;
* composition of the vapor phase: y_1 and y_2.

13.5.2 Evaluation of the Activity Coefficient

We proceed now to demonstrate how the activity coefficient for the component i of the mixture is evaluated from these data.

To develop the appropriate thermodynamic expression for γ_i we combine Eqs 11.10.3 and 11.8.2 (Section 13.4):

$$\gamma_i = \frac{y_i \hat{\phi}_i P}{x_i P_i^s \phi_i^s (Pe)_i} \tag{13.5.1}$$

The vapor pressure of pure liquid i, P_i^s, can be calculated from some appropriate relationship, such as the Antoine equation. (In practice, however, it is recommended that for improved accuracy of the calculated activity coefficients it is better to measure P_i^s directly, by running an experimental *VLE* measurement with pure i, see Example 13.1.)

Determination, thus, of the activity coefficient value requires the evaluation of the two fugacity coefficients and of the Poynting effect. To this purpose, we recall from Chapters 9 and 11 that at the low pressures of interest here:

1. Fugacity coefficients of pure saturated liquids, and of the components of a vapor phase, can be successfully evaluated through the virial equation truncated after B:

$$\ln\phi_i^s = \frac{B_{ii} P_i^s}{RT} \tag{9.11.10}$$

$$\ln\hat{\phi}_i = (2\Sigma y_j B_{ij} - B)\frac{P}{RT} \tag{11.9.5}$$

where B_{ii}, B_{ij}, and B are the second virial coefficients of the pairs i-i and i-j and the mixture respectively.

2. In the absence of experimental values for the second virial coefficients, predicted ones through the Tsonopoulos (1974; 1975) or the Hayden-O'Connell (1975) correlations, provide reliable estimates of the fugacity coefficients.

3. The Poynting effect can be determined using Eq.9.11.14.

Determination, thus, of the fugacity coefficients through the virial equation and of the Poynting effect through Eq.9.11.14 allows the evaluation of activity coefficients from the experimental data. An appropriate computer program (GAMMA) is presented in Appendix E.

13.5.3 Example 13.1

The vapor-liquid equilibrium data of Table 13.E.1 are reported (Gmehling et al, 1979) for the system: acetone(1)-cyclohexane(2) at 25°C. Calculate the activity coefficients assuming, because of the low pressures involved and for simplicity in the calculations, ideal vapor behavior.

Table 13.E.1 Vapor-Liquid Equilibrium Data and Results for the System Acetone(1)-Cyclohexane(2) at 25°C

P (mm Hg)	x_1	y_1	γ_1	γ_2	$\ln(\gamma_1/\gamma_2)$
97.45	0.0	0.0			
118.05	0.0115	0.1810	8.064	1.004	2.084
124.95	0.0160	0.2250	7.626	1.010	2.022
137.90	0.0250	0.3040	7.278	1.010	1.975
145.00	0.0300	0.3450	7.237	1.005	1.975
172.90	0.0575	0.4580	5.977	1.020	1.768
207.70	0.1125	0.5670	4.543	1.040	1.475
227.70	0.1775	0.6110	3.402	1.105	1.124
237.85	0.2330	0.6325	2.802	1.169	0.874
253.90	0.4235	0.6800	1.769	1.446	0.202
259.40	0.5760	0.7050	1.378	1.852	-0.296
261.10	0.6605	0.7170	1.230	2.233	-0.596
262.00	0.7390	0.7390	1.137	2.689	-0.860
258.70	0.8605	0.8030	1.048	3.749	-1.275
252.00	0.9250	0.8580	1.015	4.896	-1.574
243.80	0.9625	0.9160	1.007	5.604	-1.716
230.40	1.0000	1.0000			

From Eq.13.5.1:
$$\gamma_1 = \frac{y_1 P}{x_1 P_1^s} \; ; \; \gamma_2 = \frac{y_2 P}{x_2 P_2^s}$$

To calculate P_1^s and P_2^s we can use the Antoine constants given by Gmehling et al, obtaining: P_1^s = 229.9 mm Hg and P_2^s = 97.6 mm Hg, which are very close to the experimental values of 230.40 and 97.45 mm Hg respectively given in the Table.

The latter values, however, should be used - as already discussed - which give the activity coefficient values shown also in Table 13.E.1. Notice the pronounced effect of composition: as x_i decreases, the deviation of γ_i from one increases. We will discuss this further in Section 13.7.

13.5.4 Comments

1.It is apparent that vapor-liquid equilibrium measurements do not provide all the information needed for the evaluation of the activity coefficients: the vapor phase nonideality must be estimated from an equation of state. Because, however, equations of state - especially the virial - provide reliable estimates of vapor phase fugacities at low pressures, we refer to the activity coefficients calculated through Eq.13.5.1 as

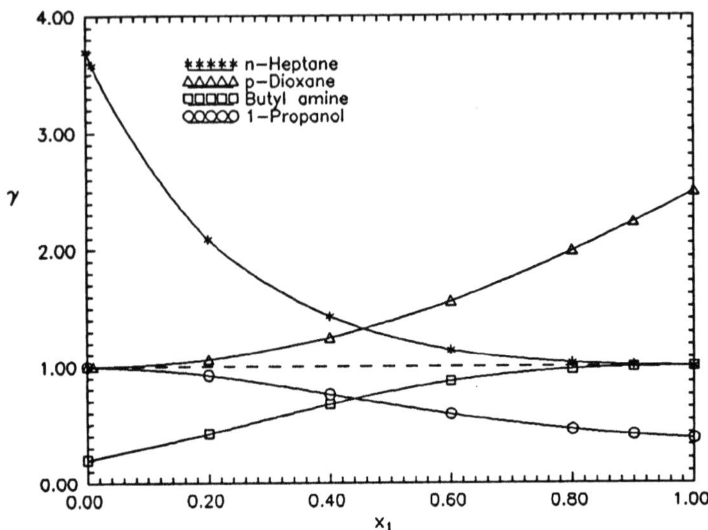

Figure 13.2 Experimentally obtained activity coefficients (lines generated from the Margules equation using infinite dilution γ's).
* Positive deviations from Raoult's law: n-heptane(1)/p-dioxane(2) at 80°C.
* Negative deviations from Raoult's law: butyl amine(1)/1-propanol(2) at 45°C.

experimental ones.

2.Typical results for two binary systems are shown in Figure 13.2. Infinite dilution activity coefficients γ_i^∞, obtained through extrapolation to $x_i = 0$ of the experimentally determined gammas, are also shown.

3.Eq.13.5.1 can be written in the following 'compact' form:

$$\gamma_i = \frac{y_i P}{x_i P_i^s} F_i \tag{13.5.2}$$

where:

$$F_i = \frac{\hat{\phi}_i}{\phi_i^s (Pe)_i} \tag{13.5.3}$$

F_i represents thus the contribution to γ_i of the nonideality in the vapor phase, $\hat{\phi}_i$ and ϕ_i^s, and of the effect of pressure on the liquid phase fugacity $(Pe)_i$. The latter is practically equal to 1.0 at the low pressures of interest here.

4.At low pressures, say up to 2-3 bar, F_i is typically in the range of 0.9 to 1.1. (This is not the case for systems containing organic acids where, as discussed in Section 11.8.4, large deviations from ideality are encountered, even at subatmospheric pressures.)

5.Activity coefficient values, on the other hand, deviate much more from one, as demonstrated in Figure 13.2. Thus, *at low pressures - typically below 10 atm - the most important, by far, quantity involved in vapor-liquid equilibrium calculations is the activity coefficient.*

6.If the pressure is low, typically 1 atm or below, and the components of the equilibrium mixture similar in chemical nature, we can safely assume ideal vapor behavior. Under these conditions the resulting error is very small, since F_i is very close to 1, and Eq.13.5.2 becomes:

$$\gamma_i = \frac{y_i P}{x_i P_i^s} \tag{13.5.4}$$

7.The experimental system mentioned in Section 13.5.1 is referred to as a 'static' one. In practice, determination of the vapor phase composition in such a system is difficult and can lead to substantial errors.

8.To avoid this problem - and also the determination of the liquid phase composition as well - specified amounts of predegassed liquids are charged into a small flask, called the 'equilibrium cell'. Since the vapor space above the liquid phase is small, the change in the liquid composition as the system reaches equilibrium is negligible and, thus, all x_i's are known. Vapor phase compositions are then estimated with the approach that will be discussed in Section 13.15.

9.Complete degassing of a liquid requires more than one freezing and thawing cycles.

10.If vapor phase compositions must also be determined experimentally, a different technique - referred to as 'dynamic' - is used. For details on experimental methods, see Hala (1967).

Before we discuss further the activity coefficient, we consider next the simple case of an ideal solution, where all gammas are equal to one.

13.6 Ideal Solutions

13.6.1 Raoult's Law

We saw in Section 11.13.2 that in an ideal solution all activity coefficients are equal to 1.0. Under these conditions Eq.13.5.1 gives:

$$\bar{P}_i = y_i P = x_i P_i^s \tag{13.6.1}$$

This represents the well-known Raoult's law, and indicates that the partial pressure, $\bar{P}_i$, of component i is simply proportional to its mole fraction in the liquid phase; and the proportionality constant is the vapor pressure of pure i at the system temperature, independently of what the other components of the mixture are.

In other words, the tendency of component i to leave the liquid state, its liquid phase fugacity, is not effected by the other components of the mixture; just by its own vapor pressure in the pure state and its mole fraction in the liquid mixture. In terms of molecular characteristics, this would require (Section 11.12.2) that:

* the forces between like and unlike molecules be the same, and
* the size and shape of the molecules be also the same.

An ideal solution represents, therefore, a hypothetical case. Because size and shape differences, however, are important only in extreme cases - such as polymer-solvent systems - we would expect that Raoult's law will be a reasonable approximation for mixtures of similar compounds, such as methanol-ethanol, hexane-heptane, etc. Indeed, the activity coefficients for the methanol-ethanol system at atmospheric pressure, for example, are to within 10% from one.

In the next six Examples we use Raoult's law in some typical *VLE* problems including applications to distillation column design.

13.6.2 Example 13.2

In Section 8.4.2 we recognized the vapor pressure of a pure liquid - at a specified temperature - as the pressure read by a manometer when this liquid is in equilibrium with the vapor above it. Thus, if in the vapor-liquid equilibrium experiment of Section 13.5.1 we introduced a pure liquid, the pressure read by the manometer represents the vapor pressure of this liquid at the bath temperature.

Similarly, the value read in the case of a mixture represents the vapor pressure of this liquid, at the corresponding composition and bath temperature. It is referred to as the *total* pressure because it represents the contribution to pressure by all components of the mixture.

Let us now assume that the mixture is one of n-hexane(1) and n-heptane(2), and that after equilibrium is reached at a bath temperature of 60°C, analysis of the liquid phase gives: $x_1 = x_2 = 0.5$. Since experimental measurements are time-consuming and costly, they are avoided whenever possible. To this purpose, let us then consider the following two questions:

1. What information would be needed to calculate the pressure that would be read by the manometer - i.e. the vapor pressure of this liquid - without resorting to an experimental measurement?

2. The same, but for the vapor phase composition.

1. The pressure over the solution results from the collision of the vapor molecules with the walls of the flask and the surface of the mercury in the manometer. Since the contribution of each component is given by its partial pressure $\bar{P}_i$, the total pressure P is:

$$P = \bar{P}_1 + \bar{P}_2$$

How are we now to evaluate $\bar{P}_1$ and $\bar{P}_2$? The answer is provided by Eq. 13.5.2:

$$\bar{P}_i = y_i P = \frac{x_i \gamma_i P_i^s}{F_i} \tag{A}$$

Thus:

$$P = \frac{x_1 \gamma_1 P_1^s}{F_1} + \frac{x_2 \gamma_2 P_2^s}{F_2} \tag{B}$$

Since x_1 and x_2 are given, we need the values for the:
* two activity coefficients,
* vapor phase fugacity coefficients, and
* vapor pressures of pure liquids (1) and (2).

For the activity coefficients we notice that because of the similarity between hexane and heptane we can assume ideal solution behavior. It follows then (why?) that the vapor phase can be assumed to behave as an ideal gas, and:

$$P = x_1 P_1^s + x_2 P_2^s \tag{C}$$

The solution pressure can be thus determined if the vapor pressures of the pure components are available. To this purpose we use the Antoine equation, with the parameter values from Appendix C, obtaining: P_1^s = 572.8 mm Hg; and P_2^s = 210.6 mm Hg, and, from Eq.(C):

$$P = (0.5)(572.8) + (0.5)(210.6) = 391.7 \text{ mm Hg}$$

2. From Eq.(A):

$$y_i = \frac{x_i \gamma_i P_i^s}{F_i P} \tag{D}$$

$y_1 = x_1 P_1^s/P = 0.731$, and $y_2 = 0.269$.

Comments

1. The calculation we have just carried out determines also the pressure at which this liquid mixture - at the specified temperature - will start boiling, i.e. where the first 'bubble' will appear. (What is the bubble's composition?) It is referred to, appropriately, as *bubble point* (B.P.) pressure calculation; and the liquid, as was the case with a pure compound, as *saturated*.

2. In the next Example, we consider a bubble point temperature calculation.

13.6.3 Example 13.3

Let us now consider the following case with reference to the system n-hexane(1) -n-heptane(2) of Example 13.2: the liquid phase composition of the equilibrium system is again: $x_1 = x_2 = 0.5$, but the pressure is equal to 300 mm Hg.

What will the equilibrium temperature of the system and the vapor phase composition be, again without resorting to an experimental measurement?

Using Eq.(C) of Example 13.2 - but expressing the two vapor pressures through the Antoine equation - we obtain by trial-and-error: $t = 52.5°C$.

Evaluation of the vapor phase composition is then straightforward:

$$y_1 = (0.5)(443.0)/300 = 0.738; \quad y_2 = 0.262$$

Comments

1. In this case, our calculation is referred to as a bubble point temperature one.
2. Our two bubble point calculations required only knowledge of the pure component vapor pressures. No mixture data were needed.
3. This, however, resulted from the assumption of ideal solution, valid here because of the similar nature of the two components, which led to activity coefficient values equal to one. We will see in Section 13.16 how bubble point calculations are carried out in real solutions.
4. In these two Examples, the liquid phase composition was specified, and:
   a. for a given T, we calculated P and y's; or
   b. for a given P, we calculated T and y's.

In the next Example, we consider the reverse case.

13.6.4 Example 13.4

Let us now assume that in the hexane-heptane system of Example 13.2 the temperature of the equilibrium system is 60°C again, but this time we know the composition of the vapor phase: $y_1 = 0.731; y_2 = 0.269$:

a. what is the system pressure? and
b. what is the liquid composition?

a. From Eq.(A), Example 13.2:

$$x_1 = \frac{y_1 P}{P_1^s} \quad \text{and} \quad x_2 = \frac{y_2 P}{P_2^s} \tag{A}$$

But since: $x_1 + x_2 = 1.0$, it follows that:

$$\frac{y_1}{P_1^s} + \frac{y_2}{P_2^s} = \frac{1}{P} \tag{B}$$

Hence: $P = 1/[0.731/572.8 + 0.269/210.6] = 391.6$ mm Hg.

b. For the liquid phase composition: $x_1 = y_1\, P/P_1^s = 0.5$; $x_2 = 0.5$

Comments

1. The fact that the values for the pressure and the liquid phase composition are the same with those of Example 13.2, should not be surprising. The one case is the reverse of the other.

2. In this Example, we have also calculated the pressure at which this vapor mixture - at the specified temperature - will start condensing, i.e. where the first liquid 'droplet' or 'dew' will appear. (What will the droplet's composition be?) It is referred to, appropriately, as *dew point* (D.P.) pressure calculation; and the vapor at these conditions, again as with the case of a pure compound, as *saturated*.

3. Assume now that the pressure and vapor phase composition are specified, say: $P = 300$ mm Hg; $y_1 = 0.738$ and $y_2 = 0.262$, and we want to calculate:

a. the temperature at which the first droplet will appear, i.e. the dew point temperature, and

b. the composition of this droplet.

In this case, we express the vapor pressures in Eq.(B) through the Antoine equation and solve - by trial-and-error - for the temperature, obtaining $t = 52.5$°C. And: $x_1 = y_1\, P/P_1^s = (0.738)(300)/443 = 0.5$.

4. The fact that the values for t and x are the same with those of Example 13.3 should not be, again, surprising. The one case is the reverse of the other.

5. It should be apparent from our discussion so far that, if a liquid and a vapor phases are in equilibrium with each other, then:

a. each phase is *saturated*,

b. the liquid phase is at its *bubble point*, and

c. the vapor phase is at its *dew point*.

6. Calculation of dew points for real solutions will be discussed in Section 13.16.

We proceed now to use bubble and dew point calculations in the following Examples that represent typical applications of vapor-liquid equilibrium to distillation column design.

13.6.5 Example 13.5

In distillation operations it is desirable, for economic reasons, that the condenser operate with water as cooling medium. Assuming, thus, that during the summer months the water from the cooling tower enters the condenser at 45°C - and using a 10°C minimum 'approach' temperature - we arrive at a condensate temperature of 55°C.

Consider now an equimolar mixture of n-hexane(1)and n-heptane(2), that is to be separated through distillation into products of 90% (mole) purity. What is the minimum pressure that the condenser should operate at?

In order that all the vapor entering the condenser be converted into liquid, the operating pressure must be at least equal to the saturation pressure of the distillate at 55°C, i.e. its bubble point value:

$$P = x_1 P_1^s + x_2 P_2^s = (0.9)(483.4) + (0.1)(173.5) = 452.4 \text{ mm Hg}$$

If the pressure is lower, we will not have complete condensation; if it is higher, there is no problem for we will end up with a *subcooled* liquid. The latter is actually desirable in practice, to avoid problems in pumping a saturated liquid back into the column as reflux, and the remaining to storage.

13.6.6 Example 13.6

From the economic point of view it is also desirable to operate distillation systems at atmospheric pressure. (Vacuum costs money and low operating pressure in the column results in large vapor volume and, consequently, large diameter). Assuming:

* that the condenser of Example 13.5 operates at 1 atm, and
* that due to pressure drop between the top of the column and the condenser the pressure there is 800 mm Hg,

consider the following question: If the vapor leaving the column is in equilibrium with the liquid at the top tray, what is the temperature there and the liquid phase composition?

The vapor has, of course, a composition of $y_1 = 0.9$ and $y_2 = 0.1$ and, since it is assumed to be in equilibrium with the liquid, it must be at its dew point temperature. Thus:

$$y_1 / P_1^s + y_2 / P_2^s = 1/800$$

and by trial-and-error, $t = 75.1$°C.

Also: $x_1 = (800)(0.9)/(924.1) = 0.779$ and $x_2 = 0.221$.

13.6.7 Example 13.7

Let us now assume that due to pressure drop through the column the boiler operates at 900 mm Hg.

a. What is the temperature in the boiler?

b. What is the composition of the vapor leaving the boiler?

a. Assuming that the liquid and vapor phases in the boiler are in equilibrium, the liquid is at its bubble point temperature:

$$x_1 P_1^s + x_2 P_2^s = 900 \text{ mm Hg}$$

where: $x_1 = 0.10$ and $x_2 = 0.90$, from Example 13.5 (why?). By trial-and-error, $t = 100°C$.

b. It can be easily seen that: $y_1 = (0.1)(1845.2)/900 = 0.205$ and $y_2 = 0.795$.

13.6.8 Summary

Let us now summarize our work with the distillation column calculations for the separation of the hexane-heptane mixture, Examples 13.5, 13.6 and 13.7.

1. The economic requirement that water be used as a cooling medium in the condenser set the temperature of the condensate.

2. A bubble point pressure calculation on the distillate determined the minimum operating pressure in the condenser, which was below the atmospheric one.

3. Economic, again, considerations led to atmospheric pressure operation for the condenser.

4. Using a reasonable pressure drop between the condenser and the top of the column, we estimated the pressure there. A dew point temperature calculation then on the vapor leaving the top of the column determined:
   * the temperature of the vapor and liquid of the top tray, and
   * the composition of the liquid leaving it.

5. Using again a reasonable pressure drop along the column, we estimated the pressure in the boiler. A bubble point temperature calculation then on the bottoms stream determined:
   * the temperature in the boiler, and
   * the composition of the vapor stream leaving it.

6. We know, thus:
   * the temperature and pressure at the two ends of the column, and
   * the compositions of the two phases at these ends.

This information is sufficient to start the so-called 'tray-to-tray' calculation, that provides the information needed for the design of the column.

7. Unfortunately, the typical mixtures encountered in practice are not ideal, and calculation of bubble and dew points requires - as we have seen - knowledge of the activity coefficients of the components involved.

We proceed, therefore, with a qualitative and quantitative consideration of activity coefficients and later, Section 13.16, we will see how they are used in carrying out bubble and dew point calculations for real solutions.

13.7 Activity Coefficients: Qualitative Remarks

Considering the importance of activity coefficients in *VLE* calculations - and the high cost of the experimental measurements required for their evaluation - a qualitative understanding of the factors that effect their values:

a. composition,
b. molecular nature of the components in the mixture, and
c. pressure and temperature,

can be very helpful in avoiding unnecessary work.

We consider the first two here and the third one, in the next Section.

13.7.1 The Effect of Composition

Consider a binary system with $x_1 = 0.95$ and $x_2 = 0.05$. The molecules of component 1 are surrounded mostly by their own, being oblivious of the existence of the molecules of component 2. We would expect, hence, its activity coefficient to be practically equal to one, i.e. to follow Raoult's law. This is indeed the case as Figure 13.2 demonstrates.

On the other hand, the molecules of component 2 are surrounded mainly by molecules of 1, i.e. they find themselves in a 'strange' environment. We would expect, therefore, that its activity coefficient would be very different from one; and, using the same physical argument, that this difference would decrease with increasing values of x_2.

This is indeed the case, as shown again in Figure 13.2. [There are, however, a few rare cases where - due to a combination of physical and chemical interactions - the activity coefficient's dependency on concentration passes through a maximum or minimum (see, for example, the system acetone-methanol in Severns et al, 1955).]

These qualitative arguments about deviations from ideality (Raoult's law) should remind us of the Lewis-Randall rule (Section 11.8.3). Indeed Raoult's law, even though it was proposed before the Lewis-Randall rule, represents a special case of the latter.

13.7.2 The Effect of Molecular Nature

Considering the requirements for ideal solution behavior, Section 13.6.1, we examine next the effect of the molecular characteristics of the mixture components on its deviation from such behavior. More specifically, we will examine their effect on:

a. the type, and
b. the magnitude,

of this deviation.

a. Types of Deviations from Ideal Solution (Raoult's Law) Behavior

Consider a binary mixture where the intramolecular (1-1, and 2-2) forces are stronger than the intermolecular (1-2) ones. As a result, the presence of component 2 increases the 'escaping' tendency of component 1, as compared to its value at the same composition, but in an ideal solution where all forces are equal; and the same is the case with component 2. Both activity coefficients are, thus, greater than one. In such systems therefore (assuming ideal vapor behavior):

* the partial pressure of component i: $\bar{P}_i(real) = x_i \gamma_i P_i^s$
* is greater than its counterpart: $\bar{P}_i(ideal) = x_i P_i^s$, had i been in an ideal solution.

We refer to them, appropriately, as demonstrating *positive* deviations from Raoult's law and an example is the system presented in the upper part of Figure 13.2. These systems represent the typical - by far - case. (And, they claim, that segregation is a human invention!).

There are a few systems, however, where 'integration' does occur, i.e. the 1-2 forces are stronger than the 1-1 and 2-2 ones. Such systems exhibit strong hydrogen bonding between the 1 and 2 molecules (i.e. association, Section 7.7), and their activity coefficient values are less than one. They are referred to, consequently, as demonstrating *negative* deviations from Raoult's law. A typical example is presented in the lower part of Figure 13.2.

These positive and negative deviations from Raoult's law are better demonstrated in Figure 13.3, where the total pressure over an equilibrium solution is plotted versus the liquid composition for isothermal conditions.

Starting with ideal solution behavior, we note that since the total pressure is the sum of the two partial pressures:

$$P(ideal) = x_1 P_1^s + x_2 P_2^s = P_2^s + \left(P_1^s - P_2^s\right) x_1$$

P(*ideal*) thus is a linear function of x_1 as shown in the Figure, line 1.

Turning to positive deviations from Raoult's law, we note that the total pressure:

$$P(real) = x_1 \gamma_1 P_1^s + x_2 \gamma_2 P_2^s$$

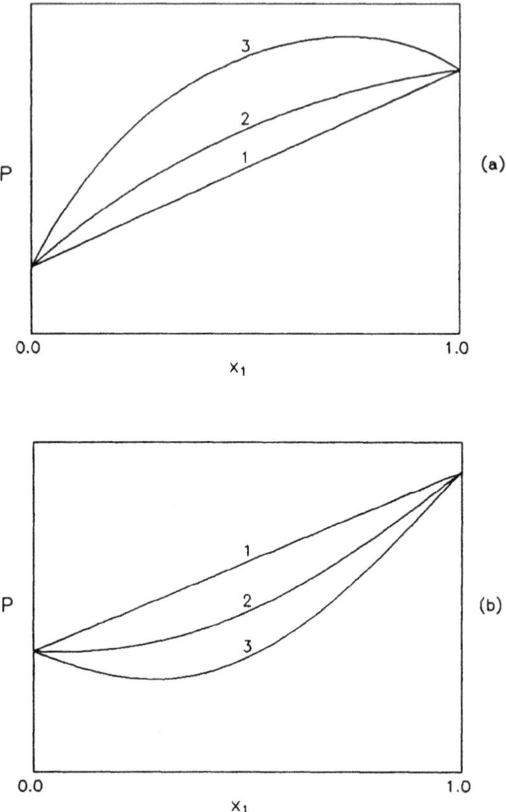

Figure 13.3 Positive (a) and negative (b) deviations from Raoult's law.

is larger than that corresponding to the ideal solution value, as shown at the top of the Figure. Notice that in the case of the upper curve - line 3 - the deviation is so large, that P goes through a maximum. At this point we have an *azeotrope*, where $x_i = y_i$.

Negative deviations are demonstrated in the lower part of the Figure. It is easily seen that here *P*(*real*) is less than *P*(*ideal*), and there is an azeotrope for curve (3) but not for (2).

Similar curves, but with opposite direction obviously, are obtained if we plot T versus x_1 for isobaric conditions. This is demonstrated with the system propanol-water at 1 atm (Gmehling et al, 1981) presented in Figure 13.4, where the phases present are also shown. Notice the presence of azeotrope at the minimum temperature. From the practical point of view, this azeotrope represents the maximum composition of propanol that can

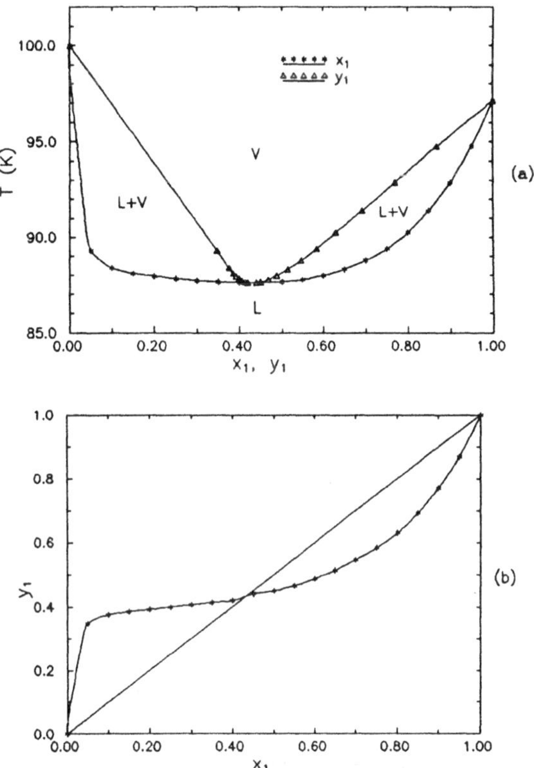

Figure 13.4 *T-x-y* (a) and *y-x* (b) diagrams for the 1-propanol(1) - water(2) system at 1 atm. Lines for identification only. L: Liquid; V: Vapor.

be obtained by distillation at this pressure.

So far we have considered deviations from ideality resulting from differences among inter-(i-j) and intra-(i-i and j-j) molecular forces only. Another factor that can play an important role is the difference in *free volume* of the mixture components. The simplest way to define this quantity (Elbro et al, 1990) is in terms of what Bondi (1968) calls 'empty volume':

$$V_f = V - V^*$$

where V is the molar volume and V^* is the molar hard-core volume, calculated from the van der Waals volumes given by Bondi.

Such differences are more pronounced in systems where there are large differences in molar volumes, such as polymer-solvent ones, and lead to negative deviations from Raoult's law as shown with a typical case in Figure 13.5 (Iwai and Arai, 1989).

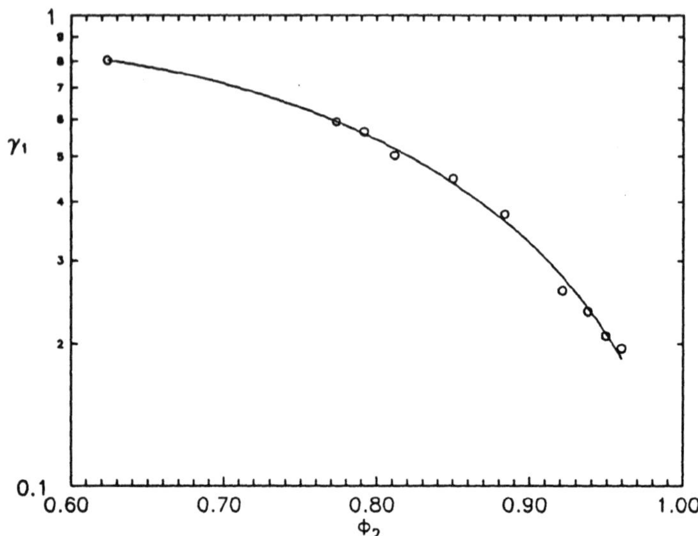

Figure 13.5 Activity coefficient of n-nonane(1) in its mixture with polybutadiene(2) at 100°C. (For polymer solutions the volume fraction Φ rather than the mole fraction is used.)

b. Magnitude of Deviations from Raoult's Law

The effect of chemical dissimilarity of the mixture components on the size of their activity coefficients is demonstrated with the infinite dilution values presented in Figure 13.6 (Pierotti et al, 1959).

Consider, for example, the binary systems of aliphatic ketones with n-heptane. The polar character of ketones - originating from the presence of the CO group - decreases with increasing values of N_c, the number of carbon atoms in the ketone: they become progressively more hydrocarbon-like, which explains the observed decrease in the infinite dilution activity coefficient values of ketones in heptane as N_c increases.

The same phenomenon, but with the opposite effect, occurs in the case of n-alcohols in water, where the main factor is the hydrogen bonding among the OH groups. Again, as the chain length increases, hydrocarbon behavior is approached and the hydrogen bonding effect diminishes. Here however the alcohols become progressively more dissimilar to water, which explains the observed increase of the infinite dilution activity coefficient values of the latter in the former, as N_c increases. Actually, this dissimilarity leads to partial miscibility with higher alcohols.

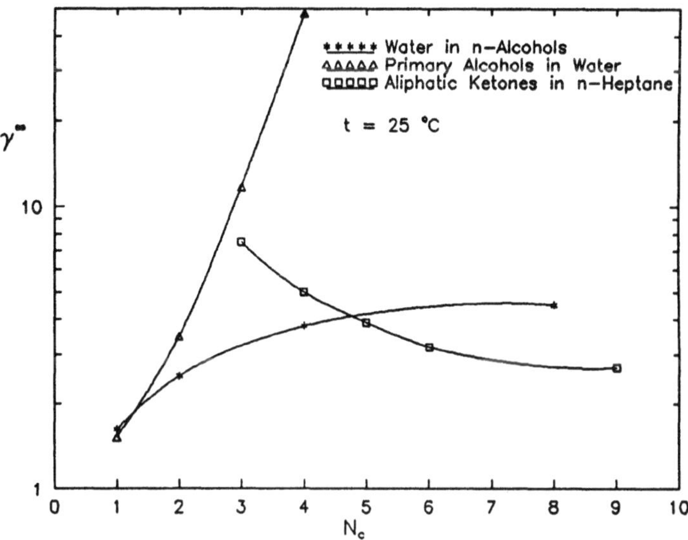

Figure 13.6 The effect of chemical dissimilarity between the components of a mixture on the activity coefficients (lines for identification only).

13.7.3 Summary

We summarize next our qualitative observations about the activity coefficient up to this point:

1. The activity coefficient is a measure of the deviation from ideal solution behavior, the 'correction' factor to Raoult's law.
2. As the composition of a given component decreases, its activity coefficient's deviation from one increases.
3. As the dissimilarity among the components of a mixture increases, so does the deviation from ideality (and, from the value of one, for γ's).
4. Dissimilarity in the physical and chemical forces has the most pronounced effect, followed by dissimilarity in free volumes. The latter becomes dominant in solvent-polymer solutions.
5. Most systems - by far - demonstrate positive deviations from Raoult's law (except solvent-polymer ones).
6. Typical maximum values of gamma for systems with positive deviations from Raoult's law:

* completely miscible: 1.1 - 10
* partially miscible: 20 - 200
* very immiscible: 200 - very high values. (The infinite dilution activi-

ty coefficient, for example, of benzene in water is of the order of 20,000 with a solubility in the parts per million.)

7.For negative deviations from Raoult's law, typical minimum values for gamma are between 0.9 and 0.5. Still smaller minimum values can be observed for the solvent activity coefficient in solvent-polymer systems (Figure 13.5).

13.8 Activity Coefficients: The Effect of Pressure and Temperature

13.8.1 Pressure

The effect of pressure on the activity coefficient - at a constant temperature and composition - is given by:

$$\left(\frac{\partial \ln \gamma_i}{\partial P}\right)_{T,x} = \frac{\bar{V}_i^E}{RT} \tag{13.8.1}$$

where $\bar{V}_i^E$ is the partial molar excess volume of component i in the liquid phase.

At the low pressures of interest here, the partial molar excess volume is rather small as we saw in Example 11.1 and, consequently, so is the effect of pressure on the activity coefficient.

Consider for example the case of water(1) in its mixture with methanol(2) at the conditions of Example 11.1: $x_1 = 0.7$; $T = 25°C$; $P = 1$ atm, where its partial molar excess volume is 0.3 cm^3/mol. Assuming this value to be independent of pressure for small pressure ranges, then the activity coefficient of water at some pressure P_2, say 5 atm, relative to its value at 1 atm is given by:

$$\gamma_1[5 \text{ atm}] = 1.00006\, \gamma_1[1 \text{ atm}]$$

a negligible effect. (It may become important, however, at high temperatures and pressures.)

13.8.2 Temperature

The effect of temperature on the activity coefficient - at a constant pressure and composition - is given by:

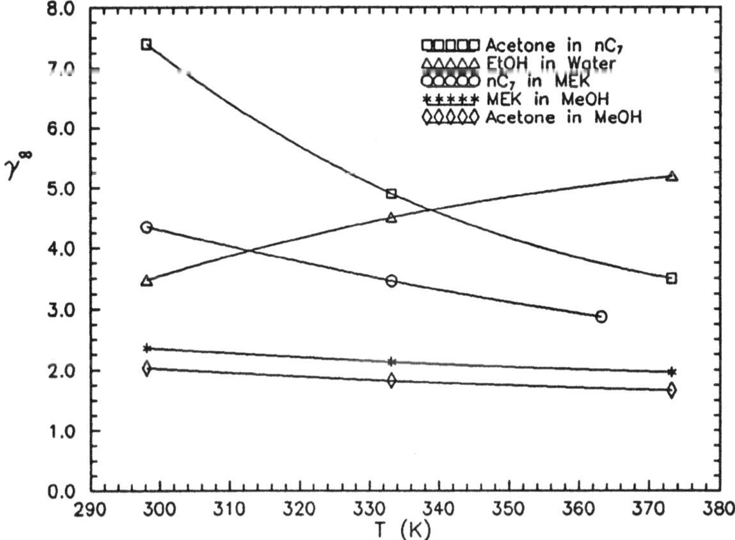

Figure 13.7 Variation of infinite dilution activity coefficients with temperature (Pierotti et al, 1959). Lines for identification only.

$$\left(\frac{\partial \ln \gamma_i}{\partial T}\right)_{P,x} = \frac{-\bar{H}_i^E}{RT^2} \tag{13.8.2}$$

The variation thus of the activity coefficient with temperature depends on the excess enthalpy of mixing, which reflects the difference between inter- and intra-molecular forces. As a result, the activity coefficient of a mixture component can be a strong function of temperature as shown in Figure 13.7, the effect being determined by the other components present. This explains, for example, the large difference in the variation of γ^∞ with temperature for acetone in nC_7, as compared to that in methanol.

We conclude, therefore, that - in the typical nonideal solution - the activity coefficient:

* is a strong function of composition,
* can be a strong function of temperature, and
* is a very weak function of pressure.

We will discuss later (Sections 13.10 through 13.14) the approach used in developing analytical relationships of the activity coefficient as a function of composition and temperature. But first, let us examine how the accuracy of experimental activity coefficients is evaluated.

13.9 Thermodynamic Consistency of Experimental Data

13.9.1 The Interrelationship Among Activity Coefficients

We have seen in Chapter 11 that all partial molar properties must satisfy the Gibbs-Duhem equation, Eq.11.6.3, which when applied to the molar excess Gibbs free energy yields (Problem 13.52):

$$RT\,\Sigma x_i d\ln\gamma_i = V^E dP - S^E dT \tag{13.9.1}$$

where V^E and S^E are the excess molar volume and entropy of the mixture respectively at the specified temperature, pressure and composition.

The activity coefficients, thus, of the components of an equilibrium mixture cannot all vary independently of each other: they must satisfy Eq.13.9.1, if they are to be thermodynamically consistent.

We examine next how this interrelationship is used to test the consistency of experimental binary data.

13.9.2 Binary Systems

Integration of Eq.13.9.1 for a binary system, from $x_1 = 0.0$ to $x_1 = 1.0$, gives (Prausnitz, 1969):

$$\int_0^1 \ln\frac{\gamma_1}{\gamma_2}dx_1 = \int_0^1 \frac{H^E}{RT^2}\frac{dT}{dx_1}dx_1 \quad (P\!:constant) \tag{13.9.2}$$

$$\int_0^1 \ln\frac{\gamma_1}{\gamma_2}dx_1 = -\int_0^1 \frac{V^E}{RT}\frac{dP}{dx_1}dx_1 \quad (T\!:constant) \tag{13.9.3}$$

Evaluation therefore of the thermodynamic consistency of binary vapor-liquid equilibrium data - at constant temperature (isothermal) or constant pressure (isobaric) conditions - requires knowledge of the values of the corresponding excess property as a function of composition.

In the typical case such experimental information is not available, and the right-hand side of Eqs 13.9.2 and 13.9.3 is set equal to zero.

While this is a reasonable assumption for Eq.13.9.3, it may not always be so for Eq.13.9.2, especially for very nonideal mixtures. For isobaric data, therefore, the semiempirical test proposed by Herington (1951) should be used. For more reliable results, the excess enthalpy can be estimated with a modified UNIFAC model (see Section 13.18.3).

13.9.3 Comments

1. Usually the integral in the left-hand side of Eqs 13.9.2 and 13.9.3 is evaluated graphically, as shown in Figure 13.8 for the system acetone-cyclohexane. As a result, this method for evaluating the thermodynamic consistency of binary data is referred to as the 'area' test.

2. For isothermal data the 'Consistency Index' (CI), defined by:

$$CI = \left| \frac{A_p - A_n}{A_p + A_n} \right| \times 100 \tag{13.9.4}$$

where A_p and A_n are the areas above and below $\ln(\gamma_1/\gamma_2) = 0.0$ respectively, becomes the measure of their quality.

For isobaric data, the measure of their quality is the quantity (CI-J) with J given by:

$$J = 150 \frac{\Delta T}{T_{min}}; \quad \Delta T = T_{max} - T_{min}$$

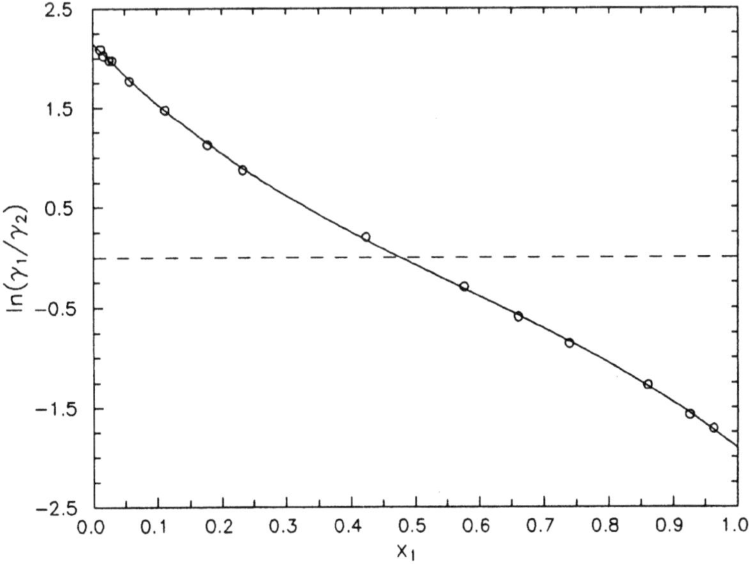

Figure 13.8 The area test for the system acetone-cyclohexane at 25°C.

Table 13.2 Classification of Typical Literature Binary *VLE* Data According to the Area Thermodynamic Consistency Test (Silverman)

Type	Isothermal (*CI*)	Isobaric (*CI-J*)	Number	%
Good	< 5	< 10	147	60
Fair	5-15	10-20	47	19
Poor	> 15	> 20	37	15
Unacceptable: area test not possible			16	6
Total			247	100

T_{max} and T_{min} are the temperature extrema in the binary system including the boiling points of the pure components and the azeotrope, if present, while the coefficient 150 is empirical, recommended by Herington on the basis of typical heat-of-mixing data (Prausnitz).

3.A value of *CI* less than 5 for isothermal data, and of (*CI-J*) less than 10 for isobaric ones, suggests good data.

4.The quality of typical literature data is demonstrated in Table 13.2 (Silverman, 1984), which indicates that about 40% of the binary systems tested do not meet the aforementioned test.

5.For more reliable results $\ln(\gamma_1/\gamma_2)$ is fitted to a polynomial, which is used to evaluate the *CI* value, by integration, and the values of the infinite dilution gammas, by extrapolation to $x_i = 0.0$.

6.It should be kept in mind, also, that the area test is a necessary but not sufficient requirement for good quality data.

7.For limitations of the area test and other techniques of testing the consistency of *VLE* data see Prausnitz et al, 1986.

In the next Example, we examine the quality of the acetone-cyclohexane data of Table 13.E.1.

13.9.4 Example 13.8

For the system: acetone-cyclohexane, Table 13.E.1, determine:

a. the consistency index, and

b. the infinite dilution activity coefficients.

a. The values of $\ln(\gamma_1/\gamma_2)$ are reported in the same Table, and are plotted versus x_1 in Figure 13.8. Drawing the 'best' line through them with a french curve yields the following approximate areas: $A_p = 47$; $A_n = -46$, both in the same units. Then, $CI = 1.1$, which suggests excellent data.

b. From the same Figure:

$$\ln\gamma_1^{\infty} = 2.13; \quad \gamma_1^{\infty} = 8.4$$

$$\ln\gamma_2^{\infty} = 1.90; \quad \gamma_2^{\infty} = 6.7$$

Comments

1. Since the data are isothermal, the *CI* index can be used with more confidence in establishing the quality of the data than the (*CI-J*) quantity, if they were isobaric.

2. Isothermal data are desirable for an other very important reason. Pressure, as we have seen, has a very small effect on low pressure activity coefficients and, consequently, such data reflect only the nonideality of the solution. Isobaric data, on the other hand, because of the significant temperature effect that may be involved, reflect this effect as well. They are obviously more desirable for distillation calculations.

3. Isothermal data, therefore, are useful in the study of solution nonideality; isobaric, for distillation calculations.

4. The reliability of the area test is facilitated, in this Example, by the relatively large number of available data. In general, it is recommended that - at least - 15 data points are taken.

5. There is obviously an uncertainty in the gamma infinity values. It would be probably larger, however, if plots of gamma versus composition were used.

13.10 Activity Coefficients and Composition: Analytical Expressions

We have already seen that the activity coefficient is a strong function of composition (Section 13.7.1, Table 13.E.1 and Figure 13.2). We consider therefore in this Section the procedure followed in developing analytical expressions (or models) for the activity coefficient as a function of composition. Several such expressions, essential for modelling of separation processes, are then considered in Sections 13.11 through 13.14.

13.10.1 Procedure

The procedure involved consists of two basic steps:

1. Using phenomenological arguments, from purely empirical to those based on statistical mechanics, an expression for the molar excess Gibbs

free energy G^E as a function of composition is developed.
2.The exact relationship:

$$RT\ln\gamma_i = \left(\frac{\partial(NG^E)}{\partial N_i}\right)_{T,P,N_j} \tag{11.13.6}$$

is then used to arrive at an expression for the activity coefficient.

13.10.2 Comments

1.Development of a model for G^E requires an understanding of the different effects involved in the mixing of molecules and, of course, an analytical representation of these effects. As a starting point we note that, since:

$$G^E = H^E - TS^E = U^E + PV^E - TS^E$$

where H^E, S^E, U^E and V^E are respectively the molar excess enthalpy, entropy, internal energy and volume of mixing, the mixing process at constant temperature T and pressure P involves three effects:
a.an energetic one, resulting from the difference in intermolecular energies between like and unlike molecules;
b.an entropic one, resulting from a lack of complete randomness in the distribution of the molecules in the mixture, caused mainly by differences in the size and shape of the molecules; and
c.one caused by an increase or decrease in volume as a result of the mixing process, caused by differences in both, intermolecular energies and molecular sizes and shapes.

2.The first effect tends to be the dominating factor in mixtures of molecules that are not very different in size and the early investigators in the field: van Laar, Hildebrand, Scatchard, Wohl, etc., assumed that: $S^E = V^E = 0$. This leads to the concept of 'regular' solution, first introduced by Hildebrand.

3.On the other hand, for mixtures involving molecules of very different size, such as polymer-solvent ones, H^E can be much smaller than S^E. The assumption of $H^E = 0$, leads to the concept of 'athermal' solution and resulted in the Flory-Huggins expression for polymeric solutions (Prausnitz et al), where $V^E = 0$ was also assumed.

4.We will classify the models for G^E and, hence, for the activity coefficient, into two types:
* the 'Wohl-type' that includes the traditional expressions of van Laar,

Margules, and Redlich-Kister and are considered purely empirical; and
* the 'local composition' type, that includes the Wilson, NRTL, and UNIQUAC expressions and are considered semi-empirical.

The Wohl-type models find only limited application today. They are presented, however, because they represent a simple example of how expressions for G^E are developed. The obtained expressions for the activity coefficient are also easier to use than those of the local composition models.

5.None of these expressions accounts for differences in free volumes (Section 13.7.2), which can be important for mixtures of molecules that are very different in size. They have been included, thus, for polymer-solvent systems (Oishi and Prausnitz, 1978; Iwai and Arai; Elbro et al); and for gas-solvent ones (Antunes, 1983; Nocon et al, 1983).

13.11 The Wohl Type Expressions

13.11.1 The General Expression for a Binary System

Wohl (1946) expressed the molar excess Gibbs free energy in a binary solution in terms of a power series of the 'effective' volume fractions of the two components, z_1 and z_2 (Prausnitz):

$$\frac{G^E}{RT(q_1x_1+q_2x_2)} = 2a_{12}z_1z_2+3a_{112}z_1^2z_2+3a_{122}z_1z_2^2+\dots \tag{13.11.1}$$

where:

$$z_1 = \frac{x_1q_1}{x_1q_1+x_2q_2}; \quad z_2 = \frac{x_2q_2}{x_1q_1+x_2q_2}$$

q_i: measure of the size of molecule i, or of its 'sphere of influence', in the solution.

a: interaction parameter, with a physical significance similar to that of the virial coefficients. For example, a_{12} reflects the interaction between two molecules, one of component 1 and one of 2; a_{122}, among three molecules, one of 1 and two of 2, etc.

The following two expressions, that of van Laar and that of Margules, can be obtained as special cases of Eq.13.11.1, even though they were developed earlier.

13.11.2 The van Laar Equation

For a binary system whose components are not too dissimilar chemically but have different molar volumes, we can assume that interaction coefficients involving more than two molecules can be neglected. Then, the Wohl expression becomes:

$$\frac{G^E}{RT} = \frac{2a_{12}x_1x_2q_1q_2}{x_1q_1+x_2q_2} \tag{13.11.2}$$

From this equation, the following expressions are obtained for γ_1 and γ_2:

$$\ln\gamma_1 = \frac{A}{[1+(Ax_1)/(Bx_2)]^2} \tag{13.11.3}$$

$$\ln\gamma_2 = \frac{B}{[1+(Bx_2)/(Ax_1)]^2} \tag{13.11.4}$$

where: $A = 2q_1 a_{12}$; $B = 2q_2 a_{12}$

Note. Actually van Laar - a student and, later, colaborator of van der Waals - developed his expression early in the 20th century starting with the van der Waals equation of state (see Prausnitz).

13.11.3 The Margules Equation

If the molecular sizes of the components of a binary system are not too different, we can assume that $q_1 = q_2 = q$ and, neglecting interactions involving more than three molecules, obtain the following expressions for the activity coefficients:

$$\ln\gamma_1 = A'x_2^2 + B'x_2^3 \tag{13.11.5}$$

$$\ln\gamma_2 = \left(A' + \frac{3}{2}B'\right)x_1^2 - B'x_1^3 \tag{13.11.6}$$

where: $A' = q(2a_{12} + 6a_{112} - 3a_{122})$; $B' = q(6a_{122} - 6a_{112})$

Eqs 13.11.5 and 13.11.6 are referred to as the 3-suffix Margules equation (two-parameter expression).

For a detailed discussion of the van Laar and Margules expressions, and a comparison of their performance in correlating binary data, see Prausnitz.

13.11.4 Evaluation of the Parameters

Evaluation of the parameter values in the van Laar and Margules expressions requires, of course, experimental activity coefficient data for the binary system involved.

To this purpose all data are regressed to obtain the parameter values that 'best fit' them. They can, of course, be also obtained from a pair of activity coefficient data by solving a system of two equations with two unknowns. Thus, if for a given x_1 the corresponding activity coefficients are γ_1 and γ_2, we obtain the following expressions for the van Laar parameters:

$$A = \left(1 + \frac{x_2}{x_1}\frac{\ln\gamma_2}{\ln\gamma_1}\right)^2 \ln\gamma_1 \tag{13.11.7}$$

$$B = \left(1 + \frac{x_1}{x_2}\frac{\ln\gamma_1}{\ln\gamma_2}\right)^2 \ln\gamma_2 \tag{13.11.8}$$

Expressions for the Margules parameters can be developed in a similar fashion.

The two equations, in spite of the different assumptions involved in their development from Wohl's expression (Eq.13.11.1), give very similar results, because of their purely empirical character and the fact that they contain the same number of adjustable parameters.

To develop some familiarity in working with nonideal systems consider the following Example.

13.11.5 Example 13.9

Using the single experimental datum for the system acetone(1)-cyclohexane(2), Table 13.E.1: $P = 253.9$ mm Hg; $x_1 = 0.4235$; $y_1 = 0.6800$ and $t = 25°C$, calculate the equilibrium pressure and vapor composition values at: $x_1 = 0.0115$; 0.1775; 0.5760; 0.7390; 0.9625, all at 25°C, and compare them to the experimental data. Assume, as in Example 13.1, ideal vapor behavior and use the pure compound vapor pressures given there (instead of the Antoine equation).

In this case:

$$P = x_1\gamma_1 P_1^s + x_2\gamma_2 P_2^s \tag{A}$$

$$y_1 = \frac{x_1 \gamma_1 P_1^s}{P} \tag{B}$$

Since x_1 and the pure compound vapor pressures are given, the values for the activity coefficients at each composition are needed. To this purpose:

1. From the given datum point calculate the corresponding γ_1 and γ_2 values using Eq. 13.5.1: $\gamma_1 = 1.769$; $\gamma_2 = 1.446$ (see Example 13.1).
2. Use these values to calculate the van Laar parameters from Eqs 13.11.7 and 13.11.8: $A = 2.016$; $B = 1.683$.
3. Calculate γ_1 and γ_2 at each composition, using these parameter values and Eqs 13.11.3 and 13.11.4.

Table 13.E.9 presents the obtained γ_1 and γ_2 values and the corresponding errors. Values of P and y_1, calculated from Eqs (A) and (B) respectively, and the corresponding % errors and deviations Δy_1 and ΔP are also included.

Comments

1. Eq.(A) determines the pressure at which the liquid at the given composition will start boiling at 25°C, i.e. the bubble point pressure; and Eq.(B) provides the composition of this bubble.
2. The most important quantity for distillation design purposes is the predicted vapor phase composition.
3. Errors in y_1 are usually reported in terms of $\Delta y = y_{(calc)} - y_{(exp)}$, because the percent error depends on whether it is based on the component with the smallest or largest y. In this case, for example, and at $x_1 = 0.0115$, while the error in y_1 is -9.6%, it is only 2.1% for y_2. Similarly to avoid masking the error at higher

Table 13.E.9 Calculated Values for the System Acetone-Cyclohexane and Comparison with the Experimental Ones at 25°C

x_1	γ_1	Error %	γ_2	Error %	P mm Hg	Error %	y_1	Error %
0.0115	7.109	-11.8	1.000	0.4	115.17	-2.4	0.1635	-9.6
0.1775	3.571	5.0	1.074	-2.8	232.12	1.9	0.6292	3.0
0.5760	1.339	2.8	1.907	3.0	256.53	-1.1	0.6928	-1.7
0.7390	1.110	-2.4	2.729	1.5	258.41	-1.4	0.7314	-1.0
0.9625	1.002	-0.5	4.848	15.6	239.92	-1.6	0.9262	1.1

x_1	Δy_1x1000	ΔP (mm Hg)
0.0115	-18	-2.88
0.1775	18	4.42
0.5760	-12	-2.87
0.7390	-8	-3.59
0.9625	10	-3.88
Overall (Abs.)	13	3.53

P values, the difference: $\Delta P = P_{(calc)} - P_{(exp)}$ is often reported.

4. The overall average absolute deviations in y and P of 0.013 and 3.53 mm Hg respectively are typical for this type of predictions, i.e. from a single experimental datum with the van Laar or the Margules equation. Typical results in predicting binary activity coefficients with the Margules equation are shown in Figure 13.2.

5. When a single datum point is used, it is better to be chosen from somewhere in the middle of the composition range as done here. In this case the impact of the uncertainties in the experimental measurements, used for the determination of the activity coefficient values, has the smallest effect (see Problem 13.6).

6. Notice, however, that this choice leads to larger errors in the low concentration activity coefficients and corresponding y values. This is typical with all two-parameter models, especially for very nonideal systems, and could create problems if products of high purity are expected from a distillation system.

13.12 The Local Composition Expressions

13.12.1 The Concept of Local Compositions

In 1964 Wilson suggested that on the molecular level a given solution, say a binary one, comprises two types of cells (Figure 13.9):

* one formed by the molecules surrounding a molecule of component 1;
* the other, by those surrounding one of component 2.

Furthermore, due to differences in the inter- and intra-molecular forces, the potential energy parameters $\lambda_{12} = \lambda_{21}$, λ_{11}, and λ_{22} are different, leading to local compositions that are not the same as the overall ones.

If, for example, the 1-1 and 2-2 interactions are stronger than the 1-2 ones, the molecules would prefer to surround themselves with their own kind. As a result, the cell mole fractions x_{ij}, referred to as 'local' mole fractions with i representing the molecule at the cell center, are not the same as the overall ones. For example, with reference to Figure 13.9, we have:

$$x_{12} = \frac{3}{7};\quad x_{11} = \frac{4}{7};\quad x_{21} = \frac{3}{7};\quad x_{22} = \frac{4}{7}$$

as compared to the overall composition of $x_1 = x_2 = 0.5$.

The concept of local compositions revolutionized the whole approach to developing expressions for the excess Gibbs free energy and, consequently, the activity coefficient. Wilson used it in developing the equa-

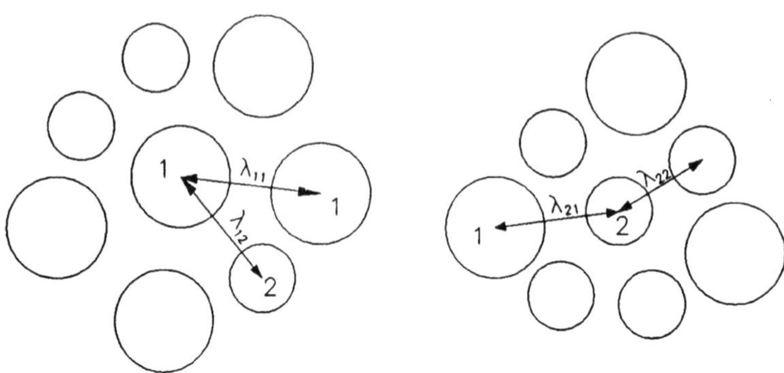

Figure 13.9 Fluid structure on the local level.

tion that carries his name; and Prausnitz and co-workers in developing the NRTL and UNIQUAC equations.

In the interest of space, only the final expressions will be presented. Their development can be found in the original articles.

13.12.2 The Wilson Equation

Wilson postulated that the ratio of local compositions are related to the overall mole fractions through a Boltzmann type expression:

$$\frac{x_{ij}}{x_{ii}} = \frac{x_j \exp(-\lambda_{ij}/RT)}{x_i \exp(-\lambda_{ii}/RT)}$$

Using this relationship he arrived at expressions for the local 'volume' fractions, which he combined with the Flory-Huggins equation for polymer solutions (Section 13.10.2) to develop an expression for the excess Gibbs free energy.

From the latter and Eq. 11.13.6 he obtained the following expression for the activity coefficient of component i in a multicomponent mixture:

$$\ln\gamma_i = 1 - \ln(\Sigma x_j \Lambda_{ij}) - \sum \frac{x_k \Lambda_{ki}}{\Sigma x_j \Lambda_{kj}} \qquad (13.12.1)$$

The summation is carried out over all the components of the mixture and:

$$\Lambda_{mn} = \frac{V_n}{V_m} \exp\left(\frac{-(\lambda_{mn} - \lambda_{mm})}{RT}\right) = \frac{V_n}{V_m} \exp\left(\frac{-\Delta\lambda_{mn}}{RT}\right) \qquad (13.12.2)$$

where V_k is the liquid molar volume of pure k. For a binary system, Eq.13.12.1 gives:

$$\ln\gamma_1 = -\ln(x_1 + x_2\Lambda_{12}) + x_2\left(\frac{\Lambda_{12}}{x_1 + x_2\Lambda_{12}} - \frac{\Lambda_{21}}{x_2 + x_1\Lambda_{21}}\right)$$

$$\ln\gamma_2 = -\ln(x_2 + x_1\Lambda_{21}) + x_1\left(\frac{\Lambda_{21}}{x_2 + x_1\Lambda_{21}} - \frac{\Lambda_{12}}{x_1 + x_2\Lambda_{12}}\right) \tag{13.12.3}$$

13.12.3 The NRTL Equation

Renon and Prausnitz (1968 and 1969) modified Wilson's equation for the local mole fractions by introducing the parameter α to account for the non-randomness of liquid solutions:

$$\frac{x_{ij}}{x_{ii}} = \frac{x_j}{x_i}\exp\left(\frac{-\alpha_{ij}(g_{ij} - g_{ii})}{RT}\right)$$

where:

g_{ij} = residual Gibbs free energy, and

$\alpha_{ij} = \alpha_{ji}$ = non-randomness parameter

Making use of this expression - and Scott's two-liquid theory - they developed the "Non-Random Two-Liquid" (NRTL) expression for the excess Gibbs free energy, from which the following equation is obtained for the activity coefficient of component i in a multicomponent mixture:

$$\ln\gamma_i = \frac{\Sigma x_j \tau_{ji} G_{ji}}{\Sigma x_k G_{ki}} + \sum \frac{x_j G_{ij}}{\Sigma x_k G_{kj}}\left(\tau_{ij} - \frac{\Sigma x_m G_{mj}\tau_{mj}}{\Sigma x_k G_{kj}}\right) \tag{13.12.4}$$

where:

$$\tau_{ij} = (g_{ij} - g_{jj})/RT = \Delta g_{ij}/RT \quad ; \quad G_{ij} = \exp(-\alpha_{ij}\tau_{ij})$$

For a binary system, Eq.13.12.4 gives:

$$\ln\gamma_1 = x_2^2\left(\tau_{21}\left(\frac{G_{21}}{x_1 + x_2 G_{21}}\right)^2 + \frac{\tau_{12}G_{12}}{(x_2 + x_1 G_{12})^2}\right) \tag{13.12.5}$$

The expression for $\ln\gamma_2$ is obtained by rotating the subscripts.

13.12.4 The UNIQUAC Equation

Abrams and Prausnitz (1975) combined Guggenheim's quasi-chemical theory with the concept of local compositions to develop the Universal Quasi-Chemical (UNIQUAC) expression for the excess Gibbs free energy. [The equation can be also developed from the two-fluid theory (Maurer and Prausnitz, 1978).]

In addition to the molecular interactions, the resulting equation also accounts - approximately - for the effects of molecular size through structural parameters obtainable from pure compound data.

For a multicomponent mixture, the activity coefficient of component *i* is given by:

$$\ln\gamma_i = \ln\gamma_i^C + \ln\gamma_i^R \tag{13.12.6}$$

γ_i^C: combinatorial part, reflecting the difference in size and shape of the molecules of the mixture, given by:

$$\ln\gamma_i^C = \ln\frac{\Phi_i}{x_i} + \frac{z}{2}q_i\ln\frac{\Theta_i}{\Phi_i} + l_i - \frac{\Phi_i}{x_i}\Sigma x_j l_j \tag{13.12.7}$$

γ_i^R: residual part, reflecting the difference in energy of interaction between the molecules in the mixture, given by:

$$\ln\gamma_i^R = -q_i\ln(\Sigma\Theta_j\tau_{ji}) + q_i - q_i\sum\frac{\Theta_j\tau_{ij}}{\Sigma\Theta_k\tau_{kj}} \tag{13.12.8}$$

where, Θ_i and Φ_i are the surface area and volume fractions respectively:

$$\Theta_i = \frac{q_i x_i}{\Sigma q_j x_j}; \quad \Phi_i = \frac{r_i x_i}{\Sigma r_j x_j} \tag{13.12.9}$$

$$\tau_{ij} = \exp\frac{-(u_{ij}-u_{jj})}{RT} = \exp\frac{-\Delta u_{ij}}{RT}$$

$$l_j = \frac{z}{2}(r_j - q_j) - (r_j - 1)$$

u_{ij}: energy parameters

z: coordination number, set equal to 10;

r, q: size and surface area parameters obtained from Bondi's method. The values involved here, however, are normalized and can be obtained for each species as the sum of the group volume (R_i) and surface area (Q_i) parameter values given by Fredenslund et al (1977), that can be found in the file VLEGPHI.HLP of Program VLEGPHI (Appendix E).

For a binary system, Eq.13.12.6 becomes:

$$\ln\gamma_1 = \ln\frac{\Phi_1}{x_1} + \frac{z}{2}q_1\ln\frac{\Theta_1}{\Phi_1} + \Phi_2\left(l_1 - \frac{r_1}{r_2}l_2\right) - q_1\ln(\Theta_1 + \Theta_2\tau_{21}) + q_1\Theta_2\left(\frac{\tau_{21}}{\Theta_1 + \Theta_2\tau_{21}} - \frac{\tau_{12}}{\Theta_2 + \Theta_1\tau_{12}}\right) \tag{13.12.10}$$

The expression for $\ln\gamma_2$ is obtained by rotating the subscripts.

13.12.5 Comments

1.Number of adjustable parameters per binary system:
 * Wilson, two: $\Delta\lambda_{12} = \lambda_{12} - \lambda_{11}$; $\Delta\lambda_{21} = \lambda_{21} - \lambda_{22}$
 * NRTL, three: $\Delta g_{12} = g_{12} - g_{22}$; $\Delta g_{21} = g_{21} - g_{11}$; $\alpha_{12} = \alpha_{21}$
 * UNIQUAC, two: $\Delta u_{12} = u_{12} - u_{22}$; $\Delta u_{21} = u_{21} - u_{11}$

2.The value of alpha in the NRTL equation can be set a priori according to the rules presented by Renon and Prausnitz which, however, are sometimes difficult to apply.

3.Marina (1973) demonstrated that a single value of alpha = -1 gives equally good results, thus eliminating the uncertainty in choosing the proper value of alpha.

A negative value for alpha is not, however, compatible with the development of the NRTL equation for it must be positive, since it is proportional to $1/z$, z being the coordination number of the liquid (i.e. the number of molecules per cell). The negative value of alpha was explained, therefore, through the concept of local 'effective' mole fractions (LEMF).

4.In practice, the NRTL/LEMF equation is often used with alpha as a third adjustable parameter to provide additional flexibility in fitting binary data. Care should be exercised, however, to make sure that alpha is used to provide improved correlation of the data rather than to fit the errors in them. Values greater than 0.5 or less than -1.5 are very questionable.

5.The NRTL/LEMF and UNIQUAC equations are applicable to both, miscible and partially miscible systems; the Wilson equation, to miscible systems only.

6.In regressing binary *VLE* data to determine the corresponding parameter values, the following observations should be kept in mind:

a.For positive deviations from Raoult's law:

i.one pair of roots is obtained for the Wilson equation;

ii.the same for the NRTL equation, except for alpha = 0.47, where two sets of roots may be obtained;

iii.more than one pair of roots for the LEMF equation may be obtained.

b.For negative deviations from Raoult's law multiplicity of roots occurs with all three equations.

c.The multiplicity of roots for the UNIQUAC equation has not been studied. Since, however, setting $q = r = l$ in the UNIQUAC equation leads to the Wilson equation, the same behavior as that of the Wilson equation should be expected.

d.If multiplicity of roots occurs, the pair with the smallest in absolute value parameters seems to give the best results, and it is obtained by using zero for the initial estimate of the parameters in the regression subroutine. (For details on items (a), (b), and (d), see Silverman, 1977.)

7.Wilson and UNIQUAC parameter values can be determined using the Program GAMPAR of Appendix E.

13.12.6 Example 13.10

Using the vapor-liquid equilibrium data of Table 13.E.10 for the system benzene(1) - i-propyl alcohol(2) at 760 mm Hg (Gmehling et al, 1982) evaluate the UNIQUAC parameters.

Table 13.E.10 Information for Example 13.10

x_1	y_1	t (°C)	P (mm Hg)	γ_1	γ_2
0.053	0.140	79.9	760.0	2.7187	0.9944
0.084	0.208	78.5	760.0	2.6493	1.0009
0.126	0.276	77.1	760.0	2.4376	1.0145
0.199	0.371	75.3	760.0	2.1834	1.0356
0.240	0.410	74.4	760.0	2.0540	1.0628
0.291	0.451	73.6	760.0	1.9074	1.0965
0.357	0.493	73.0	760.0	1.7291	1.1458
0.440	0.535	72.4	760.0	1.5492	1.2386
0.556	0.583	72.2	760.0	1.3424	1.4152
0.624	0.612	72.0	760.0	1.2625	1.5698
0.685	0.638	72.1	760.0	1.1944	1.7429
0.762	0.673	72.4	760.0	1.1210	2.0614
0.887	0.760	73.8	760.0	1.0393	3.0212
0.972	0.901	77.5	760.0	1.0025	4.3630

Using the Program GAMMA, Appendix E, the activity coefficient values presented also in Table 13.E.10 are obtained. Use of these values into Program GAMPAR also of Appendix E for the evaluation of the UNIQUAC parameters requires values for the parameters r and q. From file VLEGPHI.HLP: $r(1)$ = 3.19, $q(1)$ = 2.40; $r(2)$ = 2.78, $q(2)$ = 2.51. Use of these q and r values with Program GAMPAR gives:

$$\Delta u_{12}/R = 231.5; \quad \Delta u_{21}/R = -10.6$$

both in K, with average absolute errors of 1.0% for γ_1 and 1.1% for γ_2.

Comments

1.Evaluation of r and q for the mixture components.

* Benzene: 6 ACH groups, which is group No 9.
  Thus r = 6x0.5313 = 3.1878; q = 6x0.400 = 2.400
* i-Propyl alcohol: 1 $CHOHCH_3$ group (No 15) plus 1 CH_3 group (No 1).
  Thus r = 1.8780 + 0.9011 = 2.7791; q = 1.66 + 0.848 = 2.508

2.For improved correlation results in systems containing alcohols, Anderson and Prausnitz (1978) recommend special q values for them, equal to 0.89 for i-propyl alcohol, which must be used in the residual part only. (See Problem 13.70).

13.13 Advantages of the Local Composition Models

Activity coefficient models are judged with respect to the:

* quality of binary vapor-liquid equilibrium data correlation,
* sensitivity of parameters to temperature, and
* prediction of multicomponent vapor-liquid equilibrium behavior from binary data alone.

13.13.1 Correlation of Binary Data

Local composition models provide good correlation of binary vapor-liquid equilibrium data similar to, or somewhat better than, that of the Wohl-type models.

Typical results are presented in Table 13.3 for the average absolute deviations in P (ΔP) and y (Δy) - obtained by fitting all the experimental data with the two types of models - for the system n-butanol - toluene (Gmehling et al, 1982).

Table 13.3 Quality of Data Representation for the System n-Butanol - Toluene at 70.25°C with Different Equations

Equation	ΔP (mm Hg)	$\Delta y_1 \cdot 1000$
Margules	1.9	9.0
van Laar	1.5	7.7
Wilson	1.6	5.5
NRTL (α=0.2999)	1.6	7.4
UNIQUAC	1.6	7.3

13.13.2 Temperature Dependency of the Parameters

The parameters in the local composition models can be considered temperature independent, at least for modest temperature ranges of about 30-40°C, without a significant loss of accuracy in the obtained activity coefficient values; those of the Wohl-type models, however, cannot. This is extremely important for distillation calculations, where a single set of parameters can successfully describe the *VLE* behavior for the whole temperature range, from the top to the bottom of the column.

Actually the parameters are functions of temperature as demonstrated in Figure 13.10 for the system ethanol-toluene with the Wilson equation. Here, the binary parameters have been evaluated by regressing isothermal data at each temperature.

Table 13.4 Correlation of Binary Isobaric Data (1)

System	Temp. Range (K)	rms dev. in $y \cdot 1000$ Wilson A	Wilson B	NRTL A	NRTL B
Me Acetate-Benzene	331-350	9	6	10	9
Me Acetate-Cyclohexane	329-348	8	8	9	9
Acetone-Water	331-362	24	17	20	20
Acetonitrile-Water	350-364	21	21	20	19
Acetone-Acetonitrile	331-352	13	13	12	14

(1) Cukor and Prausnitz (1969)

A: T independent parameters

B: Parameters are linear functions of temperature

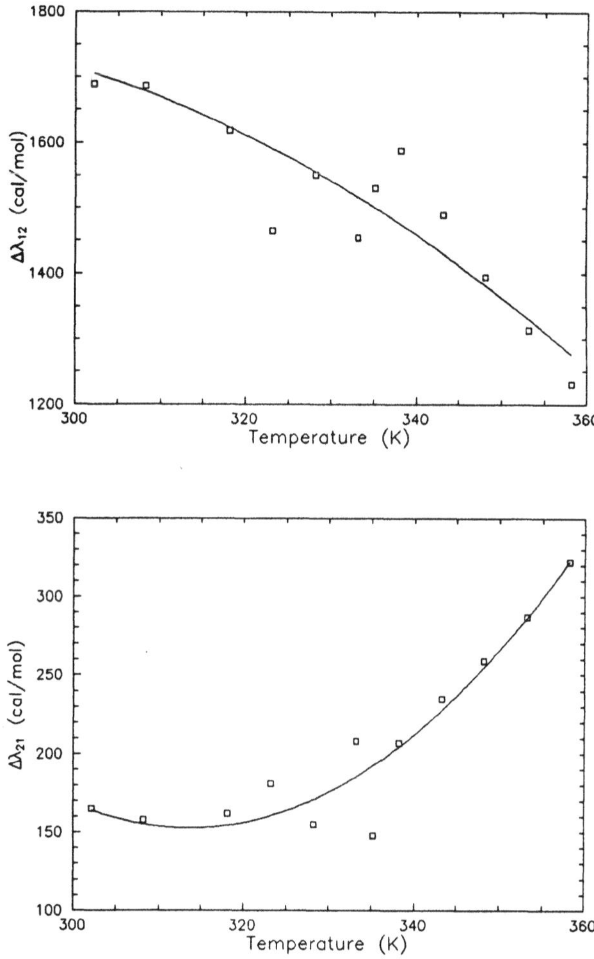

Figure 13.10 Temperature dependency of the Wilson parameters for the system ethanol(1)-toluene(2). Lines for identification only.

In spite of this variation, probably due to compensating effects, the assumption of temperature independency provides good results as suggested in Table 13.4. Notice that use of temperature-dependent parameters does not provide a significant improvement in the correlation of the binary data.

Note, finally, that even though the G^E model parameters are assumed to be temperature independent, the calculated activity coefficients are not, since the parameters are divided by the term (RT).

13.13.3 Prediction of Multicomponent *VLE* Behavior

Local composition models, as suggested by the expressions for the activity coefficient in multicomponent systems, provide description of multicomponent *VLE* behavior using binary parameters only. These parameters, in turn, can be evaluated from the corresponding binary data and, consequently, multicomponent *VLE* information - that is very scarce in the literature - can be predicted. The Wohl-type models, on the other hand, require at least ternary parameters, as Eq.13.11.1 indicates.

Typical results obtained by using binary parameters are shown graphically in Figure 13.11, indicating excellent agreement between calculated and experimental vapor phase mole fractions. On the other hand, use of binary parameters only with the Wohl-type models for multicomponent prediction gives often poor results (see Orye and Prausnitz, 1965).

This predictive capability of the local composition models - considering the large number of available binary data - is extremely important in practice, because industrial applications require multicomponent data in the typical case.

We shall return to this very important subject in Section 13.17. Undergraduate students can proceed now to Section 13.16.

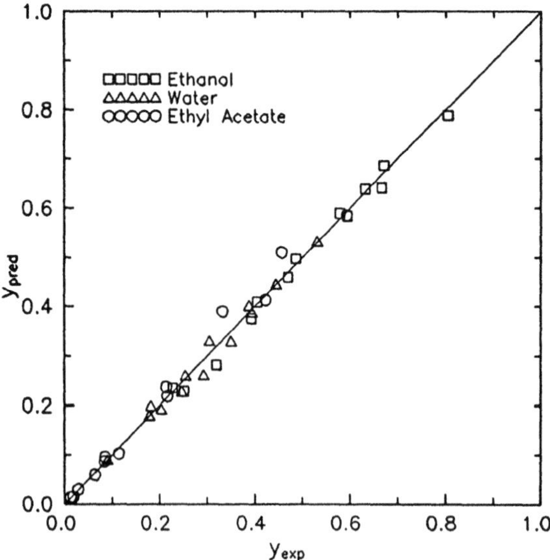

Figure 13.11 Comparison between predicted from binary data and experimental y's for the system ethanol-water-ethyl acetate with the NRTL equation at 50°C.

13.14 Single-Parameter Expressions

The simplest single-parameter expression for the molar excess Gibbs free energy of a binary system is the two-suffix Margules equation used already in Section 12.10:

$$G^E = Ax_1x_2 \tag{13.14.1}$$

where A is a function of temperature. Then from Eq.11.13.6:

$$\ln\gamma_1 = \frac{A}{RT}x_2^2; \quad \ln\gamma_2 = \frac{A}{RT}x_1^2 \tag{13.14.2}$$

Eq.13.14.1 is applicable to systems where the activity coefficients of the two components are symmetrical. It can thus correlate successfully only systems similar in structure and molar volumes, such as benzene-cyclohexane.

In 1969 this author proposed a modification of the Wilson equation, which leads to a single-parameter expression. To this purpose, λ_{kk} in Eq.13.12.2 - which according to Wilson is a constant proportional to the energy of interaction between molecules of component k - is evaluated from:

$$\lambda_{kk} = -\Delta U_k \tag{13.14.3}$$

where ΔU_k is the energy of vaporization of pure k determined at the system temperature T. It is calculated from the enthalpy of vaporization, ΔH_k, using the following thermodynamic relationship:

$$\Delta U_k = \Delta H_k - z_k RT \tag{13.14.4}$$

where z_k is the compressibility factor of saturated vapor of k at T.

Evaluation of λ_{ii} and λ_{jj} in the original expression through Eqs 13.14.3 and 13.14.4, leaves only one adjustable parameter, λ_{ij}, for the pair of components i and j. This, in turn, reduces to half the number of parameters needed to describe a system. Thus for a ternary mixture only three parameters are needed, as compared to six for the original Wilson equation.

Correlation of binary vapor-liquid equilibrium data, and prediction of multicomponent *VLE* behavior from binary ones, using this single-parameter expression is discussed by this author (1969; 1971) and by Hankinson et al (1972). Typical multicomponent predictions are shown in Figure 13.12. In general, the results obtained are reliable and close to those obtained by the two-parameter expression, even for four- and five-component systems (Hankinson et al), except for very nonideal systems.

The expression is very useful in cases where only limited data are available. An interesting such application is the case where only one infinite dilution activity coefficient for a binary system - obtained usually through gas-liquid chromatography (GLC) - is available. A combination of such GLC measurements and the single-parameter Wilson equation in predicting the solubility of SO_2 as a function of its partial pressure in several solvents is demonstrated by Bogeatzes(1973).

Other applications involve systems where very few, and of questionable accuracy, data are available. In such cases, the uncertainty in the data may be reflected more in the two-parameter expression than in the single-parameter one, which tends to 'average' it out.

In 1971, Wong and Eckert suggested that in order that the molecular meaning of the Wilson parameters be maintained:

$$\lambda_{kk} = -\frac{z}{2} \Delta U_k$$

where z, the coordination number of the liquid, is set - somewhat arbitra-

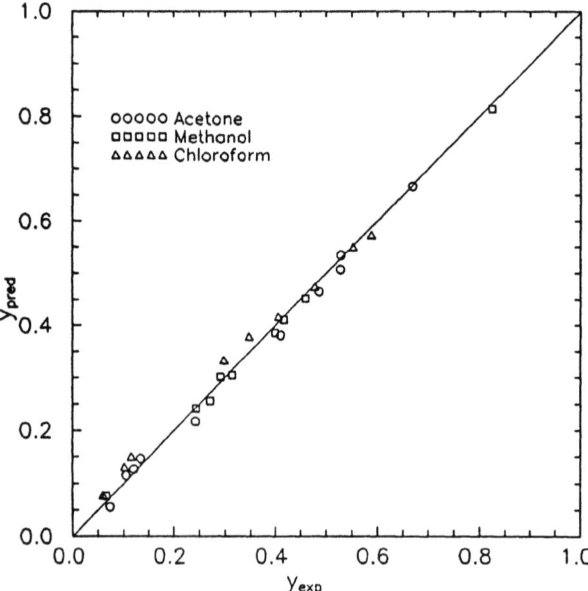

Figure 13.12 Comparison of predicted and experimental results for the *VLE* behavior of the system acetone-methanol-chloroform at 50°C using the single-parameter Wilson equation.

rily - equal to 10. This modification, however, does not affect the performance of the single-parameter Wilson equation.

Single-parameter expressions for the NRTL, LEMF, and UNIQUAC equations have also been developed. Krummins et al (1980) evaluated several of them and concluded that:

a. when all data are regressed, the single-parameter Wilson equation - with the ratio of the molar volumes arbitrarily removed - gives the best results;

b. when only one infinite dilution activity coefficient is used, the single-parameter UNIQUAC equation gives the best results;

c. reliable results are obtained up to fairly nonideal systems, say γ_i^∞ less than 10. Wong and Eckert reached the same conclusion for the single-parameter Wilson equation;

d. for strongly nonideal systems, unreliable results can be obtained.

13.15 Evaluation of Vapor Phase Composition from Liquid Composition-Total Pressure Data

In Section 13.5.4 we discussed that use of the so-called 'static' apparatus in vapor-liquid equilibrium measurements provide only the temperature, pressure, and liquid phase composition values of the equilibrium system. Let us now examine how the vapor phase composition is determined from such measurements, by considering the case of a binary system.

It is apparent that Eq. 13.5.1, obtained from the equality of fugacities, cannot provide the answer, even if we assume ideal vapor behavior. We still have:

* two equations, one for component 1 and one for 2, and
* three unknowns: the two activity coefficients and y_1.

But we have also seen in Section 13.9, that the two activity coefficients are interrelated through the Gibbs-Duhem equation, thus providing an additional equation. We examine next how this relationship is used in determining y_1.

13.15.1 The Approach

Let us assume for simplicity that the vapor phase behaves as an ideal gas. Then the total pressure P is given by:

$$P = x_1\gamma_1 P_1^s + x_2\gamma_2 P_2^s$$

If we now express the two activity coefficients through, say, the Wilson equation, we still have one equation with two unknowns: the parameters $\Delta\lambda_{12}$ and $\Delta\lambda_{21}$.

But the same parameter values must also satisfy the other T-P-x_i data as well. Thus, if we use two data points, we have a system of two equations with two unknowns. Still better, we can use all available data points and regress them for the best values of the parameters.

The method is demonstrated with a simple case in the following Example (see also Problem 13.61).

13.15.2 Example 13.11

Hermsen and Prausnitz (1963) present the P-x_i data for the system benzene(1)-cyclopentane(2) at 25°C shown in Table 13.E.11. Calculate the values of the vapor phase composition assuming, because of the low pressures involved and the similarity of the two components, ideal vapor behavior. The following Antoine constants are given:

Comp.	A	B	C
1	15.9008	2788.51	-52.36
2	15.8574	2588.48	-41.79

Again because of the similarity of the two components, consider as a first approximation the single-parameter Redlich-Kister expression, Eq. 13.14.2:

$$\ln\gamma_1 = A'x_2^2; \quad \ln\gamma_2 = A'x_1^2$$

where $A' = A/RT$. The corresponding expression for the total pressure is:

$$P = x_1 P_1^s \exp(A'x_2^2) + x_2 P_2^s \exp(A'x_1^2)$$

Values of A' for each of the experimental data points are presented in Table 13.E.11. They are close enough to suggest the validity of this one-parameter expression for this system. This is further supported by the close agreement between calculated values for P, P_{calc}, using the average value $A' = 0.457$, and the experimental ones. The average absolute deviation is only 0.29 mm Hg.

Calculation of the y_1 values is now straightforward:

$$y_1 = \frac{x_1 P_1^s \exp(A'x_2^2)}{P_{calc}}$$

and they are also presented in Table 13.E.11.

Table 13.E.11 Information and Results for Example 13.11.

x_1	P_{exp} (mm Hg)	A'	P_{calc} (mm Hg)	$P_{calc} - P_{exp}$ (mm Hg)	y_1
0.1417	294.11	0.469	293.87	-0.24	0.0643
0.2945	268.36	0.461	268.21	-0.15	0.1312
0.4362	243.29	0.458	243.25	-0.04	0.1974
0.5166	228.33	0.461	228.07	-0.26	0.2399
0.5625	218.98	0.458	218.93	-0.05	0.2669
0.8465	148.07	0.437	149.05	0.98	0.5464
	Avg.	0.457		0.29	

Comments

1. Use of a two-parameter expression follows the same principle, and the values are usually obtained by regression of the data.

2. Inclusion of the vapor phase nonideality leads to a trial-and-error calculation.

3. Hermsen and Prausnitz, for example, used a two-parameter Redlich-Kister expression, and also took into account the vapor-phase nonideality. The obtained fit of the data is, of course, somewhat better.

4. For the evaluation of vapor compositions from isothermal P-x_i data without the use of expressions for G^E, see Smith and Van Ness (1987).

13.16 Bubble and Dew Point Calculations

We have concluded, thus, our discussion of the activity coefficient and are ready to consider its use in solving vapor-liquid equilibrium problems.

The typical such problems are presented in Table 13.1 and involve - as seen in Section 13.6 - bubble point (B.P.) and dew point (D.P.) calculations. (An other common *VLE* problem, requires the so-called *flash* calculation. It involves often high pressures and is considered, therefore, in the next Chapter.)

We discussed bubble and dew point calculations for ideal solutions in Examples 13.2 through 13.7. And in Examples 13.9 and 13.11 we carried out B.P. pressure calculations for nonideal solutions assuming, however, vapor phase ideality.

We proceed now with a discussion of the more general case: nonideal liquid and nonideal vapor phases. Notice that the type of calculation involved depends on the information given, as described in Table 13.1.

The approach to calculations in this more general case is the same with that described for ideal solutions. The presence of nonideality in the two phases just increases the complexity. The principle is the same.

We will start with a presentation of the equations involved in each case and proceed with a brief discussion of the calculational schemes used to obtain numerical answers.

13.16.1 Bubble Point

a. The Equations

The total pressure over an equilibrium solution is, by definition, the sum of the partial pressures of the components involved:

$$P = \Sigma \bar{P}_i = \Sigma y_i P \tag{13.16.1}$$

Expressing $(y_i P)$ through Eq. 13.5.2, we obtain:

$$P = \sum \frac{x_i \gamma_i P_i^s}{F_i} \tag{13.16.2}$$

where F_i is given by Eq. 13.5.3. Also from Eq. 13.5.2, but solving for y_i:

$$y_i = \frac{x_i \gamma_i P_i^s}{P F_i} \tag{13.16.3}$$

where P is given by Eq. 13.16.2.

b. Calculations

Consider first the case of the B.P. *pressure* calculation: the temperature and liquid phase mole fractions are given, and we want to calculate the total pressure and vapor phase composition of the equilibrium system. To this purpose, we will assume that:

1. At the low pressures involved here, all vapor phase fugacity coefficients can be calculated from the virial equation truncated after B, as discussed in Section 13.5.2.
2. Second virial coefficients, B_{ii} and B_{ij}, are available, or can be estimated from some empirical correlation, typically that of Tsonopoulos or that of Hayden and O'Connell.
3. Values for the Poynting effect can be determined from available or estimated liquid molar volume data.
4. Parameter values for an expression describing the vapor pressure of all components involved, such as the Antoine equation, are available.
5. Binary parameter values for the Wilson, NRTL/LEMF, or UNIQUAC equations are available.

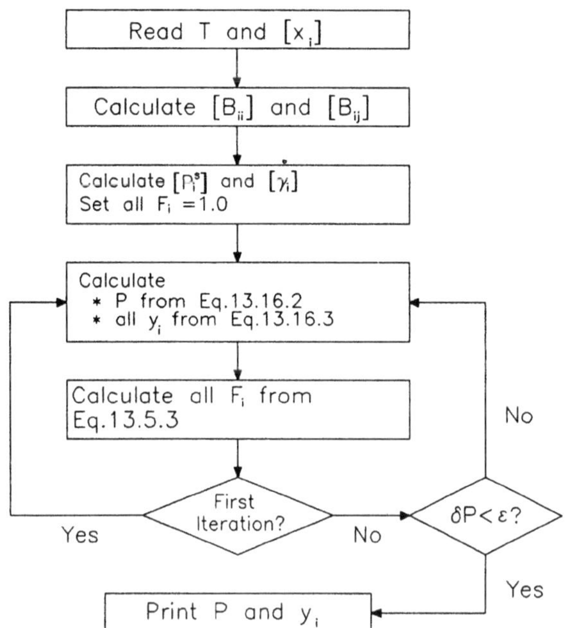

Figure 13.13 Flow diagram for a B.P. pressure calculation.

A flow diagram for a B.P. pressure calculation is presented in Figure 13.13. δP is the difference between two consecutive iterations and ϵ is the 'convergence' criterion, i.e. the value of δP below which we accept that the iterative calculation has converged.

A bubble point pressure computer program using the Wilson and UNIQUAC equations, VLEGPHI, is presented in Appendix E.

The scheme for a B.P. *temperature* calculation follows the same approach. It is, however, slightly more complex because a method for generating new temperature estimates is needed.

13.16.2 Dew Point

a. The Equations

In this case, we cannot use Eq.13.16.2 to calculate the total pressure because the x_i's are not given. Instead, we start with the observation that the sum of the calculated values must satisfy the relationship:

$$\Sigma x_i = 1.0 \qquad (13.16.4)$$

We next express x_i through Eq.13.5.2:

$$x_i = \frac{y_i P}{\gamma_i P_i^s} F_i \qquad (13.16.5)$$

Introduction into Eq.13.16.4 yields, after rearrangement:

$$P = 1/\sum \frac{y_i F_i}{\gamma_i P_i^s} \qquad (13.16.6)$$

b. Calculations

Consider first the case of a D.P. *pressure* calculation: the temperature and the vapor phase mole fractions are given, and we want to calculate the total pressure and the liquid phase compositions of the equilibrium system.

Using the same set of assumptions as with the case of the B.P. calculation, we arrive at the flow diagram shown in Figure 13.14. The approach

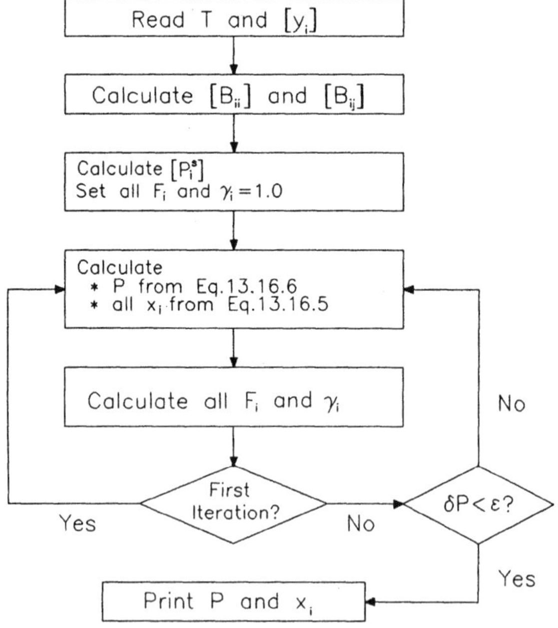

Figure 13.14 Flow diagram for a D.P. pressure calculation.

for a D.P. *temperature* calculation is similar in principle, but again slightly more complex for the same reason as with the B.P. temperature case.

Details for bubble and dew point calculations are given by Van Ness and Abbott (1982) and by Prausnitz et al (1980), who also present the appropriate computer subroutines. (The last book is a 'must' for anyone involved in such calculations). They can be also easily developed using the subroutines of Appendix E.

13.17 Prediction of Multicomponent *VLE* Behavior from Binary Data

While the typical industrial application calls for multicomponent *VLE* data, a survey of the literature reveals that well over 90% of the available data are for binary systems.

On the other hand it was pointed out in Section 13.13, that one of the main advantages of the local composition models is exactly their ability to predict multicomponent behavior from binary data alone. And this represents a commonly used approach in industrial applications.

We explore, therefore, next the factors that effect the accuracy of the obtained multicomponent predictions, more specifically:

a. the method for evaluating the necessary binary parameters;
b. the quality of the binary data used to this purpose;
c. their temperature, in relationship to that of the multicomponent system; and
d. the system type, in terms of the number and molecular nature of its components.

13.17.1 The Effect of the Method for Evaluating Binary Parameters

Binary parameters in the expressions for the activity coefficient (Wilson, NRTL, etc.) are evaluated, in the typical case, by regression of the available experimental *VLE* data. A computer subroutine is used for this purpose, which varies the parameter values until a minimum is obtained for some objective function Q, such as:

$$Q = \sum \left(\left(\frac{\gamma_{1,calc} - \gamma_{1,exp}}{\gamma_{1,exp}} \right)_i^2 + \left(\frac{\gamma_{2,calc} - \gamma_{2,exp}}{\gamma_{2,exp}} \right)_i^2 \right) \tag{13.17.1}$$

or:

$$Q = \sum \left((y_{1,calc} - y_{1,exp})_i^2 + (y_{2,calc} - y_{2,exp})_i^2 + \left(\frac{P_{calc} - P_{exp}}{P_{exp}} \right)_i^2 \right) \tag{13.17.2}$$

where 'exp' and 'calc' refer to experimental and calculated values, and the summation is over all data points. The calculated values are obtained from the appropriate expression for the activity coefficient (Wilson, NRTL, etc.) and, for obvious reasons, these functions are also referred to as 'minimization' functions.

Silverman considered ten such minimization functions (he refers to them as models), that include those most often used in the literature. Eqs 13.17.1 and 13.17.2 represent two of his models, number 4 and 5 respectively. He found that, as demonstrated in Table 13.5 for four binary systems with eight models, large variations in the obtained parameter values are possible.

How can such different parameter values provide a good description of the experimental data? The answer is suggested by the 'contour' map presented in Figure 13.15. Here, the total error Q in γ_1 and γ_2, as defined by Eq.13.17.1, for a binary system is shown as a function of the parameter values for the Wilson equation. The lines represent the loci of constant Q values. Notice, for example, that two completely different pairs of parameters [(-300,1125) and (550,690)], all in cal/mol, give the same Q value of 0.05.

This behavior is typical of all the equations that involve energy differences between like and unlike pairs of molecules for, as their single-parameter version suggests, the parameters are intercorrelatable.

Silverman examined next the effect of this variation in parameter values from the different models on the quality of the multicomponent predictions. To this purpose, he used experimental data for 247 binaries in the prediction of the *VLE* behavior of 73 multicomponent systems with the Wilson equation.

His results are summarized in Table 13.6 in terms of the average absolute deviation in predicted vapor phase compositions and total pressures for the 73 systems. The poor results for model 1, which uses the infinite dilution activity coefficients, reflect the uncertainty in evaluating them, especially when low composition data are not available. The remaining

Table 13.5 Wilson Parameter Values (cal/mol) Obtained with Different Minimization Models

Model No	Acetone-Acetonitrile		Acetone-Chloroform		Hexane-Benzene		Heptane-Toluene	
	$\Delta\lambda_{12}$	$\Delta\lambda_{21}$	$\Delta\lambda_{12}$	$\Delta\lambda_{21}$	$\Delta\lambda_{12}$	$\Delta\lambda_{21}$	$\Delta\lambda_{12}$	$\Delta\lambda_{21}$
1	-86	124	-34	-374	222	131	126	135
2	44	25	32	-424	163	194	-19	255
3	2	55	-13	-395	244	132	67	185
4	21	41	-34	-376	254	126	51	194
5	275	-118	58	-448	242	136	-21	258
6	303	-333	74	-460	297	105	-37	271
7	316	-141	31	-457	339	76	-38	272
8	227	-93	87	-464	98	233	8	234

models provide good predictions of multicomponent *VLE* behavior.

Best results are obtained with model 5 (Eq.13.17.2): the minimization of the difference between the calculated and experimental vapor phase

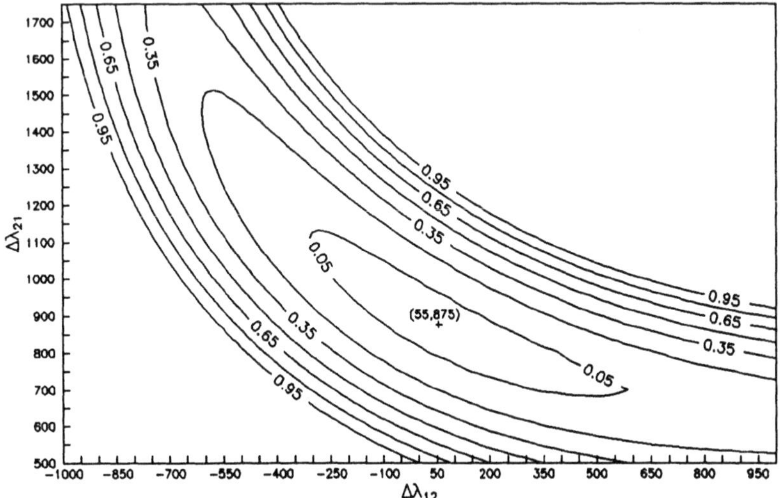

Figure 13.15 Constant Q (Eq.13.17.1) lines obtained with the Wilson equation for the system ethanol-water at 25°C

Table 13.6 Overall Average Absolute Deviation in Vapor Phase Composition (Δy) and Total Pressure (ΔP)

Model	$\Delta y \cdot 1000$	ΔP (mm Hg)
1	19.2	33.9
2	12.3	16.5
3	11.4	16.8
4	10.9	15.3
5	10.6	14.4
6	11.6	21.0
7	12.5	18.1
8	10.7	14.5
9	10.7	14.5
10	11.5	17.2

compositions plus the relative difference between calculated and experimental pressures. The required bubble point calculation, however, is computer-time consuming and can be avoided by using model 4 (Eq. 13.17.1), where the relative difference between calculated and experimental activity coefficients is minimized and gives practically the same predictions.

13.17.2 The Effect of the Quality of the Binary Data

Silverman examined also the effect of the quality of the binary data used on the accuracy of the multicomponent predictions.

The quality of the 247 binaries, as evaluated by the area test, is depicted in Table 13.2 and indicates that:

* only 60% of them can be considered of good quality;
* 19% have *CI* or (*CI-J*) values in the range 5 - 15 and 10-20 respectively and are classified as being of 'fair' quality;
* the remaining 21%, are of 'poor' or 'unacceptable' (the area test was not possible) quality.

All 247 systems were used, however, in his predictions.

The results are summarized in Table 13.7 and suggest no significant impact on the accuracy of the vapor phase composition predictions, until one of the binaries involved is of unacceptable quality. The overall deviation of 0.009 in vapor phase composition for groups A, B, and C, that include 47 multicomponent systems with no unacceptable binaries,

Table 13.7 Effect of the Quality of the Binary Data on the Prediction of Multicomponent Vapor-Liquid Equilibrium (Model 5)

Group	Number of systems	Quality of Binary Data	Overall Avg. Abs. Dev. in $y \cdot 1000$
A	13	All good	10.2
B	11	Mixture of good and fair	9.1
C	23	Mixture of good and fair and one poor	9.2
D	26	One is unacceptable	12.5

Table 13.8 Prediction of Isothermal Ternary *VLE* from Binary Data (15 Systems): The Effect of Binary Data Accuracy

	Overall Avg. Abs. Dev. in $y \cdot 1000$		
Model	A	B	C
Wilson	7.8	7.6	9.1
NRTL	9.5	8.6	11.1
NRTL/LEMF (adjustable alpha)	10.4	10.9	11.2
LEMF	10.8	9.9	11.0
UNIQUAC	10.9	10.7	12.3

A: All binary data points

B: Eliminate data points where $|y_{calc} - y_{exp}| > (0.05) \times (y_{exp})$

C: Eliminate data points where $|y_{calc} - y_{exp}| > 0.01$

y_{calc} is calculated from the experimental x values using the method of Fredenslund et al.

can be considered very good, especially if the uncertainty in the multicomponent data is taken into account.

It appears that the aforementioned interdependency between the Wilson equation parameters tends to remove some of the uncertainty in the binary data. Even for group D, the overall deviation in y is 0.013 and average deviations of over 0.025 were observed only twice, with the worst case being 0.032.

Further support for these findings is provided by Table 13.8. Here, before the binary data were regressed, they were 'screened' according to

Table 13.9 Prediction of Ternary *VLE* from Binary Data with the Wilson Equation: The Effect of Temperature of the Binary Data

System	Parameters evaluated at	Avg. Abs. Dev. in Predicted $y \cdot 1000$ at			
Hexane-cyclohexane-benzene	70°C	10°C	15°C	25°C	70°C
		6.6	6.4	5.0	2.7
Et acetate-ethanol-water	70°C	55°C	70°C		
		18.8	19.0		
Methanol-carbon tet. -benzene	55°C	34.7°C	55°C		
		2.7	2.0		

the procedure of Fredenslund et al, whereby vapor phase compositions are calculated from the experimental x's using a method based on the Gibbs-Duhem equation. These calculated y's are then used to evaluate the accuracy of the experimental ones. The results suggest that elimination of poor y data in the binary systems does not necessarily improve the obtained predictions.

13.17.3 The Effect of Temperature

Since distillation calculations require isobaric data, the temperature range of the multicomponent predictions may often be different from that of the binary systems involved.

We have already seen in Section 13.13.2 that, even though the parameters in the local composition models are functions of temperature (Figure 13.10), good correlation of binary data is obtained by considering them to be temperature-independent (Table 13.4).

This assumption can be used, also successfully, in multicomponent predictions as suggested by the quality of the results presented in Table 13.9. We should keep in mind, however, that the temperature range should be reasonable, typically below 30-40°C.

13.17.4 The Effect of System Type

Table 13.10 summarizes the prediction results of Silverman in terms of the types of multicomponent systems involved:

Table 13.10 Multicomponent Prediction Results as a Function of the System Type

Number	Type	$\Delta y \cdot 1000$
9	Systems with four(7) and five(2) components	7.6
10	Systems including hydrocarbons only	5.4
36	Systems containing alcohols	11.6
13	Systems containing water	14.5
73	All systems	10.6

* number of components, and
* degree of nonideality.

Notice that:

1.In support of the local composition models' contention that only binary parameters are needed for multicomponent predictions, the nine four- and five-component systems are predicted as well as - actually better than - the rest.

2.As expected, the degree of nonideality effects the obtained accuracy: systems containing hydrocarbons only give the best results, while those containing water, the poorest. This also explains the better prediction results for the four- and five-component systems, mostly hydrocarbon ones.

13.17.5 Example 13.12

Evaluate the performance of the UNIQUAC model in the prediction of the vapor-liquid equilibrium behavior for the system carbon tetrachloride(1)-benzene(2)-i-propyl alcohol(3) at 1.013 bar by comparing your results to the experimental data presented in Table 13.E.12 (Hala et al, 1968, pp.493-494). The required UNIQUAC binary parameters are included in the same Table and were obtained by regressing the corresponding *VLE* data: at 343.15 K for the 1-2 and 1-3 binaries (Hirata et al, 1975); and at 760 mm Hg for the 2-3 one (see Example 13.10).

The results, obtained with the Program VLEGPHI of Appendix E, are presented in the same Table. Notice the good quality of the predicted values with overall average absolute deviations of: P = 1.6%; Δy_1 = 0.004; Δy_3 = 0.007.

Table 13.E.12 Information and Results for Example 13.12

Point	T, K	x_1	x_2	$\Delta P\%$	$y_{1,exp}$	Δy_1	$y_{2,exp}$	Δy_2
1	346.6	0.032	0.362	-4.18	0.054	0.004	0.470	-0.005
2	345.1	0.071	0.430	-1.43	0.100	0.004	0.472	-0.011
3	344.4	0.301	0.054	-0.81	0.471	0.005	0.071	-0.003
4	347.0	0.429	0.528	2.03	0.416	-0.001	0.437	-0.018
5	348.6	0.424	0.555	1.21	0.420	-0.010	0.487	-0.011
6	346.0	0.092	0.734	-1.53	0.096	0.002	0.627	0.013
7	346.6	0.058	0.224	-0.79	0.112	0.004	0.348	-0.006
8	344.5	0.190	0.651	2.26	0.190	0.003	0.545	0.007
9	344.8	0.151	0.373	-1.90	0.194	0.001	0.404	0.001
10	347.8	0.106	0.143	-4.58	0.203	-0.002	0.220	-0.017
11	347.0	0.270	0.664	0.64	0.269	0.003	0.550	-0.011
12	344.6	0.256	0.425	-3.00	0.276	0.006	0.384	0.001
13	344.5	0.203	0.172	0.20	0.315	0.002	0.223	-0.006
14	346.3	0.386	0.552	1.86	0.369	-0.002	0.456	-0.009
15	343.5	0.302	0.182	0.28	0.391	0.000	0.203	-0.002
16	346.6	0.490	0.448	-0.69	0.459	-0.001	0.373	-0.003
17	343.8	0.392	0.213	-2.29	0.433	0.002	0.209	0.003
18	344.0	0.594	0.227	-1.66	0.542	0.009	0.188	0.002
19	346.0	0.713	0.236	1.15	0.641	-0.011	0.201	-0.001
20	346.4	0.830	0.125	0.10	0.741	-0.013	0.106	-0.003

$r(1) = 3.39$ $r(2) = 3.19$ $r(3) = 2.78$
$q(1) = 2.91$ $q(2) = 2.40$ $q(3) = 2.51$
$\Delta u_{12}/R = 69.99$ $\Delta u_{23}/R = 231.52$ $\Delta u_{31}/R = -76.35$, all in K
$\Delta u_{21}/R = -50.65$ $\Delta u_{32}/R = -10.61$ $\Delta u_{13}/R = 329.55$, all in K

13.18 Estimation of Vapor-Liquid Equilibrium

As expected, often not all the necessary for design purposes *VLE* data are available. Consider, for example, the case where vapor-liquid equilibrium information is needed for the design of a fractionation system to recover component A from a mixture containing, in addition, components B and C.

If experimental *VLE* data for the three binary systems: A-B, B-C and C-A are available, they can be used - as we have seen - for the prediction of the ternary behavior. Reliable results should be expected, and the availability of compilations of a large number of binary data (Hala et al;

Hirata et al; and especially the series of volumes by Gmehling and co-workers) makes this approach very attractive.

In the typical case, however, some - or even all - of the necessary binary data may not be available. In such cases we resort to estimation techniques.

13.18.1 The Group-Contribution Concept

Estimation techniques rely on the group-contribution approach that has found success with pure component properties (critical constants, heat capacities, etc.). The basic assumption behind this approach is that the property of a fluid can be approximated by the sum of the contributions of the functional groups of its molecules.

Consider, for example, the system n-hexane - acetonitrile. The deviation from Raoult's law of this system - i.e. its activity coefficients - is described according to this concept, by:

a.the size and shape of the functional groups (CH_3, CH_2, and CN) making up these molecules, and

b.the differences in interaction energies between them:

* [(CH_2/CN) - (CH_2/CH_2)], and
* [(CH_2/CN) - (CN/CN)].

(The groups: CH_3 and CH_2 are assumed, for simplicity, to have the same interactions with other groups; thus the CH_3/CN interaction is equal to the CH_2/CN one).

Notice, now, that if:

* a group-contribution model for the activity coefficient, and
* experimental *VLE* data for this system,

are available, they can be used to determine the values of the two bracketed parameters; and, consequently, the *VLE* behavior of any alkane-nitrile mixture, binary or multicomponent, can be then predicted.

Several group-contribution models have been proposed through the years (Pierotti et al, 1959; Derr and Deal, 1969; Fredenslund et al, 1975 etc.). The most recent and promising one, is the UNIFAC model of Fredenslund et al.

13.18.2 The UNIFAC Model

This model is based on the UNIQUAC equation and uses the same expression for the activity coefficient, Eq.13.12.6.

The combinatorial part of the activity coefficient is also the same, Eq.

13.12.7. Only the pure component properties r_i and q_i enter into this equation and are calculated as the sum of the group volume and area parameters, R_k and Q_k (see Section 13.12.4 for the method of evaluating these properties):

$$r_i = \sum \nu_k^{(i)} R_k \quad \text{and} \quad q_i = \sum \nu_k^{(i)} Q_k \tag{13.18.1}$$

where $\nu_k^{(i)}$, always an integer, is the number of groups of type k in molecule i.

The residual part of the activity coefficient, Eq.13.12.8, is replaced by the solution-of-groups concept:

$$\ln \gamma_i^R = \sum_{all\ groups} \nu_k^{(i)} [\ln \Gamma_k - \ln \Gamma_k^{(i)}] \tag{13.18.2}$$

where Γ_k is the group residual activity coefficient, and $\Gamma_k^{(i)}$ is the residual activity coefficient of group k in a reference solution containing only molecules of type i. In Eq.13.18.2 the term $\ln \Gamma_k^{(i)}$ is necessary to attain the normalization that activity coefficient γ_i becomes unity as $x_i = 1$. The activity coefficient for group k in molecule i depends on the molecule i in which k is situated. Consider, for example, nitriles. $\Gamma_k^{(i)}$ for the CN group in acetonitrile refers to a solution containing fifty group percent CN and fifty group percent CH_3 at the temperature of the mixture, while $\Gamma_k^{(i)}$ for the CN group in n-butyl nitrile refers to a solution containing twenty-five group percent CN, fifty group percent CH_2, and twenty-five group percent CH_3.

The group activity coefficient Γ_k is found from an expression similar to Eq.13.12.8:

$$\ln \Gamma_k = Q_k \left[1 - \ln \left(\sum \Theta_m \Psi_{mk} \right) - \sum \frac{\Theta_m \Psi_{km}}{\sum \Theta_n \Psi_{nm}} \right] \tag{13.18.3}$$

Eq.13.18.3 also holds for $\ln \Gamma_k^{(i)}$. Here, Θ_m is the area fraction of group m, and the sums are over all different groups. Θ_m is calculated in a manner similar to that for Θ_i, Eq.13.12.9:

$$\Theta_m = \frac{Q_m X_m}{\sum Q_n X_n} \tag{13.18.4}$$

where X_m is the mole fraction of group m in the mixture. The group interaction parameter Ψ_{mn} is given by:

$$\Psi_{mn} = \exp \frac{-(U_{mn} - U_{nn})}{RT} = \exp \frac{-a_{mn}}{T} \tag{13.18.5}$$

where U_{mn} is a measure of the energy of interaction between groups m and n.

Values for the group interaction parameters, two per pair, obtained by regressing available *VLE* data are given in the original article, and an update by Gmehling et al (1982a).

A computer program for the estimation of activity coefficients with the UNIFAC model is included in the book of Fredenslund et al. *VLE* calculations with the UNIFAC model can be also carried out using the Program VLEGPHI of Appendix E.

13.18.3 Comments

1.The UNIFAC model is a continuously evolving concept. For the last versions, referred to as Modified UNIFAC, see Gmehling and Weidlich (1986); and Larsen et al (1987).

2.The main limitations of the UNIFAC model are:

a.Due to lack of the necessary experimental data, values for several interaction parameters are not available. The list increases, however, as new data become available.

b.Isomeric differences are not accounted for sufficiently. This is demonstrated, for example, with the results presented in Figure 13.16, using values from the original paper of Fredenslund et al.

c.A group contribution model, by its very nature, cannot provide accuracy comparable with that of experimental data. This becomes more pronounced, of course, for mixtures involving complex molecules.

3.The UNIFAC method should be used, therefore, for preliminary calculations only; its use for final design, especially for difficult separations, is not recommended.

4.In case most of the necessary binary data for a given multicomponent prediction are available, the missing ones can be estimated with UNIFAC and then used in the prediction scheme. These binaries, however, should not include the key components of the separation.

5.For an extension of the UNIFAC model to:

a.Polymer-solvent systems, see Oishi and Prausnitz; Iwai and Arai; and Elbro et al.

b.Gas-liquid systems, see Antunes; Nocon et al; and Sander et al (1983).

c.Heats of mixing, see: Dang (1986); Gmehling and Weidlich; and Larsen et al.

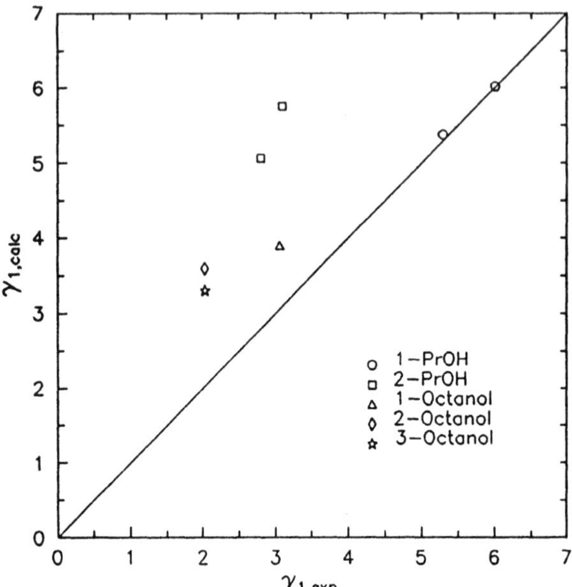

Figure 13.16 UNIFAC limitations: isomers. Experimental and calculated γ's for alcohols(1) in n-heptane(2) at $x_1 = 0.1$.

13.19 Concluding Remarks

1.The main quantity involved in low pressure vapor-liquid equilibrium is the activity coefficient, the measure of the nonideality in the liquid phase. (The same way that the fugacity coefficient reflects the nonideality in the vapor phase. Notice, however, that ideality refers to Raoult's law in the liquid phase; and to ideal gas law, in the vapor phase.)

2.The activity coefficient is a function of temperature and composition. The pressure effect is negligible at low pressures.

3.Experimental *VLE* measurements are time-consuming and costly and, consequently, prediction schemes are used whenever possible.

4.The local composition models: Wilson, NRTL/LEMF, and UNIQUAC provide successful correlation of binary data.

5.The same models provide reliable predictions of multicomponent *VLE* behavior from binary data. The accuracy becomes somewhat poorer, however, for srongly nonideal systems.

6. The obtained results are not very sensitive to:

* the quality of the data of the binary systems involved;
* their temperature, as compared to that of the predicted system, and
* the minimization function used in determining the parameter values of the model from them, but Eqs 13.17.1 and 13.17.2 are recommended.

7. The Wilson equation is recommended for completely miscible systems because of its simplicity. If a combination of miscible and partially miscible systems is involved, the UNIQUAC equation should be used.

8. In the absence of sufficient data, estimation of vapor-liquid equilibrium - mostly for preliminary design purposes, however - is often possible with the UNIFAC model.

9. The typical vapor-liquid equilibrium problem is solved through bubble or dew point calculations.

10. The number of trays and/or the reflux ratio required for a given separation depend on the relative volatility of the light (l) and heavy (k) 'key' components of the mixture:

$$\alpha_{lk} = \frac{y_l/x_l}{y_k/x_k} = \frac{\gamma_l P_l^s F_k}{\gamma_k P_k^s F_l}$$

and are thus sensitive to the accuracy of the *VLE* information, mainly activity coefficients and vapor pressures, used.

11. This sensitivity increases with decreasing values of α, as shown in Figure 13.17, where N is the number of trays and R the reflux ratio required for the desired separation. The results are based on binary mixtures using a saturated liquid feed ($x_l/x_k = 0.65/0.35$) and products of 99% purity.

Notice that in this case, if the errors in the *VLE* information cause a 2% reduction in the correct value of α, then for α equal to - say - 1.2, the number of trays is overestimated by about 40% (holding R constant); and the reflux ratio by about 60% (holding N constant). For more details on this subject, see Nelson et al (1983).

12. An excellent discussion of low pressure vapor-liquid equilibria is given by Prausnitz et al (1980, 1986). These books are a 'must' for anyone interested in the subject.

13. Compilations of *VLE* data are presented by Hala et al; by Hirata et al; and the most comprehensive one by Gmehling and co-workers.

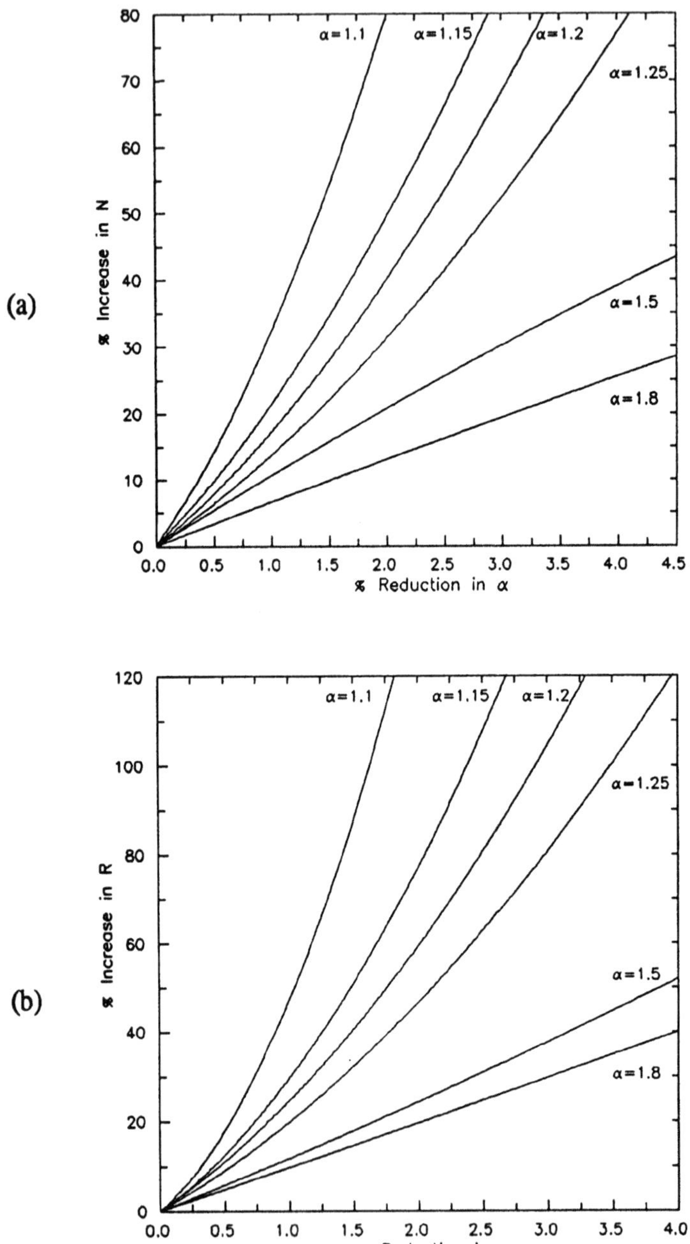

Figure 13.17 The effect of errors in *VLE* information, as reflected on the value of the relative volatility of the key components, on the number of trays (N) and reflux ratio (R) of a distillation column: (a) R = constant; (b) N = constant.

References

Abrams, D., Prausnitz, J.M., 1975. *AIChE J.*, **21**, 116.

Anderson, T.F., Prausnitz, J.M., 1978. *Ind. Eng. Chem. Process Des. Dev.*, **17**, 552.

Antunes, C., Tassios, D., 1983. *Ind. Eng. Chem. Process Des. Dev.*, **22**, 457.

Bogeatzes, A., Tassios, D., 1973. *Ind. Eng. Chem. Process Des. Dev.*, **12**, 274.

Bondi, A., 1968. *Physical Properties of Molecular Crystals, Liquids and Glasses*, Wiley, New York.

Cukor, P.M., Prausnitz, J.M., 1969. *I. Chem. Eng. Symp. Ser.* No 32 (Inst. Chem. Engrs, London), 3:88.

Dang, D., Tassios, D., 1986. *Ind. Eng. Chem. Process Des. Dev.*, **25**, 22.

Derr, E.L., Deal, C.H., 1969. *I. Chem. Eng. Symp. Ser.* No 32 (Inst. Chem. Engrs, London), 3:40.

Elbro, H.S., Fredenslund, Aa., Rasmussen, P., 1990. *Macromolecules*, **23**, 4707.

Fredenslund, Aa., Jones, R.L., Prausnitz, J.M., 1975. *AIChE J.*, **21**, 1086.

Fredenslund, Aa., Gmehling, J., Rasmussen, P., 1977. *Vapor-LiquidEquilibrium Using UNIFAC*, Elsevier, Amsterdam.

Gmehling, J., Weidlich, U., 1986. *Fluid Phase Equilibria*, **27**, 171.

Gmehling, J., and Co-workers, 1979; 1980; 1981; 1982; and ensuing years. *Vapor-Liquid Equilibrium Data Collection*, DECHEMA Chemistry Data Series, DECHEMA, Frankfurt.

Gmehling, J., Rasmussen, P., Fredenslund, Aa., 1982a. *Ind. Eng. Chem. Process Des. Dev.*, **21**, 118.

Hala, E., Pick, J., Fried, V., Vilim, O., 1967. *Vapor-Liquid Equlibrium*, Pergamon Press, New York.

Hala, E., Wichterle, I., Polak, J., Boublik, T., 1968. *Vapor-Liquid Equilibrium Data at Normal Pressures*, Pergamon Press, New York.

Hankinson, R., Langfitt, B., Tassios, D., 1972. *Can. J. Chem. Eng.*, **50**, 511.

Hayden, J.G., O'Connell, J.P., 1975. *Ind. Eng. Chem. Process Des. Dev.*, **14**, 209.

Herington, E.F., 1951. *J. Inst. Petrol.*, **37**, 457.

Hermsen, R.W., Prausnitz, J.M., 1963. *Chem. Eng. Sci.*, **61**, 1338.

Hirata, M., Ohe, S., Nagahama, K., 1975. *Computer Aided Data Book of Vapor-Liquid Equilibria*, Elsevier, Amsterdam.

Iwai, J., Arai, Y., 1989. *J. Chem. Eng. Japan*, **22**, 155.

Krummins, A., Rastogi, A., Russac, M., Tassios, D., 1980. *Can. J. Chem. Eng.*, **58**, 663.

Larsen, B.L., Rasmussen, P., Fredenslund, Aa., 1987. *Ind. Eng. Chem. Res.*, **26**, 2274.

Marina, J., Tassios, D., 1973. *Ind. Eng. Chem. Process Des. Dev.*, **12**, 67.

Maurer, G., Prausnitz, J.M., 1978. *Fluid Phase Equilibria*, **2**, 91.

Meili, A., Stuecheli, A., 1987, Feb. 16. *Chemical Engineering*, p.133.

Nelson, A.R., Olson, J.H., Sandler, S.I., 1983. *Ind. Eng. Chem. Process Des. Dev.*, **22**, 547.

Nocon, G., Weidlich, U., Gmehling, J., Onken, U., 1983. *Ber. Bunsenges. Phys. Chem.*, **87**, 17.

Oishi, T., Prausnitz, J.M., 1978. *Ind. Eng. Chem. Process Des. Dev.*, **17**, 333.

Orye, K.V., Prausnitz, J.M., 1965. *Ind. Eng. Chem.*, **57**, 5, 18.

Pierotti, G.I., Deal, C.H., Derr, E.L., 1959. *Ind. Eng. Chem.*, **51**, 95.

Prausnitz, J.M., 1969. *Molecular Thermodynamics of Fluid-Phase Equilibria*, Prentice-Hall, Englewood Cliffs, New Jersey.

Prausnitz, J.M., Lichtenthaler, R.N., de Azevedo, E.G., 1986. *Molecular Thermodynamics of Fluid-Phase Equilibria*, 2nd Ed., Prentice-Hall, Englewood Cliffs, New Jersey.

Prausnitz, J.M., Anderson, T.F., Grens, E.A., Eckert, C.A., Hsieh, R., O'Connell, J.P., 1980. *Computer Calculations for Multicomponent Vapor-Liquid and Liquid-Liquid Equilibria*, Prentice-Hall, Englewood Cliffs, New Jersey.

Renon, H., Prausnitz, J.M., 1968. *AIChE J.*, 14, 135.

Renon, H., Prausnitz, J.M., 1969. *Ind. Eng. Chem. Process Des. Dev.*, **8**, 413.

Sander, B., Skjold-Jorgensen, S., Rasmussen, P., 1983. *Fluid Phase Equilibria*, **11**, 105.

Severns, W., Sesonske, A., Perry, R., Pigford, R., 1955. *AIChE J.*, **1**, 401.

Silverman, N., Tassios, D., 1977. *Ind. Eng. Chem. Process Des. Dev.*, **16**, **13**.

Silverman, N., Tassios, D., 1984. *Ibid.*, **23**, 586. (Copyright 1984 American Chemical Society, reprinted with permission.)

Simonetty, J., Yee, D., Tassios, D., 1982. *Ind. Eng. Chem. Process Des. Dev.*, **21**, 174.

Smith, J.M., Van Ness, H.C., 1987. *Introduction to Chemical Engineering Thermodynamics*, McGraw-Hill, New York.

Tassios, D., 1969. *AIChE Annual Meeting*, Washington, D.C.

Tassios, D., 1971. *AIChE J.*, **17**, 1367.

Tsonopoulos, C., *AIChE J.*, 1974, **20**, 263; 1975, **21**, 827.

Van Ness, H.C., Abbott, M.M., 1982. *Classical Thermodynamics of Nonelectrolyte Solutions*, McGraw-Hill, New York.

Wilson, G., 1964. *J. Am. Chem. Soc.*, **86**, 127.

Wohl, K., 1946. *Trans. AIChE*, **42**, 215.

Wong, K., Eckert, C.A., 1971. *Ind. Eng. Chem. Fundam.*, **10**, 1.

Yee, D., Simonetty, J., Tassios, D., 1983. *Ind. Eng. Chem. Process Des. Dev.*, **22**, 123.

Problems

13.1 Give an industrial case where phase equilibrium is important and demonstrate its role.

13.2.a. What is the difference between intensive and extensive properties of a system?

b. What does the phase rule and the Duhem theorem state about them?

13.3 Why are mixtures, boiling at low temperatures at atmospheric pressure, separated at high pressures?

13.4 Discuss the difference in approach of calculating low and high pressure vapor-liquid equilibria.

13.5 Present your arguments in choosing the virial equation for the description of low pressure vapor phase nonideality over some cubic equation of state.

13.6 The following data for the system n-heptane(1) - p-dioxane(2) at 80°C are given:

x_1	y_1	P (mm Hg)
0.017	0.063	400.1
0.049	0.152	435.7
0.079	0.209	458.1
0.161	0.320	493.3
0.353	0.451	529.7
0.515	0.518	542.5
0.626	0.584	534.1
0.764	0.672	518.9
0.827	0.730	498.8
0.852	0.759	492.4
0.911	0.836	473.6
0.977	0.951	445.4

a. Assuming ideal vapor behavior, calculate γ_1 and γ_2 and plot them against x_1.

b. What gamma values are more accurate, those in the middle or those at the ends of the composition range? (typical errors in these experimental measurements: x_i and y_i, 0.002; T, 0.1°C; and P, 0.1 mm Hg).

c. Determine the infinite dilution activity coefficient values.

Vapor pressure data: P_1^s = 425.8, P_2^s = 380.4, in mm Hg.

13.7 With reference to the previous problem, calculate γ_1 and γ_2 at x_1 = 0.515, without assuming the vapor phase to be ideal, and compare the values with those obtained there.

13.8 Starting with $\hat{f}_i^l = \hat{f}_i^v$, develop Eq. 13.5.2.

13.9.a. Which quantities must be known from experimental measurements for the exact determination of activity coefficients?

b. How is it done typically?.

13.10 For low pressure *VLE* calculations involving a completely miscible system, give the range of typical % errors resulting from the assumption of ideality in the:

a. vapor phase, and

b. liquid phase.

13.11 Show that for an ideal solution all activity coefficients are equal to one.

13.12 In developing Eq. 13.6.1 we tacitly assumed ideal vapor behavior. Is this justified?

13.13 An equimolar liquid mixture of n-hexane and n-heptane is enclosed into a cylinder-and-piston arrangement, which in turn is placed into a constant temperature bath. The bath temperature is raised slowly and boiling starts. If the two phases are continuously in equilibrium, at what temperature will the last liquid droplet disappear if the external pressure upon the piston is one atm?

13.14 With reference to Example 13.5:

a. what are the 'economic' reasons mentioned?

b. what problems are encountered in pumping a saturated liquid?

13.15 Consider the following two mixtures of benzene(1) and toluene(2), both at 80°C:

a. an equimolar mixture at its bubble point, and

b. an equilibrium mixture of equal amounts of vapor and liquid, where in the latter: $x_1 = x_2 = 0.5$.

Calculate the system pressure and vapor composition in each case.

13.16 Separation of a mixture of benzene (40% on a molar basis) and the rest toluene into products of 95% (molar) purity is considered.

a. What is the minimum pressure that the condenser should operate at, if the water from the cooling tower is available at 40°C and an approach temperature of 10°C is used?

b. If the pressure at the top of the column is 850 mm Hg, what is the temperature and composition of the liquid leaving the top tray?

c. Assuming a pressure drop between the top of the column and the boiler of 100 mm Hg, what is the temperature and composition of the vapor leaving the boiler?

d. If the feed enters the column as saturated vapor, at a location where the pressure is 900 mm Hg, what is its temperature?

Assume in parts (b) and (c) that the two phases are in equilibrium with each other.

13.17 Three moles of benzene(1) and one mole of toluene(2) are placed in a container. The system is allowed to reach equilibrium at 90°C and 1 atm.

a. How many moles of liquid are present?

b. State and justify any assumptions you make.

13.18.a. What are positive and negative deviations from Raoult's law?

b. Explain them in terms of the molecular characteristics of the mixture components.

c. Discuss the effect of dissimilarity in chemical nature of the components of a mixture on the magnitude of the activity coefficients.

d. Explain the decrease of the infinite dilution gammas of symmetric ketones in nC_7 with increasing number of carbon atoms in them (Figure 13.6).

13.19 What is the difference between solvation and association?

13.20.a. Discuss the effect of pressure and temperature on the activity coefficient.

b. Explain in terms of molecular interactions the reasons for the weaker temperature dependency of the infinite dilution gamma of acetone in methanol, as compared to that in nC_7 (Figure 13.7).

13.21 Check the data of Problem 13.6 for thermodynamic consistency. Use the results obtained there, if justified.

13.22 A vapor mixture of acetone(1) and methanol(2) with $y_1 = 0.45$ is to be condensed at a pressure of 1 atm.

a. What is the maximum temperature that the liquid can leave the condenser at?

b. State and justify any assumptions you make.

c. Could you get a better answer? If yes, how?

The following *VLE* datum is available: $x_1 = 0.295$; $y_1 = 0.420$; $t = 59.6°C$; $P = 760$ mm Hg.

13.23 Show that for the van Laar equation: $A = \ln\gamma_1^\infty$; $B = \ln\gamma_2^\infty$.

13.24 For the system acetone (1)-cyclohexane (2) the following information is given at 25°C: $P_1^s = 230.4$ mm Hg; $P_2^s = 97.45$ mm Hg; $\gamma_1^\infty = 8.6$; and $\gamma_2^\infty = 6.6$.

a. Plot $\bar{P}_1$, $\bar{P}_2$, and P vs. x_1 assuming ideal vapor behavior.

b. Is there an azeotrope?

c. Repeat assuming ideal solution behavior for the liquid phase.

d. Plot y_1 vs. x_1 for both cases.

e. Compare in each case your results with the experimental values given in Example 13.1.

f. Draw the *P-x-y* diagram for cases (a) and (c) and identify the single- and two-phase regions.

13.25 A mixture containing 2.7 moles of methanol(1) and 1.3 of ethanol(2) is allowed to reach equilibrium at a temperature of 68°C and a pressure of 1 atm.

a. How many moles of liquid are present?

b. State and justify any assumptions you make.

c. Repeat for $t = 60°C$.

At this temperature: $\ln\gamma_1 = 0.1\,x_2^2$; $\ln\gamma_2 = 0.1\,x_1^2$.

13.26 Present three methods for the evaluation of infinite dilution activity coefficients in binary systems and comment on their accuracy.

13.27 What is the purpose of correlating binary data with:

a.a Wohl-type, and

b.a local composition,

model?

13.28 For distillation design, isobaric - for obvious reasons - *VLE* data are needed.

a.What are the 'obvious' reasons?

b.What are, to this purpose, the advantages of the Wilson equation over the van Laar one?

13.29 A vapor mixture containing 60% (mole) hexene-1(1) and the rest p-dioxane(2), is to be condensed, leaving the condenser at 60°C.

a.Estimate the minimum pressure at which the condenser should operate. State any assumptions you make and justify them.

b.Could you get more accurate results? If yes, present a step-by-step procedure of the proposed approach.

The following vapor-liquid equilibrium data are available at 80°C:

x_1	y_1	P (mm Hg)
0.885	0.930	1095
0.742	0.855	1015
0.489	0.745	992
0.285	0.629	820
0.108	0.415	602

Along with the vapor pressure data (mm Hg) below:

T (°C)	P_1^s	P_2^s
60	680	199
80	1252	380

13.30 n-Hexane(1) and ethanol(2) form an azeotrope at 760 mm Hg and 58°C with $x_1 = 0.67$. Calculate the equilibrium constant $K_1 = y_1/x_1$, at $x_1 = 0.5$ and 58°C.

13.31 With reference to bubble and dew points:

a.discuss their physical meaning;

b.what variables are specified in each case?

c.what quantities are calculated in each case?

d.what information is needed to carry out these calculations?

13.32 Why is prediction of multicomponent *VLE* from binary data important?

13.33 In predicting multicomponent *VLE* from binary data:

a.what factors effect the accuracy of the obtained results?

b.which seems to be the most important one?

c.give typical errors in y as functions of the:

* quality of the binary data
* number of components in the mixture
* degree of mixture nonideality.

13.34 Briefly discuss the concept of local compositions.

13.35 Predict the *VLE* behavior for the system acetone-cyclohexane at 25°C using the UNIFAC method and compare the results with the experimental data given in Example 13.1.

13.36 Separation of a mixture of components A, B, and C is considered.

a. Discuss the options available to you for obtaining the necessary *VLE* information.

b. Give your estimate of the anticipated accuracy.

13.37 Develop Eqs 13.11.3 and 13.11.4 from Eq. 13.11.2.

13.38 With reference to this Chapter:

a. what are the important items that you have learned?

b. give an example of an application in each case.

13.51 Develop Eqs 13.8.1 and 13.8.2.

13.52 Starting with the Gibbs-Duhem equation (Eq. 11.6.3) develop Eq. 13.9.1.

13.53 Develop Eqs 13.9.2 and 13.9.3 from Eq. 13.9.1.

13.54.a. Show that the slopes of a plot of $\ln\gamma_1$ and $\ln\gamma_2$ versus x_1, at a given point of a binary system, should satisfy the following relationship:

$$x_1\left(\frac{d\ln\gamma_1}{dx_1}\right)-x_2\left(\frac{d\ln\gamma_2}{dx_2}\right) = 0$$

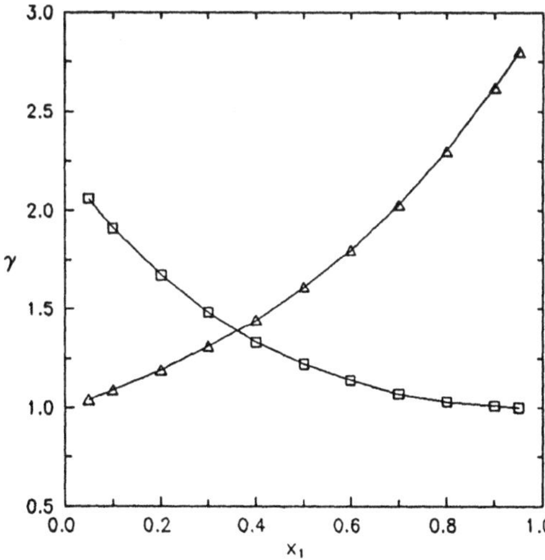

Figure 13.P.1 Isothermal activity coefficients for some binary system. (Lines for identification only.)

b.Is there any indication that the data on Figure 13.P.1 are thermodynamically inconsistent? Prove your answer.

c.Do you expect any difficulties in using this test?

13.55 Explain why the thermodynamic consistency test is a necessary but not sufficient condition for good data.

13.56 Show that if the isothermal *P-x* curve for a miscible binary system passes through a maximum, there is an azeotrope at the point of maximum.

13.57 In recent years the trend is to carry out *VLE* measurements using 'static' stills, where only the liquid phase composition is determined along with the corresponding values of *T* and *P*. Present a flow diagram demonstrating how you would use these *P-T-x* data in determining the corresponding vapor phase compositions. Do not assume ideal vapor.

13.58 A vapor mixture of methanol (1) and ethanol(2), with $y_1 = 0.95$, leaves the top of a distillation column at a pressure of 800 mm Hg. Assuming that the vapor is in equilibrium with the liquid on the top tray, calculate the temperature and composition of this liquid. Use the data of Problem 13.25 as best as you can.

13.59 Discuss, using molecular considerations, the number of parameters needed per binary system in activity coefficient models.

13.60 Discuss the advantages and disadvantages in using single-parameter G^E expressions.

13.61 Where is the Gibbs-Duhem equation used in Section 13.15?

13.62 Using the UNIFAC method with the most recent interaction parameter values, calculate γ_1 at $x_1 = 0.1$ and γ_2 at $x_2 = 0.1$ for the following alcohols (component 1) with n-hexane(2):

1-propanol; 2-propanol; 1-octanol; 2-octanol; and, 3-octanol.

Compare your results with the experimental values and those obtained from the original set of interaction parameters (see the original UNIFAC paper).

13.63.a. Explain how binary mutual solubility data can be used in the estimation of *VLE* in the completely miscible region.

b.Discuss potential problems (see Marina, 1973).

13.64.a. Explain how ternary *LLE* behavior can be predicted from binary data.

b.Discuss potential problems (see Simonetty et al, 1982).

13.65.a. Explain how ternary *VLE* behavior can be predicted from ternary *LLE* data.

b.Discuss the difficulties (see, Yee et al, 1983).

13.66 Discuss the effect of the uncertainty in the evaluation of the relative volatility of the components of a mixture on the number of trays and/or the reflux ratio required for the separation of this mixture (see Nelson et al).

13.67.a. What equation is used for the correlation of activity coefficients in polymeric solutions?

b.Correlate the data of Figure 13.5 using this method (see Prausnitz et al, 1986).

13.68 A feedstream of 60% isopropanol and the rest water is to be separated into a distillate containing 86% isopropanol and a water stream containing a maximum of 1% isopropanol (all in wt %). Two schemes are considered for this purpose:

a. Scheme I, using a single distillation column; and

b. Scheme II, using two columns A and B, operating as follows:

i. The feed enters column A, whose overhead vapor has the desired isopropanol composition.

ii. The liquid leaving the bottom of this column, containing 5%(wt) isopropanol, is split into two parts. One is completely vaporized through the boiler and returns to the column; the other becomes the reflux, and feed, to column B.

iii. The boiler of column B operates with steam and the effluent stream is water containing 1% (wt) isopropanol. The vapor from the top of the column is fed into the bottom of column A.

iv. Part of the vapor leaving the top of column A is condensed with cooling water and represents the reflux. The remaining is compressed to 94°C and becomes the heating medium for the boiler of column A.

(Scheme II uses what is referred to, appropriately, as a vapor recompression cycle, *VRC*.)

Calculate the utility: steam, cooling water, and electricity requirements of the two schemes, and compare their costs.

Given: steam, \$10/metric ton; electricity, \$0.07/kWh; cooling water, \$0.02/m^3.

Make any assumptions you consider reasonable. For details, see Meili and Stuecheli (1987).

13.69 Kojima et al (*Fluid Phase Equilibria*, 1990, **62**, 269) suggest that the three quantities: $\ln\gamma_1$, $\ln\gamma_2$ and $G^E/(x_1x_2)$ are used simultaneously in testing the thermodynamic consistency of binary data:

a. develop the test;

b. apply it to the data of Example 13.1.

13.70 Modify the Program VLEGPHI for UNIQUAC so that you can use the q value recommended by Anderson and Prausnitz in Example 13.10 and compare your results with those obtained there.

13.71 In the production of vitamin C, a stream of water at 25°C saturated with amyl acetate (1% by wt) - and some insoluble solids - is obtained.

a. Present a flow diagram for the recovery of amyl acetate.

b. Will it be an easy or difficult separation?

13.72 The preliminary process design for the production of a new antibiotic includes the presence of a chloroform - n-hexane stream. Discuss the options available for the recovery of chloroform since an azeotrope is involved. Keep in mind that hexane is used in the process only as a solvent.

13.73 Discuss the concept of free volume and its effect on the activity coefficients of the components of a liquid mixture.

14
High Pressure Vapor-Liquid Equilibrium

14.1 Introduction

We saw in the previous Chapter, Section 13.4, that high pressure vapor-liquid equilibrium (*VLE*) involves systems where some, or all, components have low boiling point temperatures at atmospheric pressure, typically mixtures of hydrocarbons - alone or with some inorganic gases: N_2, CO_2, CO, etc. (On occasion, small amounts of water are also present). This is the type of systems that we will focus our attention to.

From the thermodynamic point of view the term *high pressure* refers to pressures high enough as to have a significant effect on the thermodynamic properties of both phases, typically over 15-20 bar.

The presence, thus, of nonideality in *both* phases characterizes high pressure vapor-liquid equilibrium as compared to that at low pressure, where the main source of nonideality is in the *liquid* phase. And as a result, high pressure *VLE* calculations can be more complex than those at low pressures.

14.2 Objective and Approach

Our objective in this Chapter is to develop an understanding of vapor-liquid equilibrium behavior at high pressures and an awareness of the methods involved - and the expected accuracy - in providing a quantitative description and prediction of it.

We start with a qualitative description of mixture behavior at high pressures, and then discuss briefly the main approaches used in obtaining quantitative results:

* K-Charts;
* the Chao-Seader method; and
* Equations of State (EoS).

We proceed then to demonstrate how Bubble Point, Dew Point, and Flash calculations are carried out - in principle - at these conditions, and then apply such procedures using the simplest of the methods, the K-Charts. Use of EoS in carrying out bubble point temperature calculations is made through the computer programs presented in Appendix E.

We summarize our observations in the Concluding Remarks.

14.3 Qualitative Behavior

We will focus our qualitative description of high pressure vapor-liquid equilibrium behavior of mixtures on two very important, from the practical point of view, areas:

1. Bubble, dew, and critical points; and
2. Equilibrium compositions as functions of pressure and temperature.

14.3.1 Bubble, Dew, and Critical Points of Mixtures

The top and bottom lines in Figure 14.1 present the vapor pressure versus temperature curves for two pure compounds A and B, each terminating at the corresponding critical point, C_A and C_B.

Let us now consider a mixture of the two compounds, with a specified composition of component A equal to z_1, at a temperature T_1 and at some pressure P_1, where the mixture is in the liquid state. We reduce the pressure slowly - at T_1 - until the first bubble appears. This represents the bubble point (B.P.) pressure P_{1b} of this mixture at T_1, with a liquid composition of $x_A = z_1$. If we continue decreasing the pressure, more vapor appears until only one drop of liquid is left. This represents the dew point (D.P.) pressure P_{1d} for this mixture at T_1, with $y_A = z_1$.

We repeat now the procedure starting at a higher temperature T_2 and pressure P_2 and obtain: B.P. pressure P_{2b}, at $x_A = z_1$, and D.P. pressure P_{2d}, at $y_A = z_1$, i.e. points $[P_{2b},T_2]$ and $[P_{2d},T_2]$; and so on.

The resulting bubble and dew point curves are shown in Figure 14.1, and the 'head' - expanded - in Figure 14.2. The two curves converge in the *Critical Point* C_1, *where all thermodynamic properties of the two phases become identical.*

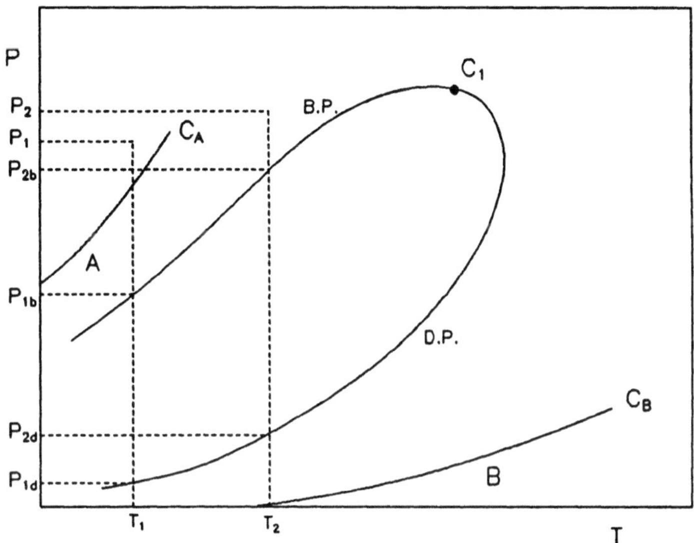

Figure 14.1 Bubble and dew point curves for a constant composition fluid.

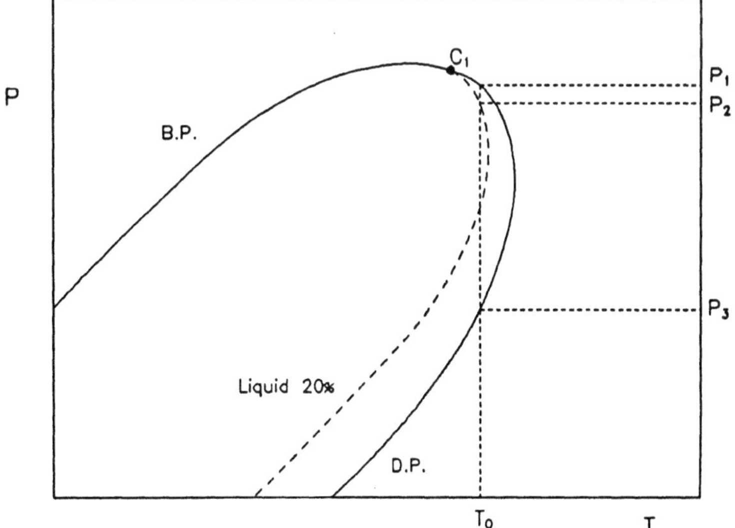

Figure 14.2 Pressure - temperature behavior of a constant composition fluid.

14.3.2 Comments

1.Notice that, contrary to our experience with pure compounds, the critical point here does not represent the maximum temperature at which the given mixture can exist in the liquid phase. It does represent, however, the highest temperature where the *whole* fluid can exist as a liquid. The maximum temperature and pressure that this envelope reaches are referred to as *cricondentherm* and *cricondenbar* respectively.

2.Notice also that at temperatures higher than the critical, up to the cricondentherm, *two* dew point pressures exist, such as P_1 and P_3 at T_0 (Figure 14.2). This is also contrary to our experience with pure fluids, and so is the next observation.

3.Between the bubble and dew point curves we have, of course, mixtures of liquid and vapor in equilibrium with each other. The dashed line in Figure 14.2, for example, represents the locus of such mixtures, where the liquid phase is 20% of the total.

If we now start with the saturated vapor at point $[P_1,T_0]$ and reduce the pressure to P_2 isothermally, a mixture of vapor and liquid appears. Further reduction, to P_3, leads to saturated vapor again, and then to superheated vapor. This would not, of course, occur with a pure compound, where isothermal reduction of the pressure of a saturated vapor leads to a superheated one. This interesting behavior is referred to as *retrograde condensation* and the region to the right of the critical temperature up to the cricondentherm, as the *retrograde* one. 'Retrograde' means moving backwards, or the reverse of normal.

4.The *P-T* phase diagram for a multicomponent system is depicted, through a typical natural gas mixture, in Figure 14.3. Notice the large retrograde region and that, here, retrograde condensation can also occur by increasing the temperature - at constant pressure.

5.Retrograde condensation is very important in gas condensate reservoirs for the reduction of pressure, as a result of gas recovery, can lead to the condensation - and thus loss - of the very valuable heavy constituents of the gas; and in gas distribution systems, where the pressure drop can lead to liquid formation and plugging of the pipes.

6.We have already encountered the phenomenon of retrograde condensation in the case of supercritical extraction (Section 11.11), where pressure reduction was used to recover the material dissolved in the gas.

7.If we repeat the procedure used in developing Figure 14.1 with the same binary system but with a different composition z_2, we obtain a new bubble and dew point curve, with critical point C_2, and so on. The locus

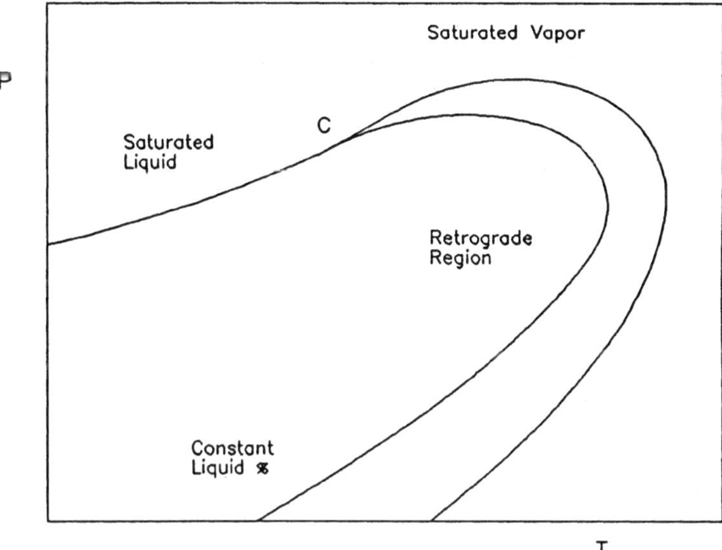

Figure 14.3 Phase diagram for a typical natural gas mixture.

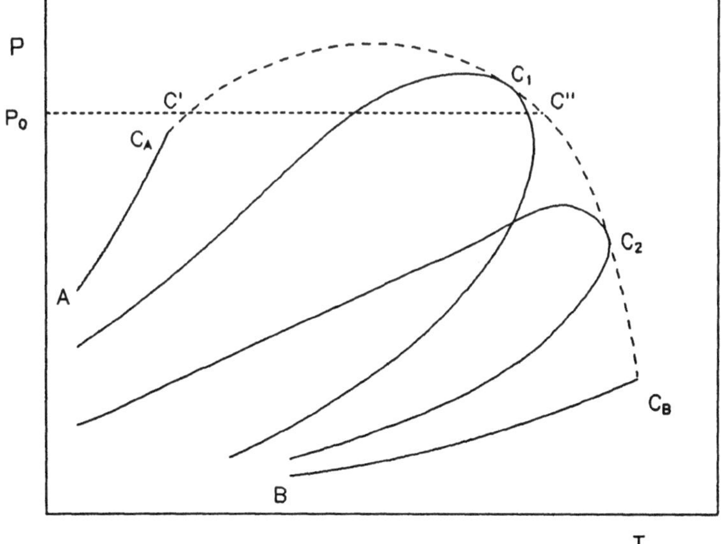

Figure 14.4 Pressure - temperature behavior for a binary fluid of varying composition.

of these critical points is shown in Figure 14.4 and is typical of hydrocarbon mixtures (see Pak and Kay, 1972, for example). Notice the presence of two critical points, such as C' and C'', at pressure P_0.

8. The locus of the critical points can also be concave upwards, typically in polar mixtures, such as methanol-benzene (McCracken and Smith, 1956). For more complex cases, see van Konynenburg and Scott (1980).

14.3.3 Equilibrium Compositions as Functions of Pressure and Temperature

We have considered so far the effect of fluid composition on the relationship between temperature and pressure of the equilibrium phases. We examine, next, the reverse case: the effect of pressure and temperature on the composition of the two equilibrium phases. Such a case is shown in Figure 14.5, which presents the effect of pressure on the *T-x-y* relationship of a binary system A-B.

At the low pressure P_1, both components can exist as liquids in their pure state, with T_A and T_B representing their boiling point temperatures: the *T-x-y* curves in this case cover the whole composition range. Moving to higher pressures we notice that, at P_2, as the composition of A increases we reach the critical point C, where $y_A = x_A$; and that at the still higher pressure P_3, two critical points are present: the *T-x-y* curves do not cover the whole composition range any more.

The practical implications of this behavior in terms of separating a mixture of A and B are apparent: at P_1, pure A and B can be recovered; at P_2, only B; and at P_3, neither.

Similar phase diagrams are obtained if we consider the effect of temperature on the *P-x-y* relationship of a binary mixture.

14.4 Quantitative Description

Consider a liquid and a vapor phase that are at equilibrium at some high pressure. The equality of fugacities for all mixture components:

$$\hat{f}_i^l = \hat{f}_i^v$$

represents, of course, the starting point for high pressure vapor-liquid equilibrium calculations, just as it did with for low pressure ones.

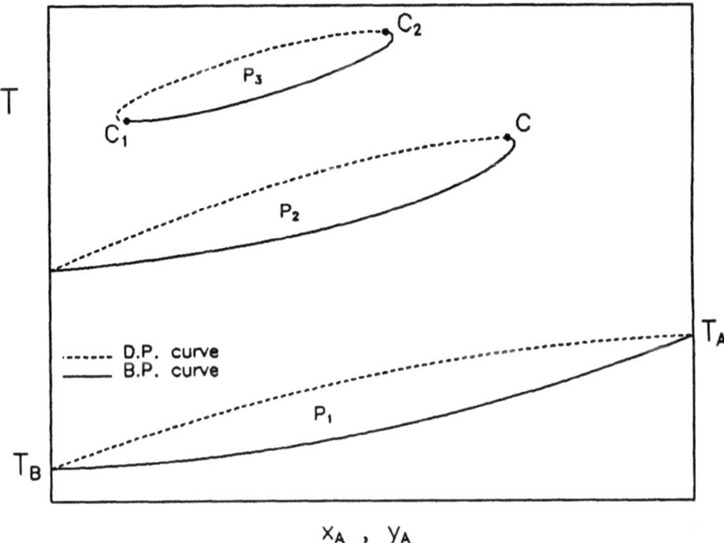

Figure 14.5 Effect of pressure on the *T-x-y* relationship of a binary fluid.

In the latter case, however, the effect of vapor phase nonideality is small - typically to within 10% of ideal vapor behavior - and it can be successfully estimated with an equation of state, such as the virial. Our main concern, therefore, lies with the liquid phase nonideality.

But in high pressure vapor-liquid equilibrium the effect of pressure is significant in both phases - becoming more pronounced close to the critical point - and, consequently, the quantitative description becomes more difficult. The only saving grace, if there is one, with high pressure vapor-liquid equilibrium calculations is that the systems involved in practice - mainly hydrocarbons alone or with some inorganic gases - typically do not deviate much from ideality.

The quantitative description of high pressure *VLE* is usually expressed through the equilibrium ratio K_i:

$$K_i = \frac{y_i}{x_i} \tag{14.4.1}$$

of the mixture components, and the methods used for its evaluation involve three broad approaches:

1. Graphical, using the so-called 'K-Charts ;
2. Separate treatment of the two phases, mainly the Chao-Seader method;
3. Simultaneous description of both phases with an equation of state.

14.5 Graphical Methods: The K-Charts

14.5.1 Charts Using *P* and *T* Only

Several such Charts have been developed in the past [Scheibel and Jenny, 1945; Winn (Hadden, 1953); DePriester, 1953]. They give the *K* values for several light hydrocarbonsas functions of the temperature and pressure of the system only. The other components of the mixture - and its composition - are not involved. Some 'average' effect is included in the *K* values, however, since experimental data were used in their construction. In addition, considering the similarity of the compounds involved and the low pressures covered in the Charts, the effect of mixture composition should not be very significant away, of course, from its critical point.

Since Charts cannot be used with computer calculations, a corresponding states type approach is developed here (Program KVALUE, Appendix E):

$$K = K^{(0)} + (\omega - \omega^{(0)})K^{(1)} \tag{14.5.1}$$

$$K^{(0)} = A_1^{(0)} + A_2^{(0)}/P_r + A_3^{(0)}P_r \tag{14.5.2}$$

$$A_i^{(0)} = a_i^{(0)} + b_i^{(0)}T_r + c_i^{(0)}T_r^2 + d_i^{(0)}T_r^3 \tag{14.5.3}$$

$$K^{(1)} = A_1^{(1)} + A_2^{(1)}/P_r + A_3^{(1)}P_r \tag{14.5.4}$$

$$A_i^{(1)} = a_i^{(1)} + b_i^{(1)}T_r + c_i^{(1)}T_r^2 + d_i^{(1)}T_r^3 \tag{14.5.5}$$

The reference compound, designated with the superscript $^{(0)}$ is n-butane, and values for the parameters in Eqs 14.5.3 and 14.5.5 are presented in Table 14.1. Methane cannot be successfully predicted by Eq.14.5.1; rather Eqs 14.5.2 and 14.5.3 should be used with the $a_{i,1}$, $b_{i,1}$, $c_{i,1}$ and $d_{i,1}$ values presented also in Table 14.1.

Eq.14.5.1, and the special expression for methane, provide reasonable estimates of *K* values for the C_1 to C_8 hydrocarbons (alkanes and alkenes) in hydrocarbon mixtures - *away* from the critical point - in the:

* P_r range 0.05 to 1.3 or about 2 to 40-50 bar;
* T_r range 0.55 to 1.1, except for C_1 where it is 1.1 to 1.7, or the approximate temperature (°C) ranges given below:

C_1 and C_2's	-70 - 50
C_3's and C_4's	-50 - 150
C_5's and C_6's	0 - 200
C_7's and C_8's	30 - 200

Under these limitations the obtained *K* values are typically to within 5% of those obtained from the DePriester charts or of those for relevant binary systems given by Knapp et al, 1982 (see Problem 14.57).

Table 14.1 Parameter Values for Eqs 14.5.3 and 14.5.5 and for CH_4

$a_1^{(0)}$ = 2.18630	$b_1^{(0)}$ = -8.80971	$c_1^{(0)}$ = 10.96486	$d_1^{(0)}$ = -3.87905
$a_2^{(0)}$ = 0.30602	$b_2^{(0)}$ = -0.64992	$c_2^{(0)}$ = -0.72423	$d_2^{(0)}$ = 1.64710
$a_3^{(0)}$ = -0.02683	$b_3^{(0)}$ = -0.28575	$c_3^{(0)}$ = 1.09824	$d_3^{(0)}$ = -0.90820
$a_1^{(1)}$ = -3.44761	$b_1^{(1)}$ = 14.55943	$c_1^{(1)}$ = -21.31054	$d_1^{(1)}$ = 11.13279
$a_2^{(1)}$ = -6.40659	$b_2^{(1)}$ = 28.27873	$c_2^{(1)}$ = -39.95181	$d_2^{(1)}$ = 17.62137
$a_3^{(1)}$ = 0.58201	$b_3^{(1)}$ = -1.28555	$c_3^{(1)}$ = 0.87828	$d_3^{(1)}$ = -0.80885
$a_{1,1}$ = 11.9200	$b_{1,1}$ = -31.1652	$c_{1,1}$ = 29.6218	$d_{1,1}$ = -9.1578
$a_{2,1}$ = 17.0379	$b_{2,1}$ = -38.4026	$c_{2,1}$ = 28.4772	$d_{2,1}$ = -6.0263
$a_{3,1}$ = -10.3018	$b_{3,1}$ = 23.1278	$c_{3,1}$ = -19.1294	$d_{3,1}$ = 5.4422

14.5.2 Charts Using the Convergence Pressure

As the critical point is approached, however, the K values converge towards the value of one in a rather pronounced fashion, as shown in Figure 14.6. To account for this effect, the concept of *convergence* pressure has been introduced. For a hydrocarbon mixture, at a specified temperature, the convergence pressure is the pressure where the K values of all components are equal to one.

Determination of K values for a given mixture according to this approach requires therefore - in addition to specification of the temperature and pressure - the evaluation of the convergence pressure. The procedure is given by Hadden and Grayson (1961), and it is applicable:

* to hydrocarbons with normal boiling points in the range from methane to n-heptane (-259°F to 209°F). Certain non-hydrocarbons (H_2, H_2S, etc.) are also included;
* in the temperature range of -260°F to 800°F; and
* for pressures up to 10,000 psia.

An extension to compounds with higher normal boiling points is given by Kesler et al (1977).

The method provides successful prediction of K values, with somewhat poorer results in the presence of H_2 and/or CH_4. [Dowling and Todd (1973), for example, report an overall error of 13.6% in K values for 253 data points in methane-containing systems; and of 16.7% for 206 points in hydrogen-containing systems.] Its major drawback is that it cannot be easily adapted for computer use.

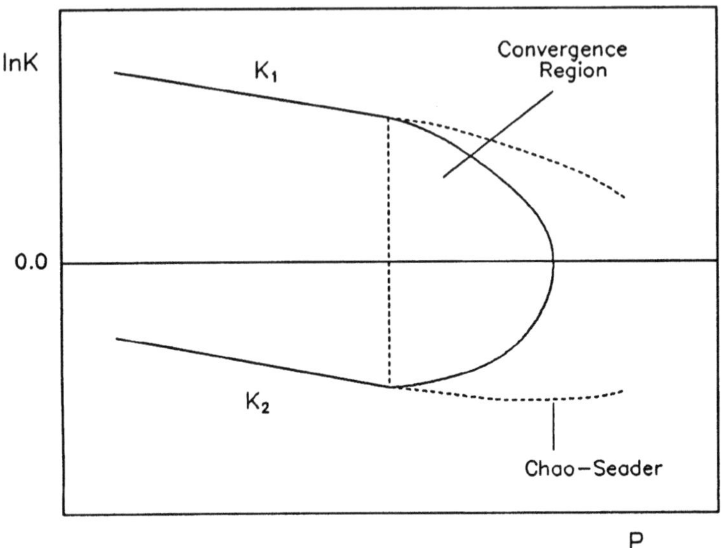

Figure 14.6 Convergence region for a binary mixture at constant temperature.

14.6 The Chao-Seader Method

14.6.1 The Method

Chao and Seader (1961) describe high pressure vapor-liquid equilibrium using the gamma-phi approach, applied to low pressure mixtures in Section 13.4:

$$K_i = \frac{y_i}{x_i} = \frac{\gamma_i}{\hat{\phi}_i} \frac{f_i^{\circ}}{P} = \frac{\gamma_i}{\hat{\phi}_i} \phi_i^{\circ} \tag{14.6.1}$$

Vapor phase fugacity coefficients $\hat{\phi}_i$ are calculated using the Redlich-Kwong equation of state, while the activity coefficients are calculated from the *regular solution* theory of Hildebrand (Hildebrand and Scott, 1964):

$$\ln\gamma_i = \frac{V_i(\delta_i - \bar{\delta})^2}{RT} \tag{14.6.2}$$

where δ_i is the *solubility parameter* of component i, defined by:

$$\delta_i = \left(\frac{\Delta H_i^v - RT}{V_i}\right)^{0.5} \tag{14.6.3}$$

ΔH_i^v is the enthalpy of vaporization and V_i the liquid molar volume, both at 298 K. The quantity $\bar{\delta}$ refers to the volumetric average value of the solubility parameter for the components of the mixture:

$$\bar{\delta} = \frac{\sum x_i V_i \delta_i}{\sum x_i V_i} \tag{14.6.4}$$

The standard state fugacity coefficient $\phi_i{}^{\circ}$ is calculated from a Pitzer-type corresponding states correlation:

$$\log\phi_i{}^{\circ} = \log\phi_i{}^{(0)} + \omega_i \log\phi_i{}^{(1)} \tag{14.6.5}$$

Expressions for $\log\phi_i{}^{(0)}$ and $\log\phi_i{}^{(1)}$ are given by the authors along with values for the acentric factor ω_i.

Through use of experimental data the correlation for the liquid phase fugacity coefficient was extended to supercritical components, and values for their 'liquid' molar volumes were established. In applying hence this method, the parameter values given by the authors should be used. The same applies to the solubility parameter values.

14.6.2 Comments

1. The use of the regular solution theory for the activity coefficient restricts the applicability of the method to nonpolar/weakly polar systems.

2. The use of the corresponding states principle approach for the evaluation of the standard state fugacity provides additional support to this restriction.

3. The evaluation, finally, of the necessary parameters from experimental data imposes restrictions on the type of mixture components.

4. The method is applicable to:

a. Hydrocarbon systems, except methane-containing ones, for:

* the reduced temperature range of 0.5 to 1.3, based on the pure component critical temperature, and

* pressures up to 2000 psia, but not to exceed 0.8 of the critical pressure of the system (see Figure 14.6).

b. Hydrogen and methane containing systems, for:

* temperatures from -100°F to about 0.93 of the pseudoreduced critical

temperature of the equilibrium liquid mixture, but not to exceed 500°F;
* pressures up to about 8000 psia;
* concentration up to 20% of other dissolved gases in the liquid.

5. Within these restrictions, the average absolute deviation for 1345 data points equals 7.9%, as compared to 7.7% for the Benedict-Webb-Rubin (BWR) equation. The method fails, however, in the convergence region as demonstrated in Figure 14.6.

6. An extension of the method, beyond the aforementioned temperature and pressure ranges, is given by Grayson and Streed (1963).

14.7 The Equation of State Approach

The methodology for solving the vapor-liquid equilibrium problem using an equation of state has been outlined in Section 13.4 and involves expressing the fugacity coefficients of the mixture components $\hat{\phi}_i$ in the liquid (l) and vapor (v) phases through an equation of state (EoS). The equilibrium ratio K_i is then calculated from:

$$K_i = \frac{y_i}{x_i} = \frac{\hat{\phi}_i^{\,l}}{\hat{\phi}_i^{\,v}} \tag{14.7.1}$$

Cubic EoS are the most commonly used equations of state for high pressure *VLE* calculations. They are relatively simple to use and provide reasonable accuracy. We confine, therefore, our discussion to such EoS and proceed to:

1. examine the information needed to carry out *VLE* calculations using them;
2. comment on their application and the obtained results;
3. comment on their application to polar systems.

14.7.1 Required Information

Consider first a two-parameter cubic EoS, say the Soave-Redlich-Kwong (SRK, Soave, 1972). Eq.11.9.6 gives the expression for the fugacity coefficient of component i in a mixture of given composition at some temperature T and pressure P with this EoS, and requires values for all:

* a_{ii}'s and b_i's, obtained through the critical properties and acentric

factor of the mixture components and the system temperature;

* a_{ij}'s$(i \neq j)$, obtained through a_{ii}, a_{jj}, and the interaction coefficients k_{ij}.

The information, thus, needed is:

* T_c, P_c, and ω for each component; and
* interaction coefficients for all the pairs of the mixture.

The same information is needed for all two-parameter EoS, such as the PR (Peng and Robinson, 1976). Three-parameter ones, such as the vdW-711 (Watson et al, 1986), require also critical compressibility factor values for all the components.

14.7.2 Comments

1.The Soave-type cubic EoS, discussed in Chapter 10, are applicable to nonpolar/weakly polar systems only.

2.Accurate prediction of pure component vapor pressures by the EoS used is essential for good multicomponent mixture results. This requires, as discussed in Section 10.6:

a.accurate T_c, P_c, and ω values, and

b.vapor pressures (P^s) that are not very low. (While the performance of the various EoS at low P^s levels varies, caution should be exercised at very low ones.)

3.Satisfactory predictions of hydrocarbon mixtures can be obtained without the use of interaction coefficients (Graboski and Daubert, 1978; Tsonopoulos and Heidman, 1986).

When significant differences in the size and type of the molecules exist, however, typically in methane-containing systems, interaction coefficients are needed for accurate results (Chao and Lin, 1986; Czerwienski et al, 1988). The last authors, for example, report that best results for the binary systems: methane with propane, cyclohexane and n-decane, are obtained using interaction coefficients of 0.020, 0.022, 0.032 respectively in the SRK EoS.

4.Interaction coefficients are also needed in the presence of non-hydrocarbons, such as H_2, CO_2, H_2S, N_2, etc. Typical results for CO_2 containing binary mixtures are presented in Table 14.2.

5.Prediction accuracy is sensitive to the value of the interaction coefficients as shown by the typical results of Figure 14.7 for the system CO_2-n-butane. For explanation, see Section 11.9.2.

6.Even though the interaction coefficients are considered temperature and pressure independent to a first approximation, they are actually functions of temperature as Figure 14.8 suggests (see also Kato et al, 1981, and Czerwienski et al).

Table 14.2 Correlation of CO_2 (1) - Solvent (2) Binaries with the SRK and PR EoS

Solvent	Range		SRK			PR		
			k_{12}	AAE%[1] in		k_{12}	AAE%[1] in	
	T(K)	P(bar)		P	K_1		P	K_1
Ethane	250	13-21	0.136	0.4	1.8	0.133	0.5	1.9
	222-289	7-57	0.138	0.9	5.2	0.135	0.9	5.3
Propane	244-266	5-26	0.132	2.0	3.2	0.129	1.8	3.0
	253-273	2-35	0.129	3.7	7.9	0.124	3.8	7.9
n-Butane	311-411	4-82	0.142	1.3	3.8	0.136	1.1	4.0
i-Pentane	277-377	1-89	0.131	2.9	5.0	0.122	2.5	3.8
n-Hexane	293-313	4-77	0.117	5.5	5.4	0.111	5.3	5.3
n-Heptane	310-477	2-133	0.110	1.9	2.5	0.100	2.3	2.5
n-Decane	277-510	1-188	0.131	5.2	3.7	0.115	5.8	3.7
Benzene	298-313	9-78	0.078	2.2	2.6	0.077	2.1	2.5
Cyclohexane	473-533	13-129	0.128	1.9	8.0	0.105	1.9	7.8
Toluene	311-477	3-152	0.113	5.0	8.7	0.106	4.8	8.2
Decalin	348-373	10-71	0.190	6.0	-	0.180	6.1	-
Overall Absolute Error %				3.0	4.8		3.0	4.7

[1] Average absolute error %.

7.For systems containing CO_2, H_2S, N_2, and CO with hydrocarbons, Graboski and Daubert (1979) give generalized correlations for the estimation of interaction coefficients in the SRK EoS; and Kato et al for CO_2 - n-alkane systems with the PR EoS.

8.For H_2/hydrocarbon systems, see: Graboski and Daubert; Lin (1980) and, especially, Tsonopoulos and Heidman.

9.For some problems encountered in using the SRK EoS, see Gray (1979).

10.The SRK and PR EoS provide similar correlation results as suggested, for example, by Table 14.2 for the CO_2 binaries.

11.The vdW-711, t-PR and t-vdW EoS give equally good results with these EoS, as Table 14.3 indicates for the first one.

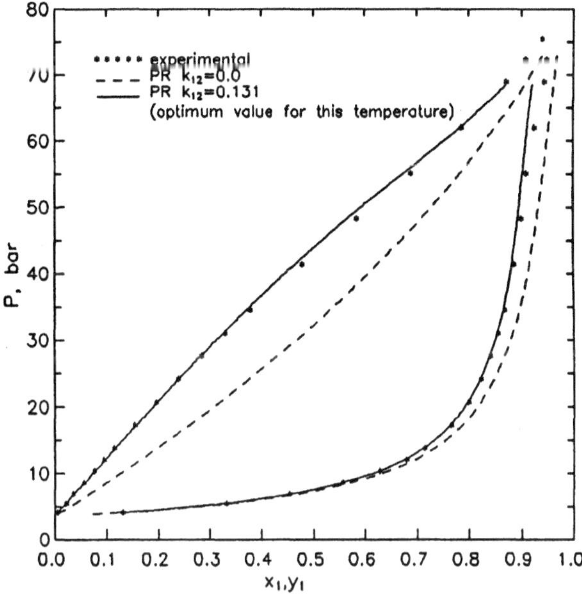

Figure 14.7 The phase envelope for the CO_2 - n-butane system at 310.9 K (exp. data from Knapp et al, 1982, p.607).

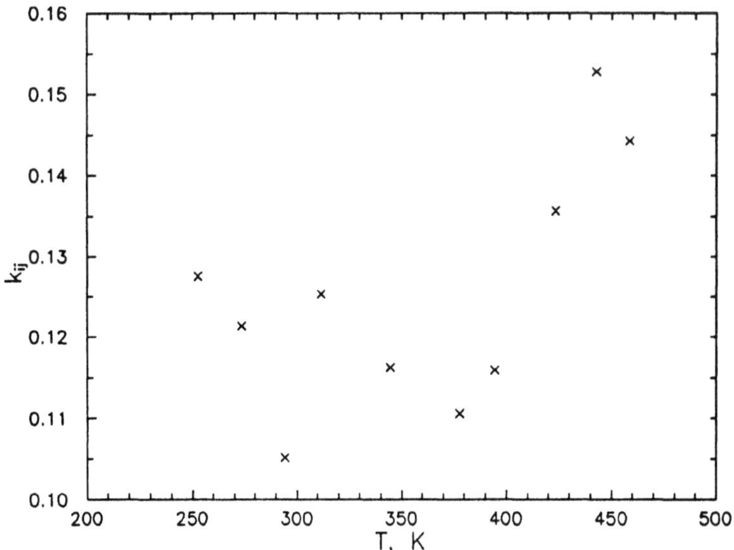

Figure 14.8 Variation of k_{ij} with temperature for the system CO_2- n-pentane and the t-PR EoS.

Table 14.3 Performance of the vdW-711, PR, and SRK EoS in the Correlation of *VLE* Data and the Prediction of Saturated Liquid Volumes (Czerwienski et al)

			$\Delta y \cdot 1000$			% Error in *P*			% Error in *V*		
System	*T*(K)	*P*(atm)	SRK	PR	vdW	SRK	PR	vdW	SRK	PR	vdW
CO_2-C_3	277/344	9/65	10	11	11	1.66	1.44	1.79	17.0	6.7	3.7
CO_2-nC_4	311/411	4/71	8	9	12	1.44	1.29	1.86	15.2	6.0	3.3
CO_2-nC_5	278/378	2/95	9	11	10	3.83	3.73	3.78	12.8	2.3	2.6
CO_2-nC_{10}	278/378	3/153	2	3	2	4.61	5.25	4.41	14.7	6.7	2.2
CO_2-nC_{20}	310/373	5/75	0	0	0	2.84	2.72	2.93	53.4	37.1	1.6
C_1-C_3	278/344	7/99	12	15	13	1.40	1.35	1.46	12.1	4.6	5.3
C_1-nC_6	294/344	7/99	6	5	6	3.92	4.16	4.10	6.5	5.2	1.1
C_1-nC_{10}	311/372	1/357	3	2	3	4.12	4.06	4.02	13.0	6.0	5.3
C_3-nC_{10}	278/478	2/68	2	2	2	2.27	2.21	2.24	17.8	4.9	3.6
Overall			6	6	7	2.9	2.9	3.0	18.1	8.8	3.2

Note. vdW stands for the vdW-711 EoS; *V* for saturated liquid volume; Δy represents the average absolute deviation in vapor phase mole fraction; % error is the average absolute one.

12.The same can be expected from other EoS, cubic and non-cubic, as suggested by the results of Chao and Lin, summarized in Table 14.4.

13.As expected from our discussion on the prediction of saturated liquid volumes of pure compounds in Section 10.6, the results for mixtures should vary among different EoS. The SRK EoS, thus, gives very poor results while the translated ones, such as the vdW-711, provide reasonable accuracy (Table 14.3).

14.The SRK and PR EoS have been also applied to water-hydrocarbon systems by adjusting the value of a_{ii} in the attractive term to reproduce water's vapor pressure.

The PR EoS gives better results for the water solubility and *K* values in the hydrocarbon-rich phase, using interaction coefficients for all hydrocarbon-water pairs. Prediction of hydrocarbon solubility in the water-rich phase however, using the same interaction coefficients, gives poor results. For details, and a method for estimating interaction coefficients in such systems, see Tsonopoulos and Heidman.

Table 14.4 Performance of Several EoS in the Correlation of Binary *VLE* Data

		Overall Avg. Abs. Error % in K_1 and K_2			
System	N(1)	SRK	PR	CCOR(2)	BACK(3)
H_2/solvent	18	6.9	6.4	5.5	---
CH_4/ "	21	4.2	---	3.8	3.5
CO_2/ "	17	4.5	4.1	4.2	4.1

(1) Number of systems
(2) Cubic chain of rotators, Kim et al (1986)
(3) Chen and Kreglewski (1977)

Table 14.5 Data Correlation of CO_2 - Solvent Binary *VLE* Systems Using the t-PR EoS with One (k_{ij}) and Two (k_{ij}, d_{ij}) Interaction Coefficients

Solvent	T, K	Avg. Abs. Error % in P	
		k_{ij}	k_{ij} and d_{ij}
nC_4	311	1.5	0.4
nC_5	294	1.7	0.7
nC_7	311	2.2	1.9
nC_{20}	323	2.3	0.4
nC_{24}	473	1.7	0.5
nC_{32}	398	5.6	1.1
nC_{44}	373	7.5	1.6

15. To improve the description of binary *VLE* data - over that obtained with the use of an interaction coefficient k_{ij} in the combining rule for a_{ij} - the typical approach involves the introduction of a second interaction coefficient by writing Eq.11.7.20 as:

$$b_{ij} = 0.5(b_i + b_j)(1 - d_{ij}) \tag{14.7.2}$$

This provides increased flexibility in fitting binary data, important when large differences in molecular size are involved, as suggested by the typical results presented in Table 14.5.

The obtained k_{ij} and d_{ij} values are not easily generalized, as indicated by Tsonopoulos and Heidman for H_2-, CO_2-, and water-hydrocarbon systems. Gasem and Robinson (1985) did obtain generalized k_{ij} and d_{ij} expressions, however, for CO_2 - n-alkanes.

16.From the non-cubic EoS, the BWR is still used for systems where sufficient data exist for the evaluation of the required parameters, mostly hydrocarbons. Good results are obtained, but the calculations are more cumbersome and computer-time consuming than those with cubic EoS.

17.Evaluation of k_{ij} values from binary *VLE* data for the equations of state discussed in Chapter 10 can be accomplished using Program EOSKIJ of Appendix E.

14.7.3 Polar Systems

Use of the equation of state approach in the correlation and prediction of *VLE* behavior of polar systems has the obvious advantage of eliminating the need for estimating the vapor phase imperfection, usually done with the virial equation.

This advantage becomes very important at high pressures. First, the virial equation truncated after B may not be applicable at these pressures. Second, even if it is applicable, the uncertainties involved in estimating second virial coefficients, especially cross ones, may lead to significant errors in the obtained vapor-phase fugacity coefficients of the mixture components at these pressures (see Problem 14.58). Finally, the two approaches may not converge in the critical region.

Application of cubic EoS to polar mixtures requires resolution of two main problems.

1.The simple Soave-type expressions for α (Eq.10.4.2) cannot provide satisfactory prediction of vapor pressures of polar compounds (why?).

2.The 'conventional' mixing rules for parameters a and b: Eqs 11.7.15 through 11.7.18, 11.7.23 and 14.7.2, are not typically sufficient for polar mixtures.

The first problem is easily dealt through use of compound-specific parameters for polar compounds, as discussed in Section 10.6.

The second problem has been the subject of extensive study in the last few years without a completely satisfactory resolution yet. The following comments represent a very brief account of this effort.

1.The introduction of new mixing rules in polar mixtures has followed two main approaches:

a.incorporation of local composition G^E models into the mixing rule for a in cubic EoS, so as to account for the non-randomness in such mixtures on the local level, as done for example by: Huron and Vidal (1979); Gupte et al (1986); Kalospiros et al (1991). This approach leads to the so-called EoS/G^E models;

b.use of density-dependent(DD) mixing rules (Mollerup, 1981; Whiting and Prausnitz, 1982; Danner and Gupte, 1986).

2.The first approach has the disadvantage that the expression for the second virial coefficient, obtained from the EoS/G^E model as the density approaches zero, does not follow the quadratic mixing rule dictated by statistical mechanics.

To overcome this, DD mixing rules have been suggested, that provide an arbitrary interpolation between this theoretically required quadratic mixing rule at zero density and some other type of mixing rule at higher densities. This approach has found limited application however.

3.Kalospiros et al have applied the methodology of Gupte et al, combining the modified vdW-711 EoS (Androulakis et al, 1989) with the UNIQUAC G^E expression, to polar systems. Excellent results were obtained in both, the correlation of binary systems and the prediction of ternary *VLE* behavior from binary data. The latter is demonstrated in Figure 14.9. Notice the improvement over the results obtained with the conventional mixing rules. The method cannot be applied in the presence of supercritical components.

4.The procedure of Michelsen (1990), using again an EoS/G^E model, appears to resolve the problem of supercritical components, however, and is expected to greatly facilitate the application of the equation of state approach to polar systems.

14.8 Bubble Point, Dew Point, and Flash Calculations

High pressure *VLE* calculations are carried out using *K* values obtained with the aforementioned methods, in a fashion similar to that used for low pressure ones.

14.8.1 Bubble and Dew Points

As we have seen in the previous Chapter in the first case, B.P., a liquid of specified composition: x_1, x_2, ... is given, and we calculate:

* the B.P. temperature at a given pressure, or

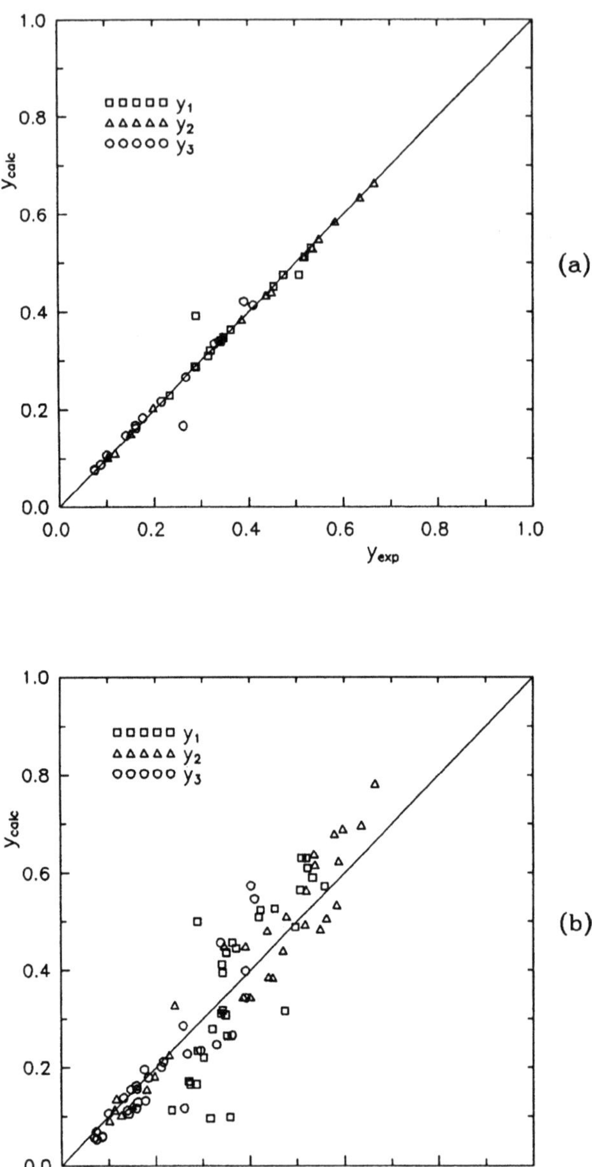

Figure 14.9 Comparison of *VLE* predictions from binary data: ethanol(1) - benzene(2) - n-heptane(3) at 180 mm Hg. (a) vdW-711/UNIQUAC model; (b) vdW-711/ conventional mixing rules (Kalospiros et al, 1991).

* the B.P. pressure, at a given temperature.

The iteration is carried out until the sum, over all mixture components, of calculated y_i's is equal to unity:

$$\sum y_i = \sum K_i x_i = 1.0 \qquad (14.8.1)$$

In the second case, D.P., a vapor of specified composition: y_1, y_2, ... is given, and we calculate:

* the D.P. temperature at a given pressure, or
* the D.P. pressure at a given temperature.

The iteration is carried out until, here, the sum of calculated x_i's is equal to unity:

$$\sum x_i = \sum \frac{y_i}{K_i} = 1.0 \qquad (14.8.2)$$

Bubble and dew point calculations are presented in the next two Examples with the simplest approach, the use of K values from Eqs 14.5.1 through 14.5.5. They can be also carried out with the computer subroutines, using the SRK EoS, given by Daubert (1985), that can be easily adapted to other EoS; and for bubble point pressure calculations, using the EoS of Chapter 10, with the Program VLEEOS presented in Appendix E, which can be easily modified to perform B.P. temperature calculations.

14.8.2 Example 14.1

Calculate the B.P. pressure at 27°C for the following mixture:

Propane(1): 0.40 moles
i-butane(2): 0.30 moles
n-butane(3): 0.30 moles.

This is a trial-and-error calculation:

1. Assume P;
2. Get K's from Eqs 14.5.1 through 14.5.5;
3. Check if: $K_1x_1 + K_2x_2 + K_3x_3 = 1.0$ (within 0.01);
4. If not, assume an other pressure and repeat.

The obtained B.P. pressure is 5.9 bar.

14.8.3 Example 14.2

Calculate the D.P. pressure for the mixture of Example 14.1, again at 27°C.

In this case we seek the pressure where:

$$\frac{y_1}{K_1}+\frac{y_2}{K_2}+\frac{y_3}{K_3} = 1.0 \quad \text{(within 0.01)}$$

Using a similar trial-and-error procedure we obtain: $P = 4.5$ bar.

14.8.4 Example 14.3

Using the data of Table 14.E.3 for the system carbon dioxide(1) - n-butane(2):
a. evaluate the optimum k_{12} value for the PR EoS;
b. using the optimum value at 310.93 K of $k_{12} = 0.131$ (Figure 14.7), calculate the B.P. pressures and vapor compositions at the given liquid compositions and temperature and compare them with those obtained in part (a).

a. Using the given data and the Program EOSKIJ of Appendix E, the value of $k_{12} = 0.148$ is obtained with an average absolute error of 0.4% in pressure; and of 0.008 in vapor phase mole fraction. Detailed results are presented in Table 14.E.3.

b. Use of $k_{12} = 0.131$ in the Program VLEEOS yields an average absolute error of 1.3% in P and 0.032 in y. The results are shown graphically in Figure 14.E.3 and demonstrate the sensitivity of *VLE* predictions to the k_{12} value (Section 14.7.2).

Table 14.E.3 Experimental and Calculated Results for the System Carbon Dioxide (1) - n-Butane (2) at 377.59 K (Knapp et al, p.607)

P (bar)	x_1	y_1	P_{calc} (bar)	Error %	$y_{1,calc}$	$y_{1,calc}-y_1$
17.237	0.004	0.026	17.381	0.83	0.025	-0.001
20.684	0.028	0.150	20.838	0.76	0.148	-0.002
24.131	0.052	0.244	24.286	0.64	0.239	-0.005
27.579	0.076	0.313	27.720	0.51	0.308	-0.005
31.026	0.100	0.367	31.137	0.36	0.361	-0.006
34.473	0.124	0.411	34.531	0.17	0.404	-0.007
41.368	0.174	0.480	41.509	0.34	0.468	-0.012
48.263	0.223	0.524	48.177	-0.18	0.510	-0.014
55.158	0.274	0.552	54.858	-0.54	0.537	-0.015
62.052	0.328	0.568	61.517	-0.86	0.553	-0.015
68.947	0.393	0.568	68.663	-0.41	0.556	-0.012
72.394	0.434	0.552	72.394	0.00	0.548	-0.004
75.152	0.498	0.498	75.152	0.00	0.498	0.000

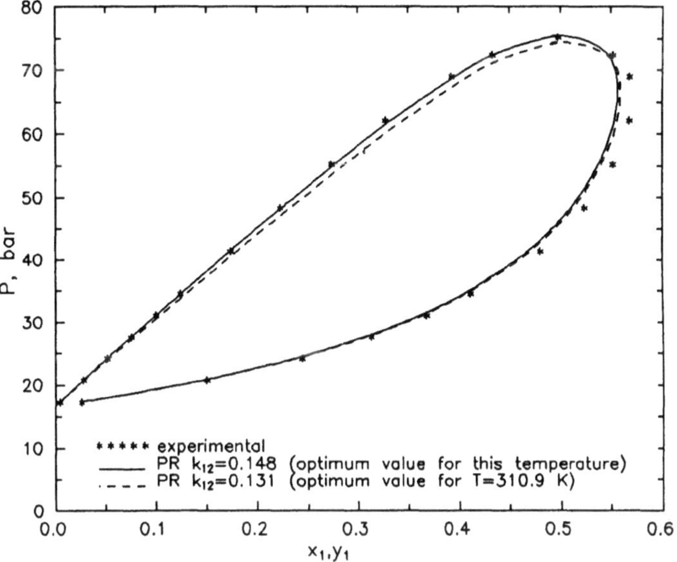

Figure 14.E.3 Bubble point pressures and compositions for the system: $CO_2(1)$ - $nC_4(2)$ at 377.59 K.

14.8.5 Flash Calculations

Consider a mixture of F moles with mole fractions of its components equal to z_1, z_2, ... available at a temperature T_f and a pressure P_f. This mixture is expanded through a throttling valve - *flashed* as it is referred to - into a vessel where the temperature is T and the pressure P (Figure 14.10). Our objective is to calculate, assuming that equilibrium has been reached:

1. the amount of liquid L and vapor V, and
2. the compositions of the two phases.

The information given is:

* F
* z_1, z_2, ...
* T and P.

The information sought is:

* L and V
* the liquid phase compositions: x_1, x_2, ...
* the vapor phase compositions: y_1, y_2, ...

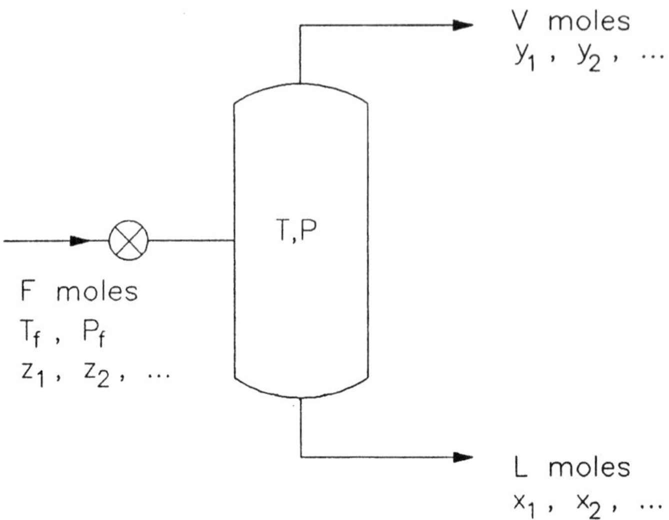

Figure 14.10 Schematic diagram for a flash process.

From a material balance on component i:

$$Fz_i = Lx_i + Vy_i \tag{14.8.3}$$

We also have:

$$y_i = K_i x_i, \text{ and}$$
$$L = F - V$$

Introduction of the last two relationships into Eq.14.8.3 and solution for x_i yields:

$$x_i = \frac{Fz_i}{F + V(K_i - 1)} \tag{14.8.4}$$

Similarly:

$$y_i = \frac{FK_i z_i}{F + V(K_i - 1)} \tag{14.8.5}$$

To evaluate V from either of the last two equations we iterate until:

$$Q_x = \sum x_i - 1 = 0 \tag{14.8.6}$$

or:

$$Q_y = \sum y_i - 1 = 0 \tag{14.8.7}$$

Determination of the x_i's and y_i's is straightforward.

Comments

1.The procedure for this type of flash calculation is demonstrated in the next Example using Eqs 14.5.1 through 14.5.5.

2.Evaluation of V can be facilitated by using instead of Eq.14.8.6 or 14.8.7, their difference Q (Rachford and Rice, 1952):

$$Q = \sum (y_i - x_i) = \sum \frac{F z_i (K_i - 1)}{F + V(K_i - 1)} = 0 \qquad (14.8.8)$$

The reason is that the derivative (dQ/dV) is always negative, which leads to rapid convergence when used with Newton's iterative procedure.

3.Two other types of flash calculations are also common and involve specification of:

* the amount of the liquid L (or vapor V) and the temperature T, to evaluate the required operating pressure P and resulting compositions, and
* L (or V) and P, to evaluate T and compositions.

4.For more details on flash calculations see, among others: King (1981); Michelsen (1982); and Henley and Seader (1981).

14.8.6 Example 14.4

Recovery of propane from the mixture of Example 14.1 is considered by flashing the mixture into a drum, operating at 27°C.

a.What is the pressure range for the existence of two phases?

b.If we operate the drum at the middle of this pressure range, what is the fraction of the obtained vapor and the mole fraction of propane in it?

a.The highest pressure for the existence of two phases at 27°C is, of course, the B.P. pressure of the mixture, equal to 5.9 bar (Example 14.1). If the pressure becomes higher, we will have subcooled liquid.

Similarly, the lowest pressure for the existence of two phases at 27°C is the D.P. one: 4.5 bar (Example 14.2). Below it, we will have superheated vapor. (See also Problems 14.25 and 14.59.)

b.The operating pressure, thus, is: (4.5 + 5.9)/2 = 5.2 bar, and we proceed to determine the value of V that satisfies Eq.14.8.6:

$$x_1 + x_2 + x_3 = 1.0$$

with x_1, x_2, and x_3 expressed through Eq.14.8.4, and:

$$F = 1 \text{ mole}; \; K_1 = 1.79; \; K_2 = 0.79; \; K_3 = 0.58.$$

A trial-and-error solution gives:

$$V = 0.485; \; x_1 = 0.289; \; x_2 = 0.334; \; x_3 = 0.377.$$

And: $y_1 = 0.518$; $y_2 = 0.264$; $y_3 = 0.218$.

The proposed flash does not produce a stream of significantly improved purity in propane as compared with the feed (see Problem 14.11).

14.9 Concluding Remarks

1.The difficulty with high pressure vapor-liquid equilibrium calculations lies in the presence of nonideality in both phases. In addition, the temperature corresponding to these pressures is often above the critical value of one, or even more, of the components of the mixture; or, may be close to that of the latter. The term 'high' is, therefore, relative: a pressure that is important for one case may not be so for another.

2.For preliminary calculations use of the *K*-charts, mainly those involving the convergence pressure, gives reasonable accuracy. The latter cannot be adapted, however, for computer calculations.

3.Within the limits of the Chao-Seader method, especially the restriction to pressures below about 0.8 of the critical value of the mixture, good results are obtained.

4.Cubic equations of state have become the main tool for high pressure *VLE* calculations. They combine simplicity with accuracy comparable to - or better than - that of other methods, including non-cubic EoS. For a comparison of the EoS approach with the Chao-Seader method, see Maddox and Erbar (1981).

5.Good representation of pure component vapor pressures is essential. It is thus recommended that this representation be checked with all mixture components, before any *VLE* calculations are carried out.

6.The EoS presented in Chapter 10, and included in the Programs of Appendix E, are recommended for *VLE* calculations of nonpolar/weakly polar systems. No interaction coefficients are needed for hydrocarbon systems, except when large differences in molecular weight are present or high degree of accuracy is required. Such coefficients, essential in the presence of non-hydrocarbons, are available for a variety of systems (Knapp et al), or can be estimated (see Section 14.7.2).

7.For saturated liquid molar volumes those EoS involving the concept of translation should be preferred.

8.Prediction of multicomponent vapor-liquid equilibrium with cubic EoS, using k_{ij} values when necessary, provides results that are qualitatively correct but not always quantitatively so, as the typical cases presented in Figures 14.11 and 14.12 demonstrate. In general:

a.Bubble points are predicted with better accuracy (typical errors in P of 2 - 5%) than dew points (2 - 10%).

b.Accuracy in the prediction of K values is rather high for the light components, but poorer for the heavy ones.

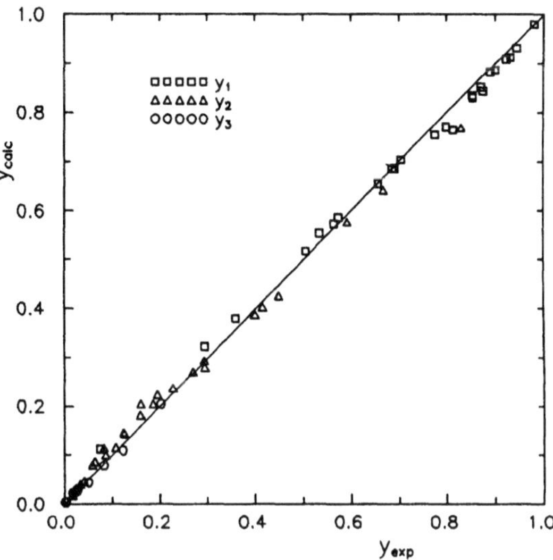

Figure 14.11 Prediction of vapor-liquid equilibrium with the PR EoS for the system: $C_2(1)$ - $C_3(2)$ - $nC_4(3)$ at 260, 270 and 280 K using k_{ij} = 0.0 for all pairs. (Data from Clark and Stead, 1988.)

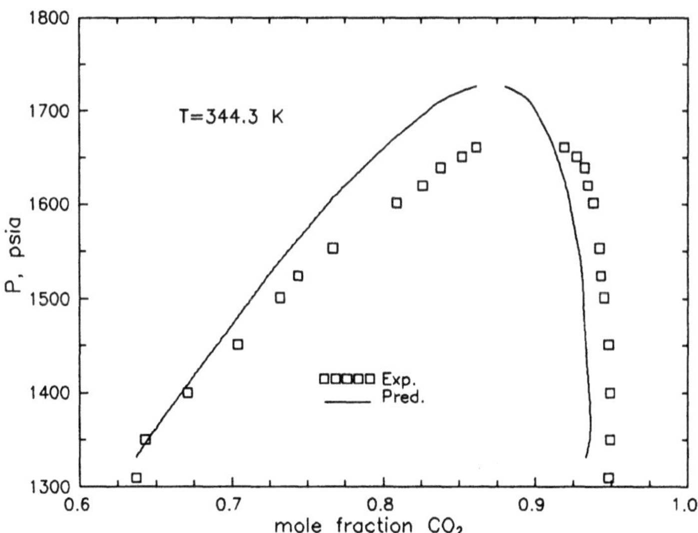

Figure 14.12 The phase envelope for the system CO_2-nC_4-nC_{10} using k_{ij} values for the two CO_2-alkane binaries, obtained from the corresponding data. Experimental data from Nagarajan et al (1990).

c. Prediction of vapor and liquid volumes from flashes, even with the translated EoS, does not always provide satisfactory results.

d. Results become poorer as the critical point is approached, even in the correlation of binary data (Figures 14.7 and 14.E.3).

9. The most extensive compilation of high pressure binary *VLE* data - and references for multicomponent ones - is presented by Knapp et al. The data are correlated with several EoS. For theoretical background, see the excellent book of Prausnitz et al (1986).

10. Successful application of equations of state to polar systems is under development in the last decade with encouraging results. They have not reached as yet, however, the level of simplicity and performance required for general acceptance.

References

Androulakis, J., Kalospiros, N., Tassios, D., 1989. *Fluid Phase Equilibria*, **45**, 135.

Chao, K.C., Lin, H.-M., 1985. *IUPAC Conference on Phase Equilibria*, Paris.

Chao, K.C., Seader, J.D., 1961. *AIChE J.*, **7**, 598.

Chen, S.S., Kreglewski, A., 1977. *Ber. Bunsenges. Phys. Chem.*, **81**, 1048.

Clark, A.Q., Stead, K., 1988. *J. Chem. Thermodynamics*, **20**, 413.

Clifford, I.L., Hunter, I.E., 1933. *J. Phys. Chem.*, **37**, 101.

Czerwienski, G., Tomasula, P., Tassios, D., 1988. *Fluid Phase Equilibria*, **42**, 63. (Reprinted by permission.)

Danner, R.P., Gupte, P.A., 1986. *Fluid Phase Equilibria*, **29**, 415.

Daubert, T.E., 1985. *Chemical Engineering Thermodynamics*, McGraw-Hill, New York.

DePriester, C.L., 1953. *Chem. Eng. Progr. Symp. Ser.*, **7**, 49.

Dowling, G.R., Todd, W.G., March 1973. *Chemical Engineering*, 115.

Gasem, K.A.M., Robinson, R.L., 1985. *National AIChE Meeting*, Houston, Texas.

Graboski, M.S., Daubert, T.E., 1978; 1979. *Ind. Eng. Chem. Process Des. Dev.*, **17**, 448; **18**, 300.

Gray, R., 1979. In *Advances in Chemistry Series*, K.C. Chao and R.L. Robinson, Editors, No **182**, ACS, Washington, D.C.

Grayson, H.G., Streed, C.W., June 25, 1963. *Paper presented at the 6th World Petroleum Congress*, Frankfurt/Main, Germany.

Gupte, P.A., Rasmussen, P., Fredenslund, Aa., 1986. *Ind. Eng. Chem. Fundam.*, **25**, 636.

Hadden, S.T., 1953. *Chem. Eng. Progr. Symp. Ser.*, No.7, 53.
Hadden, S.T., Grayson, H.G., 1961. *Hydrocarbon Processing and Petroleum Refiner*, **40**, 9, 207.
Henley, E.S., Seader, J.D., 1981. *Equilibrium-Stage Separation Operations in Chemical Engineering*, Wiley, New York.
Hildebrand, J.H., Scott, R.L., 1964. *The Solubility of Nonelectrolytes*, Dover, New York.
Huron, M.-J., Vidal, J., 1979. *Fluid Phase Equilibria*, **3**, 255.
Kalospiros, N., Misseyannis, G., Androulakis, J., Tassios, D., 1991. *Fluid Phase Equilibria*, **64**, 173. (Reprinted by permission.)
Kato, K., Nagahama, K., Hirata, H., 1981. *Fluid Phase Equilibria*, **7**, 219.
Kesler, M.G., Lee, B.J., Fish, M.I., Hadden, S.T., May 1977. *Hydrocarbon Processing*, 257.
Kim, H., Lin, H.-M., Chao, K.C., 1986. *Ind. Eng. Chem. Fundam.*, **25**, 75.
King, C.J., 1981. *Separation Processes*, McGraw-Hill, New York.
Knapp, H., Doring, R., Oellrich, L., Plocker, U., Prausnitz, J.M., 1982. *Vapor-Liquid Equilibria for Mixtures of Low Boiling Substances*, DECHEMA Chemistry Series, Vol. **VI**, Frankfurt/Main, Germany.
Lin, H.-M., 1980. *Ind. Eng. Chem. Process Des. Dev.*, **19**, 501.
Maddox, R.N., Erbar, J.H., Feb. 2, 1981. *Oil & Gas J.*, p.74.
McCracken, P.G., Smith, J.M., 1956. *AIChE J.*, **2**, 498.
Michelsen, M.L., 1982. *Fluid Phase Equilibria*, **9**, 1 and 21.
Michelsen, M.L., 1990. *Fluid Phase Equilibria*, **60**, 47 and 213.
Mollerup, J., 1981. *Fluid Phase Equilibria*, **7**, 121.
Nagarajan, N., Gasem, K.A.M., Robinson, R.L., 1990. *J.C.E.D.*, **35**, 228.
Pak, S.C., Kay, W.B., 1972. *Ind. Eng. Chem. Fundam.*, **11**, 225.
Peng, D.Y., Robinson, D.R., 1976. *Ind. Eng. Chem. Fundam.*, **15**, 59.
Prausnitz, J.M., Lichtenthaler, R.N., de Azevedo, E.G., 1986. *Molecular Thermodynamics of Fluid-Phase Equilibria*, Prentice-Hall, Englewood Cliffs, N.J.
Rachford, H.H., Jr., Rice, J.D., 1952(October). *J. Petrol. Technol.*, **4**, 19.
Scheibel, E.G., Jenny, E.F., 1945. *Ind. Eng. Chem.*, **37**, 80.
Soave, G., 1972. *Chem. Eng. Sci.*, **27**, 1197.
Tsonopoulos, C., Heidman, J.L., 1986. *Fluid Phase Equilibria*, **30**, 391.
Van Konynenburg, P.H., Scott, R.L., 1980. *Trans. Roy. Soc.*, London, **298A**, 495.
Watson, P., Cascella, M., May, D., Salerno, S., Tassios, D., 1986. *Fluid Phase Equilibria*, **27**, 35.
Whiting, W.B., Prausnitz, J.M., 1982. *Fluid Phase Equilibria*, **9**, 119.

Problems

14.1.a. Is there a difference in the definition of the critical point of a pure compound and of a mixture?

b.Discuss the differences in behavior of these two types of fluids in the critical region.

14.2 With reference to mixture behavior at high pressures:

a.can isothermal reduction of pressure for a saturated vapor lead to the formation of liquid?

b.the same, but for isobaric temperature increase.

c.what is retrograde condensation? Is it of any practical importance? Give an example.

d.why are *K* values equal to one at the critical point?

14.3 Can changing the operating pressure of a distillation column effect the purity of the obtained products? Explain.

14.4 With reference to natural gas:

a.what is its contribution to the world energy supply?

b.when is retrograde condensation important?

c.what are 'gas hydrates'?

14.5 Prepare the *P-x-y* diagram for a binary system at three temperatures and show the critical points if present.

14.6 What are the main approaches used in high pressure *VLE* calculations?

14.7 Consider the mixture C_1-nC_5 at 300 K and 50 atm. What problems do you see in using the gamma-phi approach to describe the *VLE* behavior of this system at these conditions?

14.8.a.A vapor mixture of propane(1), i-butane(2), and n-butane(3), leaving the top of a distillation column with composition:

$$y_1 = 0.7; \; y_2 = 0.2; \; y_3 = 0.1$$

is to be condensed in a total condenser. What is the minimum pressure that the condenser should operate at, so that the liquid leaves at 30°C?

b.Discuss any assumptions you make and estimate their quantitative effect on the accuracy of your answer.

c.Is there a better method? If yes, which one and what accuracy would you expect?

14.9 With reference to the previous problem, assuming a 0.2 bar pressure drop between the condenser and the top of the column and that the vapor and liquid on the top tray of the distillation column are in equilibrium with each other, determine:

a.the temperature at that tray;

b.the composition of the liquid leaving the tray.

14.10 If a partial condenser were used in Problem 14.8, determine:

a.the pressure it should operate at, so that half of the vapor is condensed;

b.the composition of the vapor leaving the condenser.

The liquid condensate is at 30°C as in Problem 14.8.

14.11 With reference to Example 14.4, plot:

a. the amount of vapor, and

b. its propane composition,

as a function of the drum operating pressure and comment on the purity of the obtained vapor. Operating temperature: 27°C.

14.12 At 40 atm and 333 K a liquid phase consisting of propane and a non-volatile hydrocarbon is in equilibrium with a vapor phase containing 85% (mole) nitrogen and the rest propane. Neglecting the solubility of nitrogen in the liquid phase, find the mole % of propane in the liquid.

14.13 The overhead vapor stream from a fractionating column going to the condenser has the following composition (molar):

ethylene = 0.20; propane = 0.30; n-butane = 0.50

It is desired to operate the condenser so that 70% (mole) of the total stream is liquified in the condenser. If the temperature in the condenser is 40°C, what is:

a. the required pressure?

b. the composition of each phase leaving the condenser?

14.14 A system consisting of methane, propane, and a non-volatile hydrocarbon is in equilibrium at 100°F and 600 psia. If the vapor phase contains 94.0% (mole) methane and the rest propane, what is the composition of the liquid phase?

14.15 The liquid feed to a fractionating column has the molar composition:

$$C_2 = 0.15;\ C_3 = 0.40;\ nC_4 = 0.45$$

If the temperature and pressure in the column at the point of introduction of the feed are respectively 50°C and 14 atm, what mole % of the feed is vaporized and what is the composition of the vapor?

14.16 A gas stream at 25°C and 1 atm, consisting of 20% (mole) nitrogen and the rest carbon dioxide, is separated into practically pure CO_2 with 95% recovery. The product streams leave the separation process at 1 atm and 25°C, the temperature of the surroundings. Find the minimum amount of work required in J/mol of CO_2 recovered.

14.17 Discuss the advantages and disadvantages in using the:

a. *K*-Charts;

b. Chao-Seader; and

c. Equations of State;

approaches in carrying out *VLE* calculations.

14.18 What is the 'solubility parameter'? How can it be calculated from vapor pressure data?

14.19.a. Could you use the regular solution theory for low pressure *VLE* calculations?

b. What information would you need?

c. What would the restrictions be?

14.20 In using cubic EoS for high pressure *VLE* calculations:

a. what information is needed?

b. when are interaction coefficients needed?

c. how are they evaluated?

14.21 Discuss the advantages and disadvantages of the EoS approach over the $\gamma - \hat{\phi}$ one in the correlation and prediction of *VLE*. Consider:

a. low pressures; and

b. high pressures, including the critical region;

for nonpolar and polar mixtures.

14.22 Experimental data for the system methanol-hexane are to be correlated with the SRK EoS.

a. What changes, if any, are necessary to this purpose in the SRK EoS?

b. How would you obtain the best results?

14.23 The feed to a butane-pentane splitter has the following composition:

$$iC_4 = 0.06; nC_4 = 0.17; iC_5 = 0.32; nC_5 = 0.45$$

For a 95% recovery of nC_4 it is estimated that the distillate will also contain: 0.0595 moles of iC_4, 0.016 moles of iC_5 and 0.0059 moles of nC_5 per mole of feed. Using a condensate temperature that you would recommend in this case, determine:

a. the operating pressure for a total condenser;

b. the composition of the liquid leaving the top tray;

c. the bottoms temperature and composition.

Make any assumptions you need, but justify them.

14.24 In Example 14.3 it is stated that use of k_{12}= 0.131 gives poorer prediction of vapor phase compositions than use of k_{12} = 0.148. Yet it appears from Figure 14.E.3 that the curve (dashed) corresponding to k_{12} = 0.131 is closer to the experimental dew point curve than that for k_{12} = 0.148. Determine which k_{12} gives the best values for K_1 and K_2.

14.25 Is the statement made in part (a) of Example 14.4 with reference to the D.P. pressure: "Below it, we will have superheated vapor", correct? Explain qualitatively.

14.26 What are the most important items you have learned in this Chapter? Give a possible industrial application in each case.

14.51 Using the data of Table 14.P.1 calculate:

a. the value of the interaction coefficient for the SRK, PR, and vdW-711 EoS at each temperature;

b. repeat using all data points together;

c. explain any differences in the interaction coefficient values.

14.52 Prepare a flow diagram for a bubble point pressure calculation with the Chao-Seader method.

14.53 Calculate the total pressure and vapor phase composition for the system $C_1(1)$ - $CO_2(2)$ at 230 K, x_1 = 0.170 and x_1 = 0.543, using the Chao-Seader method. Compare your findings with the experimental values given in Table 14.P.1.

14.54 Explain in terms of intermolecular forces the following observation: At 280 K and 70 bar total pressure, the *K* value for hexane in the methane-hexane system is much larger than that in the hydrogen - hexane one at the same tempe-

Table 14.P.1 Vapor-Liquid Equilibrium Data for Methane(1) - Carbon Dioxide(2). (Knapp et al, p.401).

T(K)	P (bar)	x_1	y_1	T (K)	P (bar)	x_1	y_1
230	8.916	0.000	0.000	250	30.397	0.053	0.361
230	20.265	0.050	0.525	250	50.662	0.166	0.575
230	40.529	0.170	0.728	250	70.926	0.326	0.615
230	55.728	0.318	0.764	250	79.539	0.405	0.564
230	62.821	0.394	0.762	270	31.947	0.000	0.000
230	68.900	0.534	0.751	270	37.013	0.018	0.108
230	69.994	0.543	0.730	270	42.140	0.040	0.190
230	71.403	0.584	0.716	270	58.575	0.113	0.353
250	20.265	0.010	0.104	270	80.633	0.260	0.411

rature and pressure. At 280 K and 1 bar, however, the two K values are about the same.

14.55 Develop the second virial coefficient expression for a binary mixture and the SRK EoS.

14.56 The mixture in Problem 14.13 is to be condensed completely at 30°C. Calculate the minimum pressure required to this purpose, using:

a. Eqs 14.5.1 through 14.5.5; and

b. an EoS of your choice.

Which answer is more accurate? Why?

14.57.a. Establish the reliability of Eqs 14.5.1 through 14.5.5 in predicting the K values of the hydrocarbons involved, using a method of your choice.

b. Demonstrate the impact of this uncertainty on a real problem of your choice again.

14.58 Correlate the data for NH_3-H_2O of Clifford and Hunter (1933) at all temperatures given, using:

a. the gamma-phi approach;

b. the EoS/G^E approach.

14.59 Answer Problem 14.25 quantitatively.

15
Chemical Reaction Equilibrium

The petrochemical industry was born in the early 1900's when Thomas Edison's electricity was becoming popular and the use of kerosene was on its way down. With a decreasing kerosene market, chemical engineers in the major oil companies began studying means for usefully converting the excess residue remaining after gasoline was distilled from petroleum.

The solution to the problem came as early as 1913, when William Burton of the Standard Oil Company (Indiana) put into operation a series of thermal "cracking" units known as "Burton Stills". This was the beginning of a method for breaking the larger carbon molecules into smaller ones through a heat process, a process for making chemicals such as ethylene, propylene, butylene, and other olefins from crude oil.

Then a new technique was developed - known as "catalytic cracking" - and before long there was an abundance of the petrochemicals needed for elastomers, plastics, and synthetic fibers. Ethylene dominates all the other petroleum building blocks and one of its first uses was as ethylene glycol, the anti-freeze used in cars. Another automotive-associated use was tetra-ethyl lead (TEL), a gasoline additive to reduce engine knock and increase fuel efficiency.

Soon after commercialization of ethylene glycol and TEL came a whole stream of ethylene derivatives such as styrene and polyethylene. Other major compounds made available from the cracking of liquid petroleum were the elastomers, synthetic fabrics, and the plastics.

("Ten Greatest Achievements of Chemical Engineers", American Institute of Chemical Engineers (AIChE), 1983.)

15.1 Introduction

Chemical reactors represent the cornerstone of the Chemical and Petroleum Industries. Thus, of the ten greatest achievements of Chemical Engineers - published by the American Institute of Chemical Engineers in celebrating its 75th anniversary - at least six involve chemical reactions:

* Synthetic rubber
* Drugs
* Plastics
* Synthetics replacing the silkworm
* Petrochemicals (see previous page)
* Fertilizers

Application of a chemical reaction for industrial purposes must consider two key questions:

1. How far will the reaction go if sufficient time to reach equilibrium is provided, i.e. what is the equilibrium conversion? and
2. How long will it take for a certain level of conversion - up to the equilibrium value, of course - to be attained?

The first represents the subject of Thermodynamics; the second, that of Kinetics, and both must be considered in the design of a reactor. Furthermore the two are often in conflict. For example in the case of the oxidation of SO_2 to SO_3 for the production of sulfuric acid, low temperatures favor high conversions but the kinetics is very slow, and the reverse is the case at high temperatures. Thus - as it is often the case in our profession - an optimum must be found, based on economic considerations.

15.2 Objective and Approach

The thermodynamic basis for calculating chemical reaction equilibrium was developed in Section 12.5. There we showed that if a system of components A_1, A_2, ..., A_k, reacting according to the equation:

$$\nu_1 A_1 + \nu_2 A_2 + \ldots + \nu_k A_k = 0$$

where ν_1, ν_2, ..., ν_k, represent the *stoichiometric* numbers, reaches equilibrium at constant temperature and pressure, then:

$$\ln K = \frac{-\Delta G^\circ}{RT} \tag{12.5.10}$$

where:

$$K = \prod \left(\frac{\hat{f}_i}{f_i^\circ} \right)^{\nu_i} = \prod \hat{a}_i^{\nu_i} \tag{12.5.11}$$

$\hat{a}_i$ is the activity of component i in the mixture, and ΔG° is the Gibbs free energy of the reaction - the free energy of the products (G_p°) minus that of the reactants (G_r°) both evaluated at the standard state, i.e. at the temperature of the reaction and a pressure of one atmosphere - referred to therefore as *standard free energy*:

$$\Delta G^\circ = G_p^\circ - G_r^\circ = \sum \nu_i G_i^\circ \tag{12.5.12}$$

Our objective in this Chapter is to demonstrate how these equations - with their rather abstract quantities - can be used in determining the equilibrium conversion of a reacting system, which - as outlined in the Introduction - represents the contribution of thermodynamics to the application of chemical reaction equilibrium in the Chemical and Petroleum Industries. More specifically our objective is to examine how:

* temperature,
* pressure, and
* feed composition

effect the equilibrium conversion of the reacting system.

We start with the evaluation of the standard free energy of the reaction as a function of the system temperature T - remember it is a function of T only - so that the equilibrium constant K can be evaluated at the temperature of interest.

We proceed with the determination of the equilibrium conversion from the known value of the equilibrium constant K, considering first single reactions taking place in a single phase, and examine the factors that effect this conversion.

We then move to the more complex cases: reactions taking place in heterogeneous systems and multiple reactions. To this purpose we consider first the Phase Rule and the Duhem theorem - as they apply to reacting systems - and discuss the methodology for identifying the number of independent reactions for the formation of a given equilibrium mixture.

Calculation of equilibrium conversions for heterogeneous systems and for multiple reactions is then outlined mostly through some typical examples.

We summarize our findings in the Concluding Remarks.

15.3 The Dependency of K on Temperature

15.3.1 The Problem

According to Eqs 12.5.10 and 12.5.12 evaluation of the equilibrium constant K for a given reaction - at a specified temperature T - requires values for the standard free energies of the components of the reaction, also at T. The latter can be replaced by the *standard free energy of formation*, ΔG_f^o, of each component from its elements, as we will see in Example 15.1.

Such values are available for a large variety of compounds, but at 298.2 K and the ideal gas state at P = 1 atm (or 1 bar). Values for selected compounds are presented in Table 15.1; for other compounds, see the same reference, Reid et al (1987), or the references given by Kyle (1984).

We proceed, therefore, to describe a method for evaluating K values at temperatures other than 298.2 K from the free energy of formation values of the components of the reaction at 298.2 K. To this purpose, we:

1. demonstrate that the variation of K with T depends on the standard enthalpy of the reaction ΔH°, which itself is a function of temperature;
2. establish the functional relationship of ΔH° to T; and
3. develop the expression of K as a function of temperature.

15.3.2 The Variation of K with the Enthalpy of the Reaction

Starting with the definition of the Gibbs free energy (Eq.1.6.5):

$$G = H - TS \tag{1.6.5}$$

we will demonstrate that:

$$\frac{d\ln K(T)}{dT} = \frac{\Delta H^\circ(T)}{RT^2} \tag{15.3.1}$$

where the symbolism (T) is introduced to emphasize the temperature dependency of K and ΔH°.

The standard enthalpy of the reaction, ΔH°, represents - like the free energy - the difference in enthalpy between products (H_p^o) and reactants (H_r^o), both evaluated at the standard state. It is calculated - again like the free energy - from the enthalpies of formation of the reaction components at the standard state, ΔH_f^o:

Table 15.1 Standard Enthalpy and Free Energy of Formation for Selected Compounds(1) at 298.2 K in the Ideal Gas State and, for ΔG_f^o, at 1 atm (kJ/mol)

Compound	ΔH_f^o	ΔG_f^o
Methane	-74.52	-50.519
Ethane	-83.82	-31.954
Ethylene	52.510	68.440
Acetylene	228.20	210.68
Propane	-104.68	-24.39
Propylene	19.71	62.15
n-Butane	-125.79	-16.70
i-Butane	-134.18	-20.76
n-Pentane	-146.71	-8.765
n-Hexane	-166.92	-0.08122
n-Heptane	-187.80	7.977
n-Octane	-208.75	16.00
Water	-241.814	-228.59
H_2S	-20.63	-33.44
NH_3	-45.898	-16.400
NO	90.250	86.567
NO_2	33.18	51.328
SO_2	-296.84	-300.12
SO_3	-395.72	-370.95
CO	-110.53	-137.15
CO_2	-393.51	-394.40
Methanol	-200.94	-162.32
Ethanol	-234.95	-167.85

(1) Daubert, T.E., Danner, R.P. (Eds.) 1992. *Physical and Thermodynamic Properties of Pure Chemicals. Data Compilation*, Hemisphere Publishing Corp., New York (extant 1992).

$$\Delta H^\circ = H_p^\circ - H_r^\circ = \sum \nu_i H_i^\circ = \sum \nu_i \Delta H_{f_i}^\circ \qquad (15.3.2)$$

Such values are available for a large variety of compounds, but at 298.2 K and the ideal gas state. Values for selected compounds are presented in Table 15.1. For other compounds, see the same reference, Reid et al, or the references given by Kyle.

Differentiation of Eq.1.6.5 yields:

$$dG = dH - TdS - SdT = dU + PdV + VdP - TdS - SdT$$

From the first law for a reversible change, $TdS = dU + PdV$, and consequently:

$$dG = VdP - SdT$$

Since we are interested in equilibrium attained at constant pressure:

$$\frac{dG}{dT} = -S$$

and, expressing S through Eq. 1.6.5, we obtain:

$$\frac{dG}{dT} = \frac{G-H}{T}$$

Rearrangement and division by (RT) yields:

$$\frac{1}{RT}\frac{dG}{dT} - \frac{G}{RT^2} = \frac{d(G/RT)}{dT} = \frac{-H}{RT^2}$$

It follows that:

$$\frac{d(\Delta G^\circ/RT)}{dT} = \frac{-\Delta H^\circ}{RT^2} \tag{15.3.3}$$

Introduction of Eq. 12.5.10 into Eq. 15.3.3 leads to the desired expression for the variation of K with ΔH°, Eq. 15.3.1.

Eq. 15.3.1 represents the starting point for establishing the temperature dependency of the equilibrium constant K. To this purpose we must first establish the temperature dependency of ΔH° on T.

15.3.3 The Temperature Dependency of ΔH°

To develop the temperature dependency of ΔH° on T we take advantage of the fact that enthalpy is a state function and, following the scheme depicted in Figure 15.1, we write Eq. 15.3.2 as:

$$\Delta H^\circ(T) = H_p^\circ(T) - H_r^\circ(T) = \left[H_p^\circ(T) - H_p^\circ(T_0)\right] + \\ + \left[H_p^\circ(T_0) - H_r^\circ(T_0)\right] + \left[H_r^\circ(T_0) - H_r^\circ(T)\right] \tag{15.3.4}$$

where $T_0 = 298.2$ K and the subscripts p and r refer to products and reactants respectively.

It can be easily shown (Problem 15.5, for example) that the first and third bracketed terms are given by:

$$\left[H_p^\circ(T) - H_p^\circ(T_0)\right] + \left[H_r^\circ(T_0) - H_r^\circ(T)\right] =$$

$$= \int_{T_0}^{T} \sum \nu_i \left[a_i + b_i T + c_i T^2 + d_i T^3 \right] dT$$

where a_i, b_i, c_i, and d_i are the constants in the temperature dependency expression of the heat capacity of component A_i (such as Eq.2.6.7 for an ideal gas).

The second bracketed term in Eq.15.3.4 represents, of course, the standard enthalpy of the reaction at $T_0 = 298.2$ K, $\Delta H^\circ(T_0)$.

After some algebra Eq.15.3.4 becomes:

$$\Delta H^\circ(T) = \Delta H_0 + S_a T + \frac{S_b}{2} T^2 + \frac{S_c}{3} T^3 + \frac{S_d}{4} T^4 \qquad (15.3.5)$$

where:

$$\Delta H_0 = \Delta H^\circ(T_0) - S_a T_0 - \frac{S_b}{2} T_0^2 - \frac{S_c}{3} T_0^3 - \frac{S_d}{4} T_0^4 \qquad (15.3.6a)$$

$$S_a = \sum \nu_i a_i \; ; \; S_b = \sum \nu_i b_i \; ; \; S_c = \sum \nu_i c_i \; ; \; S_d = \sum \nu_i d_i \qquad (15.3.6b)$$

Note. For liquids and solids a fifth term in the temperature dependency of the heat capacity is often needed (see, for example, Appendix C), and Eq.15.3.5 - and the ensuing Eqs 15.3.6 and 15.3.7 - must be modified accordingly.

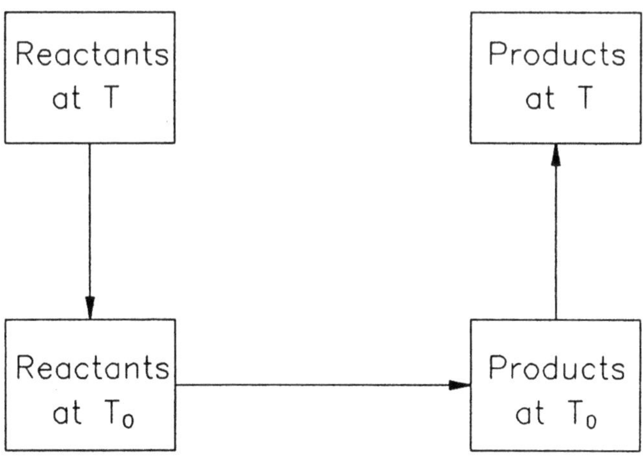

Figure 15.1 Scheme for the evaluation of the standard enthalpy of a reaction at temperature T from its value at T_0.

15.3.4 The Expression for $K(T)$

Introduction of Eq. 15.3.5 into Eq. 15.3.1 yields, after integration between T_0 and T, the expression for $K(T)$:

$$\ln K(T) = \ln K(T_0) - \frac{\Delta H_0}{R}\left(\frac{1}{T} - \frac{1}{T_0}\right) + \frac{S_a}{R}\ln\frac{T}{T_0} + \\ + \frac{S_b}{2R}\left(T - T_0\right) + \frac{S_c}{6R}\left(T^2 - T_0^2\right) + \frac{S_d}{12R}\left(T^3 - T_0^3\right) \qquad (15.3.7)$$

where $K(T_0)$ is the value of K at 298.2 K, given by Eqs 12.5.10 and 12.5.12:

$$\ln K(T_0) = \frac{-\Delta G°(T_0)}{RT_0} = \frac{-\sum \nu_i \, \Delta G_{f_i}°(T_0)}{RT_0} \qquad (15.3.8)$$

15.3.5 Comments

1. Evaluation, therefore, of the equilibrium constant for a given reaction as a function of temperature requires:

a. Standard free energy and enthalpy of formation values of the reaction components at 298.2 K (Table 15.1).

b. Heat capacities as functions of temperature for the reaction components at 1 atm (Appendix C).

2. If heat capacity data are not available, we can assume that $\Delta H°$ is independent of temperature and integrate Eq. 15.3.1:

$$\ln K(T) = \ln K(T_0) - \frac{\Delta H°}{R}\left(\frac{1}{T} - \frac{1}{T_0}\right) \qquad (15.3.9)$$

Eq. 15.3.9 gives good results for reasonable distances from T_0, as Example 15.2 will demonstrate.

3. Standard enthalpies of formation are calculated from calorimetric measurements, mostly combustion reactions, and those for the free energy through the relationship:

$$\Delta G° = \Delta H° - T\Delta S°$$

where $\Delta S°$ is the standard entropy change associated with the reaction.

4.The latter may be obtained from the absolute entropies of the pure species involved, calculated from Eq.3.12.3:

$$S(T) = S(T=0) + \int_0^{T_f} \frac{(C_P)_s}{T} dT + \frac{\Delta H_f}{T_f} + \int_{T_f}^{T_v} \frac{(C_P)_l}{T} dT + \frac{\Delta H_v}{T_v} + \int_{T_v}^{T} \frac{(C_P)_g}{T} dT \tag{15.3.10}$$

where:

* from the third law, $S(T=0) = 0$;

* the subscripts f and v stand for fusion and vaporization respectively, and s, l, g for solid, liquid and gas;

* no solid-phase transitions are involved.

Details are given by Lewis et al (1961) and by Denbigh (1981). See also Section 16.9.2 and Problem 15.58.

5.If no data on free energies and enthalpies of formation are available, determination of K at two temperatures is required to establish the temperature dependency for K, of the type given by Eq.15.3.9 (see Problem 15.7).

6.Values of the free energy of formation as a linear function of temperature can be estimated by group contribution techniques (Reid et al). The uncertainties, however, in the obtained values - combined with the sensitivity of the equilibrium conversion to the value of the standard free energy of the reaction (see Problem 15.28) - make such estimates useful for examining the feasibility of a given reaction only.

7.Free energy and enthalpy of formation values are typically available at 25°C and in the ideal gas state. Evaluation of their values at other conditions, such as in the liquid phase, is demonstrated in Example 15.3.

8.In Example 15.4 we apply our knowledge of chemical reaction equilibrium to one of the most important reactions, the combustion of fossil fuels, and examine the amount of useful energy that can be produced.

15.3.6 Example 15.1

Consider the reaction: $CH_4 + 2O_2 \rightarrow CO_2 + 2H_2O$

(1) (2) (3) (4)

Show that the standard free energy of this reaction:

$$\Delta G^\circ = G^\circ(3) + 2G^\circ(4) - G^\circ(1) - 2G^\circ(2) \tag{A}$$

can be expressed in terms of the standard free energies of formation of the compounds involved:

$$\Delta G^\circ = \Delta G_f^\circ(3) + 2\Delta G_f^\circ(4) - \Delta G_f^\circ(1) \tag{B}$$

For each compound the free energy can be calculated from the free energies of its elements and the corresponding free energy of formation, as follows:

$$G^o(1) = G^o(C) + 2G^o(H_2) + \Delta G_f^o(1)$$
$$G^o(2) = G^o(O_2) + 0.0$$
$$G^o(3) = G^o(C) + G^o(O_2) + \Delta G_f^o(3)$$
$$G^o(4) = G^o(H_2) + 0.5G^o(O_2) + \Delta G_f^o(4)$$

Introduction of these expressions into Eq.(A) gives Eq.(B). The same procedure can be used for the standard enthalpy of the reaction.

15.3.7 Example 15.2

For the reaction: $H_2 + (1/2)\,O_2 \rightarrow H_2O$, calculate K as a function of temperature from 300 K to 1500 K, using:
a. $\Delta H^o = f(T)$
b. ΔH^o = constant, equal to its value at $T_0 = 298.2$ K.
The following information is obtained from Table 15.1 and from Appendix C.

Comp.	$\Delta G_f^o(T_0)$	$\Delta H_f^o(T_0)$	a	$b\cdot 10^3$	$c\cdot 10^5$	$d\cdot 10^9$
	kJ/mol					
H_2	0	0	28.36	4.943	-0.920	6.142
O_2	0	0	28.66	-2.380	2.008	-11.500
H_2O	-228.59	-241.814	32.24	1.908	1.057	-3.602

a. From Eqs 15.3.6 and 15.3.8:

$$S_a = -10.450;\ S_b = -1.845\cdot 10^{-3};\ S_c = 0.973\cdot 10^{-5};\ S_d = -3.994\cdot 10^{-9}$$

$$\Delta H_0 = -238.69\ \text{kJ/mol, and } \ln K(T_0) = 228590/(RT_o) = 92.20$$

Hence from Eq. 15.3.7:

$$\ln K = 92.20 + 28710\left(\frac{1}{T} - \frac{1}{T_0}\right) - 1.257\ln\left(\frac{T}{T_0}\right) - 1.110\cdot 10^{-4}\left(T - T_0\right) +$$
$$+ 1.951\cdot 10^{-7}\left(T^2 - T_0^2\right) - 4.003\cdot 10^{-11}\left(T^3 - T_0^3\right) \qquad \text{(A)}$$

b. Using $\Delta H^o = -241.814$ kJ/mol = constant in Eq. 15.3.9:

$$\ln K = 92.20 + 29085\left(\frac{1}{T} - \frac{1}{T_0}\right) \qquad \text{(B)}$$

The results for the two methods are presented in Table 15.E.2. As expected, the errors are low at small distances from 298.2 K and become larger with increasing ones.

Table 15.E.2 Variation of ln*K* with Temperature, Example 15.2, Using Eqs (A) and (B)

T (K)	ln$K$(A)	ln$K$(B)	% Error
300	91.61	91.61	0.0
500	52.70	52.83	0.2
750	33.07	33.44	1.1
1000	23.17	23.75	2.4
1250	17.19	17.93	4.1
1500	13.19	14.05	6.6

15.3.8 Example 15.3

The standard free energy and enthalpy of formation of water vapor at 298.2 K are -228.59 and -241.814 kJ/mol respectively. What are the corresponding values for liquid water at the same conditions?

Notice that: $$\Delta G_f^o(l) - \Delta G_f^o(v) = G^o(l) - G^o(v)$$

where (l) and (v) stand for liquid and vapor respectively. If we now follow a scheme similar to that of Figure 15.1, noting that the superscript (°) implies P = 1 atm, i.e. $G^o(l) = G(l,1)$ and $G^o(v) = G(v,1)$, we have:

$$G(l,1) - G(v,1) = [\ G(l,1) - G(l,P^s)\ ] + [\ G(l,P^s) - G(v,P^s)\ ] + \\ + [\ G(v,P^s) - G(v,1)\ ]$$

where P^s = 3.166 kPa, is the vapor pressure of water at 25°C.

Since at constant temperature: $dG = VdP$, where V is the molar volume, we obtain for the first and third bracketed terms, respectively:
First term, V = 18 cm^3/mol:

$$G(l,1) - G(l,P^s) = V(1 - P^s) = 1.77 \text{ J/mol}$$

Third term, $V = RT/P$, valid in these low pressures:

$$G(v,P^s) - G(v,1) = RT\ln(P^s/1) = -8591.25 \text{ J/mol}$$

The second bracketed term is, of course, equal to zero (why?). Thus:

$$\Delta G_f^o(l) = -237.18 \text{ kJ/mol}$$

Similarly, the calculation of $\Delta H_f^o(l)$ leads to three terms: the first and third that represent the effect of pressure - in the range P^s to 1 atm - on liquid and vapor enthalpies, which are negligible in this pressure range; and the second, that represents the enthalpy difference between saturated liquid and vapor at 25°C, i.e. the negative of the enthalpy of vaporization of water at 25°C, equal to -44.0 kJ/mol. Hence:

$$\Delta H_f^o(l) = -285.81 \text{ kJ/mol}$$

15.3.9 Example 15.4

One of the most important chemical reactions involves the combustion of fossil fuels, since over 90% of the energy used in the world comes from them. Assuming that the fuel under consideration is propane, where the combustion reaction is:

$$C_3H_8 + 5O_2 \rightarrow 3CO_2 + 4H_2O$$

determine:

a. the maximum amount of useful energy, i.e. the exergy, that can be produced in a flow process at 25°C and 1 atm, and

b. the amount of exergy obtained as a function of temperature, from 400 K to 1500 K, by using the enthalpy of combustion also in a flow process.

a. The maximum obtainable work is given by Eq.5.6.1:

$$\Phi = W_s(id) = T_0(S_0 - S) - (H_0 - H) \qquad \text{(A)}$$

where S and H are the entropy and enthalpy of the system in its initial state, and the subscript ($_0$) refers to the final state where the temperature and pressure are those of the surroundings, taken as 298.2 K and 1 atm.

Since the combustion takes place at atmospheric pressure and 298.2 K, the initial state of the system here is the reactants at these conditions; and the final, the products also at the same conditions. Eq.(A) becomes, therefore:

$$\Phi = -\left[G_p^\circ - G_r^\circ\right] = -\Delta G^\circ$$

Using the free energy of formation data from Table 15.1, noting that water is in the liquid state (see Example 15.3):

$$\Phi = -[3(-394.40) + 4(-237.18) - (-24.39)] = 2107.5 \text{ kJ/mol}$$

Note. The exergy obtained from a chemical reaction, is referred to as *chemical* exergy (Section 5.7.4).

b. In this case water is in the vapor phase and, using enthalpies of formation and heat capacity data from Table 15.1 and Appendix C respectively, we obtain the following expression for the heat of the reaction from Eqs 15.3.5 and 15.3.6a:

$$\Delta H^\circ(T) = \Delta H^\circ(T_0) + S_a(T - T_0) + \frac{S_b}{2}(T^2 - T_0^2) + $$
$$+ \frac{S_c}{3}(T^3 - T_0^3) + \frac{S_d}{4}(T^4 - T_0^4)$$

where: $\Delta H^\circ(T_0) = 2045.8$ kJ/mol, and from Eq.15.3.6b:

$$S_a = 43.413;\ S_b = -4.139 \cdot 10^{-2};\ S_c = -9.476 \cdot 10^{-5};\ S_d = 7.159 \cdot 10^{-8}.$$

A plot of $\Delta H^\circ(T)$ vs. T is presented in Figure 15.E.4. To calculate next the exergy that can be obtained, we set up a Carnot cycle that absorbs the enthalpy of reaction at T and rejects heat at T_0 (Eq.5.6.3):

$$\Phi = -\Delta H^\circ(T)\left(1 - \frac{T_0}{T}\right)$$

A plot of exergy versus T is shown also in Figure 15.E.4, along with the chemical exergy at 298.2 K.

Comments

1. It is apparent from Figure 15.E.4 that use of the free energy gives the largest amount of useful energy (chemical exergy). This requires, however, that the reaction is carried out in a fuel cell. While research in this subject has been going on for a long time, the obtained efficiencies are still low, typically about 20 - 40%. In addition, high initial costs combined with the perceived risk element in this still-developing technology, have led to very limited application of fuel cells (Parkinson, 1986).

2. If we could use all the exergy obtained from the enthalpy of combustion at high temperatures, we could still recover about 80% of the chemical exergy at 298.2 K.

3. In practice however, even with modern power plants, we obtain only about 45% of the maximum amount based on the enthalpy of combustion (see Section 3.15).

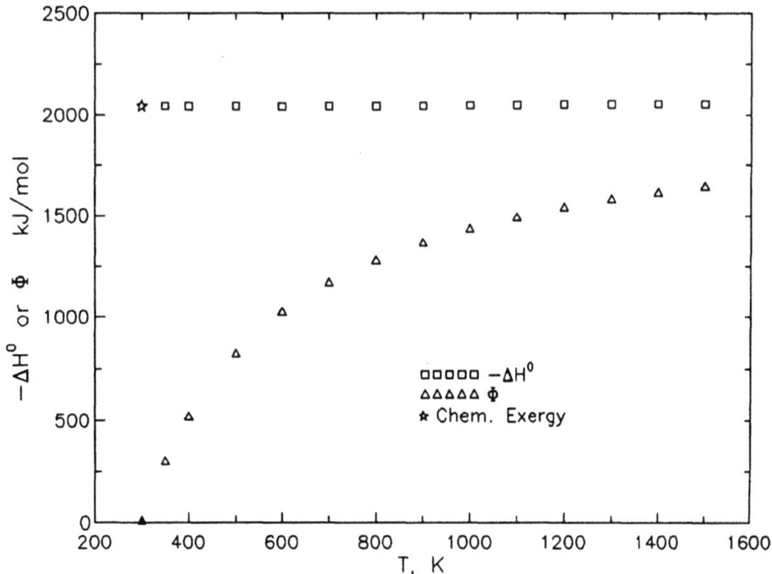

Figure 15.E.4 Enthalpy and exergy produced from the combustion of propane at atmospheric pressure.

15.4 Equilibrium Conversion: Gas-Phase Reactions

Once the value of the equilibrium constant K for the reaction under consideration - and the temperature of interest - is determined, the equilibrium conversion can be obtained. To this purpose we first express K in terms of compositions, which in turn are expressed in terms of the progress of the reaction variable (see Section 12.5.3). The latter is then calculated from the known K value and, consequently, the equilibrium conversion can be determined.

15.4.1 *K* in Terms of Composition

For a gas-phase reaction the activity of component i of the reacting mixture is given by:

$$\hat{a}_i = \frac{\hat{f}_i}{f_i^\circ} = \frac{\hat{f}_i}{1} = \hat{\phi}_i y_i P \tag{15.4.1}$$

and Eq.12.5.11 becomes:

$$K = \prod (\hat{\phi}_i y_i P)^{\nu_i} = K_\phi K_y P^\nu \tag{15.4.2}$$

where:

$$\nu = \sum \nu_i \quad ; \quad K_\phi = \prod \hat{\phi}_i^{\nu_i} \quad ; \quad K_y = \prod y_i^{\nu_i}$$

Since fugacity coefficients can be determined as functions of composition through an equation of state - such as the virial - the equilibrium constant K has thus been expressed in terms of composition.

15.4.2 Evaluation of the Equilibrium Conversion

As we have seen in Section 12.5.3, the number of moles of component i of the reaction mixture at equilibrium is given by:

$$N_i = N_i^\circ + \nu_i e \tag{15.4.3}$$

where e is the equilibrium value of the progress of the reaction variable and the superscript ($^\circ$) denotes the value of N_i at the beginning of the reaction.

Thus, the total number of moles present at equilibrium is:

$$N = \sum N_i = \sum N_i^{\circ} + e \sum \nu_i \tag{15.4.4}$$

and the corresponding mole fractions are given by:

$$y_i = \frac{N_i}{N}$$

All mole fractions are, therefore, expressed in terms of a single variable, e, which can be evaluated from the known value of K. The equilibrium conversions are then calculated from Eq.15.4.3.

Comments

1. At low pressures K_ϕ can be set equal to 1, and the fact that most gas-phase reactions take place at relatively high temperatures for kinetic purposes, provides additional support for this assumption.

2. At higher pressures K_ϕ can be substantially different from unity even at high temperatures, as we will see with the case of the ammonia synthesis (Example 15.7). In such cases K_ϕ can be estimated with an equation of state. This requires an iterative calculation which is initialized by setting $K_\phi = 1$.

In the following two Examples we demonstrate the evaluation of conversions for gas-phase reactions taking place at constant temperature and then consider the case of reactions occurring at adiabatic conditions.

15.4.3 Example 15.5

Ethylene can be produced from the vapor phase dehydration of ethanol:

$$\underset{(1)}{C_2H_5OH} \rightarrow \underset{(2)}{C_2H_4} + \underset{(3)}{H_2O}$$

Calculate the equilibrium conversion at 100°C and 1 atm. Given: K = 3.

At equilibrium, starting with one mole of ethanol:

$N_1 = 1 - e$
$N_2 = e$
$N_3 = e$
$N = 1 + e$

From Eq.15.4.2: $K = K_y K_\phi P^\nu$. Setting $K_\phi = 1.0$ - because of the low pressure involved - we arrive at the following expression for K_y:

$$K_y = 3 = \frac{y_2 y_3}{y_1} = \frac{e^2}{(1-e)(1+e)}$$

Solution gives $e = 0.866$ and the equilibrium conversion is 86.6%.

15.4.4 Example 15.6

Sulfuric acid is one of the most important inorganic chemicals and the main step in its production is the oxidation of SO_2 to SO_3:

$$\underset{(1)}{SO_2} + \underset{(2)}{0.5O_2} \rightarrow \underset{(3)}{SO_3}$$

which takes place in the presence of a catalyst - platinum in the earlier days, vanadium pentoxide more recently - and at atmospheric pressure. Determine the equilibrium conversion of SO_2 as a function of temperature from 400°C to 800°C. Since both the combustion of sulfur to SO_2 and its oxidation to SO_3 are carried out with air, assume that the feed contains 0.5 mole of O_2 and 5.6 moles of N_2 per mole of SO_2.

For calculational simplicity, use the following expression for K as a function of temperature (Daubert, 1985):

$$K = \exp\left[-\left(\frac{1}{R}\right)\left(-\frac{22630}{T} + 5.281\ln T - 0.959\cdot 10^{-2}\,T + 0.28\cdot 10^{-5}\,T^2 - 7.68\right)\right] \quad \text{(A)}$$

where R = 1.987 cal/mol K ; $[T]$ = K.

Values of K as a function of temperature are presented in Table 15.E.6. To calculate the equilibrium conversion, we note that:

$$K = K_y K_\phi P^\nu = K_y = \frac{y_3}{y_1 y_2^{0.5}}$$

assuming K_ϕ = 1.0 again.

Expressing the equilibrium amounts of the mixture components in terms of the progress of the reaction variable, e, we have:

$N_1 = 1 - e$
$N_2 = 0.5 - 0.5e$
$N_3 = e$
$N_4 = 5.6$ (Nitrogen)
$N = 7.1 - 0.5e$

Table 15.E.6 Equilibrium Constants and Conversions for the Oxidation of SO_2

T(°C)	K	Conversion %
400	440.15	96.0
500	45.13	83.4
600	7.77	57.8
700	1.90	30.0
800	0.60	13.0

Hence:

$$K = \frac{e/N}{[0.5(1-e)/N]^{0.5}[(1-e)/N]} = \frac{e(2N)^{0.5}}{(1-e)^{1.5}} \tag{B}$$

or:

$$K^2 (1 - e)^3 + e^3 - 14.2e^2 = 0$$

Solution of this cubic equation gives the value of *e* at each temperature. The corresponding conversions are presented also in Table 15.E.6. Notice that as the temperature increases the conversion drops off substantially, suggesting - from the thermodynamic point of view - operation at lower temperatures. On the other hand - from the kinetic point of view - rates are very slow at low temperatures, leading to very large reactors. This is the conflict that often occurs, as mentioned in the Introduction.

In practice, therefore, the reaction is carried out in a series of reactors, starting at higher temperatures and ending at about 450°C (Redlich, 1976). This, combined with the use of an excess in air that favors the reaction, leads to low concentration of SO_2 in the effluent gases especially important from the environmental point of view.

15.4.5 Adiabatic Reactions

Let us now consider the equilibrium conversion of a reaction that takes place adiabatically. In this case the temperature does not remain constant and it is evaluated by using an energy balance, in addition to the expression for the reaction equilibrium constant.

The approach will be demonstrated with the oxidation of SO_2 to SO_3 discussed in Example 15.6. To this purpose, let us assume that the feed enters the adiabatic reactor at 400°C = 673 K and 1 atm, and proceed to calculate the resulting:

* temperature *T*, and
* equilibrium conversion *e*,

when the mixture leaves the reactor.

At equilibrium, from Eq.(B) of Example 15.6:

$$K = \frac{e(2N)^{0.5}}{(1-e)^{1.5}} \tag{A}$$

Expressing *K* as a function of temperature through Eq.(A) - also of Example 15.6 - we obtain an equation containing two unknowns: *e* and *T*.

We need thus a second equation which will come from the fact that, since the reactor is adiabatic, the total enthalpy change ΔH will be equal to zero.

To express this enthalpy change in terms of *T*, we follow a scheme similar to that shown in Figure 15.1 and, assuming ideal gas behavior at $P = 1$ atm, we set $\Delta H = \Delta H^\circ$:

$$\Delta H^\circ = H_p^\circ(T) - H_r^\circ(673) = [H_p^\circ(T) - H_p^\circ(T_0)] +$$

$$[H_p^\circ(T_0) - H_r^\circ(T_0)] + [H_r^\circ(T_0) - H_r^\circ(673)] = 0 \tag{B}$$

The second bracketed term can be expressed now in terms of e through the enthalpy of the reaction at $T_0 = 298.2$ K, $\Delta H^\circ(T_0)$:

$$[H_p^\circ(T_0) - H_r^\circ(T_0)] = \Delta H^\circ(T_0) e \tag{C}$$

Using the enthalpy of formation values of Table 15.1 we obtain:

$$\Delta H^\circ(T_0) = -(395.72 - 296.84) = -98.88 \text{ kJ/mol}$$

For the first and third bracketed terms we have respectively:

$$[H_p^\circ(T) - H_p^\circ(T_0)] = \int_{T_0}^{T} [N_1 C_{P_1} + N_2 C_{P_2} + N_3 C_{P_3} + N_4 C_{P_4}] \, dT \tag{D}$$

$$[H_r^\circ(T_0) - H_r^\circ(673)] = \int_{673}^{T_0} [N_1^\circ C_{P_1} + N_2^\circ C_{P_2} + N_3^\circ C_{P_3} + N_4^\circ C_{P_4}] \, dT \tag{E}$$

where C_P is the ideal gas heat capacity.

Introduction of Eqs (C), (D) and (E) into Eq.(B) gives the required second equation.

To simplify the solution of this system of two equations, (A) and (B), with two unknowns, T and e, we will use the following approximate mean heat capacity values:

$$C_{P_1} = 47.9;\ C_{P_2} = 32.0;\ C_{P_3} = 68.0;\ C_{P_4} = 30.4;\ \text{all in J/mol K.}$$

The solution is found by trial-and-error. A simple method is plotting e versus T for the two equations. The intersection of the two lines gives the answer:

$$T = 890 \text{ K} = 617^\circ\text{C, and } e = 0.52.$$

15.5 Equilibrium Conversion: Liquid-Phase Reactions

To establish the relationship between conversion and the equilibrium constant in this case, we follow the same approach as for gas-phase reactions.

The activity of component i of a liquid mixture is given by:

$$\hat{a}_i = \frac{\hat{f}_i}{f_i^\circ} = \frac{x_i \gamma_i f_i^\circ(P)}{f_i^\circ(1)} \tag{15.5.1}$$

where $f_i^\circ(P)$ and $f_i^\circ(1)$ represent the fugacity of pure liquid i at the temperature of the system, but the former at the pressure of the system P, whereas the latter at the standard state of one atmosphere.

Except close to the critical point, the ratio of the two standard state fugacities can be calculated from:

$$\frac{f_i^\circ(P)}{f_i^\circ(1)} = \exp\left[\frac{V_i(P-1)}{RT}\right]$$

where V_i is the average value of the molar volume of pure liquid i in the pressure range from 1 atm to P at the temperature T. The expression thus for K, Eq.12.5.11, becomes:

$$K = K_x K_\gamma K_f \tag{15.5.2}$$

where:

$$K_x = \prod x_i^{\nu_i} \quad ; \quad K_\gamma = \prod \gamma_i^{\nu_i} \quad ; \quad K_f = \prod \left[\exp\left(\frac{V_i(P-1)}{RT}\right)\right]^{\nu_i}$$

Since:

* the activity coefficients can be expressed as a function of composition through, say, the Wilson equation, and
* the compositions can be expressed in terms of e,

the relationship between conversion and the equilibrium constant is established.

Comments

1. While K_ϕ for a gas-phase reaction can be estimated from an equation of state, evaluation of K_γ requires knowledge of the values of the necessary parameters in one of the models for the activity coefficient, that provide reliable predictions of multicomponent γ's from binary ones, such as that of Wilson.

2. If these parameters are not available, activity coefficients can be estimated with the UNIFAC method (see Problem 15.51).

3. Evaluation of equilibrium conversions requires, because of the nonlinear character of the equation, an iterative scheme which can be initialized by assuming ideal solution behavior.

4. For pressures not too far removed from 1 atm, K_f can be safely set equal to one.

15.6 Factors Effecting the Equilibrium Conversion

15.6.1 Temperature

As indicated by Eq.15.3.1, for exothermic reactions - ΔH° negative - an increase in temperature leads to a decrease in K and, consequently, in K_y or K_x. This, in turn, leads to a decrease in equilibrium conversion, as demonstrated in Example 15.6. The reverse is the case for endothermic reactions.

15.6.2 Pressure

While K is not a function of pressure, equilibrium conversions are effected by the latter.

For gas-phase reactions we obtain from Eq.15.4.2:

$$K_y = \frac{K}{K_\phi P^\nu} \tag{15.6.1}$$

Thus if $\nu < 0$, an increase in pressure increases K_y and, consequently, the equilibrium conversion. The reverse is true for $\nu > 0$, while for $\nu = 0$ pressure has no effect. A change in pressure effects K_ϕ, of course, but to a much smaller degree.

For liquid-phase reactions pressure effects K_γ and K_f, Eq.15.5.2, but to a negligible degree, unless very high pressures are involved.

15.6.3 Inert Components

Assuming that the presence of inert components increases the total number of moles N, and since:

$$K_y = K_x = \prod \frac{N_i^{\nu_i}}{N^\nu} \tag{15.6.2}$$

a negative ν leads to decreased conversion, and the reverse is the case when ν is positive. There is no effect when $\nu = 0$. The effect of inert components is very important in oxidation reactions since air, rather than pure oxygen, is often used.

15.6.4 Excess Reactant in Feed

Consider, as an example, the gas-phase reaction:

$$aA + bB \rightleftharpoons cC + dD$$

then:

$$K_y = \frac{N_C^{\,c} N_D^{\,d}}{N_A^{\,a} N_B^{\,b}} \frac{1}{N^{\nu}} \tag{15.6.3}$$

where: $\nu = c + d - a - b$

An increase in, say, the initial number of moles of component A would increase N_A and, consequently, N_C and N_D, i.e. the equilibrium conversion. While this is often the case, notice that the increase in N_A will also effect N and, depending on the value of ν, it could have an adverse effect on the equilibrium conversion (see Problem 15.19).

The effect of temperature on the equilibrium conversion has been demonstrated in Example 15.6. In the following Examples, 15.7 and 15.8, we demonstrate the effect of pressure and inerts respectively.

15.6.5 Example 15.7. The Ammonia Synthesis

Only a few short years ago, world leaders were predicting that the battle to feed humanity was over and that we could expect to undergo vast famines with hundreds of millions of people starving to death before the year 1985.

Chemical engineers took up the challenge in a variety of ways, perhaps the most significant of which was the breakthrough which permitted production of large quantities of ammonia in compact single-unit plants. Ammonia is the key building block for nitrogen fertilizers. And of all fertilizer elements, nitrogen is the most needed.

The basic process for synthesizing ammonia was developed early in this century, but what was needed was an inexpensive way to produce it in tremendous quantities. How well the chemical engineers succeeded is shown by the fact that as recently as 50 years ago, the annual production of synthesized ammonia was just over 300,000 tons. Thanks to chemical engineering breakthroughs, a single ammonia plant can today produce this with ease, and annual production is measured in the millions of tons. Many of these large-scale ammonia plants are strategically located in countries like China and India where on site production of fertilizers is helping the war against hunger.

Chemical engineering is at the heart of most other fertilizers as well.

One of the proudest accomplishments of the chemical engineering profession is the surge ahead in this round of the battle against the most constant enemy of humankind.
["Ten Greatest Achievements of Chemical Engineers", AIChE]

The synthetic production of ammonia:

$$\underset{(1)}{(1/2)\,N_2} + \underset{(2)}{(3/2)\,H_2} \rightarrow \underset{(3)}{NH_3}$$

attracted tremendous attention at the beginning of the 20th century, as the available natural fertilizers became insufficient to meet the needs of an increasing world population. More than 1000 catalysts were tried, for example, at BASF (Ludwigshafen, Germany) and are shown in the Company's museum.

Increased temperatures favor the rate of the reaction, but are not desirable from the conversion point of view. On the other hand, since $\nu = -1$, increased pressures favor conversion. Examine, thus, the effect of pressure on the equilibrium composition of ammonia, obtained from a stoichiometric mixture of H_2 and N_2 (3:1) at 450 °C, where $K = 7.739 \cdot 10^{-3}$ (Redlich).

Let z represent the mole fraction y_3 of ammonia in the equilibrium mixture. Then, since $y_2 = 3y_1$ (why?), and $y_1 + y_2 + y_3 = 1$, we have:

$$y_1 = (1 - z)\,(1/4); \;\; y_2 = (1 - z)\,(3/4)$$

From Eq.15.4.2:

$$K = K_y K_\phi P^\nu = \frac{z K_\phi P^{-1}}{\left[\dfrac{(1-z)}{4}\right]^{0.5}\left[\dfrac{3(1-z)}{4}\right]^{1.5}}$$

which leads to the following quadratic equation for z:

$$z^2 - (2+q)z + 1 = 0$$

where $q = 3.079 K_\phi / KP$.

At $P = 1$ atm, we can assume ideal gas behavior, $K_\phi = 1$ and $z = 2.5 \cdot 10^{-3}$. At $P = 10$ atm, K_ϕ cannot be assumed to be equal to 1. Rather it must be calculated from an equation of state through an iterative calculation. Redlich, for example, using the Redlich-Kwong equation of state, reports $K_\phi = 0.994$. The closeness of this value to one should not be surprising considering the high temperature involved. Using this value we obtain $z = 0.024$, a ten-fold increase over $P = 1$ atm. At higher pressures K_ϕ deviates much more from unity. Thus, it is equal to 0.304 at 1000 atm, which leads to $z = 0.707$.

Redlich presents calculated values of z - using this approach - for pressures up to 1000 atm, along with the corresponding experimental data (Larson and Dodge, 1923; Larson, 1924). Both are shown in Figure 15.E.7. The results obtained by assuming $K_\phi = 1.0$ are also included for comparison purposes, and indicate that - as expected - substantial errors are realized at very high pressures by assuming ideal gas behavior.

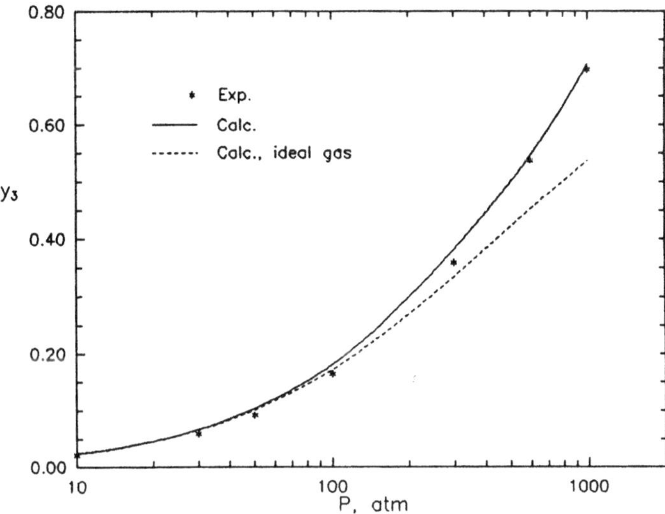

Figure 15.E.7 Equilibrium ammonia mole fraction at 450°C from a stoichiometric feed as a function of pressure.

15.6.6 Example 15.8

Repeat Example 15.6 using pure SO_2 and O_2 at their stoichiometric ratio at T = 600°C and P = 1 atm.

In this case: $N = (1 - e) + (0.5 - 0.5e) + e = 1.5 - 0.5e$, and from Eq.(B) of Example 15.6:

$$K^2(1 - e)^3 + e^3 - 3e^2 = 0$$

From Table 15.E.6, $K = 7.77$, which gives a conversion of 72.7%, substantially higher than the 57.8% one obtained in Example 15.6. This increase in the absence of the inert N_2 is in agreement with Section 15.6.3, since $\nu = -0.5$.

15.7 The Phase Rule and the Duhem Theorem for Reacting Systems

We have discussed, so far, equilibrium conversions for single reactions taking place in a single phase, mainly in the gaseous one, that represent a fraction of the industrial applications of chemical reaction equilibrium. We consider, next, reacting systems where:

* multiple reactions occur, a common case for gas-phase systems, or
* more than one phases are present, mostly gas-solid systems.

To this purpose we examine in this Section how the number of:

a. independent reactions, and
b. degrees of freedom,

are established for such systems. Evaluation of equilibrium conversions is considered in the next two Sections.

15.7.1 The Phase Rule and the Number of Independent Reactions

In Section 12.7 we showed that for a reacting system the phase rule takes the form:

$$F = k - r + 2 - \phi - SC \tag{12.7.4}$$

where:

F: number of degrees of freedom
k: number of components present
r: number of independent reactions taking place
ϕ: number of phases present
SC: number of imposed special constraints

While determination of k and ϕ is straightforward, this may not be always the case with r.

Consider, for example, an equilibrium mixture consisting of CO, CO_2, H_2, H_2O, and CH_4. How are we to determine the number of independent reactions r that lead to this equilibrium mixture?

The following procedure consisting of three steps is used to this purpose:

1. Write an equation for the formation of each compound from its elements.
2. Combine algebraically these equations to eliminate any elements not present in the equilibrium mixture. This is done by combining one equation, containing such an element, with any other equation containing the same element.
3. Repeat step 2 with the new set of equations until all elements not present in the equilibrium mixture are eliminated.

Comments

1. It is possible, of course, that two elements may be eliminated simultaneously.

2. When all elements, not present in the mixture, are eliminated we obtain the number of independent reactions r. Notice that different combinations may lead to different sets of final reactions but their number should be the same.

3. All sets of final equations should, of course, lead to the same equilibrium conversion.

Determination of the number of independent reactions and degrees of freedom is discussed in the following Examples.

15.7.2 Example 15.9

Determine the number of independent reactions for the equilibrium mixture of CO, CO_2, H_2, H_2O, and CH_4 mentioned in Section 15.7.1.

Step 1. Formation reactions:

$$C + (1/2)O_2 \rightleftharpoons CO \quad (A)$$

$$C + O_2 \rightleftharpoons CO_2 \quad (B)$$

$$H_2 + (1/2)O_2 \rightleftharpoons H_2O \quad (C)$$

$$C + 2H_2 \rightleftharpoons CH_4 \quad (D)$$

Step 2. Eliminate C, not present in the equilibrium mixture:

Subtract Eq.(A) from Eq.(B): $CO + (1/2)O_2 \rightleftharpoons CO_2$ (E)

Subtract Eq.(D) from Eq.(A): $CH_4 + (1/2)O_2 \rightleftharpoons CO + 2H_2$ (F)

The resulting, by eliminating C, new set of equations are (C), (E) and (F).

Step 3. Eliminate O_2:

Subtract Eq.(C) from Eq.(E): $CO + H_2O \rightleftharpoons CO_2 + H_2$ (G)

Subtract Eq.(C) from Eq.(F): $CH_4 + H_2O \rightleftharpoons CO + 4H_2$ (H)

The number of independent reactions is therefore two, (G) and (H). Notice, however, that a different combination procedure could lead to a different set of independent reactions, but again two, such as:

$$CO_2 + CH_4 \rightleftharpoons 2CO + 2H_2 \quad (I)$$

$$CH_4 + 2H_2O \rightleftharpoons CO_2 + 4H_2 \quad (J)$$

15.7.3 Example 15.10

Determine the degrees of freedom for the equilibrium mixture of Example 15.9, assumed to be in the vapor phase.

We note that: $k = 5$; $r = 2$; $\phi = 1$; $SC = 0$. Hence: $F = 5 - 2 + 2 - 1 = 4$.

If we specify, thus, the temperature and pressure, say the atmospheric ones of 1 atm and 298 K, we can specify only two additional variables such as the mole fractions of two of the five components in the equilibrium mixture, say those of CO and CO_2.

This conclusion appears contrary to our experience that, at atmospheric conditions, we can form mixtures of these five compounds with *any* composition desirable, provided that the composition of water is below that at the mixture's dew point. Such mixture, however, is not an equilibrium one from the thermodynamic point of view, but it appears to be so for - at this low temperature - the rate of the possible reactions is practically zero, and the mixture maintains its original composition. At higher temperatures, however, a new mixture determined by the thermodynamic equilibrium will be formed.

15.7.4 Example 15.11

Determine the degrees of freedom for the following equilibrium cases:

a. a two-phase mixture of ethylene, water, and ethanol resulting from the hydration of ethylene using the stoichiometric amounts of the reactants;

b. a mixture resulting from the decomposition of calcium carbonate in vacuum;

c. a vapor mixture of Cl_2, O_2, H_2O, and HCl resulting from the oxidation of HCl.

a. The reaction is:

$$C_2H_4 + H_2O \rightleftharpoons C_2H_5OH$$

and: $k = 3$; $r = 1$; $\phi = 2$; $SC = 1$ (specification of the C_2H_4/H_2O ratio).

Hence: $F = 3 - 1 + 2 - 2 - 1 = 1$, and only one variable, temperature or pressure for example, can be specified.

b. The reaction is:

$$CaCO_3 \rightleftharpoons CaO + CO_2$$

and: $k = 3$; $r = 1$; $\phi = 3$ (two solid phases + vapor); $SC = 0$.

Hence: $F = 1$, and only one variable, such as the system temperature or the pressure of CO_2, can be specified.

c. The reaction is:

$$2HCl + (1/2)O_2 \rightleftharpoons Cl_2 + H_2O$$

and: $k = 4$; $r = 1$; $\phi = 1$; $SC = 0$. Hence: $F = 4$.

15.7.5 The Duhem Theorem

In developing the Duhem theorem for non-reacting systems in Section 12.7, we subtracted the number of equations from the number of variables. For a reacting system the number of equations increases by r, the number of independent reactions; but so does the number of independent variables, since each reaction introduces a new variable, the equilibrium value of the progress of this reaction, e.

The Duhem theorem therefore can be stated in the following general form: *An equilibrium system, resulting from specified initial amounts of its components, is completely determined if two variables that vary independently at equilibrium are fixed, independently of the number of reactions and phases involved.* This theorem is very important in reacting systems, for we are often interested in determining the equilibrium composition of a system of specified initial amounts of reactants at constant temperature and pressure.

15.8 Equilibrium Conversion: Heterogeneous Systems

We will concentrate here on the most important type of these systems, those involving gas-solid reactions, such as combustion of solid fuels, decomposition of solids, and reduction of metal oxides. For the rather rare case of gas-liquid systems, see - among others - Sandler (1977) or Smith and Van Ness (1987).

15.8.1 The Equilibrium Constant in Terms of Composition for a Gas(g)-Solid(s) Reaction

Consider the reaction: $a\text{A(s)} + b\text{B(g)} \rightarrow c\text{C(s)} + d\text{D(g)}$

The equilibrium constant is given by:

$$K = \frac{\hat{a}_C^{\,c}\,\hat{a}_D^{\,d}}{\hat{a}_A^{\,a}\,\hat{a}_B^{\,b}} \tag{15.8.1}$$

For the solid compound i the activity is given by:

$$\hat{a}_i = \frac{x_i \gamma_i f_i^{\circ}(P)}{f_i^{\circ}(1)}$$

In the typical case, solids do not form solutions and consequently x_i and γ_i are equal to one. The activity of the solid, therefore, becomes:

$$\hat{a}_i = \exp\left[V_i(P-1)/RT\right] \tag{15.8.2}$$

where V_i is the molar volume of the solid at temperature T. With the exception therefore of very high pressures, the activities of solids are equal to one, and Eq.15.8.1 becomes:

$$K = \frac{(y_D \hat{\phi}_D P)^d}{(y_B \hat{\phi}_B P)^b} = K_y K_\phi P^{(d-b)} \tag{15.8.3}$$

In the following Examples we apply this equation to the determination of the equilibrium conversion for two typical industrial gas-solid reactions.

15.8.2 Example 15.12. Solid ⇀ Solid + Gas

Calculate the equilibrium partial pressure of CO_2 in the temperature range 400 K to 1000 K for the decomposition of calcium carbonate:

$$\underset{(1)}{CaCO_3} \rightharpoonup \underset{(2)}{CaO} + \underset{(3)}{CO_2}$$

In this temperature range, the equilibrium constant is given by the following approximate relationship: $\ln K = 18.84 - 21788/T$.

Eq.15.8.3 becomes: $K = K_y K_\phi P = y_3 P = \bar{P}_3$

A plot of $\ln\bar{P}_3$ versus $(1/T)$ is presented in Figure 15.E.12. Decomposition will take place to the left of the line only. Thus at 1000 K: $K = 0.052$, and the equilibrium partial pressure of CO_2 over the decomposing $CaCO_3$ is 0.052 atm. If $\bar{P}_3 > 0.052$ atm, no decomposition will take place.

15.8.3 Example 15.13. Solid + Gas ⇀ Solid + Gas

A mixture containing 20% CO and 80% N_2 passes over FeO reducing it to metallic iron:

$$\underset{(1)}{FeO} + \underset{(2)}{CO} \rightharpoonup \underset{(3)}{Fe} + \underset{(4)}{CO_2}$$

Assuming that equilibrium is reached, calculate the amount of iron produced per minute - at 1000 K and a pressure of 1 atm - given a gas mixture flow rate of 100 moles per minute. $K = 0.52$.

Eq.15.8.3 becomes: $K = K_y = \dfrac{y_4}{y_2}$

where: $y_2 = (20 - e)/N$; $y_4 = e/N$; $N = (20 - e) + e + 80 = 100$

Hence: $K = \dfrac{e}{20-e}$, $e = 6.84$, and 6.84 moles of Fe are formed per minute.

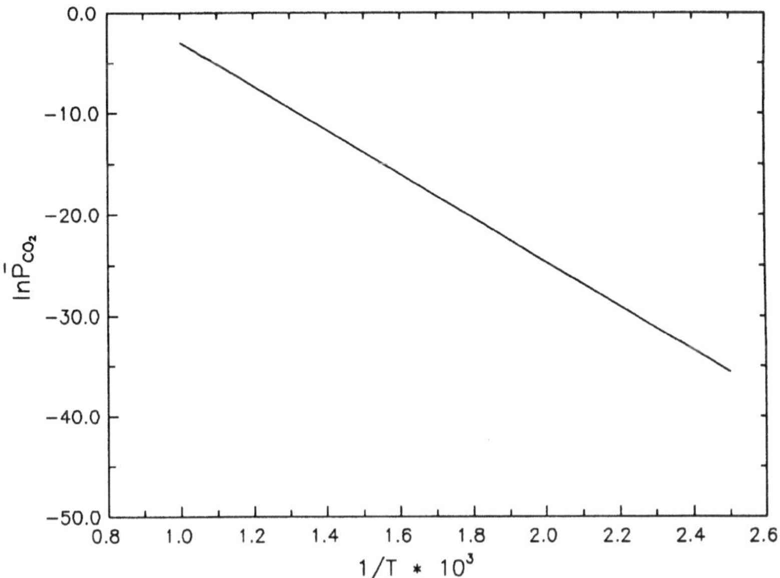

Figure 15.E.12 Equilibrium partial pressure of CO_2 from the decomposition of $CaCO_3$ ($\bar{P}$, atm; T, K).

15.9 Multiple Reactions

15.9.1 The Problem

We consider next systems where more than one reactions take place, which is often the case in gas-phase reactions. The thermodynamic analysis of such systems is straightforward for, as we saw in Section 12.6, each reaction is in equilibrium. Thus for a system involving r reactions, there are r progress of the reaction variables with equilibrium values of: $e_1, e_2, \ldots, e_r$, through which all compositions are expressed.

Determination thus of the equilibrium conversion requires the calculation of the r values of e, by solving a system of r equations with r unknowns. And this is where the problem lies. The equations are very nonlinear and solution of the resulting system becomes difficult. For r = 2, a trial-and-error or a graphical solution can be easily determined. For r = 3, a graphical solution is possible, but somewhat tedious. If repeated calculations are required, therefore, this approach becomes impractical. In such cases, or for $r > 3$, a solution using Lagrange multipliers must be used.

We will demonstrate the graphical method for systems involving two and three reactions in the following three Examples. These cover the majority of industrial applications, since systems involving more than three reactions are relatively rare. For the approach that uses Lagrange multipliers, see among others: Smith and Van Ness; Daubert; or Oliver et al (1962).

15.9.2 Example 15.14

One of the most important industrial reactions is the catalytic production of the 'synthesis gas':

$$CH_4 + H_2O \rightleftharpoons CO + 3H_2$$

An additional reaction takes place to a significant extent:

$$CO + H_2O \rightleftharpoons CO_2 + H_2$$

Starting with a ratio of 5 moles of steam to 1 mole of methane, i.e. with an excess of steam for reasons that will be discussed in Example 15.16, calculate the equilibrium composition of the resulting mixture at 600°C and 1 atm. At this temperature, the equilibrium constants for the first and second reaction are respectively (Hougen et al, 1959): $K_1 = 0.574$; $K_2 = 2.21$.

Let component 1, CH_4; 2, H_2O; 3, CO; 4, H_2; and 5, CO_2; and e_1, e_2 the values of the progress of the reaction variables for the first and second reaction respectively at equilibrium. Then:

$N_1 = 1 - e_1$
$N_2 = 5 - e_1 - e_2$
$N_3 = e_1 - e_2$
$N_4 = 3e_1 + e_2$
$N_5 = e_2$
$N = 6 + 2e_1$

At $P = 1$ atm, $K = K_y$ for both reactions. Hence:

$$K_1 = \frac{y_3 y_4^3}{y_1 y_2} = \frac{(e_1 - e_2)(3e_1 + e_2)^3}{N^2(1 - e_1)(5 - e_1 - e_2)} = 0.574 \qquad \text{(A)}$$

$$K_2 = \frac{y_5 y_4}{y_3 y_2} = \frac{e_2(3e_1 + e_2)}{(e_1 - e_2)(5 - e_1 - e_2)} = 2.21 \qquad \text{(B)}$$

We have thus a system of two equations with two unknowns, which we will solve with the graphical method as follows:

1. Assume a value for e_1 and determine the corresponding value of e_2 from Eq.(A) by trial-and-error, since it is a fourth degree equation.
2. For the same value of e_1, calculate e_2 by solving Eq.(B) which is quadratic.
3. Repeat, using a new value for e_1.
4. Plot the obtained values of e_2 vs. e_1 for the two equations. The intersection is the solution, giving:

$$e_1 = 0.912 \text{ and } e_2 = 0.633.$$

Hence: $y_1 = 0.0112, y_2 = 0.4416, y_3 = 0.0357, y_4 = 0.4306, y_5 = 0.0809$.

15.9.3 Example 15.15

With reference to the case presented in Example 15.14, a third possible reaction is:

$$2CO \rightleftharpoons C + CO_2$$

with a value of $K_3 = 8.14$ at 600°C. This reaction is undesirable because the carbon deposits on the catalyst reducing its activity. Considering the conditions of Example 15.14, will there be any carbon deposition?

The value of K_3 for this reaction, calculated from the results of Example 15.14 using Eq.15.8.3, is:

$$K_3 = y_5/y_3^2 = 63.5$$

which is much larger than the equilibrium value of 8.14. No carbon, therefore, will be formed(why?).

15.9.4 Example 15.16

An examination of the second reaction of Example 15.14:

$$CO + H_2O \rightleftharpoons CO_2 + H_2$$

indicates that an excess of water shifts the reaction to the right, increasing the amount of CO_2. This increase, in turn, shifts the third reaction:

$$2CO \rightleftharpoons C + CO_2$$

to the left, prohibiting the undesirable formation of carbon. At what steam to methane ratio will carbon be formed?

Following Example 15.14, we replace the 5 moles of H_2O by z, the unknown number of moles of steam per mole of methane, and introduce an additional equation resulting from the formation of carbon. Thus we have the following system of three equations with three unknowns:

$$K_1 = \frac{(e_1-e_2)(3e_1+e_2)^3}{N^2(1-e_1)(z-e_1-e_2)} = 0.574 \quad \text{(A)}$$

$$K_2 = \frac{e_2(3e_1+e_2)}{(e_1-e_2)(z-e_1-e_2)} = 2.21 \quad \text{(B)}$$

$$K_3 = \frac{Ne_2}{(e_1-e_2)^2} = 8.14 \quad \text{(C)}$$

where: $N = z + 1 + 2e_1$.

This system can be solved graphically for the three unknowns: e_1, e_2, and z, by finding the point of intersection of three lines in space. To this purpose:

1. Assume a value of z;
2. Plot e_2 versus e_1 for the three Eqs (A), (B), and (C);
3. Repeat until the obtained from the three lines triangle becomes a point.

This is demonstrated in Figure 15.E.16 and gives:

$z = 1.38$; $e_1 = 0.539$; and $e_2 = 0.227$.

15.10 Concluding Remarks

1. The equilibrium conversion for a given reaction is determined from knowledge of its equilibrium constant K, which in turn is evaluated in the typical case from data on the:

a. standard free energy and enthalpy of formation at 298.2 K; and

b. heat capacity as a function of temperature at $P = 1$ atm;

of the reactants and products.

2. In addition, for gas-phase reactions an equation of state is needed to estimate the fugacity coefficients of the mixture components; and for liquid-phase ones, parameter values of some model for the activity coefficient.

3. Because of the logarithmic relationship involved the value of K, and therefore that of the equilibrium conversion, for a reaction is very sensitive to the values used for the free energy and enthalpy of formation (Problem 15.28).

4. Since $\ln K = -\Delta G^\circ/RT$, the feasibility of a reaction is determined by the value of ΔG°. The following approximate criteria can be used:

$\Delta G^\circ < 0$:	Very promising, high conversion;
$0 < \Delta G^\circ < 4$:	Promising, but low conversion;

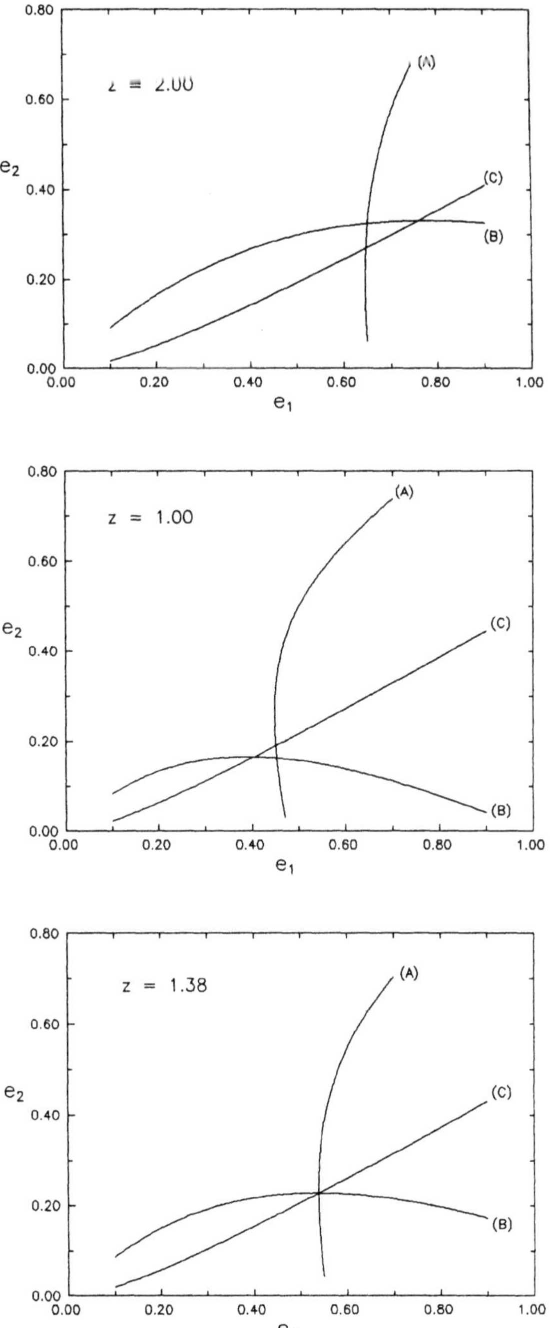

Figure 15.E.16 Graphical solution of Example 15.16.

$4 < \Delta G^{o} < 40$: Doubtful, but it warrants further examination;
$\Delta G^{o} > 40$: Very unfavorable;
where ΔG^{o} is in kJ/mol.

The term 'feasible' is used here in the practical sense for, of course, every reaction is feasible even if the equilibrium conversion is of the order of 10^{-20}.

5.The fact that a reaction is thermodynamically feasible does not necessarily mean that it will indeed take place. A catalyst is often needed for the reaction to attain significant conversions.

6.Keep in mind, however, that while equilibrium conversions can be determined - or at least estimated - in the majority of cases without resort to experimental work, this is not the case with finding the appropriate catalyst. Even though great progress has been made in identifying promising catalysts - since the days when 1000 catalysts were tried for the ammonia synthesis - extensive experimental work is still required.

7.Catalysts can be also used to suppress undesirable, even highly favored thermodynamically, side-reactions. Thus in the methanol synthesis, as an example:

$$CO + 2H_2 \rightarrow CH_3OH$$

a second reaction:

$$CH_3OH + H_2 \rightarrow CH_4 + H_2O$$

takes place.

With values of K at 600°C in the order of 10^{-4} and 10^{10} respectively for the first and second reaction, the second one would not allow the formation of any methanol. Yet, by using the proper catalyst and high pressures, production of methanol is realized.

8.Heterogeneous and multiple reactions represent a complex topic that has been covered very briefly here. For a more extensive discussion, see Kyle (1984) and, especially, Smith and Misseu (1982). The last reference includes also appropriate computer subroutines. Finally, for reactions involving electrolytic solutions, see Sandler.

References

Colombo, F., Cori, L., Dalloro, L., Delogu, P., 1983. *Ind. Eng. Chem. Fundam.*, **22**, 219.

Daubert, T.E., 1985. *Chemical Engineering Thermodynamics*, McGraw-Hill, New York.

Denbigh, K., 1981. *The Principles of Chemical Equilibrium*, Cambridge University Press, Cambridge.

Hougen, O.A., Watson, K.M., Ragatz, R.A., 1959. *Chemical Process Principles, Part II, Thermodynamics*, Wiley, New York.

Kyle, B.G., 1984. *Chemical and Process Thermodynamics*, Prentice-Hall, Englewood Cliffs, N.J.

Larson, A.T., Dodge, R.L., 1923. *J. Am. Chem. Soc.*, **45**, 2918.

Larson, A.T., 1924. *Ibid.*, **46**, 367.

Lewis, G.N., Randall, M., Pitzer, K.S., Brewer, L., 1961. *Thermodynamics*, McGraw-Hill, New York.

Oliver, R.C., Stefanou, S.E., Baier, R.W., Feb. 19, 1962. *Chemical Engineer*, p. 21.

Parkinson, G., Dec. 8, 1986. *Chemical Engineering*, p.14.

Redlich, O., 1976. *Thermodynamics: Fundamentals and Applications*, Elsevier, New York.

Reid, R.C., Prausnitz, J.M., Poling, B.E., 1987. *The Properties of Gases and Liquids*, 4th Ed., McGraw-Hill, New York.

Sandler, S., 1977. *Chemical and Engineering Thermodynamics*, Wiley, New York.

Smith, W.R., Misseu, R.W., 1982. *Chemical Reaction Equilibrium Analysis*, Wiley, New York.

Smith, J.M., Van Ness, H.C., 1987. *Introduction to Chemical Engineering Thermodynamics*, McGraw-Hill, New York.

Problems

15.1 Consider the reaction: $A + B \rightarrow C$
Discuss the type of information that can be obtained about this reaction from:

a. thermodynamics, and

b. kinetics.

15.2 For the gas-phase reaction: $C_2H_4 + H_2O \rightarrow C_2H_5OH$
the equilibrium constant at 145°C and 1 bar is $K = 6.8 \cdot 10^{-2}$. How would you calculate K at $P = 10$ atm?

15.3.a. Have you used a scheme of the type depicted in Figure 15.1 for any other thermodynamic calculations? Explain.

b. Discuss the information needed for the evaluation of the standard state fugacity for a gaseous, liquid, or solid compound involved in a reaction.

15.4 For the reaction: $A + B \rightarrow C + D$
what information is needed in order to establish the dependency of the equilibrium constant K on temperature?

15.5 For the gas phase reaction: $aA + bB \rightarrow cC + dD$
show that, with reference to Eq. 15.3.4:

$$[H_p°(T)-H_p°(T_0)]+[H_r°(T_0)-H_r°(T)] = \sum \int_{T_0}^{T} \nu_i C_{P_i}^{*} dT$$

15.6 Develop the temperature dependency of the equilibrium constant for the reaction: $SO_2 + (1/2)O_2 \rightarrow SO_3$.

15.7.a. Show that knowledge of K at two temperatures can be used to establish the temperature dependency of K.

b. State any assumptions you make and discuss the resulting limitations.

15.8 Present a schematic diagram for determining:

a. the standard free energy of formation in Example 15.3;

b. the enthalpy of the adiabatic reaction of Section 15.4.5 as a function of the temperature T of the products.

15.9 For the gas-phase reaction: A $\rightarrow$ 2B taking place at 400 K and 2 bar, K = 45.6. Calculate the equilibrium conversion, starting with pure A. Begin with:

$$K = \frac{\hat{a}_B^{\,2}}{\hat{a}_A}$$

15.10 Consider the gas-phase reaction: A + B $\rightarrow$ C + D
Starting with one mole of A and one of B, 0.75 moles of C are found at equilibrium at 400 K and 1 bar. How many moles of C will be present at equilibrium, under the same conditions, if we start with three moles of A per mole of B?

15.11 Determine the equilibrium conversion for the production of methanol from a stoichiometric ratio of CO and H_2 at 275°C and 100 atm. Assume the mixture to be an ideal one of real gases.

15.12 What is the maximum temperature that can be attained by burning CH_4 at 1 atm with the stoichiometric amount of O_2, both available at 25°C and 1 atm?

15.13 Consider the feasibility of producing ethanol through the vapor-phase hydration of ethylene:

$$C_2H_4 + H_2O \rightarrow C_2H_5OH$$

Determine the equilibrium conversion for an equimolar feed at 200°C and:

a. 1 atm;

b. 3 atm.

15.14 Consider the combustion of methane: $CH_4 + 2O_2 \rightarrow CO_2 + 2H_2O$ for the production of electricity.

a. What is the maximum amount of electricity that can be produced?

b. If we were to use the heat of combustion, what is the maximum amount of electricity that would be produced as a function of temperature?

c. Explain why in practice the amount of electricity produced is only about half of that in your answer in part (b).

15.15 Consider a power plant that absorbs 95% of the heat of combustion of methane and operates as follows:

a. steam at 100 bar and 500°C leaves the boiler/superheater;

b. the turbine is adiabatic with an efficiency of 0.9;

c. the condenser operates at 1 mm Hg and the water leaving it is saturated.

How much electricity is produced per mole of methane? Compare your answer with the results of the previous problem.

15.16 The reaction: $A + B \rightleftharpoons C$
takes place in the vapor phase at 200°C and 2 bar with $K = 15$.

a. What is the equilibrium conversion for a stoichiometric ratio of reactants?

b. If we add to the reactants one mole of an inert gas per mole of A (and B), will the conversion increase or decrease?

c. Prove your answer by calculating the actual conversion.

15.17 Calculate ΔG_f^o and ΔH_f^o for saturated liquid n-hexane at 350 K.

15.18.a. What is the LeChatelier principle?

b. What does it predict about the effect of temperature, pressure, and excess reactant in the feed, on the equilibrium conversion? Compare the predictions with the findings in Section 15.6.

15.19 For the ammonia synthesis at 450°C and 5 atm:

a. plot the equilibrium conversion as a function of the initial ratio: moles of H_2 per mole of N_2, from 1 to 5, making any assumptions you consider appropriate;

b. estimate the effect of these assumptions on the accuracy of the obtained results;

c. do your results satisfy the LeChatelier principle?

15.20 Under what thermodynamic conditions does the equilibrium constant K have a maximum or a minimum value?

15.21 What is the minimum information needed to determine the equilibrium conversion of a:

a. liquid, or

b. vapor,

phase reaction?

15.22 Consider a gaseous mixture composed of: NH_3, NO_2, NO, H_2O, O_2, and N_2. Determine the number of independent reactions.

15.23 What is the dew point composition of water in the mixture of Example 15.9 at 298 K and 1 bar (make any assumptions you deem appropriate)?

15.24 At 600 K the equilibrium constant for the reaction:

$$CO + 2H_2 \rightleftharpoons CH_3OH$$

is: $K = 9.0 \cdot 10^{-5}$. Starting with one mole of CO and 3 moles of H_2, what pressure is required to convert 85% of the CO to methanol at 600 K, assuming equilibrium conversion?

Consider ideal solution behavior, and for the purpose of calculating fugacity coefficients, assume the required pressure to be 300 atm.

15.25 Mobil has developed a technology whereby methanol can be converted into gasoline. Show how coal can be a starting material for the production of synthetic gasoline. What other chemicals can be produced from coal? (See Chem. & Eng. News, August 14, 1989, p.25.)

15.26 Determine the number of independent reactions and degrees of freedom for an equilibrium gas mixture of CO_2, H_2, CH_3OH, CO and H_2O.
15.27 The following simultaneous reactions:

$$A + 2B \rightarrow 3C + D \quad K = 5.0$$
$$A + C \rightarrow D + E \quad K = 0.68$$

take place in the vapor phase at 1 bar. Starting with one mole of A and one of B, how many moles of C, D, and E are present at equilibrium.
15.28 To investigate the sensitivity of equilibrium conversions to the standard free energy and enthalpy of formation values at 298.2 K, determine the amounts of NO, NO_2, and NH_3 formed at 1 atm from stoichiometric feeds at 500 K, 1000 K, 1500 K and 2000 K using:

a. the values given in Table 15.1;
b. the same ΔG_f^o values but increase ΔH_f^o by 1%;
c. the reverse of (b).

Assume, for simplicity, temperature independent enthalpy of formation.
15.29 In Section 15.10 the comment is made that if $4 < \Delta G^o$ (kJ/mol) < 40, the reaction is doubtful, but warrants further consideration. What factors would you consider?
15.30 Discuss the desirable and undesirable effects of using excess steam in the production of synthesis gas (Examples 15.14, 15.15 and 15.16).
15.31 What is the reason for the presence of two standard states in Eq.15.5.1?
15.32.a. Develop Eqs (I) and (J) in Example 15.9.
b. Prove with any method you want that this set of reactions and the set (G) and (H) lead to the same composition for the resulting equilibrium system.
15.33 Calculate the equilibrium conversion of SO_2 in the adiabatic reaction of Section 15.4.5 using the information given in Table 15.1 and in Appendix C.
15.34 Consider the reaction: C(graphite) → C(diamond). At what pressure would graphite be converted to diamond at 298.2 K?
Given(Sandler): Standard state free energy change: 685 cal/mol; $d_{(graphite)} = 2.25$ g/cm^3; $d_{(diamond)} = 3.51$ g/cm^3. Assume both substances to be incompressible.
15.35 Starting with Eq.15.8.1 develop Eq.15.8.3.
15.36 Consider a gas-phase and a liquid-phase reaction of your choice and describe, in each case, the modifications that you would use in order to increase the equilibrium conversion.
15.37 With reference to this Chapter:

a. what are main objectives?
b. discuss briefly the most important items that you have learned, and give an example in each case.

15.51 Methyl t-butyl ether has good antiknock properties and is becoming thus progressively an important gasoline additive as tetra-ethyl lead is phased out for environmental reasons. It is produced industrially by the liquid-phase reaction of i-butene with methanol:

$$i\text{-}C_4H_8 + CH_3OH \rightarrow t\text{-}C_4H_9OCH_3$$

Colombo et al (1983) report the following expression for K:

$$K = K_x K_\gamma = -10.0982 + \frac{1251.05}{T} + 0.2667 \ln T \quad (T, \text{K})$$

In addition to i-butene, the feed contains 1-butene.

For a feed containing: 1 mole of methanol, 1 mole of i-butene, and 1.22 moles of 1-butene, calculate the equilibrium conversion at 60°C. Use the UNIFAC model for activity coefficients. For details and experimental results, see Colombo et al.

15.52 Assuming that the pyrolysis of propane involves the following two reactions:

$$C_3H_8 \rightleftharpoons C_3H_6 + H_2$$
$$C_3H_8 \rightleftharpoons C_2H_4 + CH_4$$

calculate the composition of the equilibrium mixture at 1000 K and 1 atm.

15.53 A mixture of $BaCO_3$ and C is heated to 1482 K in an evacuated batch container, where the following reactions occur:

$$BaCO_3 \rightleftharpoons BaO + CO_2 \qquad K = 0.1 \text{(Daubert)}$$
$$CO_2 + C \rightleftharpoons 2CO$$

What pressure will be developed at equilibrium?

15.54 Butene-1 can be produced by the catalytic dehydrogenation of n-butane:

$$nC_4H_{10} \rightleftharpoons C_4H_8 + H_2$$

Further dehydrogenation of butene-1 to 1,3-butadiene is also possible:

$$C_4H_8 \rightleftharpoons C_4H_6 + H_2$$

Calculate the equilibrium conversion at 1 atm and:

a. 800 K; and

b. 1000 K.

Given for 1,3-butadiene at 298.2 K; $\Delta G_f^\circ = 36.01$; $\Delta H_f^\circ = 26.33$, both in kcal/mol.

15.55 A coal gasifier is fed with steam and air in the molar ratio of 1:2.5 and produces a stream containing H_2, CO, O_2, H_2O, CO_2, and N_2. Assuming the coal bed to be pure carbon, calculate the equilibrium composition of the stream at 10 atm and 1000 K.

The following standard free energy of formation data, in kJ/mol, are given at 1000 K: H_2O: -197; CO: -200; CO_2: -396.

15.56 What are the environmental reasons mentioned in Problem 15.51?

15.57 In evaluating the first integral in Eq. 15.3.10, what happens as $T_f \rightarrow 0$?

15.58.a. What is actually meant by 'absolute' entropy in connection with the third law?

b. What are its implications on calculating entropy of reactions?

15.59 Write an essay on the following argument, advanced by some historians: The development of the process for the synthetic manufacture of NH_3 prolonged World War I.

16
Elements of Statistical Mechanics

16.1 Introduction

Thermodynamics, as we have studied it so far, is referred to as *Classical* Thermodynamics and is built upon its empirical laws. They are called empirical because they are the results of observations through the years (Chapter 4) and their validity lies with the fact that they have never failed.

Using these laws and the three concepts (properties) introduced by them: temperature, internal energy, and entropy, along with the auxiliary ones: enthalpy, free energies, heat capacity, etc., the whole building of thermodynamics is constructed.

Yet, classical thermodynamics provides no clues about the origin of these laws, especially of the more complex second one, or any methods for determining these properties.

We might, therefore, ask at this point:

1. Is there a rigorous, from first principles, development of the second law? (and perhaps something deeper, or more 'profound' philosophically, in it); and
2. Is there an avenue for the evaluation of various thermodynamic properties without having to resort to experimental measurements, as is the case with the classical approach?

To look for possible answers to these questions we recall that classical thermodynamics deals with the bulk, or macroscopic, properties of matter. It does not consider its molecular structure. Internal energy and entropy, for example, are defined completely independently of any molecular considerations. (The reason why we have used, on certain occasions, such considerations was to gain some insight, not because they were needed per se).

It is reasonable, therefore, to expect that consideration of the molecular structure of matter might provide some answers to the aforementioned questions. This link between molecular and macroscopic behavior is the task of *Statistical Mechanics*.

16.2 Objective and Approach

To carry out this task, statistical mechanics combines two major fields of knowledge:

1. Classical and quantum mechanics, that provide the *description* of the molecular behavior; and
2. Statistical theory, that provides the means for *averaging* this molecular behavior, so that the experimentally observed macroscopic properties of matter can be calculated.

Statistical mechanics is a broad field that cannot be covered in the limited space available here. On the other hand, its contributions to chemical engineering thermodynamics - practically minimal a few years ago - are becoming progressively more and more important.

We will attempt, therefore, to provide a simple, elementary, introduction into the subject in the hope that it will help the reader comprehend these - and future - contributions. More specifically, our objective in this Chapter is to develop some understanding of:

1. the approaches used for the description of molecular behavior;
2. the methodology for determining macroscopic properties by averaging this behavior; and
3. the practical capabilities of statistical mechanics by applying the first two items to the case of an ideal gas.

Application to real fluids is considered in the next Chapter.

We will start with a brief discussion of the two methods used for describing molecular behavior: classical and quantum mechanics, and then consider the averaging process: first the principle behind it and then the results obtained for the special case of a constant-volume closed system in equilibrium with a heat reservoir.

This will lead us to the most important concept in statistical mechanics: the *partition function* (p.f.), from which - if available - all thermodynamic properties can be evaluated.

We proceed then to apply our findings to the simple case of an ideal gas. To this purpose we express the p.f. of an ideal gas in terms of the molecular p.f., which is further simplified by expressing it through the p.f. of the different molecular energy modes.

Finally, we develop the p.f. for the translational energy mode, and apply it to the evaluation of the properties of an ideal monatomic gas.

We close with some remarks about the accomplishments of statistical mechanics with respect to the two questions posed in the Introduction.

For a more detailed discussion of the material covered in this Chapter, see Denbigh (1981), a reference on which we rely a great deal. For an in depth coverage of the material - and that of the next Chapter - see among others: Hill (1960); Hatsopoulos and Keenan (1965); Knuth (1966); Reed and Gubbins (1973); and Gray and Gubbins (1984).

The Magnitude of the Atomic World

The diameter of an atom is about one hundred-millionth of a centimeter. In order to visualize this diminutive size, imagine an orange blown up to the size of the earth. The atoms of the orange will then have the size of cherries. Myriads of cherries, tightly packed into a globe of the size of the earth - that's a magnified picture of the atoms in an orange.

An atom, therefore, is extremely small compared to macroscopic objects, but it is huge compared to the nucleus in its center. In our picture of cherry-sized atoms, the nucleus of an atom will be so small that we will not be able to see it. If we blew up the atom to the size of a football, or even to room size, the nucleus would still be too small to be seen by the naked eye. To see the nucleus we would have to blow up the atom to the size of the biggest dome in the world, the dome of St. Peter's Cathedral in Rome. In an atom of that size, the nucleus would have the size of a grain of salt! A grain of salt in the middle of the dome of St. Peter's, and specks of dust whirling around it in the vast space of the dome - this is how we can picture the nucleus and electrons of an atom. (Capra, 1983, p.75)

16.3 Description of the Molecular Behavior

In describing molecular behavior we must consider the molecules themselves as well as the particles that constitute them, the electrons and nuclei. Two models are available to this purpose:

* Classical Mechanics; and
* Quantum Mechanics.

16.3.1 Classical Mechanics

The classical model of Nature is based on Newton's laws, which govern the motion of particles with respect to a fixed coordinates system.

For a particle with mass m, and a location described by vector r, the fundamental law is:

$$\textit{force} = \textit{mass} \times \textit{acceleration}, \text{ or}$$

$$F = m \frac{d^2 r}{dt^2} \tag{16.3.1}$$

where F is the force acting on the particle and t represents time. If we know, thus:

1. the force acting upon a particle; and
2. the location and velocity of the particle at some time t_0;

we can then follow its trajectory with respect to time.

To this purpose assume, for simplicity, that a particle of mass m moves in the x-direction only, under the influence of a constant force F_0, and that at time $t = 0$ its location is given by x_0 and its velocity by u_0. Solution of Eq.16.3.1 yields:

$$x = \frac{1}{2} \frac{F_0}{m} t^2 + u_0 t + x_0 \tag{16.3.2}$$

Complete knowledge, therefore, of the location and kinetic energy of the particle at any time t is possible.

16.3.2 Quantum Mechanics

Particles, Waves, and Quanta

The concepts of quantum theory were not easy to accept even after their mathematical formulation had been completed. Their effect on the physicist's imagination was truly shattering. Rutherford's experiments had shown that atoms, instead of being hard and indestructible, consisted of vast regions of space in which extremely small particles moved, and now quantum theory made it clear that even these particles were nothing like the solid objects of classical physics. The subatomic units of matter are very abstract entities which have a dual aspect. Depending on how we look at them, they appear sometimes as particles, sometimes as waves; and this dual nature is also exhibited by light which can take the form of electromagnetic waves or of particles.

This property of matter and of light is very strange. It seems impossible to accept that something can be, at the same time, a particle - i.e. an entity confined to a very small volume - and a wave, which is spread out over a large region of space. This contradiction gave rise to most of the koan-like paradoxes which finally led to the formulation of quantum theory. The whole development started when Max Planck discovered that the energy of heat radiation is not emitted continuously, but appears in the form of 'energy packets'. Einstein called these energy packets 'quanta' and recognized them as a fundamental aspect of nature. He was bold enough to postulate that light and every other form of electromagnetic radiation can appear not only as electromagnetic waves, but also in the form of these quanta. The light quanta, which gave quantum theory its name, have since been accepted as bona fide particles and are now called photons. They are particles of a special kind, however, massless and always travelling with the speed of light. (Capra, p.76)

a. Introductory Remarks

The domination of science through the 19th century by the Newtonian model came to an end with the discovery of the subatomic particles in the turn of this century. Physicists soon realized that, in this fine scale, nature portrays a dual behavior: sometimes particle-like, other times wave-like, the latter demonstrated in electron and proton diffraction.

Conversely, it was also realized that waves behave some times like particles as in the case of the photoelectric effect. Here, they were forced to create the *photon* as the hypothetical particle corresponding to a certain amount, a *quantum*, of electromagnetic energy. The very existence, of course, of these quanta of energy contrasted the Newtonian notion of continuity of energy.

Modern physics, where the concept of quantum plays a pivotal role, was thus born. Here is a small sample of the difficulties that arose in describing this dual-nature of the very small particles:

"It would have been obliging on the part of nature if it were possible to attribute to the smallest pieces of nature (electrons and protons) the same Newtonian properties - in particular, definite positions and velocities - as are normally associated with the notion of a particle. Experiment shows, however, that these minute objects do not have such clear-cut properties." (Denbigh, 1981, p.365)

"The complete definiteness of nature as described by classical mechanics was destroyed by the necessity of assigning an inertia to electromagnetic radiation (photoelectric effect).....We can do without particles but not without waves in nature." (Reed and Gubbins, p.8)

As for the conceptual difficulties that arose from this dual character of nature in the fine atomic scale, here is what two of the pioneers of the field said, as quoted by Capra, p.42:

"All my attempts to adapt the theoretical foundation of physics to this [new type of] knowledge failed completely. It was as if the ground had been pulled out from under one, with no firm foundation to be seen anywhere, upon which one could build." (Einstein)

"The great extension of our experience in recent years has brought to light the insufficiency of our simple mechanical conceptions and, as a consequence, has shaken the foundation on which the customary interpretation of observations was based." (Niels Bohr)

Matter at the Subatomic Level

The apparent contradiction between the particle and the wave picture was solved in a completely unexpected way which called in question the very foundation of the mechanistic world-view - the concept of the reality of matter. At the subatomic level, matter does not exist with certainty in definite places, but rather shows 'tendencies to exist', and atomic events do not occur with certainty at definite times and in definite ways, but rather show 'tendencies to occur'. In the formalism of quantum theory, these tendencies are expressed as probabilities and are associated with mathematical quantities which take the form of waves. This is why particles can be waves at the same time. They are not real three dimensional waves like sound or water waves. They are 'probability-waves', abstract mathematical quantities with all the characteristic properties of waves which are related to the probabilities of finding the particles at particular points in space and at particular times. All the laws of atomic physics are expressed in terms of these probabilities. We can never predict an atomic event with certainty; we can say only how likely it is to happen. (Capra, p.77)

b. The Schrödinger Equation

To describe this dual character of nature, we introduce the following postulate: *the mechanical behavior of matter on the atomic scale is described by the Schrödinger equation* which, for N particles:

- * having masses m_1, m_2, ..., m_N; and
- * moving in a force field described by the potential energy Φ and with a total energy E;

has the following form:

$$\sum \frac{1}{m_i}\left[\frac{\partial^2 \Psi}{\partial x_i^2} + \frac{\partial^2 \Psi}{\partial y_i^2} + \frac{\partial^2 \Psi}{\partial z_i^2}\right] = \frac{-8\pi^2}{h^2}[E - \Phi]\Psi \qquad (16.3.3)$$

where h is the Planck constant and the summation is over all the N particles.

Note that in terms of classical mechanics this system would be described through the coordinates x_i, y_i, and z_i of the particles and the corresponding values of their velocities, along with the values of E and Φ.

For an interpretation of the wave function Ψ, the following observations, from the eloquent presentation of the Schrödinger equation by Denbigh, provide some insight into its mathematical properties:

1."Its numerical solution, for appropriate conditions, expresses the observable properties without contravening the principle that it is impossible to make an exact and simultaneous specification of position and velocities. It may be remarked that it is because of this principle of uncertainty that wave mechanics seek to describe the state of a system by means of the function Ψ, whose purpose is to describe probabilities and not certainties." (Denbigh, 1981, p.365)

2."The wave function Ψ has the following property: the probability that the various particles have their coordinates in the range x_1 to x_1+dx_1, ..., z_N to z_N+dz_N, is given by:

$$\Psi^2 \, dx_1 \, dy_1 \dots dz_N \tag{16.3.4}$$

3.The essential requirement of Ψ, if it is to fulfil its wave-mechanical purpose, is that it shall describe a probability, in accordance with Eq. 16.3.4. For this purpose it must be *single-valued, continuous and finite* throughout the range of the coordinates. For example, if it is known that all of the particles are contained within a certain range of the coordinates (e.g. if they are in a box), then Ψ^2 must have the above properties within this range but must be zero at all points outside it. Otherwise Ψ^2 does not have the required properties as a probability density.

In brief, not all possible solutions of the differential Eq.16.3.3 lead necessarily to values of Ψ which correspond to physical reality; *it may occur that significant solutions are obtained only for particular values* E_0, E_1, *etc., of the energy*. These are called *eigenvalues* and the corresponding values of Ψ are Ψ_0, Ψ_1, etc., and are called *eigensolutions*.

4.Each eigensolution specifies a possible quantum state of the system, and their successive numbering gives rise to a quantum number. Instead of numbering the quantum states, an alternative procedure is to number the permissible energy levels (eigenvalues) successively. If it occurs that there are Ω_i different quantum states of the system all of which correspond to the same energy E_i, this level is said to be Ω_i-fold *degenerate*." (Denbigh, 1981, pp.366-367)

16.3.3 Comments

1.Both, the classical and quantum mechanical approach for describing molecular behavior are used in statistical mechanics. The first is only applicable, however, when the temperature is high enough for the thermal energy kT, where k is the Boltzmann constant, to be very large as compared to the spacing of the lowest quantum energy levels, and when the particle masses are not too small. Thus, the translational motion of molecules can be treated classically above about 50 K; but not the vibrational and electronic motions.

2.In applying classical mechanics, it is more convenient to replace the fundamental second-order differential Eq.16.3.1 by the two Hamilton's equations of motion that are first-order differential equations (see, for example, Reed and Gubbins).

3.The terms 'eigenvalues' and 'eigensolutions' are not as abstract as it may appear at first. This is demonstrated with a system of two simultaneous equations in Example 16.1.

4.The Schrödinger equation can be only solved under simple conditions, especially where there are 'separable' energy states. As a result of this separability, the equation can be broken down into a number of simpler differential equations, each containing a smaller number of independent variables (see Problem 16.4). The importance of this energy 'separability' will become apparent in discussing the translational wave function of a single particle in Section 16.4.

5.Independently of the degree of our understanding of molecular behavior, one thing we can accept: the microscopic picture of a fluid is not a static but a dynamic one, with the particles moving continuously among the vast number of quantum states, compatible with the available energy levels (Example 16.2 and Section 16.4).

Some averaging procedure is needed, therefore, so that the macroscopic or bulk properties of the fluid be obtained. It is considered in Section 16.5.

16.3.4 Example 16.1

Identify the values of E for which the system of simultaneous equations:

$$(1 - E)x + 3y = 0$$
$$3x + (1 - E)y = 0$$

has solutions, other than the trivial one: $x = y = 0$.

Such solutions can be obtained only when $E = 4$ or $E = -2$. For any other value of E the equations are inconsistent and, consequently, cannot refer to the same physical phenomenon described by the variables x and y. For example, if $E = 3$: $x = (3/2)y$ from the first equation, while $x = (2/3)y$ from the second one. On the other hand:

* for $E = 4$, $x = y$ from both equations; and
* for $E = -2$, $x = -y$, again from both equations.

Thus:

* the eigenvalue $E = 4$ has as its solution, or eigenfunction: $x = y$; and
* the eigenvalue $E = -2$, the eigenfunction: $x = -y$.

(This is similar to an example given by Margenau and Murphy, *The Mathematics of Physics and Chemistry*, § 8.1, Van Nostrand, New York, 1944, as quoted by Denbigh, 1981, p.367. See also Problem 16.3.)

16.3.5 Example 16.2

Consider four particles moving freely into a given box with three quantized translational energies available to them: e_0, e_1, and e_2. If we assume, for simplicity, that: $e_1 = 2e_0$, and $e_2 = 3e_0$, and that the total energy is constant, equal to $6e_0$:

a. how many distributions of this energy among the particles are possible?
b. how many quantum states are available to them?

a. We could have:

* two particles in e_0 and two in e_1; or
* one in e_2 and three in e_0;

both compatible with the total energy of the system.

b. These two possibilities represent two different quantum states of the system, and we do not know in which one it is at a given moment. Notice, furthermore, that the number of such states increases drastically with increasing system energy (Problem 16.5). From the classical thermodynamics point of view, however, the system is completely defined, since the volume, number of moles, and energy are specified.

We need, thus, a method for evaluating the average properties of the system as it moves through all the available quantum states, which is considered in Section 16.5. But first, to develop some familiarity with the Schrödinger equation - and the relevant terms: energy separability, quantum numbers, permissible energy levels and quantum states - we apply it to the translational motion of a single particle.

16.4 The Translational Wave Function of a Single Particle

16.4.1 The General Solution

For a single particle of mass m moving freely in the absence of any potential energy, the Schrödinger equation becomes:

$$\frac{\partial^2\Psi}{\partial x^2} + \frac{\partial^2\Psi}{\partial y^2} + \frac{\partial^2\Psi}{\partial z^2} = \frac{-8\pi^2 m}{h^2} e\Psi \tag{16.4.1}$$

where e is the kinetic energy of the particle.

This energy can be broken down into three independent parts corresponding to the velocity components u_x, u_y and u_z along the three axes:

$$e = \frac{1}{2}m\,[u_x^2 + u_y^2 + u_z^2] = e_x + e_y + e_z \tag{16.4.2}$$

Assume now that Ψ can be written as the product of three independent functions, each containing only one independent variable, $\Psi_x = f_x(x)$, $\Psi_y = f_y(y)$, $\Psi_z = f_z(z)$:

$$\Psi = \Psi_x \Psi_y \Psi_z \tag{16.4.3}$$

It can be easily shown that introduction of Eqs 16.4.2 and 16.4.3 into Eq. 16.4.1 leads to three easy to solve differential equations, each containing only one independent variable (Problem 16.4):

$$\frac{1}{\Psi_x}\frac{d^2\Psi_x}{dx^2} = \frac{-8\pi^2 m}{h^2} e_x \tag{16.4.4}$$

$$\frac{1}{\Psi_y}\frac{d^2\Psi_y}{dy^2} = \frac{-8\pi^2 m}{h^2} e_y \tag{16.4.5}$$

$$\frac{1}{\Psi_z}\frac{d^2\Psi_z}{dz^2} = \frac{-8\pi^2 m}{h^2} e_z \tag{16.4.6}$$

The solution of Eq. 16.4.4 is:

$$\Psi_x = A_x \sin\left[\frac{\pi x}{h}\sqrt{8me_x} + A_1\right] \tag{16.4.7}$$

where A_x and A_1 are constants. Similar solutions are obtained for Eqs 16.4.5 and 16.4.6.

16.4.2 Quantum Numbers, Permissible Energy Levels, and the Wave Function

Assume now that the particle moves inside a box which is in the origin of the coordinates systems with dimensions a, b, and c, along the x, y, and z axes respectively. Then Ψ_x (Eq.16.4.7) must be zero for values of x less than zero and greater than a, from which we obtain:

$$A_1 = 0 \tag{16.4.8a}$$

$$\frac{\pi a}{h}\sqrt{8me_x} = \pi p_x \tag{16.4.8b}$$

where p_x is an integer, different from zero, for Ψ_x would be equal to zero throughout. From the last equation:

$$e_x = \frac{h^2}{8m}\left(\frac{p_x}{a}\right)^2 \tag{16.4.9a}$$

Similarly:

$$e_y = \frac{h^2}{8m}\left(\frac{p_y}{b}\right)^2 \tag{16.4.9b}$$

$$e_z = \frac{h^2}{8m}\left(\frac{p_z}{c}\right)^2 \tag{16.4.9c}$$

The integers p_x, p_y, and p_z are the *quantum* numbers.

Introduction of Eqs 16.4.8 into Eq.16.4.7 yields:

$$\Psi_x = A_x \sin\left[\frac{p_x \pi x}{a}\right] \tag{16.4.10}$$

The solution of Eq.16.4.1, i.e. the *translational* wave function of the particle, is obtained by introducing Eq.16.4.10, and the similar expressions for Ψ_y and Ψ_z, into Eq.16.4.3:

$$\Psi = A_x A_y A_z \sin\left[\frac{p_x \pi x}{a}\right] \sin\left[\frac{p_y \pi y}{b}\right] \sin\left[\frac{p_z \pi z}{c}\right] \tag{16.4.11}$$

The *permissible* energy levels are obtained by introducing Eqs 16.4.9 into Eq.16.4.2:

$$e = \frac{h^2}{8m}\left[\frac{p_x^2}{a^2} + \frac{p_y^2}{b^2} + \frac{p_z^2}{c^2}\right] \tag{16.4.12}$$

16.4.3 Comments

1. Notice the importance of energy separability, Eq.16.4.2, in solving Eq.16.4.1. (Could it be solved if this were not the case?)
2. The origin of the quantum numbers and quantized energy levels is the imposition of boundary conditions. In the absence of confinement, the energy is continuous.
3. The 'ground state', the state with the lowest energy level, corresponds to $p_x = p_y = p_z = 1$. Notice that this energy is not equal to zero.
4. To every e_x corresponds only one quantum state. Thus e_x is not degenerate. The same applies to e_y and e_z. The total translational energy, however, can be degenerate. If, for example, $a = b = c$ the three quantum numbers: $q_x = 10$, $q_y = 5$, $q_z = 3$ and $q_x = 11$, $q_y = 2$, $q_z = 3$ give the same energy but two different quantum states.
5. To develop a sense about the size of quantum numbers, see Problem 16.20.

16.5 The Averaging Process

16.5.1 The Problem

Consider a system of fixed values of mass and volume, which is in equilibrium with a heat reservoir. When we say in classical thermodynamics that a system is in equilibrium, we mean that all its properties - temperature, pressure etc. - do not vary with time. But from the molecular point of view, the equilibrium state is dynamic and not static for, as a result of the continuous collisions of the molecules, the system moves continuously from one quantum state to an other.

Let us assume that we want to calculate some property Θ for this system by determining its average value over all possible quantum states that are available to it. To this purpose we observe the system over a long time interval τ, and let t_i be the time spent in state i. The probability P_i, then, that the system is at a given moment in state i, is expressed by:

$$P_i = \lim_{\tau \to \infty} (t_i/\tau) \qquad (16.5.1)$$

The sum of all these probabilities is, of course, equal to 1.

The average value of the property O, is then given by.

$$\Theta = \sum P_i \Theta_i \tag{16.5.2}$$

where the summation is over all accessible quantum states (q.s.), and Θ_i is the property value in q.s. i.

Now, we know from experience that measurements of a system in equilibrium give values that do not change from time to time. Actually, the system passes through all q.s. a great many times over the period of a measurement and, consequently, the value of the property measured experimentally corresponds to the average value Θ.

Assuming that all the Θ_i values can be obtained from quantum or classical - if applicable - mechanics, we need the values of the probabilities P_i, if we are to determine the average and, thus, the observable quantities. We consider next the principle on which the determination of these probabilities is based.

16.5.2 The Statistical Postulate

In discussing classical mechanics in Section 16.3, we noted that if the position and velocity of the molecules were known at some time t_0, along with the forces acting upon them, we could then follow their paths and be able to determine their positions and velocities at any time t. The average values could then be calculated.

Leaving aside the computational problems associated with following such a large number of molecules - of the order of 10^{23} - simultaneous knowledge of their original positions and velocities is impossible by virtue of the *uncertainty principle*. Any measurement of the position of the molecules will effect their velocity and vice versa.

Knowledge of the state of the system in complete detail is thus unattainable and this represents one of the bases of the molecular interpretation of the second law (Denbigh, 1981, p.336; see also Section 3.13). It is impossible, therefore, to determine the average values of the properties of a given system by applying Newtonian mechanics to its individual molecules.

We resort rather to a postulate, in addition to those of mechanics, which is of statistical character and asserts that: *For a system in internal equilibrium the probability of a given quantum state is a function of its energy only.*

16.5.3 Ensembles

The totality of quantum states available to a system is referred to as an *ensemble*. On the basis of the constraints imposed on the system, three types of ensembles are used:

1. *Microcanonical*: system of fixed values of mass, volume, and energy;
2. *Canonical*: system of fixed values of mass and volume that is in thermal equilibrium with a heat reservoir;
3. *Grand canonical*: system of fixed value of volume in equilibrium with:
   * a heat reservoir,
   * reservoirs of the various substances it contains through semipermeable membranes (open system).

It is apparent that in the case of a microcanonical ensemble each quantum state i has, according to the aforementioned statistical postulate, the same probability:

$$P_i = \frac{1}{\Omega}$$

where Ω is the number of quantum states available to the system under the given constraints (fixed values of mass, volume and energy).

The canonical ensemble is considered in the next Section, while for the grand canonical ensemble the reader is referred to more comprehensive texts on the subject (Reed and Gubbins, for example).

16.6 The Canonical Ensemble

16.6.1 The Characteristic Property of the Probability Function

Consider a system of fixed values of mass and volume, which is in equilibrium with a heat reservoir. In this case, the energy of the system is not constant. Rather, through interactions with the molecules of the heat reservoir, the system assumes a range of energy values around its mean one. These values are, of course, quantized and we label them E_0, E_1,..., starting from the lowest one. How are we to evaluate the probability of each of the quantum states (q.s.) of the system?

According to the statistical postulate, the probability of q.s. i with energy E_i is a function of this energy only:

$$P_i = f(E_i) \tag{16.6.1a}$$

Assume next that a second system is also brought in equilibrium with the reservoir. The probability that the second system is in the kth q.s. with energy e_k is given by:

$$P_k = f(e_k) \tag{16.6.1b}$$

Let us now consider the probability P_{ik}, that the first system is in q.s. i while, simultaneously, the second one is in q.s. k. According to the postulate:

$$P_{ik} = f(E_i + e_k) \tag{16.6.1c}$$

At the same time, since the reservoir is large, there is no interaction between the two systems. Thus, the probability that the first system will be in q.s. i and the second one in k is given by the product of the two probabilities, P_i and P_k:

$$P_{ik} = P_i \, P_k \tag{16.6.2a}$$

or:

$$f(E_i + e_k) = f(E_i) \, f(e_k) \tag{16.6.2b}$$

We must find, therefore, a function that has the following property: its value for the sum of two variables equals the product of the function values of these two variables taken separately (Example 16.3).

16.6.2 Example 16.3

Show that if: $f(x)f(y) = f(x+y)$, then: $f(x) = A \exp(bx)$; $f(y) = B \exp(by)$; where A, B and b are constants (Denbigh, 1981, p.344).

Partial differentiation with respect to y gives:

$$f(x)\,\frac{df(y)}{dy} = \frac{df(x+y)}{dy} = \frac{df(x+y)}{d(x+y)}\,\frac{d(x+y)}{dy} = \frac{df(x+y)}{d(x+y)}$$

Similarly, partial differentiation with respect to x gives:

$$f(y)\,\frac{df(x)}{dx} = \frac{df(x+y)}{d(x+y)}$$

Thus:

$$f(x)\,\frac{df(y)}{dy} = f(y)\,\frac{df(x)}{dx}$$

or:

$$\frac{1}{f(y)}\,\frac{df(y)}{dy} = \frac{1}{f(x)}\,\frac{df(x)}{dx}$$

Since x and y are independent variables, the left-hand side of the equation does not depend on x, neither does the right-hand one on y. Thus, each side must be equal to a constant b, the same for both:

$$\frac{1}{f(x)}\frac{df(x)}{dx} = b \rightarrow df(x) = bf(x)dx$$

and: $f(x) = A\exp(bx)$. Similarly: $f(y) = B\exp(by)$.

16.6.3 The Partition Function

It follows from Example 16.3 that if two systems of constant mass and volume are in equilibrium with a heat reservoir, the probability of any of their quantum states is an exponential function of its energy:

$$P_i = C_1 \exp(bE_i) \tag{16.6.3a}$$

$$P_k = C_2 \exp(be_k) \tag{16.6.3b}$$

Then, from Eq.16.6.2a:

$$P_{ik} = C_1 C_2 \exp[b(E_i + e_k)] \tag{16.6.3c}$$

C_1 and C_2 are constants characteristic of the first and second system respectively; b is also a constant, but the same for both systems, if Eq. 16.6.2b is to be satisfied. In addition b must be negative, otherwise the probability P_i would become infinite for infinite values of the energy of the q.s. Finally, we can speculate that since both systems are in equilibrium with the heat reservoir, b must be related to the temperature of it. This is indeed demonstrated by Denbigh (1981, p.350), who shows that:

$$b = -\frac{1}{kT} \tag{16.6.4}$$

where k is the Boltzmann constant.

Returning now to the system under consideration - and using C instead of C_1 - we note that since $\Sigma P_i = 1.0$:

$$C = \frac{1}{Q}$$

where:

$$Q = \sum \exp(bE_i) \tag{16.6.5}$$

The probability, thus, of quantum state i with energy E_i is given by:

$$P_i = \frac{\exp(bE_i)}{\sum \exp(bE_i)} = \frac{\exp(bE_i)}{Q} \tag{16.6.6}$$

The summation is taken over all the available quantum states, and Q is called the *partition function* (p.f.) of the system at constant volume and temperature.

Comments:

1. The totality of q.s. available to a closed system of constant volume which is in equilibrium with a heat reservoir, is referred to as a *canonical ensemble*, and Eq.16.6.6 describes the *canonical* distribution of probabilities.

2. Eq.16.6.6 indicates that, since b is negative, the higher the energy level of a quantum state the lower its probability.

3. If the energy level E_i is Ω_i-fold degenerate, i.e. there are Ω_i q.s. that correspond to this energy level, then the partition function can be written as:

$$Q = \sum_{\text{q.s.}} \exp(bE_i) = \sum_{\text{e.l.}} \Omega_i \exp(bE_i) \tag{16.6.7}$$

where the first summation is over all quantum states and the second one over all energy levels.

16.7 Thermodynamic Properties from the Partition Function

The relationships between the thermodynamic properties of a system and its partition function Q are developed very eloquently by Denbigh (1981, p.345). We present here the final results only:

$$A = -kT\ln Q \tag{16.7.1}$$

$$U = kT^2\left(\frac{\partial \ln Q}{\partial T}\right)_{V,\,\text{comp.}} \tag{16.7.2}$$

$$S = k\ln Q + kT\left(\frac{\partial \ln Q}{\partial T}\right)_{V,\,\text{comp.}} \tag{16.7.3}$$

Finally, since from thermodynamics:

$$P = -\left(\frac{\partial A}{\partial V}\right)_{T,\,\text{comp.}}$$

then:

$$P = kT\left(\frac{\partial \ln Q}{\partial V}\right)_{T,\,\mathrm{comp.}} \tag{16.7.4}$$

We conclude, therefore, that *if the partition function Q of a constant composition fluid is known as a function of temperature and volume, all thermodynamic properties of the system can be determined.* This is demonstrated next with the case of an ideal gas.

16.8 The Partition Function of an Ideal Gas

16.8.1 In Terms of the Molecular Partition Function

In the case of a pure ideal gas of volume V and temperature T, because there are no interactions among the molecules, the expression for the partition function is greatly simplified for it can be expressed in terms of the partition functions of the individual molecules, q:

$$Q = \frac{1}{N!}q^N \tag{16.8.1}$$

where N is the number of molecules in the gas, and:

$$q = \sum_{\mathrm{q.s.}} \exp[\frac{-e_j}{kT}] = \sum_{\mathrm{e.l.}} \omega_j \exp[\frac{-e_j}{kT}] \tag{16.8.2}$$

Comments:

1. The lower case symbols refer to molecules, ω_j is the degeneracy of molecular energy level e_j, while the capital symbols Ω_i and E_i, refer to the whole gas.

2. The development of Eq. 16.8.1 is given, among others, by Denbigh and by Reed and Gubbins. The basic assumption involved in its development is that each molecule is in a different q.s., and it is valid at temperatures above about 20 K. The kind of molecular statistics under these conditions is referred to as Maxwell-Boltzmann. At lower temperatures, the possibility of several molecules being in one quantum state must be considered, and Bose-Einstein or Fermi-Dirac statistics must be used (Reed and Gubbins).

3. The term $N!$ arises from the fact that the particles are identical and free to move throughout the volume V.

4.In the case of an ideal gas mixture containing N_1 molecules of type 1 and N_2 of type 2:

$$Q = \frac{1}{N_1!\,N_2!} q_1^{N_1} q_2^{N_2} \tag{16.8.3}$$

Extension to ideal gas mixtures of more than two components is straightforward.

The problem, thus, of evaluating the partition function of an ideal gas has been reduced to that of determining the molecular partition function. We will see next that this is further simplified through the factoring of the molecular partition function.

16.8.2 Factoring of the Molecular Partition Function

For an ideal gas the molecular energy is the sum of two independent contributions:

$$e = e(tr) + e(int) \tag{16.8.4}$$

where:

$e(tr)$: translational energy; and

$e(int)$: internal energy, comprising the rotational, $e(r)$, vibrational, $e(v)$, internal rotation, $e(ir)$, electronic $e(e)$, and nuclear spin $e(ns)$. (Notice that in classical thermodynamics the term internal energy, Section 1.9, refers to *all* energy possessed by the molecules of the system).

Consider now a translational energy level $e_j(tr)$ with degeneracy $\omega_j(tr)$ and an internal energy level $e_k(int)$ with degeneracy $\omega_k(int)$. The degeneracy of the state with energy:

$$e_i = e_j(tr) + e_k(int) \tag{16.8.5a}$$

is:

$$\omega_i = \omega_j(tr)\,\omega_k(int) \tag{16.8.5b}$$

This arises from the fact that anyone of the $\omega_j(tr)$ translational quantum states can be combined with all $\omega_k(int)$ internal quantum states, to give a state of energy e_i.

Introduction of Eqs 16.8.5a and b into Eq.16.8.2 yields:

$$q = \sum_{j,k} \omega_j(tr)\,\omega_k(int)\exp[-\frac{e_j(tr)+e_k(int)}{kT}]$$

It follows that:

$$q = q(tr)\,q(int) \tag{16.8.6}$$

where:

$$q(tr) = \sum \omega_j(tr) \exp[\frac{-e_j(tr)}{kT}]$$

$$q(int) = \sum \omega_k(int) \exp[\frac{-e_k(int)}{kT}]$$

The summations are over all available translational or internal energy levels and $q(tr)$ and $q(int)$ are the translational and internal partition functions respectively.

If we further assume that:

$$e(int) = e(r) + e(v) + e(ir) + e(e) + e(ns) \tag{16.8.7}$$

then:

$$q(int) = q(r)\, q(v)\, q(ir)\, q(e)\, q(ns) \tag{16.8.8}$$

The validity of the last expression is limited by the fact that the rotational and vibrational motions are not independent: the centrifugal force from the rotational motion stretches the molecule, thus effecting its vibrational motion; and the change in nuclear distances due to vibration, effects the moment of inertia and thus the rotation.

Evaluation of the internal p.f. is rather complex, involving a combination of spectroscopic measurements and theory (see, for example, Reed and Gubbins), and will not be considered here. On the other hand, evaluation of the translational p.f. is simpler and is presented next.

16.8.3 The Translational Partition Function

Consider an ideal gas molecule moving inside a box, which is in the origin of the coordinates system with dimensions a, b and c along the x, y, and z axes respectively (Section 16.4). According to the separability of the translational energy, Eq.16.4.2, and following the reasoning of the previous Section:

$$q(tr) = q(x)\, q(y)\, q(z) \tag{16.8.9}$$

where:

$$q(x) = \sum_{p_x} \exp[\frac{-h^2}{8mkTa^2} p_x^2] = \sum_{p_x} \exp(-g_a p_x^2) \tag{16.8.10}$$

$$g_a = \frac{h^2}{8mkTa^2}$$

Eq.16.8.10 is obtained by introducing Eq.16.4.9a into Eq.16.8.2, and similar expressions are obtained for $q(y)$ and $q(z)$.

To evaluate this summation, we note that use of reasonable values of T and a gives very small values for g_a. This, in turn, leads to terms that, starting with a value of practically equal to one - for $p_x = 1$, decrease extremely slowly. We can replace thus this summation with the integral:

$$q(x) = \sum_{1}^{\infty} \exp[-g_a p_x^2] = \int_0^{\infty} \exp[-g_a p_x^2] dp_x \qquad (16.8.11)$$

where p_x is not limited anymore to integer values (see also Problem 16.13).

Evaluation of the integral gives:

$$q(x) = \frac{1}{2}\left(\frac{\pi}{g_a}\right)^{1/2} = \frac{a}{h}(2\pi mkT)^{1/2} \qquad (16.8.12)$$

Similar expressions are obtained for $q(y)$ and $q(z)$ and, consequently:

$$q(tr) = V\left[\frac{2\pi mkT}{h^2}\right]^{3/2} \qquad (16.8.13)$$

where: $V = abc$, is the volume of the box.

The replacement of the summation with an integral in Eq.16.8.11 is the result of the fact that the spacing of translational energies is very small as compared to kT (Problem 16.14.d). Under these conditions the partition function can be evaluated through classical mechanics (see, for example, Reed and Gubbins) and it is referred to as *classical* treatment. This approximation is valid for molecular gases at temperatures above about 50 K, but not for very low temperature solids, or for liquid He, H_2 and Ne.

The quantum mechanical treatment was chosen here to enhance our understanding of the concepts of quantum numbers and energy levels, which are 'masked' in the classical one, and to demonstrate the conditions under which the classical treatment is possible. These conditions, for example, are not met by the vibrational energy levels and, consequently, $q(v)$ cannot be treated classically.

16.9 The Description of an Ideal Gas

16.9.1 Equation of State

We have arrived, finally, to the partition function of an ideal gas:

$$Q = \frac{1}{N!}[q(tr)\,q(int)]^N = \frac{1}{N!}\left[V\left(\frac{2\pi mkT}{h^2}\right)^{3/2} q(int)\right]^N \tag{16.9.1}$$

Using Eq.16.7.4, since $q(int)$ is volume-independent, we obtain:

$$P = kT\left[\frac{\partial \ln Q}{\partial V}\right]_{T,N} = \frac{NkT}{V} = \frac{nRT}{V} \tag{16.9.2}$$

where n is the number of moles, $N = nN_A$ (N_A is the Avogadro number), and $R = N_A k$ is the ideal gas constant. Eq.16.9.2 is the ideal gas equation we have already used.

16.9.2 Thermodynamic Properties

For monatomic gases it turns out that, with few exceptions (Denbigh, 1981, p.379), $q(int) = 1.0$. Thus, on a molar basis following Section 16.7:

$$U = \frac{3}{2}RT \tag{16.9.3}$$

$$H = U + PV = \frac{5}{2}RT \tag{16.9.4}$$

and:

$$C_V = (3/2)R \tag{16.9.5}$$

$$C_P = (5/2)R \tag{16.9.6}$$

The development of the expressions for the Helmholtz free energy A, the Gibbs free energy G, and the entropy S is straightforward (Problems 16.16 and 16.17).

For polyatomic molecules $q(int)$ must, of course, be determined by considering the different contributions to it, as shown in Eqs 16.8.7 and 16.8.8. This is done by a combination of spectroscopic measurements for the various energy levels, and theoretical calculations for the degeneracy of the energy levels (see, for example, Reed and Gubbins). Ideal gas energies can then be calculated but only relative to an arbitrarily chosen value of zero, adopted by convention for the lowest energy level (ground state).

Ideal gas entropies can also be calculated from statistical mechanics (for a monatomic gas, see Problem 16.17). Such entropy values can be used in calculating the standard entropies of reactions, because the absolute entropies of the elements drop out by virtue of the stoichiometry of the reaction. Implied also here is the validity of the third law, i.e. that the entropy change of the reaction at 0 K is zero. Thus, even though absolute entropies cannot be known - for example because of the possibility of randomness within the nucleus - entropies of reaction can be determined.

The same, however, does not apply to the enthalpies of reaction. Here, the enthalpy of reaction at 0 K - due to the fact that the enthalpies of the compounds are relative to their values at the lowest energy level (ground state), which are not the same among different compounds - must be accounted for.

Application of statistical mechanics to real fluids becomes difficult because of the complications that arise from the presence of intermolecular forces. Through simplifications and approximations, however, more and more understanding of the properties of such fluids is achieved through statistical mechanics, as outlined in the next Chapter.

16.10 Concluding Remarks

In the beginning of the Chapter we set out to examine if statistical mechanics could provide:

1. A rigorous, from first principles, development of the second law of thermodynamics; (and perhaps a deeper, or more profound philosophically, insight into it); and
2. An avenue for evaluating thermodynamic properties.

The degree to which the second objective is accomplished is discussed briefly for the case of an ideal gas in Section 16.9. For details, see the books mentioned in Section 16.2. Real fluids are considered in Chapter 17.

We turn thus to the issues raised in the first objective.

16.10.1 A Rigorous Development of the Second Law

Statistical mechanics attributes the origin of the second law to the molecular nature of matter (see Sections 3.13 and 16.5.2, and for a detailed discussion Denbigh, 1981, p.56). It also provides a quantitative relationship between entropy and molecular behavior through a comparison of statistical analogies with thermodynamic functions (this is eloquently done by Denbigh, 1981, p.350; see also Hatsopoulos and Keenan, Part II). Accepting, for the sake of argument, that this does represent a rigorous development of the second law, let us now examine the foundation of statistical mechanics itself, i.e. the two postulates: the statistical one; and the quantum mechanical one, the Schrödinger equation.

In support of the first one, Denbigh (1981, p.338) writes: "It may be remarked that the assumption that each quantum state (for a system of constant energy) has an 'equal a priori probability' is less arbitrary than any alternative assumption, in the absence of any information to the contrary." And then he proceeds to quote David Hume: "In this connexion the following remarks from David Hume are very much to the point: 'as chance is nothing real in itself, and, properly speaking, is merely the negation of a cause, its influence on the mind is contrary to that of causation; and it is essential to it to leave the imagination perfectly indifferent, either to consider the existence or non-existence of the object (event) which is regarded as contingent ... Since, therefore, an entire indifference is essential to chance, no one chance can possibly be superior to another, otherwise than as it is composed of a superior number of equal chances. For if we affirm that one chance can, after any other manner, be superior to another, we must at the same time affirm, that there is something which gives it the superiority, and determines the event rather to that side than the other: that is, in other words, we must allow of a cause, and destroy the supposition of chance, which we had before established' (A Treatise of Human Nature (1740), Book I, Part III, §XI)".

In support of the second postulate we have the fact that whenever the Schrödinger equation can be solved (the ideal gas for example), the results agree with the experimental observations. This, of course, is the same type of support given to the second law.

Have we, thus, indeed a rigorous development of the second law? Are we closer to the 'ultimate' truth? Or, have we just fallen victims of the famous slogan of theoretical mathematicians: "The road to complete rigor leads to meaninglessness"?

16.10.2 A Philosophical Insight into the Second Law

We have associated knowledge of molecular behavior with the second law in Sections 3.13 and 16.5. Let us now compare the ideas about this knowledge from the classical and quantum mechanics points of view.

For the first, here is what the famous French mathematician Pierre Laplace, who set out to refine and perfect Newton's calculations, writes

(Denbigh, 1975, p.119): "Consider an intelligence which, at any instant, could have a knowledge of all forces controlling nature together with the momentary conditions of all the entities of which nature consists. If this intelligence were powerful enough to submit all this data to analysis it would be able to embrace in a single formula the movements of the largest bodies in the universe and those of the lightest atoms; for it nothing would be uncertain; *the future and the past would be equally present to its eyes.*" (Emphasis added.)

Contrast this mechanistic view, which can be considered as the basis of 'determinism', with the quantum mechanics point of view, discussed in Section 16.5.2: *"Knowledge of the state of a system in complete detail is, because of the uncertainty principle, unattainable"*; and: "at the subatomic level, matter does not exist with certainty at definite places, but rather shows *'tendencies to exist'*, and atomic events do not occur with certainty at definite times and in definite ways, but rather show *'tendencies to occur.'*" (Insert, Section 16.3.2).

For a fascinating discussion of determinism, see Denbigh (1975), who writes, p.120: "If determinism is correct it would seem that nothing essentially new ever comes into existence - i.e. in the sense of not been contained, so to say, in what already existed. Creativeness, whether in man or nature, appears, therefore, as if it were an illusion." And farther down, in distinguishing between scientific laws and determinism: "What scientists recognize as being the normal regularity of nature has been pressed into the service of an all embracing philosophical thesis."

For a still broader perspective on this subject, let us consider the classical and quantum mechanics view of the separation between the individual and the world around him or her. In the classical (mechanistic) point of view, the individual is defined through his mind as epitomized by Descartes' famous statement: "Cogito ergo sum", "I think, therefore, I exist". This mind, thus, is completely separated from the world around it; it is the 'intelligence' that, according to Laplace's aforementioned statement, could - as an outside observer - make the future and the past "equally present to its eyes".

It should be noted that this separation between individual and nature, that has dominated western thought for the last few centuries, has been blamed by several individuals - from philosophers to environmentalists - for the human isolation and the predicament, in general, of the western civilization.

Let us now contrast this separation with the interconnection between the observer and the 'observed' phenomena in atomic physics. Capra writes: "Quantum theory has thus demolished the classical concepts of solid objects and of strictly deterministic laws of nature. At the subatomic level, the solid material objects of classical physics dissolve into wave-like patterns of probabilities, and these patterns, ultimately, do not represent probabilities of things, but rather *probabilities of interconnections*. A careful analysis of the process of observation in atomic physics has shown that the subatomic particles have no meaning as isolated entities but can only be understood as interconnections between the preparation of an experiment and the subsequent measurement. Quantum theory thus reveals a basic oneness of the universe. It shows that we cannot decompose the world into independently existing smallest units. As we penetrate into matter, nature does not show us any isolated 'basic building blocks', but rather appears as a complicated web of relations between the various parts of the whole. These relations always include the observer in an essential way. The human observer constitutes the final link in the chain of observational processes, and the properties of any atomic object can be understood only *in terms of the object's interaction with the observer*. This means that the classical ideal of an objective description of nature is no longer valid. The Cartesian partition between the I and the world, between the observer and the observed, cannot be made when dealing with atomic matter. In atomic physics, we can never speak about nature without, at the same time, speaking about ourselves." (Emphasis added, Capra, p.77.)

Note: Capra, actually, explores the parallels between quantum mechanics, and modern physics in general, and the mysticism of some of the classic Greeks (Milesians and Heraclitus) and of the eastern religions (Hinduism, Buddhism, and Taoism). You will probably consider such an attempt rather farfetched. But none other than the famous physicist Bohr - when knighted in 1947 for his contributions to the Danish cultural life - chose for his coat-of-arms the Chinese symbol of 't'ai-chi', representing the complementary relationship of the archetypic opposites 'yin' and 'yang' and the inscription "Contraria Sunt Complementa", "Opposites Are Complementary" (Capra, p.175). What could be a better description of the particle-wave dualism? Anyway, the book provides an excellent qualitative presentation of quantum mechanics - and modern physics in general - and Capra's aforementioned connection of atomic phenomena to some classic greek philosophies and eastern religions makes for fascinating reading.

References

Capra, F., 1983. *The Tao of Physics*, Harpercollins Ltd., London (reprinted with permission).

Denbigh, K.G., 1981. *The Principles of Chemical Equilibrium*, Cambridge University Press, Cambridge.

Denbigh, K.G., 1975. *An Inventive Universe*, George Braziller, New York.

Gray, C.G., Gubbins, K.E., 1984. *Theory of Molecular Fluids*, Vol. 1: *Fundamentals*, Oxford University Press, Oxford.

Hatsopoulos, G.N., Keenan, J.H., 1965. *Principles of General Thermodynamics*, Wiley, New York.

Hill, T.L., 1960. *Introduction to Statistical Thermodynamics*, Addison-Wesley, Reading, Mass.

Knuth, E.L., 1966. *Introduction to Statistical Thermodynamics*, McGraw-Hill, New York.

Reed, T.M., Gubbins, K.E., 1973. *Applied Statistical Mechanics*, McGraw-Hill, New York.

Problems

16.1 With reference to statistical mechanics:

a. what is its objective?

b. how does it accomplish it, in principle?

c. how does it provide, in principle, values of macroscopic properties?

16.2 What are the limitations of classical mechanics in describing molecular behavior?

16.3 Develop a system of two equations involving the variables x, y, and z and identify the values of x and y that satisfy it.

16.4.a. Consider a particle that moves in a box with kinetic energy E and no potential energy. Show that if $E = E_x + E_y + E_z$, the eigenfunction Ψ of the Schrödinger equation can be written as the product of three eigenfunctions, each containing only one independent variable.

b. Has this been used in the development of the partition function of the ideal gas? Explain.

Hint. Assume that: $\Psi = \Psi_x \Psi_y \Psi_z$, and proceed to demonstrate that it satisfies the Schrödinger equation.

16.5 With reference to Example 16.2 assume that the total energy is $10e_0$.

a. How many quantum states are possible?

b. Would it make any difference if the particles were distinguishable?

16.6 What are, qualitatively, the reasons behind the uncertainty principle?

16.7 Discuss the molecular origin of the second law (see Denbigh, p.56).

16.8.a. What is the statistical postulate?

b. How is it used in developing the partition function for the canonical ensemble?

16.9 In using statistical mechanics to determine macroscopic properties, what information do we obtain from quantum mechanics and what from statistical theory?

16.10 What information is contained in the partition function? What is its practical importance?

16.11 In factoring the partition function of an ideal gas:

a. what assumptions are involved?

b. discuss the validity of these assumptions.

16.12 Develop Eq.16.8.10 for $q(y)$.

16.13 Using your own numbers, demonstrate that Eq.16.8.10 can be replaced by an integral and proceed to evaluate it. Identify any limitations, if they exist.

16.14 With reference to the translational motion of an ideal gas molecule and using your own input data:

a. give an eigenvalue and corresponding eigenfunction;

b. demonstrate that a given energy level can be degenerate.

c. show that if the molecule is in a box of infinite size, its energy is not quantized.

d. Compare the spacing of the energy levels with kT and comment on its importance.

16.15 Develop the Maxwell-Boltzmann distribution law:

$$\bar{N}_i = \frac{\omega_i \exp(-e_i/kT)}{\sum \omega_j \exp(-e_j/kT)} N$$

where $\bar{N}_i$ is the mean number of molecules in the ith energy level and N the total number of molecules of the ideal gas. (See Denbigh or any text on Statistical Mechanics.)

16.16 Show that for one mole of an ideal monatomic gas:

$$A = -RT[\ln V + \frac{3}{2}\ln(2\pi MkT) + 1 - \ln(h^3 N_A^{5/2})]$$

$$G = -RT[\ln V + \frac{3}{2}\ln(2\pi MkT) - \ln(h^3 N_A^{5/2})]$$

where M is the molecular weight and N_A the Avogadro number.

Hint. Use the Stirling approximation: $\ln x! = x\ln x - x$. (Is it valid here?)

16.17 For the same gas:

a. show that the molar entropy is given by:

$$S = R[\ln(\frac{M^{3/2}\,T^{5/2}}{P}) - 1.16]$$

which is known as the Sackur-Tetrode equation, and P is in atmospheres.

b. This equation demonstrates that as $T \to 0$, $S \to \infty$, which violates the third law. What is the reason behind this paradox?

16.18 Consider water in the ideal gas state at some temperature T. Discuss the contributions of statistical mechanics in determining its:

a. equation of state;

b. internal energy;

c. heat capacity;

d. entropy; and

e. free energy of formation.

16.19 What is a:

a. microcanonical;

b. canonical; and

c. grand canonical

ensemble? Which one(s) would you consider more important in practice?

16.20 The mean translational energy of an ideal gas molecule may be taken equal to $(3/2)kT$. Using an ideal gas of your choice, examine the magnitude of the quantum numbers as a function of the volume and temperature of the gas.

16.21.a. What is Maxwell's 'demon'?

b. Does it violate the second law?

16.22 How do you explain the fact that Gibbs and Boltzmann derived practically all statistical mechanics before Einstein started publishing?

16.23 What useful information, from the practical point of view, do we obtain from statistical mechanics?

16.24 Comment on the degree to which the questions posed in the Introduction are answered by statistical mechanics.

16.25 Contrast the ideas about knowledge from the classical and the quantum mechanics point of view.

17 Statistical Mechanics: Application to Real Fluids

17.1 Introduction

Direct solution of the Schrödinger equation for real fluids is not possible. The potential energy of interaction among the molecules is mutual and, consequently, a separate energy cannot be attributed to each molecule. The lack of this energy separability does not allow the simplification of the partition function of the fluid in terms of that of the individual molecules. (Remember that this was the case with the ideal gas, because there the system energy is not mutual.)

Through approximations and the use of molecular simulation, however, application of statistical mechanics to real fluids has provided enhanced understanding of their behavior and improved description of it through equations of state. Furthermore, the molecular approach of statistical mechanics, as compared to the macroscopic one of traditional thermodynamics, combined with molecular simulation renders it a very useful tool in new applications: biochemical processes; prediction of matter behavior at and near surfaces, for example in porous materials or thin films; ceramic materials; polymers; etc.

17.2 Objective and Approach

Our objective in this Chapter is to review very briefly some of the approximations involved in applying statistical mechanics to real fluids. Apparently this is a complex subject with continuously growing applications and its detailed discussion is beyond the scope of this book. It is hoped, however, that the interested reader can get the assistance and direction needed for further involvement in this field.

We start by assuming that the internal motions of the molecules are independent of density, which allows us to separate the partition function of a real fluid into two parts: one that involves the intermolecular forces - i.e. the potential energy of the fluid - and is expressed in terms of the *configuration integral*; and one that involves the internal motions in the ideal gas state.

We proceed then with a second assumption, that of *pairwise additivity*, according to which the potential energy of the fluid can be approximated by the sum of interactions between all molecular pairs. This assumption allows us to determine the thermodynamic properties of a real fluid in terms of the intermolecular potential energies and the *radial distribution function*, which describes the position of the molecules in the space they occupy.

Unfortunately, the radial distribution function cannot be determined with sufficient accuracy at all densities. For low density fluids, however, solution is possible through a Taylor's series expansion around zero density, which will lead to the virial equation.

We turn then to dense fluids and outline the main approaches used for them, which we classify into three main categories: Molecular Simulation, Perturbation, and Semiempirical. More emphasis is given to the last one which, using results from the other two approaches, leads to semiempirical equations of state.

We close with some concluding remarks.

17.3 The Partition Function of a Real Fluid

17.3.1 The Configuration Integral

Consider a fluid consisting of N molecules at temperature T and volume V. Following Reed and Gubbins (1973), we assume that the total energy of the fluid is the sum of two independent contributions:

a.that involving the center-of-mass (*cm*) degrees of freedom (translational), and

b.that involving only internal (*int*) degrees of freedom (vibrational, rotational, etc.), which are considered independent of density.

Under this assumption, the partition function (p.f.) of the fluid Q can be written - as discussed in the previous Chapter - as the product of the two corresponding partition functions:

$$Q = Q_{(cm)} Q_{(int)} \tag{17.3.1}$$

For a fluid where the potential energy ϕ depends only on the positions $r_1,\ldots, r_N$ of the molecular centers of mass and not their orientation:

$$\phi = f(r_1,\ldots,r_N) \tag{17.3.2}$$

where r is a position vector.

Then, following Reed and Gubbins:

$$Q_{(cm)} = \frac{L\Lambda^{-3N}}{N!} \tag{17.3.3}$$

where:

$$L = \int\ldots\int \exp\left(\frac{-\phi}{kT}\right) dr_1 \ldots dr_N \tag{17.3.4}$$

is called the *configuration integral*, because its value depends on the relative position (configuration) of the molecules (the integrations are performed over the volume of the system); and,

$$\Lambda = \frac{h}{(2\pi mkT)^{1/2}} \tag{17.3.5}$$

is called the *de Broglie wavelength*.

17.3.2 Comments

1.The physical significance of the configuration integral is that it represents the contribution of intermolecular forces to the partition function of a fluid.

2.Notice that for an ideal gas, $\phi = 0$, the configuration integral is equal to V^N.

3.If we introduce Eq.17.3.3 into Eq.17.3.1 we obtain:

$$Q = \left[\frac{1}{N!}\left(\frac{V}{\Lambda^3}\right)^N Q_{(int)}\right]\left(\frac{L}{V^N}\right) \tag{17.3.6}$$

i.e. the expression for the p.f. of a real fluid.

4.The first bracketed term in Eq.17.3.6 represents the partition function in the ideal gas state, $Q_{(id)}$, at the same T, V, and N as the real fluid; the second one, the effect of non-ideal gas behavior, i.e. of the non-zero intermolecular forces. Thus Eq.17.3.6 becomes:

$$Q = Q_{(id)}\left(\frac{L}{V^N}\right) \qquad (17.3.7)$$

5.The main assumption involved here is that the internal partition function is independent of density. It is valid for monatomic molecules, but can lead to significant errors for polyatomic molecules, where their rotation and vibration is effected by density, especially at high values of it.

6.Two more assumptions are also implicit in the development of Eq. 17.3.3, but are of minor importance from the practical point of view, and are discussed in Chapter 16:

a.Applicability of Maxwell-Boltzmann statistics; and

b.Classical treatment of the translational p.f.

7.If the potential energy depends not only on the location of the centers of the molecules but also on their orientation, Eqs 17.3.2 and 17.3.4 must be modified accordingly.

8.It is apparent from Eq.17.3.7 that, if the configuration integral can be evaluated, all thermodynamic properties of the fluid can be determined (See Example 17.1) - provided, of course, that $Q_{(int)}$ be available (Section 16.8) - and within the limitations of the aforementioned assumptions.

9.Evaluation of the configuration integral requires, as Eq.17.3.4 indicates, knowledge of the:

a.intermolecular potential; and

b.distribution (location) of the molecules in space, described by what are referred to as *distribution functions*.

Intermolecular potentials have been discussed in Chapter 7, so we focus our attention on the distribution functions.

10.The general case, where the potential energy is calculated by considering pairs, triplets, and so on, of molecular combinations, is discussed by Reed and Gubbins and leads to 'high-order distribution functions' that cannot be evaluated.

11.A commonly made assumption about the potential energy of a system is that of *pairwise additivity*, i.e. that it is equal to the sum of pair potentials $\Gamma(r_{ij})$, where r_{ij} is the distance between molecules i and j:

$$\phi = \sum_{1 \le i} \sum_{< j \le N} \Gamma(r_{ij}) \qquad (17.3.8)$$

In this case only the second-order distribution function, known as the *radial distribution function*, is required for the evaluation of the thermodynamic properties of the fluid (Reed and Gubbins, p.174). It is discussed in Section 17.4.

12. At very low densities, such as at nearly ideal gas conditions, Eq. 17.3.8 is valid because the probability that three or more molecules are close enough to interact is very small, compared with that of only two molecules. At higher densities, however, the assumption of pairwise additivity will cause some error. Thus, it is estimated that the nonadditive contributions to the internal energy of liquid argon at its triple point is of the order of 5 to 10 percent (Reed and Gubbins, p.97).

17.3.3 Example 17.1. Thermodynamic Properties from the Configuration Integral

One way to isolate the effect of intermolecular forces on the properties of a fluid is to use the residual properties. For any property X, the residual property $X_{(r)}$ is defined by:

$$X_{(r)}[T,V,N] = X(T,V,N) - X_{(id)}[T,V,N]$$

where (id) refers to ideal gas behavior.

Develop expressions for the residual pressure $P_{(r)}$, internal energy $U_{(r)}$, and Helmholtz free energy $A_{(r)}$, in terms of the configuration integral.

From Eq.16.7.4:

$$P_{(r)} = kT\left(\frac{\partial \ln Q}{\partial V}\right)_{T,N} - kT\left(\frac{\partial \ln Q_{(id)}}{\partial V}\right)_{T,N}$$

which by virtue of Eq.17.3.7 becomes:

$$P_{(r)} = kT\left(\frac{\partial \ln L}{\partial V}\right)_{T,N} - P_{(id)} \tag{A}$$

With the same approach:

$$U_{(r)} = kT^2\left(\frac{\partial \ln L}{\partial T}\right)_{V,N} \tag{B}$$

$$A_{(r)} = -kT\ln(L/V^N) \tag{C}$$

If the configuration integral is known, therefore, all properties of the real fluid can be determined, provided that the ideal gas properties at the same conditions be available (from the previous Chapter).

Following Comments 9, 10, and 11 of Section 17.3.2, we consider next the radial distribution function.

17.4 The Radial Distribution Function and Thermodynamic Properties

17.4.1 The Radial Distribution Function *g(r)*

Assume that we have a very fast camera that can take a picture of the spacial arrangement of molecules of a given system. We focus now our attention on a specific molecule j in the picture and proceed to determine the number of molecules whose centers lie in a volume element dV at distance r from j (Figure 17.1). This number is equal to $\rho(r,t)dV$, where $\rho(r,t)$ is the local number density (molecules per unit volume) at distance r from j at the time of observation t, and $dV = 4\pi r^2 dr$. We repeat this measurement over a time interval so that we obtain the time-average value of the local number density $\rho(r)$.

We notice that if R_o is the radius of molecule j, then because of the strong repulsive forces it is unlikely that there will be an other molecule, whose center lies in a sphere of radius equal to $2R_o$.

If we now move just beyond $2R_o$, we will find more molecules (per unit volume) than the bulk average, because of the effect of attractive forces. A little farther out, however, this effect diminishes and the number of molecules present becomes equal to the bulk average.

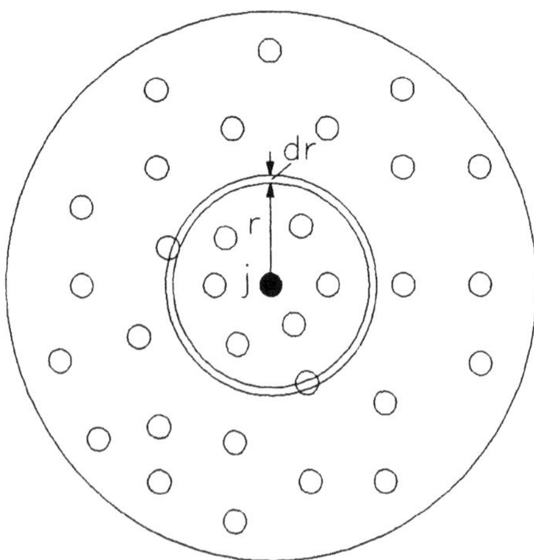

Figure 17.1 Schematic diagram of a fluid for the determination of its radial distribution function, pressure, and internal energy.

If we divide the local densities at various distances from molecule j by the bulk average $\rho = N/V$, we obtain the *radial*, or *pair*, *distribution function*, $g(r)$. In addition to r, g is a strong function of density ρ, and a weak function of T:

$$g = \frac{\rho(r)}{\rho} = g(r,\rho,T) \tag{17.4.1}$$

Typically, however, we use the notation $g(r)$ with an implied dependency on temperature and density.

Radial distribution functions of gases and liquids can be obtained from X-ray and neutron scattering experiments, in the same manner as the crystal structure is studied with X-ray diffraction (see, for example, Pings, 1967). Typical shapes for the three types of matter are presented in Figure 17.2, and are described by Mansoori and Haile (1983) as follows: "At low densities there is essentially no local structure in the fluid and $g(r)$ quickly assumes its long range limit. In crystalline solids, structure persists to larger r, and $g(r)$ is a series of Gaussian functions, corresponding to molecular vibrations about lattice sites. At liquid densities, the fluid exhibits short range structure similar to the solid and long range disorder characteristic of the gas phase."

They can also be estimated theoretically if the intermolecular potential is known, but the accuracy declines with increasing density.

17.4.2 Thermodynamic Properties from *g(r)* and Γ*(r)*

We will demonstrate next that the thermodynamic properties of a fluid can be determined from knowledge of the radial distribution function, $g(r)$, and the pair potential, $\Gamma(r)$. To this purpose we will develop expressions for its equation of state and internal energy.

a. Equation of State

Following Example 17.1, we write: $(VP) = (VP)_{(id)} + (VP)_{(r)}$. The first term represents the ideal gas contribution, and is equal to NkT; the second, the residual contribution due to the intermolecular forces. To calculate the latter we note that, with reference to Figure 17.1:

* the force between the central molecule and one at a distance r is:

$$-\frac{d\Gamma(r)}{dr} = -\Gamma'(r)$$

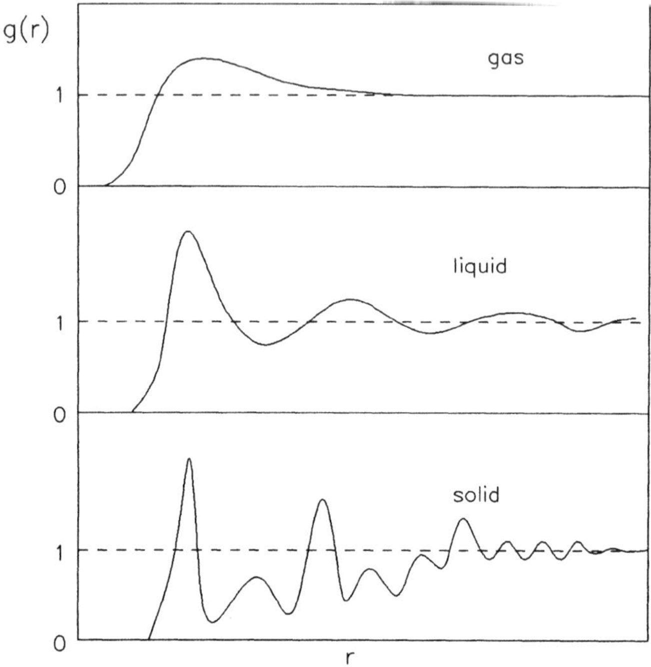

Figure 17.2 Local structure of matter.

* the number of molecules in the shell of width dr at a distance r is given by: $\rho g(r) 4\pi r^2 dr$, where ρ is the number density, molecules per volume unit;

* the total force acting between the central molecule and all others in the shell is: $-\Gamma'(r) 4\pi r^2 \rho g(r) dr$ coressponding to pressure $-\Gamma'(r) \rho g(r) dr$.

Multiplication of this by the volume of the sphere of radius r, $(4/3)\pi r^3$, gives the contribution to $(VP)_{(r)}$ of the central molecule interacting with all molecules in the shell. Integration then over all r yields:

$$(VP)_{(r)} = \left(\frac{N}{2}\right)\frac{-4\pi}{3}\int_0^\infty \rho\Gamma'(r) g(r) r^3 dr \tag{17.4.2}$$

where the factor (1/2) is introduced to avoid counting each molecule twice, once as central and once as peripheral.

Hence:

$$P = \rho kT - \frac{2\pi\rho^2}{3}\int_0^\infty \Gamma'(r) g(r) r^3 dr \tag{17.4.3}$$

b. Internal Energy

Following the same approach we arrive at:

$$U_{(r)} = 2\pi N\rho . \int_0^{\infty} \Gamma(r) g(r) r^2 dr \tag{17.4.4}$$

and consequently:

$$U = \frac{3}{2} NkT + U_{(int)} + U_{(r)} \tag{17.4.5}$$

It is apparent that only the residual internal energy depends on the intermolecular potential and, as Eq.17.4.5 suggests, it also represents the potential energy of the fluid based on pairwise additivity.

Expressions for the Helmholtz free energy and the chemical potential in terms of $g(r)$ and $\Gamma(r)$ can also be developed (see Reed and Gubbins).

17.4.3 Comments

1. We conclude that if the assumption of pairwise additivity for the potential energy, Eq.17.3.8, is valid and the radial distribution function is available, the thermodynamic properties of a fluid can be determined from Eqs 17.4.3 through 17.4.5. (Provided, of course, that the pair potential be known.)

2. The assumption of pairwise additivity is valid at low densities, but it will lead to errors at high densities (Section 17.3.2).

3. The quality of the obtained results depends, in addition to the validity of Eq.17.3.8, on how accurately the pair potential is described. While our knowledge in this field is increasing continuously - see, for example, Reed and Gubbins - it is still limited.

4. The assumption that the rotational and vibrational contributions are independent of density can also lead, as discussed in Section 17.3.2, to errors for polyatomic molecules, especially at higher densities.

5. Experimental determination of $g(r)$ is rather cumbersome. Furthermore, it is of sufficient accuracy for the determination of internal energy, but not of pressure.

6. Use of such measurements would defeat, anyway, the objective of predicting thermodynamic properties without resorting to laboratory measurements.

7. Theoretical determination of $g(r)$, on the other hand, based on knowledge of the pair potential as function of the intermolecular distance, is

possible and yields good results at low densities, but poorer ones with increasing density.

8. Approximations are, hence, necessary if statistical mechanics is to provide results for real fluids. As expected, these approximations are more successful at lower densities than at higher ones, as we will see next.

17.5 Low Densities: The Virial Equation

17.5.1 The Virial Equation

Consider the expression for the compressibility factor z:

$$z = \frac{PV}{NkT} = \frac{P}{\rho kT}$$

We know that as the density ρ approaches zero at a constant temperature T, the product (PV) remains finite and becomes equal to NkT at the limit $\rho = 0$, i.e. $z = 1$, at $\rho = 0$.

A Maclaurin's series expansion in density around $\rho = 0$ yields:

$$z = 1 + \sum_{2}^{\infty} B_n(T)\rho^{n-1} \tag{17.5.1}$$

where:

$$B_n(T) = \frac{1}{(n-1)!}\left[\frac{\partial^{n-1}(P/\rho kT)}{\partial \rho^{(n-1)}}\right]_{\rho=0} \tag{17.5.2}$$

are known as the virial coefficients, B_2 the second, B_3 the third, and so on.

17.5.2 Comments

1. In Chapter 8 we used B for the second virial coefficient, C for the third, and so on. Here we use, for generality, B_2, B_3, etc. B_1 is the unity that appears before the summation.

2. We also stated that the virial coefficients of pure compounds are functions of temperature only. This is apparent here, because they are defined at the limit $\rho = 0$, at a specified temperature.

3. The virial equation was first proposed by Kammerlingh Onnes in 1901 on a purely empirical basis; he also suggested the name *virial*.

4. In the 1930's, Joseph Mayer showed that the virial equation has a rigorous derivation from statistical mechanics and that the virial coefficients are related to the intermolecular potentials: B_2 to two-body interactions, B_3 to three-body ones, and so on.

5. The procedure involves a series expansion of $g = g(r,\rho,T)$ around $\rho = 0$, and is a very lengthy one (see, for example, Reed and Gubbins). We present here only the final result for the second virial coefficient:

$$B = B_2 = 2\pi N_A \int_0^\infty \left[1 - \exp\left(-\frac{\Gamma(r)}{kT}\right)\right] r^2\,dr \qquad (17.5.3)$$

where N_A is the Avogadro number.

6. Because of this development of the virial equation from statistical mechanics it is the only one of the equations of state with a solid theoretical basis.

7. When applied to real fluids the virial equation has two main disadvantages:

a. Since it is not applicable to condensed phases (why?), it cannot be used for the simultaneous description of all phases at equilibrium.

b. Pair potentials for such fluids are typically not available and, consequently, virial coefficients are determined from *PVT* data. These data rarely allow the reliable evaluation of virial coefficients beyond the second one, however (see Section 8.9.4).

The usefulness, thus, of the virial equation is restricted to low pressure vapors, where coefficients beyond the second one are not needed.

17.5.3 Example 17.2. The Second Virial Coefficient of the Hard-Sphere Fluid

The simplest fluid is one composed of hard spheres. The intermolecular potential is:

$$\Gamma(r) = 0, \quad r > \sigma$$
$$\Gamma(r) = \infty, \quad r \leq \sigma$$

where σ is the sphere diameter. In other words, the fluid is composed of tiny marbles which do not interact with each other until they touch: then a repulsion of infinite size occurs. What is the second virial coefficient of this fluid?

From Eq.17.5.3:

$$B = 2\pi N_A\left[\int_0^{\sigma}(1-e^{-\infty})r^2dr+\int_{\sigma}^{\infty}(1-e^{0})r^2dr\right] = 2\pi N_A\int_0^{\sigma}r^2dr =$$

$$= \frac{2\pi N_A}{3}\sigma^3$$

Notice that for this fluid the second virial coefficient is positive and independent of temperature.

17.6 High Densities: The Approaches

Uncertainties in the prediction of the radial distribution function at high densities from knowledge of the pair potential do not allow for the direct evaluation of thermodynamic properties in such densities through Eqs 17.4.3 and 17.4.5. Three general approaches have been used to circumvent this problem:

1. Molecular Simulation;
2. Perturbation; and
3. Semiempirical.

In the first approach, we start with a molecular shape (spheres, dumbbells, etc.) and an expression for the intermolecular potential (hard-sphere, HS, Lennard-Jones, LJ, etc.) and proceed to calculate the macroscopic properties of this hypothetical fluid through numerical methods employing the appropriate - for the methodology used - equations.

In the second approach, a 'perturbation' around some known reference fluid is involved. This reference fluid is a hypothetical one, such as the HS fluid, whose particles interact with the hard-sphere potential, or the LJ fluid, whose particles interact with a Lennard-Jones potential, and its properties become known through molecular simulation.

In the third approach, we use our qualitative understanding of the structure of fluids and intermolecular forces in formulating the partition function. To account for the molecular volume, for example, we introduce in the partition function, Eq.17.3.6, the so-called 'free volume'. By understanding the limitations of the approximations made, this procedure provides avenues for the development of improved equations of state. We will place more emphasis on this approach because it is simpler and has found considerable application in the literature.

17.7 Molecular Simulation

Assume that we place, in some random fashion, several hundred particles (molecules) in a box. We refer to them as the system and specify it by defining:

a. The structure of the molecules (spheres, dumbbells, etc.);

b. The state of the system in terms of its density, molecules per unit volume, and the average kinetic energy or temperature of them; and

c. The nature of forces that are exerted among the molecules (hard-sphere, Lennard-Jones, etc.).

We now allow the molecules to move around and from their positions we can determine the radial distribution function for each one and, by averaging them, that for the whole system. When the latter remains constant, the system is in equilibrium. The movement of the molecules is carried out through two main techniques:

* Monte Carlo; and
* Molecular Dynamics.

An excellent qualitative discussion of these techniques is given by Alder and Wainwright (1959a), who are among the pioneers in the field, and only a brief description - based on their article - is presented here. For the quantitative aspects of the techniques, see Allen and Tildesley (1987) and the references given by: Gubbins et al (1983); Haile (1986); Panagiotopoulos et al (1986); and Quirke (1986).

17.7.1 Monte Carlo (Metropolis et al, 1953)

Here, the molecules are displaced - one at a time - using randomly generated numbers from a uniform distribution. Thus, molecule 1 moves from position $x_1(0)$, $y_1(0)$, $z_1(0)$ to position $x_1(1)$, $y_1(1)$, $z_1(1)$ using three numbers generated randomly (such as through a roulette - hence the name Monte Carlo). The same is done with the other molecules. After each move, the corresponding properties of the molecules are determined using their relative positions. In making these movements two factors must be considered.

First, the move may be prohibited, for the molecule arrives at a position occupied by an other one. In this case the molecule is returned to its previous location and the corresponding distribution is used twice in the averaging, over all configurations, process. In other words, the more often a given distribution leads to an impossible configuration, the more probable this distribution is.

Second, due to the forces that are exerted among the molecules, a certain move is more likely to occur than an other. Thus two molecules within the range of attraction are more likely to approach each other than move apart. This is taken into account by assigning probabilities reflecting the force that tends to encourage or inhibit a certain movement.

17.7.2 Molecular Dynamics (Alder and Wainwright, 1959b)

The Monte Carlo technique provides the relative location of the molecules and, hence, the equilibrium properties of the system. It does not provide, however, a description of the actual molecular motions, which is accomplished through the technique of molecular dynamics.

Here, once the initial location and velocities of the molecules are set - along with the appropriate analytical description of the intermolecular forces - no element of chance is involved. The molecules move according to Newton's laws of motion (see Section 16.3.1).

17.7.3 Comments

1.The random moves of the molecules in the Monte Carlo method are not the same as the motions that carry the molecules from one position to an other in real matter. As a result, the order in which the configurations occur is also unrealistic. (In a system at equilibrium, however, this order is not important and the obtained average is independent of it.)

2.Molecular dynamics, on the other hand, is not restricted to systems at equilibrium. It can be used, for example, to determine how fast, i.e. after how many molecular movements, a system will reach equilibrium. Once equilibrium is reached, all properties can be determined from the relative positions of the molecules.

3.In addition, since the molecular configurations obtained with the molecular dynamics method are in the right order, the time-dependent behavior of the system, such as diffusion, can be studied.

4.If the correct intermolecular potential of a fluid is known, all its properties can be obtained by simulation. Because of limitations in the description of such potentials for real fluids, however, most simulation work has dealt with hypothetical fluids, such as the hard-sphere (Section 17.7.4) or the Lennard-Jones ones.

5. While early simulations dealt with pure fluids only, more emphasis is given progressively to mixtures containing molecules:
* of different sizes and shapes; and
* interacting with various potentials.

6. More emphasis is also given to the very important area of phase equilibria, where special simulation techniques are needed, since the conventional Monte Carlo and molecular dynamics techniques do not enable direct calculations. For a description of these techniques, see Gubbins (1989b) and especially Panagiotopoulos and Stapleton (1989).

7. The most successful such technique is the 'Gibbs ensemble' of Panagiotopoulos et al, which was used by Harismiadis et al (1991) to generate phase diagrams for binary Lennard-Jones fluids of varying values for the size and energy parameters.

8. Incorporation of more realistic models for the molecular forces in simulations has yielded results that are closer to real fluid behavior. Such results have been used either for preliminary testing of theories, or for the approximate description of the behavior of real fluids. [For more on this subject, see: Panagiotopoulos et al; Quirke; Haile; and de Pablo and Prausnitz (1989).]

9. An important contribution of simulation studies may be in the establishment of mixing rules, especially density-dependent ones, which allow an equation of state to cover the whole density range (see, for example, Nakanishi et al, 1986, and Lee et al, 1986); or in the development of new models for the activity coefficient (Sandler and Lee, 1986).

10. For a review of the current and future role of molecular simulation in industrial applications, see Gubbins (1989a).

17.7.4 The Hard-Sphere Fluid

The simplest fluid to be studied through molecular simulation is the hard-sphere one, whose intermolecular potential is described in Example 17.2. The radial distribution function and pressure for this fluid have been established as functions of density, the latter from molecular dynamics calculations. The most successful analytical expression for the pressure of a hard-sphere fluid (P^{hs}) is that proposed by Carnahan and Starling (1969; 1972):

$$P^{hs} = \frac{RT}{V}\left[\frac{1+y+y^2-y^3}{(1-y)^3}\right] \tag{17.7.1}$$

where:
$$y = \frac{\pi\sqrt{2}}{6}\left(\frac{V_0}{V}\right) = \frac{\pi}{6}\left(\frac{N_A}{V}\right)\sigma^3 \tag{17.7.2}$$

V is the molar volume and V_0 is the hard-sphere molar volume at closest packing, equal to:

$$V_0 = \frac{N_A\sigma^3}{\sqrt{2}}$$

This is demonstrated in Figure 17.3, where the compressibility factor values of the hard-sphere fluid obtained from molecular simulation and from Eq.17.7.1 are shown.

The usefulness of the hard-sphere fluid is based on the evidence that for dense fluids, such as liquids, the attractive forces between molecules serve primarily to determine the volume, i.e. the density. Otherwise they play a minor role, as compared to the strong repulsive forces, in determining its structure, i.e. the radial distribution function (Reed and Gubbins, p.249). The hard-sphere fluid represents, therefore, an attractive choice as a base on which to build semiempirical equations of state, or as a reference fluid for perturbation calculations.

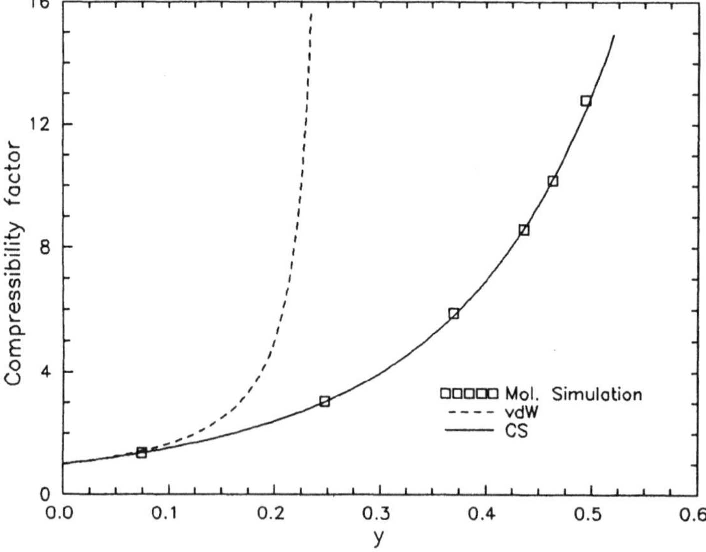

Figure 17.3 The compressibility factor for the hard-sphere fluid.

17.8 Perturbation

In this approach the properties (e.g. distribution function or free energy) of a system with a given intermolecular potential Γ, are related to those of a reference system with potential Γ_0, usually by an expansion in powers of the perturbation potential $\Gamma_1 = \Gamma - \Gamma_0$. The perturbation terms, first-order, second-order, etc., involve both Γ and the distribution function of the reference system (see, among others, Gubbins et al, and, especially, Gray and Gubbins, 1984).

It is apparent that the closer the actual fluid is to the reference one, the easier the series will converge. Thus, hard spheres would be a better reference fluid for one whose molecules interact with a square-well (Section 7.6.3) potential, than with a Lennard-Jones one.

The requirements for the reference fluid and the corresponding theories are summarized by Gubbins et al as follows: "The reference fluid used in perturbation theory should meet two requirements: its properties should be well understood, and it should be close to the fluid of interest ... For atomic fluids it is possible to satisfy both of these requirements by using the repulsive part of the potential as the reference fluid potential. For molecular fluids the situation is more complex, and no reference fluid is yet available which meets both of the above requirements. Two types of theory have been proposed: those that adopt a reference fluid of spherical molecules, and those in which the reference molecules are nonspherical. The spherical reference fluid is well understood - the equation of state and structure are known - but it is not close to the fluid of interest, so that the series convergence is often poor. Reference fluids of nonspherical molecules are generally closer to the real fluid, but are more poorly understood."

A main application of the perturbation approach is to describe hypothetical fluids without resorting to simulation work that consumes large amounts of computer time. Thus, a fluid interacting with a square-well potential can be studied by using a perturbation around the well-known hard-sphere fluid, with the attractive part of the potential representing the perturbation. The success is determined by comparison to the simulation results for such a fluid.

For details, as well as an extensive discussion of perturbation theories, see Reed and Gubbins; and also Gray and Gubbins.

17.9 Equations of State with the Semiempirical Approach

We present first a generalized partition function for the canonical ensemble, referred to as the van der Waals one (Vera and Prausnitz, 1972; Anderson and Prausnitz, 1980a), and then proceed to demonstrate:

a. the assumptions involved in developing the empirical van der Waals Equation of State (EoS); and

b. how improved EoS may be developed by using information from molecular simulation.

17.9.1 The Generalized van der Waals Partition Function

The essential assumption of the generalized van der Waals (vdW) theory is that the partition function of a real fluid is of the form:

$$Q = \frac{1}{N!}\left(\frac{V}{\Lambda^3}\right)^N\left(\frac{V_f}{V}\right)^N\left[\exp\left(\frac{-\phi}{2kT}\right)\right]^N q_{(int)}^N \qquad (17.9.1)$$

Here V is the volume of the system, V_f is the 'free volume', i.e. the part of the total volume that is available to the N molecules, if we take into account their own finite size; $(\phi/2)$ is the potential energy associated with one molecule; and $q_{(int)}$ is the internal molecular partition function. (See also Problem 17.12.) A comparison of Eq. 17.9.1 with Eq. 17.3.6 indicates that:

a. the term: $(1/N!)[V/\Lambda^3]^N$, and the last one, represent respectively the contributions of the translational and internal motions of the molecules in the ideal gas state; and

b. the second and third bracketed terms represent the contribution of intermolecular forces.

Using next the expression for P in terms of Q (Eq. 16.7.4):

$$P = kT\left(\frac{\partial \ln Q}{\partial V}\right)_{T,N}$$

we obtain:

$$P = P_1 + P_2 + P_3 \qquad (17.9.2)$$

where:

$$P_1 = NkT\left(\frac{\partial \ln V_f}{\partial V}\right)_{T,N} \tag{17.9.3}$$

$$P_2 = \frac{-N}{2}\left(\frac{\partial \phi}{\partial V}\right)_{T,N} \tag{17.9.4}$$

$$P_3 = NkT\left(\frac{\partial \ln q_{(int)}}{\partial V}\right)_{T,N} \tag{17.9.5}$$

P_1 represents the contributions to pressure from the repulsive forces; P_2 those from the attractive forces; and P_3 those from the internal motions of the molecules.

We demonstrate next the development of the simple empirical vdW EoS starting with Eq.17.9.1.

17.9.2 The van der Waals Equation of State

Consider a pure fluid of volume V containing N molecules, each with a hard-core diameter σ. To develop an expression for the free volume, we subtract from the total volume, the volume not accessible to the molecules, which is referred to as the 'excluded' volume.

To calculate the latter, consider a molecule moving around in a fluid of low density so that it has only one closest neighbor. Since the closest distance between the centers of these two molecules is σ, the space that 'belongs' to their centers is a sphere of radius σ with a volume equal to $(4\pi/3)\sigma^3$. The free volume is, thus:

$$V_f = V - \left(\frac{N}{2}\right)\left(\frac{4\pi}{3}\right)\sigma^3 = V - 4N\frac{\pi}{6}\sigma^3 \tag{17.9.6}$$

van der Waals used Eq.17.9.6 for V_f in the form:

$$V_f = V - \frac{N}{N_A}b \tag{17.9.7}$$

where: $b = 4b_0$, and $b_0 = N_A(\pi/6)\sigma^3$, is the actual volume of one mole of hard-core molecules.

Thus, on a molar basis:

$$P_1 = RT\left[\frac{\partial \ln(V-b)}{\partial V}\right]_{T,N} = \frac{RT}{V-b} \tag{17.9.8}$$

where V stands, now, for the molar volume and N, for the Avogadro number.

To calculate the potential energy ($\phi/2$) associated with each molecule we use Eq. 17.4.4 and replace the lower limit by σ, since $g(r)$ is zero for $r < \sigma$. van der Waals, in effect, assumed that $g(r)$ is independent of temperature and volume for all values of r; as a result, the integral is a constant, which van der Waals expressed as $(-2a/N^2)$, where a is a positive constant (Vera and Prausnitz). Thus, on a molar basis:

$$\phi = \frac{-2a}{VN_A} \tag{17.9.9}$$

and:

$$P_2 = \frac{-N_A}{2}\frac{\partial}{\partial V}\left[\frac{-2a}{VN_A}\right]_{T,N_A} = \frac{-a}{V^2}$$

If we now assume $P_3 = 0$, we obtain the van der Waals EoS:

$$P = \frac{RT}{V-b} - \frac{a}{V^2}$$

Considering the shortcomings of the vdW EoS, it is apparent that the simplifying assumptions made in developing the partition function are rather limiting. Indeed van der Waals and his coworkers recognized, as Vera and Prausnitz comment, that "Eq. 17.9.7 for V_f is only valid at low densities. As the density rises, three-body, four-body, etc., collisions become increasingly important and, because of overlap, the excluded volume calculated on the basis of two-body collisions alone is too large". Other expressions for the free volume are given by Vera and Prausnitz.

Similarly, the assumption that $g(r)$ is independent of volume and temperature is also restrictive, and Vera and Prausnitz demonstrate how relaxing this assumption can lead, for example, to the Redlich-Kwong EoS.

We discuss, next, how use of molecular simulation results leads to a new generation of equations of state.

17.9.3 The Carnahan-Starling-vdW and -RK Equations of State

Considering the aforementioned limitations of the free volume expression given by Eq. 17.9.7 and the comments on the hard-sphere fluid (Section

Table 17.1 Performance of the vdW and RK EoS, and the Carnahan-Starling Versions of Them

Equation	Error in Enthalpy (J/mol)	%Error in Density
vdW	24.4	4.5
CSvdW	17.6	2.9
RK	8.8	1.5
CSRK	4.2	1.1

17.7.4), it is reasonable to use Eq.17.7.1 for the contribution of repulsive forces to the EoS. Thus Carnahan and Starling (1972) used this expression for P_1 in Eq.17.9.2, and for the attractive term those of van der Waals and of Redlich and Kwong. They also set $P_3 = 0$. Thus:

$$P(\text{CSvdW}) = \frac{RT}{V}\left[\frac{1+y+y^2-y^3}{(1-y)^3}\right] - \frac{a}{V^2}$$

and a similar expression is obtained for P(CSRK).

The improvement over the original equations is demonstrated in Table 17.1. The results are averages over several pure fluids, mostly light gases and hydrocarbons up to C_6 (Carnahan and Starling).

Anderson and Prausnitz (1980a; 1980b) and Dimitrelis and Prausnitz (1986) have also used the CS-vdW combination, but with temperature-dependent parameters, for the correlation and prediction of phase equilibria in nonpolar mixtures.

17.9.4 Rotational and Attractive Contributions

We have neglected, so far, rotational contributions to the equation of state by assuming that $q_{(int)}$ is independent of density. In addition, for the attractive forces, the simple vdW and RK terms have been used.

For small spherical molecules like argon, $q_{(int)}$ is independent of density. For polyatomic molecules, however, this assumption is not valid. The ability of a molecule to rotate depends on whether or not the molecule is alone or surrounded by other molecules. The extent, thus, to which rotation can be exercised depends on density. Furthermore, the ability of the atoms in a molecule to vibrate is also effected by the presence of near neighbors, in other words, by density. This density effect, however, is much smaller for vibration than it is for rotation.

Turning to molecular attractions, the molecular dynamics results of Alder et al (1972) indicate that, even the simple square-well interaction potential, leads to a much more complex expression for the attractive term, than those used in the van der Waals and Redlich-Kwong EoS.

To account for the effect of density on molecular rotation, Prigogine (1957) suggested the incorporation of this effect into the translational degrees of freedom. This idea was used, among others, by Flory (1970) but the resulting equation is only applicable to high densities. It does not approach, for example, the ideal gas law as density goes to zero. Beret and Prausnitz (1975), however, used Prigogine's idea with Alder's molecular dynamics calculations in developing the Perturbed-Hard-Chain-Theory (PHCT) equation of state that covers the whole density range.

Chao and coworkers (Chien et al, 1983) avoided Prigogine's assumption by using for the rotational contributions the equation of state for hard dumbbell molecules developed by Boublik and Nezbeda (1977). For the attractive term, they also used the results of Alder et al. They refer to the resulting equation of state as the Chain-of-Rotators (COR) EoS.

These rather complex equations, the PHCT and the COR, give good results but cannot be solved analytically for density. As a result, numerical solutions for density become more computer-time consuming - as compared to cubic EoS - for applications requiring repetitive iterative calculations, such as of fugacities in phase equilibrium calculations. [The study of Mathias and Benson (1986), however, seems to question this.] To eliminate this problem, Chao and coworkers (Lin et al, 1983 and Kim et al, 1986) developed the Cubic Chain of Rotators (CCOR) EoS (Chapter 14), that gives similar results with the COR EoS; and Kim et al (1988) have presented a simplified version of the PHCT EoS (SPHCT).

17.9.5 Comments

1.Results from molecular simulation reflecting repulsive and attractive forces, mainly through the hard-sphere and square-well potentials, have been used in the development of semiempirical equations of state.

2.The presence of only repulsive forces in the hard-sphere fluid, and the successful description of its equation of state by the CS expression, makes the latter a very attractive choice for the repulsive term of an EoS. On the other hand, the vdW repulsive term gives very poor results in describing the hard-sphere fluid as shown in Figure 17.3, where the curve labeled (vdW) is obtained from Eq.17.9.8. (See, also, Prausnitz, 1985; and Lin et al.)

3.In terms of describing real fluid behavior, Dimitrelis and Prausnitz indicate an improved performance of the CS over the vdW repulsive term in the correlation of hydrogen-solvent vapor-liquid equilibrium (*VLE*). To this purpose the Boublik-Mansoori mixing rules (Mansoori et al, 1971), not applicable to the vdW term, were used. A similar conclusion is suggested by the results of Table 17.1.

4.The detailed study of Adachi et al (1983), however, suggests that use of a rigorous repulsive term (such as the CS one) - instead of the simple vdW one - combined with a traditional attractive term (for example the vdW or RK), does not provide improvement in predicting pure fluid properties. A similar conclusion has been reached for binary mixtures by Dohrn and Prausnitz (1990), who compared an EoS, composed of a CS reference term plus a simple perturbation term, with some cubic ones.

5.It could be argued of course that a rigorous, but also simple for practical application purposes, attractive term is needed, which represents the objective of substantial effort in the last few years.

6.Gasem and Robinson (1989) report that the SPHCT gives poor vapor pressures for hydrocarbons below 5 mm Hg where, as we have seen in Section 10.6, the simpler EoS: PR, SRK, t-vdW, and t-PR, give good to excellent results. (This could probably be due to the method of parameter evaluation used, rather than the EoS itself.) They also report that in the correlation of ethane/n-alkane vapor-liquid equilibrium data, the Soave-Redlich-Kwong (SRK) and SPHCT EoS give similar results up to C_{20}; for larger alkanes, the SPHCT gives somewhat better results. (This could probably be due to the estimated T_c, P_c and ω values used in the SRK EoS, rather than the EoS itself.)

7.For polar systems the CCOR EoS provides successful correlation of binary *VLE* behavior using conventional mixing rules but with two interaction coefficients (Lin et al). Empirical EoS, however, can also provide successful correlation in this case using two interaction coefficients, except for systems close to partial miscibility (Kalospiros et al, 1991).

8.For polar fluids Bryan and Prausnitz (1987) modified the CS term to account - in addition to the repulsive forces - for orientation-averaged attractive (dipole-dipole) forces; this reference system is then combined with a vdW-type perturbation term to account for the additional attractive forces arising from dispersion and induction.

9.The real test, of course, for these EoS with polar systems will be in their ability to predict multicomponent *VLE* behavior from binary data. This, in turn, may depend more on the mixing rules than on the EoS themselves. Thus, no prediction results are given with the CCOR EoS, and empirical EoS fail (Kalospiros et al).[Remember that, as discussed in Section 14.7.3, empirical EoS can meet successfully this test when combined with local-composition G^E models. The resulting EoS, however, do not provide the required quadratic mixing rule for the mixture second virial coefficient.]

10. Efforts to introduce local-composition, density-dependent mixing rules, obtained from molecular simulation results, into equations of state have not been applied to real systems yet. (See, for example, Lee and Sandler, 1987).

It appears, thus, that molecular simulation results will play an important role in future development of EoS especially for polar systems.

17.10 Concluding Remarks

1. The presence of potential energy in real fluids does not allow the simplification of the partition function in terms of the molecular ones, as is the case with the ideal gas.

2. A certain degree of simplification is achieved by assuming pairwise additivity for the potential energy. This allows for the calculation of thermodynamic properties through the radial distribution function and the pair potential.

3. At high densities, the radial distribution function is determined primarily by the repulsive forces, while the attractive ones have little effect. The latter do make large contributions, however, to the thermodynamic properties; thus the attractive forces are entirely responsible for fluid-fluid (gas-liquid, liquid-liquid and 'gas-gas') phase transitions, which are absent when only repulsive forces are present. For less dense liquids (e.g. near the critical point), and particularly for gases, the attractive forces have a larger effect on $g(r)$ (Gray and Gubbins).

4. Unfortunately, theory cannot provide accurate determination of the radial distribution function for a given intermolecular potential, especially at higher densities. The problem is further exacerbated by limitations in the description of realistic (real-fluid) intermolecular potentials. Approximations are, therefore, necessary.

5. At lower densities, such approximations lead to the virial equation; at higher ones, computer simulation of hypothetical fluids contributes to enhanced understanding of real fluid behavior and, when combined with

semiempirical approaches, may provide improved - over traditional ones - equations of state (Section 17.9).

6. Application of molecular simulation to real systems, which gives qualitative results so far, will provide more quantitative answers as more realistic intermolecular potentials are used (see, for example de Pablo and Prausnitz). The computer cost is very high as yet, however (for cost estimates, see Gubbins, 1989a).

7. Determination of intermolecular potentials for complex molecules with sufficient for this purpose accuracy, however, is very difficult and slow. Rowlinson (1990) states: "It is salutary to remember that until about twenty years ago the accepted figure for so simple an energy of interaction as that between a pair of argon atoms was about 20% in error. Spectroscopists and workers in quantum mechanics are working on the problem for more interesting molecules but their progress is slow."

8. An other approach to determining such potentials may become possible in the future as the computing power increases. From Rowlinson again: "If we can look forward to an increase of, say, a thousand in computer speed and memory in the next few years then, one can envisage trial-and-error attacks on certain problems. By this I mean that one could guess the intermolecular forces for a set of molecules, compute the thermodynamic properties, compare these with known experiments, and then adjust the parameters until there is good agreement. In this way one could envisage building up a library of adequate intermolecular potentials, which would be complicated functions of molecular coordinates, but which would serve as a sound basis for predicting the properties of unknown systems. Such a use of computers would be expensive if not impossible today, but may be feasible tomorrow."

References

Adachi, Y., Lu, B.C.-Y., Sugie, H., 1983. *Fluid Phase Equilibria*, **13**, 133.

Alder, B.J., Young, D.A., Mark, M.A., 1972. *J. Chem. Phys.*, **56**, 3013.

Alder, B.J., Wainwright, T.E., Oct. 1959a. *Scientific American*, 3.

Alder, B.J., Wainwright, T.E., 1959b. *J. Chem. Phys.*, **31**, 459.

Allen, M.P., Tildesley, D.J., 1987. *Computer Simulation of Liquids*, Oxford University Press, Oxford.

Anderson, T.F., Prausnitz, J.M., 1980a; 1980b. *Ind. Eng. Chem. Process Des. Dev.*, **19**, 1; **19**, 9.

Beret, S., Prausnitz, J.M., 1975. *AIChE J.*, **21**, 1123.

Boublik, T., Nezbeda, I., 1977. *Chem. Physics Letters*, **46**, 315.

Bryan, P.F., Prausnitz, J.M., 1987. *Fluid Phase Equilibria*, **38**, 201.
Carnahan, N.F., Starling, K.E., 1969. *J. Chem. Phys.*, **51**, 635.
Carnahan, N.F., Starling, K.E., 1972. *AIChE J.*, **18**, 1184.
Chien, M.C., Greenkorn, R.A., Chao, K.C., 1983. *AIChE J.*, **29**, 560.
de Pablo, J.J., Prausnitz, J.M., 1989. *Fluid Phase Equilibria*, **53**, 177.
Dimitrelis, D., Prausnitz, J.M., 1986. *Fluid Phase Equilibria*, **31**, 1.
Dohrn, R., Prausnitz, J.M., 1990. *Fluid Phase Equilibria*, **61**, 53.
Flory, P.J., 1970. *Disc. Faraday Soc.*, **49**, 7.
Gasem, K.A.M., Robinson, R.L., Jr., 1989. *AIChE Spring National Meeting*, Houston.
Gray, C.G., Gubbins, K.E., 1984. *Theory of Molecular Fluids: Vol.1: Fundamentals*, Oxford University Press, Oxford.
Gubbins, K.E., Shing, K.S., Street, W.B., 1983. *J. Phys. Chem.*, **87**, 4573.
Gubbins, K.E., 1989a. *Chem. Eng. Progress*, February, 38.
Gubbins, K.E., 1989b. *Molecular Simulation*, **2**, 223.
Haile, J.H., 1986. *Fluid Phase Equilibria*, **29**, 307.
Harismiadis, V.I., Koutras, N.K., Tassios, D.P., Panagiotopoulos, A.Z., 1991. *Fluid Phase Equilibria*, **65**, 1.
Hirschfelder, J.O., Curtiss, C.F., Bird, R.B., 1954. *Molecular Theory of Gases and Liquids*, Wiley, New York.
Kalospiros, N., Misseyannis, G., Androulakis, I., Tassios, D., 1991. *Fluid Phase Equilibria*, **64**, 173.
Kim, H.Y., Lin, H.M., Chao, K.C., 1986. *Ind. Eng. Chem. Fundam.*, **25**, 75.
Kim, C.-H., Vilmalchand, P., Donohue, M.D., Sandler, S.I., 1988. *AIChE J.*, **34**, 834.
Lee, K.-H., Sandler, S.I., Patel, N.C., 1986. *Fluid Phase Equilibria*, **25**, 31.
Lee, K.-H., Sandler, S.I., 1987. *Fluid Phase Equilibria*, **34**, 113.
Lin, H.M., Kim, H.Y., Guo, T.M., Chao, K.C., 1983. *Fluid Phase Equilibria*, **13**, 143.
Mansoori, G.A., Carnahan, N.F., Starling, K.E., Leland, T.W. Jr., 1971. *J. Chem. Phys.*, **54**, 1523.
Mansoori, G.A., Haile, J.M., 1983. In *Molecular-Based Study of Fluids*, Haile, J.M., Mansoori, G.A., Eds, *Advances in Chemistry Series 204*, *Am. Chem. Soc.*, Washington, D.C.
Mathias, P.M., Benson, M.S., 1986. *AIChE J.*, **32**, 2087.
Metropolis, N.A., Rosenbluth, A.W., Rosenbluth, M.N., Teller, A.H., Teller, E., 1953. *J. Chem. Phys.*, **21**, 1087.
Nakanishi, K., Adachi, Y., Fujihara, I., 1986. *Fluid Phase Equilibria*, **29**, 347.
Panagiotopoulos, A.Z., Suter, U.W., Reid, R.C., 1986. *Ind. Eng. Chem. Fundam.*, **25**, 525.
Panagiotopoulos, A.Z., Stapleton, M.R., 1989. *Fluid Phase Equilibria*, **53**, 133.
Pings, C.J., 1967. *Disc. Faraday Soc.*, **43**, 89.
Prausnitz, J.M., 1985. *Fluid Phase Equilibria*, **24**, 63.
Prigogine, I., 1957. *Molecular Theory of Solutions*, North-Holland, Amsterdam.
Quirke, N., 1986. *Fluid Phase Equilibria*, **29**, 283.

Reed, T.M., Gubbins, K.E., 1973. *Applied Statistical Mechanics*, McGraw-Hill, New York.

Rowlinson, J.W., 1990. *The Prediction of Physical Properties - A Challenge to Chemical Engineers*, The Inaugural T.W. Leland, Jr. Lecture in Chemical Engineering, Rice University, Houston.

Sandler, S.I., Lee, K.-H., 1986. *Fluid Phase Equilibria*, **30**, 135.

Vera, J.H., Prausnitz, J.M., 1972. *Chem. Eng. J.*, **3**, 1.

Problems

17.1 With reference to Eq.17.3.1:

a.what assumptions are involved in its development?

b.what are the resulting limitations?

17.2 Show that for an ideal gas of N molecules in a volume V, the configuration integral is equal to V^N.

17.3 Develop Eqs (A), (B), (C) of Example 17.1.

17.4 Discuss the physical meaning of the radial distribution function.

17.5.a. Develop Eq.17.4.4 and state the assumptions involved.

b.Develop expressions for the Helmholtz free energy and for the enthalpy of a fluid in terms of $g(r)$ and $\Gamma(r)$.

17.6.a. Show that the equation of state of a pure fluid can be obtained from knowledge of the radial distribution function and of the pair potential.

b.Discuss the assumptions involved and the resulting limitations.

17.7 Discuss the advantages and disadvantages of the virial equation as compared with other EoS.

17.8 Discuss the difficulties involved in evaluating second and third virial coefficients of pure fluids and mixtures:

a.through statistical mechanics

b.from *PVT* data.

17.9 With reference to the second virial coefficient:

a.how do you explain in terms of intermolecular forces the fact that B_2 for the hard-sphere fluid is positive?

b.develop the expression for the second virial coefficient of a Lennard-Jones fluid. (See Hirschfelder et al, 1954.)

17.10.a. Discuss briefly the Monte Carlo and Molecular Dynamics simulation techniques;

b.Present the advantages of the one over the other.

17.11 How is the pressure of a fluid with a specified pair potential determined in molecular simulation?

17.12 What assumptions are involved in developing Eq.17.9.1 from Eq.17.3.6?

17.13 Develop Eqs 17.9.3 through 17.9.5.

17.14 Explain how Eq. 17.9.9 is obtained from Eq. 17.4.4.
17.15 What assumptions are involved in developing the vdW EoS from Eq. 17.9.1?
17.16 Develop the Soave-Redlich-Kwong EoS starting with Eq. 17.9.1. (For guidance, see Vera and Prausnitz.).
17.17 What requirements should a fluid meet so that it can be used as a reference fluid in a perturbation calculation?
17.18 Discuss briefly the capabilities and limitations of the perturbation method.
17.19 Does the rotational motion of the molecules lead to an increase or a decrease in the fluid pressure?
17.20 Would the CSRK EoS be more valid for methane or for n-butane? Why?
17.21 Evaluate the performance of the CSvdW, CSRK, and t-vdW EoS (see Chapter 10 for the latter) in describing the compressibility factor of Ar along the critical isotherm. Could you get better results by changing the value of the translation in the t-vdW EoS? At what cost?
17.22.a. Modify Eq. 17.9.1 to reflect Prigogine's assumption about rotational contributions to the p.f.;
b. Develop the corresponding expression for the CSvdW EoS;
c. Compare its performance with the original one in the prediction of the volumetric behavior of a real fluid of your choice.
17.23 An equation of state is to be developed for propane through statistical mechanics:
a. identify the starting point(s);
b. discuss the limitations involved;
c. recommend the best - in your judgement - approach and the experimental data you will need;
d. discuss the capabilities and limitations of the resulting EoS.

Note. Problems 17.10, 17.11, 17.18, 17.21, 17.22, and 17.23 can be extended to projects.

APPENDIX

A

Conversion Factors

Length	1 m	= 100 cm
		= 3.28084 ft
		= 39.3701 in
Mass	1 kg	= 2.20462 lb_m
Force	1 N	= 1 kg m sec^{-2}
		= 10^5 dyne
		= 0.224809 lb_f
Pressure	1 Pa	= 1 N m^{-2}
		= $0.986923 \cdot 10^{-5}$ atm
		= 10^{-5} bar
		= $14.5038 \cdot 10^{-5}$ psia
Energy	1 J	= 1 N m = 1 Pa m^3
		= 10^7 dyne cm = 10^7 erg
		= 0.239006 cal
		= $9.47831 \cdot 10^{-4}$ Btu
		= 0.737562 ftlb_f
Power	1 Watt	= 1 J s^{-1}
		= $1.34102 \cdot 10^{-3}$ horsepower (hp)
		= 0.737562 ftlb_f s^{-1}
		= 0.239006 cal s^{-1}

APPENDIX

B

Steam Tables

Table B.1 Properties of Saturated Steam (SI Units)

Table B.2 Properties of Superheated Steam (SI Units)

All tables are generated by computer from programs, supplied by Professor H.C. Van Ness, based on "The 1976 IFC (International Formulation Committee) Formulation for Industrial Use: A Formulation of the Thermodynamic Properties of Ordinary Water Substance," as published in the "ASME Steam Tables,", 4th ed., App.I, pp.11-29, The Am. Soc. Mech. Engrs., New York, 1979.

TABLE B.1. SATURATED STEAM SI UNITS

V = SPECIFIC VOLUME cu cm/g
U = SPECIFIC INTERNAL ENERGY kJ/kg
H = SPECIFIC ENTHALPY kJ/kg
S = SPECIFIC ENTROPY kJ/(kg K)

TEMPERATURE		ABS PRESS	SPECIFIC VOLUME V		INTERNAL ENERGY U		ENTHALPY H		ENTROPY S	
C	K	kPa	SAT LIQUID	SAT VAPOR	SAT LIQUID	SAT VAPOR	SAT LIQUID	SAT VAPOR	SAT LIQUID	SAT VAPOR
0	273.15	0.611	1.000	206300	-0.04	2375.6	-0.04	2501.6	0.0000	9.1578
0.01	273.16	0.611	1.000	206200	0.0	2375.6	0.00	2501.6	0.0	9.1575
1	274.15	0.657	1.000	192600	4.17	2376.9	4.17	2503.4	0.0153	9.1311
2	275.15	0.705	1.000	179900	8.39	2378.3	8.39	2505.2	0.0306	9.1047
4	277.15	0.813	1.000	157300	16.80	2381.1	16.80	2508.9	0.0611	9.0526
6	279.15	0.935	1.000	137800	25.21	2383.8	25.21	2512.6	0.0913	9.0014
8	281.15	1.072	1.000	121000	33.60	2386.6	33.60	2516.2	0.1213	8.9513
10	283.15	1.227	1.000	106400	41.99	2389.3	41.99	2519.9	0.1510	8.9020
12	285.15	1.401	1.000	93840	50.38	2392.1	50.38	2523.6	0.1805	8.8536
14	287.15	1.597	1.001	82900	58.75	2394.8	58.75	2527.2	0.2098	8.8060
16	289.15	1.817	1.001	73380	67.12	2397.6	67.13	2530.9	0.2388	8.7593
18	291.15	2.062	1.001	65090	75.49	2400.3	75.50	2534.5	0.2677	8.7135
20	293.15	2.337	1.002	57840	83.86	2403.0	83.86	2538.2	0.2963	8.6684
22	295.15	2.642	1.002	51490	92.22	2405.8	92.23	2541.8	0.3247	8.6241
24	297.15	2.982	1.003	45930	100.6	2408.5	100.6	2545.5	0.3530	8.5806
25	298.15	3.166	1.003	43400	104.8	2409.9	104.8	2547.3	0.3670	8.5592
26	299.15	3.360	1.003	41030	108.9	2411.2	108.9	2549.1	0.3810	8.5379
28	301.15	3.778	1.004	36730	117.3	2414.0	117.3	2552.7	0.4088	8.4959
30	303.15	4.241	1.004	32930	125.7	2416.7	125.7	2556.4	0.4365	8.4546
32	305.15	4.753	1.005	29570	134.0	2419.4	134.0	2560.0	0.4640	8.4140
34	307.15	5.318	1.006	26600	142.4	2422.1	142.4	2563.6	0.4913	8.3740
35	308.15	5.622	1.006	25240	146.6	2423.5	146.6	2565.4	0.5049	8.3543
36	309.15	5.940	1.006	23970	150.7	2424.8	150.7	2567.2	0.5184	8.3348
38	311.15	6.624	1.007	21630	159.1	2427.5	159.1	2570.8	0.5453	8.2962
40	313.15	7.375	1.008	19550	167.4	2430.2	167.5	2574.4	0.5721	8.2583
42	315.15	8.198	1.009	17690	175.8	2432.9	175.8	2577.9	0.5987	8.2209
44	317.15	9.100	1.009	16040	184.2	2435.6	184.2	2581.5	0.6252	8.1842
45	318.15	9.582	1.010	15280	188.3	2436.9	188.4	2583.3	0.6383	8.1661
46	319.15	10.09	1.010	14560	192.5	2438.3	192.5	2585.1	0.6514	8.1481
48	321.15	11.16	1.011	13230	200.9	2440.9	200.9	2588.6	0.6776	8.1125
50	323.15	12.34	1.012	12050	209.2	2443.6	209.3	2592.2	0.7035	8.0776
52	325.15	13.61	1.013	10980	217.6	2446.2	217.6	2595.7	0.7293	8.0432
54	327.15	15.00	1.014	10020	226.0	2448.9	226.0	2599.2	0.7550	8.0093
55	328.15	15.74	1.015	9578.9	230.2	2450.2	230.2	2601.0	0.7677	7.9925
56	329.15	16.51	1.015	9158.7	234.3	2451.5	234.4	2602.7	0.7804	7.9759
58	331.15	18.15	1.016	8380.8	242.7	2454.1	242.7	2606.2	0.8058	7.9431
60	333.15	19.92	1.017	7678.5	251.1	2456.8	251.1	2609.7	0.8310	7.9108
62	335.15	21.84	1.018	7043.7	259.4	2459.4	259.5	2613.2	0.8560	7.8790
64	337.15	23.91	1.019	6469.0	267.8	2462.0	267.8	2616.6	0.8809	7.8477
65	338.15	25.01	1.020	6202.3	272.0	2463.2	272.0	2618.4	0.8933	7.8322
66	339.15	26.15	1.020	5948.2	276.2	2464.5	276.2	2620.1	0.9057	7.8168
68	341.15	28.56	1.022	5475.6	284.6	2467.1	284.6	2623.5	0.9303	7.7864
70	343.15	31.16	1.023	5046.3	292.9	2469.7	293.0	2626.9	0.9548	7.7565
72	345.15	33.96	1.024	4655.7	301.3	2472.2	301.4	2630.3	0.9792	7.7270
74	347.15	36.96	1.025	4300.0	309.7	2474.8	309.7	2633.7	1.0034	7.6979
75	348.15	38.55	1.026	4134.1	313.9	2476.0	313.9	2635.4	1.0154	7.6835
76	349.15	40.19	1.027	3975.7	318.1	2477.3	318.1	2637.1	1.0275	7.6693

TABLE B.1. SATURATED STEAM SI UNITS (Continued)

TEMPERATURE		ABS PRESS	SPECIFIC VOLUME V		INTERNAL ENERGY U		ENTHALPY H		ENTROPY S	
C	K	kPa	SAT LIQUID	SAT VAPOR	SAT LIQUID	SAT VAPOR	SAT LIQUID	SAT VAPOR	SAT LIQUID	SAT VAPOR
78	351.15	43.65	1.028	3679.6	326.5	2479.8	326.5	2640.4	1.0514	7.6410
80	353.15	47.36	1.029	3409.1	334.9	2482.3	334.9	2643.8	1.0753	7.6132
82	355.15	51.33	1.031	3161.6	343.3	2484.8	343.3	2647.1	1.0990	7.5858
84	357.15	55.57	1.032	2935.0	351.7	2487.3	351.7	2650.4	1.1225	7.5587
85	358.15	57.80	1.033	2828.8	355.9	2488.5	355.9	2652.0	1.1343	7.5454
86	359.15	60.11	1.033	2727.2	360.1	2489.7	360.1	2653.6	1.1460	7.5321
88	361.15	64.95	1.035	2536.5	368.5	2492.2	368.5	2656.9	1.1693	7.5058
90	363.15	70.11	1.036	2361.3	376.9	2494.6	376.9	2660.1	1.1925	7.4799
92	365.15	75.61	1.038	2200.2	385.3	2497.0	385.4	2663.4	1.2156	7.4543
94	367.15	81.46	1.039	2051.9	393.7	2499.4	393.8	2666.6	1.2386	7.4291
95	368.15	84.53	1.040	1982.2	397.9	2500.6	398.0	2668.1	1.2501	7.4166
96	369.15	87.69	1.041	1915.3	402.1	2501.8	402.2	2669.7	1.2615	7.4042
98	371.15	94.30	1.042	1789.3	410.5	2504.1	410.6	2672.9	1.2842	7.3796
100	373.15	101.33	1.044	1673.0	419.0	2506.5	419.1	2676.0	1.3069	7.3554
110	383.15	143.27	1.052	1209.9	461.2	2518.0	461.3	2691.3	1.4185	7.2388
120	393.15	198.54	1.061	891.5	503.5	2529.0	503.7	2706.0	1.5276	7.1293
130	403.15	270.13	1.070	668.1	546.0	2539.4	546.3	2719.9	1.6344	7.0261
140	413.15	361.38	1.080	508.5	588.7	2549.3	589.1	2733.1	1.7390	6.9284
150	423.15	476.00	1.091	392.4	631.6	2558.6	632.1	2745.4	1.8416	6.8358
160	433.15	618.06	1.102	306.8	674.8	2567.1	675.5	2756.7	1.9425	6.7475
170	443.15	792.02	1.114	242.6	718.2	2574.9	719.1	2767.1	2.0416	6.6630
180	453.15	1002.7	1.128	193.8	762.0	2581.9	763.1	2776.3	2.1393	6.5819
190	463.15	1255.1	1.142	156.3	806.1	2588.1	807.5	2784.3	2.2356	6.5036
200	473.15	1554.9	1.156	127.2	850.6	2593.2	852.4	2790.9	2.3307	6.4278
210	483.15	1907.7	1.173	104.2	895.5	2597.3	897.7	2796.2	2.4247	6.3539
220	493.15	2319.8	1.190	86.04	940.9	2600.3	943.7	2799.9	2.5178	6.2817
230	503.15	2797.6	1.209	71.45	986.9	2602.1	990.3	2802.0	2.6102	6.2107
240	513.15	3347.8	1.229	59.65	1033.5	2602.5	1037.6	2802.2	2.7020	6.1406
250	523.15	3977.6	1.251	50.04	1080.8	2601.4	1085.8	2800.4	2.7935	6.0708
260	533.15	4694.3	1.276	42.13	1129.0	2598.6	1134.9	2796.4	2.8848	6.0010
270	543.15	5505.8	1.303	35.59	1178.1	2593.9	1185.2	2789.9	2.9763	5.9304
280	553.15	6420.2	1.332	30.13	1228.3	2587.0	1236.8	2780.4	3.0683	5.8586
290	563.15	7446.1	1.366	25.54	1279.8	2577.5	1290.0	2767.6	3.1611	5.7848
300	573.15	8592.7	1.404	21.65	1333.0	2565.0	1345.1	2751.0	3.2552	5.7081
310	583.15	9870.0	1.448	18.33	1388.1	2549.1	1402.4	2730.0	3.3512	5.6278
320	593.15	11289.1	1.500	15.48	1445.7	2528.9	1462.6	2703.7	3.4500	5.5423
330	603.15	12862.5	1.561	12.99	1506.4	2503.1	1526.5	2670.2	3.5528	5.4490
340	613.15	14605.2	1.639	10.78	1571.5	2468.7	1595.5	2626.2	3.6616	5.3427
350	623.15	16535.1	1.741	8.799	1643.0	2422.2	1671.8	2567.7	3.7801	5.2177
360	633.15	18675.1	1.896	6.940	1728.8	2355.8	1764.2	2485.4	3.9210	5.0600
370	643.15	21054.0	2.214	4.973	1843.6	2238.1	1890.2	2342.8	4.1108	4.8144
374.15	647.30	22120.0	3.170	3.170	2037.3	2037.3	2107.4	2107.4	4.4429	4.4429

TABLE B.2. SUPERHEATED STEAM SI UNITS

ABS PRESS KPA (SAT TEMP) DEG C		SAT WATER	SAT STEAM	TEMPERATURE, DEG C (TEMPERATURE, K) 75 (348.15)	100 (373.15)	125 (398.15)	150 (423.15)	175 (448.15)	200 (473.15)	225 (498.15)	250 (523.15)
	V	1.000	129200	160640	172180	183720	195270	206810	218350	229890	241430
1	U	29.334	2385.2	2480.8	2516.4	2552.3	2588.5	2624.9	2661.7	2698.8	2736.3
(6.98)	H	29.335	2514.4	2641.5	2688.6	2736.0	2783.7	2831.7	2880.1	2928.7	2977.7
	S	0.1060	8.9767	9.3828	9.5136	9.6365	9.7527	9.8629	9.9679	10.0681	10.1641
	V	1.010	14670	16030	17190	18350	19510	20660	21820	22980	24130
10	U	191.822	2438.0	2479.7	2515.6	2551.6	2588.0	2624.5	2661.4	2698.6	2736.1
(45.83)	H	191.832	2584.8	2640.0	2687.5	2735.2	2783.1	2831.2	2879.6	2928.4	2977.4
	S	0.6493	8.1511	8.3168	8.4486	8.5722	8.6888	8.7994	8.9045	9.0049	9.1010
	V	1.017	7649.8	8000.0	8584.7	9167.1	9748.0	10320	10900	11480	12060
20	U	251.432	2456.9	2478.4	2514.6	2550.9	2587.4	2624.1	2661.0	2698.3	2735.8
(60.09)	H	251.453	2609.9	2638.4	2686.3	2734.2	2782.3	2830.6	2879.2	2928.0	2977.1
	S	0.8321	7.9094	7.9933	8.1261	8.2504	8.3676	8.4785	8.5839	8.6844	8.7806
	V	1.022	5229.3	5322.0	5714.4	6104.6	6493.2	6880.8	7267.5	7653.8	8039.7
30	U	289.271	2468.6	2477.1	2513.6	2550.2	2586.8	2623.6	2660.7	2698.0	2735.6
(69.12)	H	289.302	2625.4	2636.8	2685.1	2733.3	2781.6	2830.0	2878.7	2927.6	2976.8
	S	0.9441	7.7695	7.8024	7.9363	8.0614	8.1791	8.2903	8.3960	8.4967	8.5930
	V	1.027	3993.4	....	4279.2	4573.3	4865.8	5157.2	5447.8	5738.0	6027.7
40	U	317.609	2477.1	....	2512.6	2549.4	2586.2	2623.2	2660.3	2697.7	2735.4
(75.89)	H	317.650	2636.9	....	2683.8	2732.3	2780.9	2829.5	2878.2	2927.2	2976.5
	S	1.0261	7.6709	....	7.8009	7.9268	8.0450	8.1566	8.2624	8.3633	8.4598
	V	1.030	3240.2	....	3418.1	3654.5	3889.3	4123.0	4356.0	4588.5	4820.5
50	U	340.513	2484.0	....	2511.7	2548.6	2585.6	2622.7	2659.9	2697.4	2735.1
(81.35)	H	340.564	2646.0	....	2682.6	2731.4	2780.1	2828.9	2877.7	2926.8	2976.1
	S	1.0912	7.5947	....	7.6953	7.8219	7.9406	8.0526	8.1587	8.2598	8.3564
	V	1.037	2216.9	....	2269.8	2429.4	2587.3	2744.2	2900.2	3055.8	3210.9
75	U	384.374	2496.7	....	2509.2	2546.7	2584.2	2621.6	2659.0	2696.7	2734.5
(91.79)	H	384.451	2663.0	....	2679.4	2728.9	2778.2	2827.4	2876.6	2925.8	2975.3
	S	1.2131	7.4570	....	7.5014	7.6300	7.7500	7.8629	7.9697	8.0712	8.1681
	V	1.043	1693.7	....	1695.5	1816.7	1936.3	2054.7	2172.3	2289.4	2406.1
100	U	417.406	2506.1	....	2506.6	2544.8	2582.7	2620.4	2658.1	2695.9	2733.9
(99.63)	H	417.511	2675.4	....	2676.2	2726.5	2776.3	2825.9	2875.4	2924.9	2974.5
	S	1.3027	7.3598	....	7.3618	7.4923	7.6137	7.7275	7.8349	7.9369	8.0342

TABLE B.2. SUPERHEATED STEAM SI UNITS (Continued)

ABS PRESS KPA (SAT TEMP) DEG C		SAT WATER	SAT STEAM	TEMPERATURE, DEG C (TEMPERATURE, K) 75 (348.15)	100 (373.15)	125 (398.15)	150 (423.15)	175 (448.15)	200 (473.15)	225 (498.15)	250 (523.15)
	V	1.044	1673.0	....	1673.0	1792.7	1910.7	2027.7	2143.8	2259.3	2374.5
101.325	U	418.959	2506.5	....	2506.5	2544.7	2582.6	2620.4	2658.1	2695.9	2733.9
(100.00)	H	419.064	2676.0	....	2676.0	2726.4	2776.2	2825.8	2875.3	2924.8	2974.5
	S	1.3069	7.3554	....	7.3554	7.4860	7.6075	7.7213	7.8288	7.9308	8.0280
	V	1.049	1374.6	....	....	1449.1	1545.6	1641.0	1735.6	1829.6	1923.2
125	U	444.224	2513.4	....	....	2542.9	2581.2	2619.3	2657.2	2695.2	2733.3
(105.99)	H	444.356	2685.2	....	....	2724.0	2774.4	2824.4	2874.2	2923.9	2973.7
	S	1.3740	7.2847	....	....	7.3844	7.5072	7.6219	7.7300	7.8324	7.9300
	V	1.053	1159.0	....	....	1204.0	1285.2	1365.2	1444.4	1523.0	1601.3
150	U	466.968	2519.5	....	....	2540.9	2579.7	2618.1	2656.3	2694.4	2732.7
(111.37)	H	467.126	2693.4	....	....	2721.5	2772.5	2822.9	2872.9	2922.9	2972.9
	S	1.4336	7.2234	....	....	7.2953	7.4194	7.5352	7.6439	7.7468	7.8447
	V	1.057	1003.34	....	....	1028.8	1099.1	1168.2	1236.4	1304.1	1371.3
175	U	486.815	2524.7	....	....	2538.9	2578.2	2616.9	2655.3	2693.7	2732.1
(116.06)	H	487.000	2700.3	....	....	2719.0	2770.5	2821.3	2871.7	2921.9	2972.0
	S	1.4849	7.1716	....	....	7.2191	7.3447	7.4614	7.5708	7.6741	7.7724
	V	1.061	885.44	....	....	897.47	959.54	1020.4	1080.4	1139.8	1198.9
200	U	504.489	2529.2	....	....	2536.9	2576.6	2615.7	2654.4	2692.9	2731.4
(120.23)	H	504.701	2706.3	....	....	2716.4	2768.5	2819.8	2870.5	2920.9	2971.2
	S	1.5301	7.1268	....	....	7.1523	7.2794	7.3971	7.5072	7.6110	7.7096
	V	1.064	792.97	....	....	795.25	850.97	905.44	959.06	1012.1	1064.7
225	U	520.465	2533.2	....	....	2534.8	2575.1	2614.5	2653.5	2692.2	2730.8
(123.99)	H	520.705	2711.6	....	....	2713.8	2766.5	2818.2	2869.3	2919.9	2970.4
	S	1.5705	7.0873	....	....	7.0928	7.2213	7.3400	7.4508	7.5551	7.6540
	V	1.068	718.44	....	....	....	764.09	813.47	861.98	909.91	957.41
250	U	535.077	2536.8	....	....	....	2573.5	2613.3	2652.5	2691.4	2730.2
(127.43)	H	535.343	2716.4	....	....	....	2764.5	2816.7	2868.0	2918.9	2969.6
	S	1.6071	7.0520	....	....	....	7.1689	7.2886	7.4001	7.5050	7.6042
	V	1.073	605.56	....	....	....	633.74	675.49	716.35	756.60	796.44
300	U	561.107	2543.0	....	....	....	2570.3	2610.8	2650.6	2689.9	2729.0
(133.54)	H	561.429	2724.7	....	....	....	2760.4	2813.5	2865.5	2916.9	2967.9
	S	1.6716	6.9909	....	....	....	7.0771	7.1990	7.3119	7.4177	7.5176

TABLE B.2. SUPERHEATED STEAM SI UNITS (Continued)

ABS PRESS KPA (SAT TEMP) DEG C		SAT WATER	SAT STEAM	TEMPERATURE, DEG C (TEMPERATURE, K) 300 (573.15)	350 (623.15)	400 (673.15)	450 (723.15)	500 (773.15)	550 (823.15)	600 (873.15)	650 (923.15)
1	V	1.000	129200	264500	287580	310660	333730	356810	379880	402960	426040
(6.98)	U	29.334	2385.2	2812.3	2889.9	2969.1	3049.9	3132.4	3216.7	3302.6	3390.3
	H	29.335	2514.4	3076.8	3177.5	3279.7	3383.6	3489.2	3596.5	3705.6	3816.4
	S	0.1060	8.9767	10.3450	10.5133	10.6711	10.8200	10.9612	11.0957	11.2243	11.3476
10	V	1.010	14670	26440	28750	31060	33370	35670	37980	40290	42600
(45.83)	U	191.822	2438.0	2812.2	2889.8	2969.0	3049.8	3132.3	3216.6	3302.6	3390.3
	H	191.832	2584.8	3076.6	3177.3	3279.6	3383.5	3489.1	3596.5	3705.5	3816.3
	S	0.6493	8.1511	9.2820	9.4504	9.6083	9.7572	9.8984	10.0329	10.1616	10.2849
20	V	1.017	7649.8	13210	14370	15520	16680	17830	18990	20140	21300
(60.09)	U	251.432	2456.9	2812.0	2889.6	2968.9	3049.7	3132.3	3216.5	3302.5	3390.2
	H	251.453	2609.9	3076.4	3177.1	3279.4	3383.4	3489.0	3596.4	3705.4	3816.2
	S	0.8321	7.9094	8.9618	9.1303	9.2882	9.4372	9.5784	9.7130	9.8416	9.9650
30	V	1.022	5229.3	8810.8	9581.2	10350	11120	11890	12660	13430	14190
(69.12)	U	289.271	2468.6	2811.8	2889.5	2968.7	3049.6	3132.2	3216.5	3302.5	3390.2
	H	289.302	2625.4	3076.1	3176.9	3279.3	3383.3	3488.9	3596.3	3705.4	3816.2
	S	0.9441	7.7695	8.7744	8.9430	9.1010	9.2499	9.3912	9.5257	9.6544	9.7778
40	V	1.027	3993.4	6606.5	7184.6	7762.5	8340.1	8917.6	9494.9	10070	10640
(75.89)	U	317.609	2477.1	2811.6	2889.4	2968.6	3049.5	3132.1	3216.4	3302.4	3390.1
	H	317.650	2636.9	3075.9	3176.8	3279.1	3383.1	3488.8	3596.2	3705.3	3816.1
	S	1.0261	7.6709	8.6413	8.8100	8.9680	9.1170	9.2583	9.3929	9.5216	9.6450
50	V	1.030	3240.2	5283.9	5746.7	6209.1	6671.4	7133.5	7595.5	8057.4	8519.2
(81.35)	U	340.513	2484.0	2811.5	2889.2	2968.5	3049.4	3132.0	3216.3	3302.3	3390.1
	H	340.564	2646.0	3075.7	3176.6	3279.0	3383.0	3488.7	3596.1	3705.2	3816.0
	S	1.0912	7.5947	8.5380	8.7068	8.8649	9.0139	9.1552	9.2898	9.4185	9.5419
75	V	1.037	2216.9	3520.5	3829.4	4138.0	4446.4	4754.7	5062.8	5370.9	5678.9
(91.79)	U	384.374	2496.7	2811.0	2888.9	2968.2	3049.2	3131.8	3216.1	3302.2	3389.9
	H	384.451	2663.0	3075.1	3176.1	3278.6	3382.7	3488.4	3595.8	3705.0	3815.9
	S	1.2131	7.4570	8.3502	8.5191	8.6773	8.8265	8.9678	9.1025	9.2312	9.3546
100	V	1.043	1693.7	2638.7	2870.8	3102.5	3334.0	3565.3	3796.5	4027.7	4258.8
(99.63)	U	417.406	2506.1	2810.6	2888.6	2968.0	3049.0	3131.6	3216.0	3302.0	3389.8
	H	417.511	2675.4	3074.5	3175.6	3278.2	3382.4	3488.1	3595.6	3704.8	3815.7
	S	1.3027	7.3598	8.2166	8.3858	8.5442	8.6934	8.8348	8.9695	9.0982	9.2217

TABLE B.2. SUPERHEATED STEAM SI UNITS (Continued)

ABS PRESS KPA (SAT TEMP) DEG C		SAT WATER	SAT STEAM	TEMPERATURE, DEG C (TEMPERATURE, K) 300 (573.15)	350 (623.15)	400 (673.15)	450 (723.15)	500 (773.15)	550 (823.15)	600 (873.15)	650 (923.15)
	V	1.044	1673.0	2604.2	2833.2	3061.9	3290.3	3518.7	3746.9	3975.0	4203.1
101.325	U	418.959	2506.5	2810.6	2888.5	2968.0	3048.9	3131.6	3215.9	3302.0	3389.8
(100.00)	H	419.064	2676.0	3074.4	3175.6	3278.2	3382.3	3488.1	3595.6	3704.8	3815.7
	S	1.3069	7.3554	8.2105	8.3797	8.5381	8.6873	8.8287	8.9634	9.0922	9.2156
	V	1.049	1374.6	2109.7	2295.6	2481.2	2666.5	2851.7	3036.8	3221.8	3406.7
125	U	444.224	2513.4	2810.2	2888.2	2967.7	3048.7	3131.4	3215.8	3301.9	3389.7
(105.99)	H	444.356	2685.2	3073.9	3175.2	3277.8	3382.0	3487.9	3595.4	3704.6	3815.5
	S	1.3740	7.2847	8.1129	8.2823	8.4408	8.5901	8.7316	8.8663	8.9951	9.1186
	V	1.053	1159.0	1757.0	1912.2	2066.9	2221.5	2375.9	2530.2	2684.5	2838.6
150	U	466.968	2519.5	2809.7	2887.9	2967.4	3048.5	3131.2	3215.6	3301.7	3389.5
(111.37)	H	467.126	2693.4	3073.3	3174.7	3277.5	3381.7	3487.6	3595.1	3704.4	3815.3
	S	1.4336	7.2234	8.0280	8.1976	8.3562	8.5056	8.6472	8.7819	8.9108	9.0343
	V	1.057	1003.34	1505.1	1638.3	1771.1	1903.7	2036.1	2168.4	2300.7	2432.9
175	U	486.815	2524.7	2809.3	2887.5	2967.1	3048.3	3131.0	3215.4	3301.6	3389.4
(116.06)	H	487.000	2700.3	3072.7	3174.2	3277.1	3381.4	3487.3	3594.9	3704.2	3815.1
	S	1.4849	7.1716	7.9561	8.1259	8.2847	8.4341	8.5758	8.7106	8.8394	8.9630
	V	1.061	885.44	1316.2	1432.8	1549.2	1665.3	1781.2	1897.1	2012.9	2128.6
200	U	504.489	2529.2	2808.8	2887.2	2966.9	3048.0	3130.8	3215.3	3301.4	3389.2
(120.23)	H	504.701	2706.3	3072.1	3173.8	3276.7	3381.1	3487.0	3594.7	3704.0	3815.0
	S	1.5301	7.1268	7.8937	8.0638	8.2226	8.3722	8.5139	8.6487	8.7776	8.9012
	V	1.064	792.97	1169.2	1273.1	1376.6	1479.9	1583.0	1686.0	1789.0	1891.9
225	U	520.465	2533.2	2808.4	2886.9	2966.6	3047.8	3130.6	3215.1	3301.2	3389.1
(123.99)	H	520.705	2711.6	3071.5	3173.3	3276.3	3380.8	3486.8	3594.4	3703.8	3814.8
	S	1.5705	7.0873	7.8385	8.0088	8.1679	8.3175	8.4593	8.5942	8.7231	8.8467
	V	1.068	718.44	1051.6	1145.2	1238.5	1331.5	1424.4	1517.2	1609.9	1702.5
250	U	535.077	2536.8	2808.0	2886.5	2966.3	3047.6	3130.4	3214.9	3301.1	3389.0
(127.43)	H	535.343	2716.4	3070.9	3172.8	3275.9	3380.4	3486.5	3594.2	3703.6	3814.6
	S	1.6071	7.0520	7.7891	7.9597	8.1188	8.2686	8.4104	8.5453	8.6743	8.7980
	V	1.073	605.56	875.29	953.52	1031.4	1109.0	1186.5	1263.9	1341.2	1418.5
300	U	561.107	2543.0	2807.1	2885.8	2965.8	3047.1	3130.0	3214.5	3300.8	3388.7
(133.54)	H	561.429	2724.7	3069.7	3171.9	3275.2	3379.8	3486.0	3593.7	3703.2	3814.2
	S	1.6716	6.9909	7.7034	7.8744	8.0338	8.1838	8.3257	8.4608	8.5898	8.7135

TABLE B.2. SUPERHEATED STEAM SI UNITS (Continued)

ABS PRESS KPA (SAT TEMP) DEG C		SAT WATER	SAT STEAM	TEMPERATURE, DEG C (TEMPERATURE, K) 150 (423.15)	175 (448.15)	200 (473.15)	220 (493.15)	240 (513.15)	260 (533.15)	280 (553.15)	300 (573.15)
350 (138.87)	V	1.079	524.00	540.58	576.90	612.31	640.18	667.75	695.09	722.27	749.33
	U	583.892	2548.2	2567.1	2608.3	2648.6	2680.4	2712.0	2743.4	2774.8	2806.2
	H	584.270	2731.6	2756.3	2810.3	2863.0	2904.5	2945.7	2986.7	3027.6	3068.4
	S	1.7273	6.9392	6.9982	7.1222	7.2366	7.3226	7.4045	7.4828	7.5581	7.6307
400 (143.62)	V	1.084	462.22	470.66	502.93	534.26	558.85	583.14	607.20	631.09	654.85
	U	604.237	2552.7	2563.7	2605.8	2646.7	2678.8	2710.6	2742.2	2773.7	2805.3
	H	604.670	2737.6	2752.0	2807.0	2860.4	2902.3	2943.9	2985.1	3026.2	3067.2
	S	1.7764	6.8943	6.9285	7.0548	7.1708	7.2576	7.3402	7.4190	7.4947	7.5675
450 (147.92)	V	1.088	413.75	416.24	445.38	473.55	495.59	517.33	538.83	560.17	581.37
	U	622.672	2556.7	2560.3	2603.2	2644.7	2677.1	2709.2	2741.0	2772.7	2804.4
	H	623.162	2742.9	2747.7	2803.7	2857.8	2900.2	2942.0	2983.5	3024.8	3066.0
	S	1.8204	6.8547	6.8660	6.9946	7.1121	7.1999	7.2831	7.3624	7.4384	7.5116
500 (151.84)	V	1.093	374.68	....	399.31	424.96	444.97	464.67	484.14	503.43	522.58
	U	639.569	2560.2	....	2600.6	2642.7	2675.5	2707.8	2739.8	2771.7	2803.5
	H	640.116	2747.5	....	2800.3	2855.1	2898.0	2940.1	2981.9	3023.4	3064.8
	S	1.8604	6.8192	....	6.9400	7.0592	7.1478	7.2317	7.3115	7.3879	7.4614
550 (155.47)	V	1.097	342.48	....	361.60	385.19	403.55	421.59	439.38	457.00	474.48
	U	655.199	2563.3	....	2598.0	2640.6	2673.8	2706.4	2738.6	2770.6	2802.6
	H	655.802	2751.7	....	2796.8	2852.5	2895.7	2938.3	2980.3	3022.0	3063.5
	S	1.8970	6.7870	....	6.8900	7.0108	7.1004	7.1849	7.2653	7.3421	7.4158
600 (158.84)	V	1.101	315.47	....	330.16	352.04	369.03	385.68	402.08	418.31	434.39
	U	669.762	2566.2	....	2595.3	2638.5	2672.1	2705.0	2737.4	2769.6	2801.6
	H	670.423	2755.5	....	2793.3	2849.7	2893.5	2936.4	2978.7	3020.6	3062.3
	S	1.9308	6.7575	....	6.8437	6.9662	7.0567	7.1419	7.2228	7.3000	7.3740
650 (161.99)	V	1.105	292.49	....	303.53	323.98	339.80	355.29	370.52	385.56	400.47
	U	683.417	2568.7	....	2592.5	2636.4	2670.3	2703.5	2736.2	2768.5	2800.7
	H	684.135	2758.9	....	2789.8	2847.0	2891.2	2934.4	2977.0	3019.2	3061.0
	S	1.9623	6.7304	....	6.8004	6.9247	7.0162	7.1021	7.1835	7.2611	7.3355
700 (164.96)	V	1.108	272.68	....	280.69	299.92	314.75	329.23	343.46	357.50	371.39
	U	696.285	2571.1	....	2589.7	2634.3	2668.6	2702.1	2735.0	2767.5	2799.8
	H	697.061	2762.0	....	2786.2	2844.2	2888.9	2932.5	2975.4	3017.7	3059.8
	S	1.9918	6.7052	....	6.7598	6.8859	6.9784	7.0651	7.1470	7.2250	7.2997

TABLE B.2. SUPERHEATED STEAM SI UNITS (Continued)

ABS PRESS KPA (SAT TEMP) DEG C		SAT WATER	SAT STEAM	TEMPERATURE, DEG C (TEMPERATURE, K) 325 (598.15)	350 (623.15)	400 (673.15)	450 (723.15)	500 (773.15)	550 (823.15)	600 (873.15)	650 (923.15)
	V	1.079	524.00	783.01	816.57	883.45	950.11	1016.6	1083.0	1149.3	1215.6
350	U	583.892	2548.2	2845.6	2885.1	2965.2	3046.6	3129.6	3214.2	3300.5	3388.4
(138.87)	H	584.270	2731.6	3119.6	3170.9	3274.4	3379.2	3485.4	3593.3	3702.7	3813.9
	S	1.7273	6.9392	7.7181	7.8022	7.9619	8.1120	8.2540	8.3892	8.5183	8.6421
	V	1.084	462.22	684.41	713.85	772.50	830.92	889.19	947.35	1005.4	1063.4
400	U	604.237	2552.7	2844.8	2884.5	2964.6	3046.2	3129.2	3213.8	3300.2	3388.2
(143.62)	H	604.670	2737.6	3118.5	3170.0	3273.6	3378.5	3484.9	3592.8	3702.3	3813.5
	S	1.7764	6.8943	7.6552	7.7395	7.8994	8.0497	8.1919	8.3271	8.4563	8.5802
	V	1.088	413.75	607.73	633.97	686.20	738.21	790.07	841.83	893.50	945.10
450	U	622.672	2556.7	2844.0	2883.8	2964.1	3045.7	3128.8	3213.5	3299.8	3387.9
(147.92)	H	623.162	2742.9	3117.5	3169.1	3272.9	3377.9	3484.3	3592.3	3701.9	3813.2
	S	1.8204	6.8547	7.5995	7.6840	7.8442	7.9947	8.1370	8.2723	8.4016	8.5255
	V	1.093	374.68	546.38	570.05	617.16	664.05	710.78	757.41	803.95	850.42
500	U	639.569	2560.2	2843.2	2883.1	2963.5	3045.2	3128.4	3213.1	3299.5	3387.6
(151.84)	H	640.116	2747.5	3116.4	3168.1	3272.1	3377.2	3483.8	3591.8	3701.5	3812.8
	S	1.8604	6.8192	7.5496	7.6343	7.7948	7.9454	8.0879	8.2233	8.3526	8.4766
	V	1.097	342.48	496.18	517.76	560.68	603.37	645.91	688.34	730.68	772.96
550	U	655.199	2563.3	2842.4	2882.4	2963.0	3044.7	3128.0	3212.8	3299.2	3387.3
(155.47)	H	655.802	2751.7	3115.3	3167.2	3271.3	3376.6	3483.2	3591.4	3701.1	3812.5
	S	1.8970	6.7870	7.5043	7.5892	7.7500	7.9008	8.0433	8.1789	8.3083	8.4323
	V	1.101	315.47	454.35	474.19	513.61	552.80	591.84	630.78	669.63	708.41
600	U	669.762	2566.2	2841.6	2881.7	2962.4	3044.3	3127.6	3212.4	3298.9	3387.1
(158.84)	H	670.423	2755.5	3114.3	3166.2	3270.6	3376.0	3482.7	3590.9	3700.7	3812.1
	S	1.9308	6.7575	7.4628	7.5479	7.7090	7.8600	8.0027	8.1383	8.2678	8.3919
	V	1.105	292.49	418.95	437.31	473.78	510.01	546.10	582.07	617.96	653.79
650	U	683.417	2568.7	2840.9	2881.0	2961.8	3043.8	3127.2	3212.1	3298.6	3386.8
(161.99)	H	684.135	2758.9	3113.2	3165.3	3269.8	3375.3	3482.1	3590.4	3700.3	3811.8
	S	1.9623	6.7304	7.4245	7.5099	7.6712	7.8224	7.9652	8.1009	8.2305	8.3546
	V	1.108	272.68	388.61	405.71	439.64	473.34	506.89	540.33	573.68	606.97
700	U	696.285	2571.1	2840.1	2880.3	2961.3	3043.3	3126.8	3211.7	3298.3	3386.5
(164.96)	H	697.061	2762.0	3112.1	3164.3	3269.0	3374.7	3481.6	3589.9	3699.9	3811.4
	S	1.9918	6.7052	7.3890	7.4745	7.6362	7.7875	7.9305	8.0663	8.1959	8.3201

TABLE B.2. SUPERHEATED STEAM SI UNITS (Continued)

ABS PRESS KPA (SAT TEMP) DEG C		SAT WATER	SAT STEAM	TEMPERATURE, DEG C (TEMPERATURE, K) 175 (448.15)	200 (473.15)	220 (493.15)	240 (513.15)	260 (533.15)	280 (553.15)	300 (573.15)	325 (598.15)
750 (167.76)	V	1.112	255.43	260.88	279.05	293.03	306.65	320.01	333.17	346.19	362.32
	U	708.467	2573.3	2586.9	2632.1	2666.8	2700.6	2733.7	2766.4	2798.9	2839.3
	H	709.301	2764.8	2782.5	2841.4	2886.6	2930.6	2973.7	3016.3	3058.5	3111.0
	S	2.0195	6.6817	6.7215	6.8494	6.9429	7.0303	7.1128	7.1912	7.2662	7.3558
800 (170.41)	V	1.115	240.26	243.53	260.79	274.02	286.88	299.48	311.89	324.14	339.31
	U	720.043	2575.3	2584.0	2629.9	2665.0	2699.1	2732.5	2765.4	2797.9	2838.5
	H	720.935	2767.5	2778.8	2838.6	2884.2	2928.6	2972.1	3014.9	3057.3	3109.9
	S	2.0457	6.6596	6.6851	6.8148	6.9094	6.9976	7.0807	7.1595	7.2348	7.3247
850 (172.94)	V	1.118	226.81	228.21	244.66	257.24	269.44	281.37	293.10	304.68	319.00
	U	731.080	2577.1	2581.1	2627.7	2663.2	2697.6	2731.2	2764.3	2797.0	2837.7
	H	732.031	2769.9	2775.1	2835.7	2881.9	2926.6	2970.4	3013.4	3056.0	3108.8
	S	2.0705	6.6388	6.6504	6.7820	6.8777	6.9666	7.0503	7.1295	7.2051	7.2954
900 (175.36)	V	1.121	214.81	....	230.32	242.31	253.93	265.27	276.40	287.39	300.96
	U	741.635	2578.8	....	2625.5	2661.4	2696.1	2729.9	2763.2	2796.1	2836.9
	H	742.644	2772.1	....	2832.7	2879.5	2924.6	2968.7	3012.0	3054.7	3107.7
	S	2.0941	6.6192	....	6.7508	6.8475	6.9373	7.0215	7.1012	7.1771	7.2676
1000 (179.88)	V	1.127	194.29	....	205.92	216.93	227.55	237.89	248.01	257.98	270.27
	U	761.478	2581.9	....	2620.9	2657.7	2693.0	2727.4	2761.0	2794.2	2835.2
	H	762.605	2776.2	....	2826.8	2874.6	2920.6	2965.2	3009.0	3052.1	3105.5
	S	2.1382	6.5828	....	6.6922	6.7911	6.8825	6.9680	7.0485	7.1251	7.2163
1100 (184.07)	V	1.133	177.38	....	185.92	196.14	205.96	215.47	224.77	233.91	245.16
	U	779.878	2584.5	....	2616.2	2653.9	2689.9	2724.7	2758.8	2792.2	2833.6
	H	781.124	2779.7	....	2820.7	2869.6	2916.4	2961.8	3006.0	3049.6	3103.3
	S	2.1786	6.5497	....	6.6379	6.7392	6.8323	6.9190	7.0005	7.0778	7.1595
1200 (187.96)	V	1.139	163.20	....	169.23	178.80	187.95	196.79	205.40	213.85	224.24
	U	797.064	2586.9	....	2611.3	2650.0	2686.7	2722.1	2756.5	2790.3	2832.0
	H	798.430	2782.7	....	2814.4	2864.5	2912.2	2958.2	3003.0	3046.9	3101.0
	S	2.2161	6.5194	....	6.5872	6.6909	6.7858	6.8738	6.9562	7.0342	7.1266
1300 (191.61)	V	1.144	151.13	....	155.09	164.11	172.70	180.97	189.01	196.87	206.53
	U	813.213	2589.0	....	2606.4	2646.0	2683.5	2719.4	2754.3	2788.4	2830.3
	H	814.700	2785.4	....	2808.0	2859.3	2908.0	2954.7	3000.0	3044.3	3098.8
	S	2.2510	6.4913	....	6.5394	6.6457	6.7424	6.8316	6.9151	6.9938	7.0859

TABLE B.2. SUPERHEATED STEAM SI UNITS (Continued)

ABS PRESS KPA (SAT TEMP) DEG C		SAT WATER	SAT STEAM	TEMPERATURE, DEG C (TEMPERATURE, K) 350 (623.15)	375 (648.15)	400 (673.15)	450 (723.15)	500 (773.15)	550 (823.15)	600 (873.15)	650 (923.15)
	V	1.112	255.43	378.31	394.22	410.05	441.55	472.90	504.15	535.30	566.40
750	U	708.467	2573.3	2879.6	2920.1	2960.7	3042.9	3126.3	3211.4	3298.0	3386.2
(167.76)	H	709.301	2764.8	3163.4	3215.7	3268.3	3374.0	3481.0	3589.5	3699.5	3811.0
	S	2.0195	6.6817	7.4416	7.5240	7.6035	7.7550	7.8981	8.0340	8.1637	8.2880
	V	1.115	240.26	354.34	369.29	384.16	413.74	443.17	472.49	501.72	530.89
800	U	720.043	2575.3	2878.9	2919.5	2960.2	3042.4	3125.9	3211.0	3297.7	3386.0
(170.41)	H	720.935	2767.5	3162.4	3214.9	3267.5	3373.4	3480.5	3589.0	3699.1	3810.7
	S	2.0457	6.6596	7.4107	7.4932	7.5729	7.7246	7.8678	8.0038	8.1336	8.2579
	V	1.118	226.81	333.20	347.29	361.31	389.20	416.93	444.56	472.09	499.57
850	U	731.080	2577.1	2878.2	2918.8	2959.6	3041.9	3125.5	3210.7	3297.4	3385.7
(172.94)	H	732.031	2769.9	3161.4	3214.0	3266.7	3372.7	3479.9	3588.5	3698.6	3810.3
	S	2.0705	6.6388	7.3815	7.4643	7.5441	7.6960	7.8393	7.9754	8.1053	8.2296
	V	1.121	214.81	314.40	327.74	341.01	367.39	393.61	419.73	445.76	471.72
900	U	741.635	2578.8	2877.5	2918.2	2959.0	3041.4	3125.1	3210.3	3297.1	3385.4
(175.36)	H	742.644	2772.1	3160.5	3213.2	3266.0	3372.1	3479.4	3588.1	3698.2	3810.0
	S	2.0941	6.6192	7.3540	7.4370	7.5169	7.6689	7.8124	7.9486	8.0785	8.2030
	V	1.127	194.29	282.43	294.50	306.49	330.30	353.96	377.52	400.98	424.38
1000	U	761.478	2581.9	2876.1	2917.0	2957.9	3040.5	3124.3	3209.6	3296.4	3384.9
(179.88)	H	762.605	2776.2	3158.5	3211.5	3264.4	3370.8	3478.3	3587.1	3697.4	3809.3
	S	2.1382	6.5828	7.3031	7.3864	7.4665	7.6190	7.7627	7.8991	8.0292	8.1537
	V	1.133	177.38	256.28	267.30	278.24	299.96	321.53	342.98	364.35	385.65
1100	U	779.878	2584.5	2874.7	2915.7	2956.8	3039.6	3123.5	3208.9	3295.8	3384.3
(184.07)	H	781.124	2779.7	3156.6	3209.7	3262.9	3369.5	3477.2	3586.2	3696.6	3808.5
	S	2.1786	6.5497	7.2569	7.3405	7.4209	7.5737	7.7177	7.8543	7.9845	8.1092
	V	1.139	163.20	234.49	244.63	254.70	274.68	294.50	314.20	333.82	353.38
1200	U	797.064	2586.9	2873.3	2914.4	2955.7	3038.6	3122.7	3208.2	3295.2	3383.8
(187.96)	H	798.430	2782.7	3154.6	3208.0	3261.3	3368.2	3476.1	3585.2	3695.8	3807.8
	S	2.2161	6.5194	7.2144	7.2983	7.3790	7.5323	7.6765	7.8132	7.9436	8.0684
	V	1.144	151.13	216.05	225.46	234.79	253.28	271.62	289.85	307.99	326.07
1300	U	813.213	2589.0	2871.8	2913.2	2954.5	3037.7	3121.9	3207.5	3294.6	3383.2
(191.61)	H	814.700	2785.4	3152.7	3206.3	3259.7	3366.9	3475.0	3584.3	3695.0	3807.1
	S	2.2510	6.4913	7.1751	7.2594	7.3404	7.4940	7.6385	7.7754	7.9060	8.0309

TABLE 3.2. SUPERHEATED STEAM SI UNITS (Continued)

ABS PRESS KPA (SAT TEMP) DEG C		SAT WATER	SAT STEAM	200 (473.15)	225 (498.15)	250 (523.15)	275 (548.15)	300 (573.15)	325 (598.15)	350 (623.15)	375 (648.15)
				TEMPERATURE, DEG C (TEMPERATURE, K)							
1400 (195.04)	V	1.149	140.72	142.94	153.57	163.55	173.08	182.32	191.35	200.24	209.02
	U	828.465	2590.8	2601.3	2651.7	2698.6	2743.2	2786.4	2828.6	2870.4	2911.9
	H	830.074	2787.8	2801.4	2866.7	2927.6	2985.5	3041.6	3096.5	3150.7	3204.5
	S	2.2837	6.4651	6.4941	6.6285	6.7477	6.8560	6.9561	7.0499	7.1386	7.2233
1500 (198.29)	V	1.154	131.66	132.38	142.53	151.99	161.00	169.70	178.19	186.53	194.77
	U	842.933	2592.4	2596.1	2647.7	2695.5	2740.8	2784.4	2826.9	2868.9	2910.6
	H	844.663	2789.9	2794.7	2861.5	2923.5	2982.3	3038.9	3094.2	3148.7	3202.8
	S	2.3145	6.4406	6.4508	6.5885	6.7099	6.8196	6.9207	7.0152	7.1044	7.1894
1600 (201.37)	V	1.159	123.69	....	132.85	141.87	150.42	158.66	166.68	174.54	182.30
	U	856.707	2593.8	....	2643.7	2692.4	2738.3	2782.4	2825.2	2867.5	2909.3
	H	858.561	2791.7	....	2856.3	2919.4	2979.0	3036.2	3091.9	3146.7	3201.0
	S	2.3436	6.4175	....	6.5503	6.6740	6.7852	6.8873	6.9825	7.0723	7.1577
1700 (204.31)	V	1.163	116.62	....	124.31	132.94	141.09	148.91	156.51	163.96	171.30
	U	869.866	2595.1	....	2639.6	2689.3	2735.8	2780.3	2823.5	2866.0	2908.0
	H	871.843	2793.4	....	2851.0	2915.3	2975.6	3033.5	3089.6	3144.7	3199.2
	S	2.3713	6.3957	....	6.5138	6.6398	6.7526	6.8557	6.9516	7.0419	7.1277
1800 (207.11)	V	1.168	110.32	....	116.69	124.99	132.78	140.24	147.48	154.55	161.51
	U	882.472	2596.3	....	2635.5	2686.1	2733.3	2778.2	2821.8	2864.5	2906.7
	H	884.574	2794.8	....	2845.5	2911.0	2972.3	3030.7	3087.3	3142.7	3197.5
	S	2.3976	6.3751	....	6.4787	6.6071	6.7214	6.8257	6.9223	7.0131	7.0993
1900 (209.80)	V	1.172	104.65	....	109.87	117.87	125.35	132.49	139.39	146.14	152.76
	U	894.580	2597.3	....	2631.2	2682.8	2730.7	2776.2	2820.1	2863.0	2905.4
	H	896.807	2796.1	....	2840.0	2906.7	2968.8	3027.9	3084.9	3140.7	3195.7
	S	2.4228	6.3554	....	6.4448	6.5757	6.6917	6.7970	6.8944	6.9857	7.0723
2000 (212.37)	V	1.177	99.536	....	103.72	111.45	118.65	125.50	132.11	138.56	144.89
	U	906.236	2598.2	....	2626.9	2679.5	2728.1	2774.0	2818.3	2861.5	2904.1
	H	908.589	2797.2	....	2834.3	2902.4	2965.4	3025.0	3082.5	3138.6	3193.9
	S	2.4469	6.3366	....	6.4120	6.5454	6.6631	6.7696	6.8677	6.9596	7.0466
2200 (217.24)	V	1.185	90.652	....	93.067	100.35	107.07	113.43	119.53	125.47	131.28
	U	928.346	2599.6	....	2617.9	2672.7	2722.7	2769.7	2814.7	2858.5	2901.5
	H	930.953	2799.1	....	2822.7	2893.4	2958.3	3019.3	3077.7	3134.5	3190.3
	S	2.4922	6.3015	....	6.3492	6.4879	6.6091	6.7179	6.8177	6.9107	6.9985

TABLE B.2. SUPERHEATED STEAM SI UNITS (Continued)

ABS PRESS KPA (SAT TEMP) DEG C		SAT WATER	SAT STEAM	TEMPERATURE, DEG C (TEMPERATURE, K) 400 (673.15)	425 (698.15)	450 (723.15)	475 (748.15)	500 (773.15)	550 (823.15)	600 (873.15)	650 (923.15)
	V	1.149	140.72	217.72	226.35	234.95	243.50	252.02	268.98	285.85	302.66
1400	U	828.465	2590.8	2953.4	2994.9	3036.7	3078.7	3121.1	3206.8	3293.9	3382.7
(195.04)	H	830.074	2787.8	3258.2	3311.8	3365.6	3419.6	3473.9	3583.3	3694.1	3806.4
	S	2.2837	6.4651	7.3045	7.3828	7.4585	7.5319	7.6032	7.7404	7.8710	7.9961
	V	1.154	131.66	202.92	211.01	219.05	227.06	235.03	250.89	266.66	282.37
1500	U	842.933	2592.4	2952.2	2993.9	3035.8	3077.9	3120.3	3206.0	3293.3	3382.1
(198.29)	H	844.663	2789.9	3256.6	3310.4	3364.3	3418.4	3472.8	3582.4	3693.3	3805.7
	S	2.3145	6.4406	7.2709	7.3494	7.4253	7.4989	7.5703	7.7077	7.8385	7.9636
	V	1.159	123.69	189.97	197.58	205.15	212.67	220.16	235.06	249.87	264.62
1600	U	856.707	2593.8	2951.1	2992.9	3034.8	3077.0	3119.4	3205.3	3292.7	3381.6
(201.37)	H	858.561	2791.7	3255.0	3309.0	3363.0	3417.2	3471.7	3581.4	3692.5	3805.0
	S	2.3436	6.4175	7.2394	7.3182	7.3942	7.4679	7.5395	7.6770	7.8080	7.9333
	V	1.163	116.62	178.55	185.74	192.87	199.97	207.04	221.09	235.06	248.96
1700	U	869.866	2595.1	2949.9	2991.8	3033.9	3076.1	3118.6	3204.6	3292.1	3381.0
(204.31)	H	871.843	2793.4	3253.5	3307.6	3361.7	3416.1	3470.6	3580.5	3691.7	3804.3
	S	2.3713	6.3957	7.2098	7.2887	7.3649	7.4388	7.5105	7.6482	7.7793	7.9047
	V	1.168	110.32	168.39	175.20	181.97	188.69	195.38	208.68	221.89	235.03
1800	U	882.472	2596.3	2948.8	2990.8	3032.9	3075.2	3117.8	3203.9	3291.5	3380.5
(207.11)	H	884.574	2794.8	3251.9	3306.1	3360.4	3414.9	3469.5	3579.5	3690.9	3803.6
	S	2.3976	6.3751	7.1816	7.2608	7.3372	7.4112	7.4830	7.6209	7.7522	7.8777
	V	1.172	104.65	159.30	165.78	172.21	178.59	184.94	197.57	210.11	222.58
1900	U	894.580	2597.3	2947.6	2989.7	3031.9	3074.3	3117.0	3203.2	3290.8	3380.0
(209.80)	H	896.807	2796.1	3250.3	3304.7	3359.1	3413.7	3468.4	3578.6	3690.0	3802.8
	S	2.4228	6.3554	7.1550	7.2344	7.3109	7.3851	7.4570	7.5951	7.7265	7.8522
	V	1.177	99.536	151.13	157.30	163.42	169.51	175.55	187.57	199.50	211.36
2000	U	906.236	2598.2	2946.4	2988.7	3031.0	3073.5	3116.2	3202.5	3290.2	3379.4
(212.37)	H	908.589	2797.2	3248.7	3303.3	3357.8	3412.5	3467.3	3577.6	3689.2	3802.1
	S	2.4469	6.3366	7.1296	7.2092	7.2859	7.3602	7.4323	7.5706	7.7022	7.8279
	V	1.185	90.652	137.00	142.65	148.25	153.81	159.34	170.30	181.19	192.00
2200	U	928.346	2599.6	2944.1	2986.6	3029.1	3071.7	3114.5	3201.1	3289.0	3378.3
(217.24)	H	930.953	2799.1	3245.5	3300.4	3355.2	3410.1	3465.1	3575.7	3687.6	3800.7
	S	2.4922	6.3015	7.0821	7.1621	7.2393	7.3139	7.3862	7.5249	7.6568	7.7827

TABLE B.2. SUPERHEATED STEAM SI UNITS (Continued)

ABS PRESS KPA (SAT TEMP) DEG C		SAT WATER	SAT STEAM	TEMPERATURE, DEG C (TEMPERATURE, K) 225 (498.15)	250 (523.15)	275 (548.15)	300 (573.15)	325 (598.15)	350 (623.15)	375 (648.15)	400 (673.15)
	V	1.193	83.199	84.149	91.075	97.411	103.36	109.05	114.55	119.93	125.22
2400	U	949.066	2600.7	2608.6	2665.6	2717.3	2765.4	2811.1	2855.4	2898.8	2941.7
(221.78)	H	951.929	2800.4	2810.6	2884.2	2951.1	3013.4	3072.8	3130.4	3186.7	3242.3
	S	2.5343	6.2690	6.2894	6.4338	6.5586	6.6699	6.7714	6.8656	6.9542	7.0384
	V	1.201	76.856	....	83.205	89.220	94.830	100.17	105.32	110.33	115.26
2600	U	968.597	2601.5	....	2658.4	2711.7	2760.9	2807.4	2852.3	2896.1	2939.4
(226.04)	H	971.720	2801.4	....	2874.7	2943.6	3007.4	3067.9	3126.1	3183.0	3239.0
	S	2.5736	6.2387	....	6.3823	6.5110	6.6249	6.7281	6.8236	6.9131	6.9979
	V	1.209	71.389	....	76.437	82.187	87.510	92.550	97.395	102.10	106.71
2800	U	987.100	2602.1	....	2650.9	2705.9	2756.3	2803.7	2849.2	2893.4	2937.0
(230.05)	H	990.485	2802.0	....	2864.9	2936.0	3001.3	3062.8	3121.9	3179.3	3235.8
	S	2.6106	6.2104	....	6.3331	6.4659	6.5824	6.6875	6.7842	6.8746	6.9601
	V	1.216	66.626	....	70.551	76.078	81.159	85.943	90.526	94.969	99.310
3000	U	1004.7	2602.4	....	2643.2	2700.0	2751.6	2799.9	2846.0	2890.7	2934.6
(233.84)	H	1008.4	2802.3	....	2854.8	2928.2	2995.1	3057.7	3117.5	3175.6	3232.5
	S	2.6455	6.1837	....	6.2857	6.4228	6.5422	6.6491	6.7471	6.8385	6.9246
	V	1.224	62.439	....	65.380	70.721	75.593	80.158	84.513	88.723	92.829
3200	U	1021.5	2602.5	....	2635.2	2693.9	2746.8	2796.0	2842.7	2887.9	2932.1
(237.45)	H	1025.4	2802.3	....	2844.4	2920.2	2988.7	3052.5	3113.2	3171.8	3229.2
	S	2.6786	6.1585	....	6.2398	6.3815	6.5037	6.6127	6.7120	6.8043	6.8912
	V	1.235	57.025	....	58.693	63.812	68.424	72.710	76.776	80.689	84.494
3500	U	1045.4	2602.4	....	2622.7	2684.5	2739.5	2790.0	2837.8	2883.7	2928.4
(242.54)	H	1049.8	2802.0	....	2828.1	2907.8	2979.0	3044.5	3106.5	3166.1	3224.2
	S	2.7253	6.1228	....	6.1732	6.3221	6.4491	6.5611	6.6626	6.7563	6.8443
	V	1.242	53.888	....	54.812	59.814	64.282	68.410	72.311	76.055	79.687
3700	U	1060.6	2602.1	....	2614.0	2678.0	2734.4	2786.0	2834.4	2880.8	2926.0
(245.75)	H	1065.2	2801.4	....	2816.8	2899.3	2972.3	3039.1	3102.0	3162.2	3220.8
	S	2.7547	6.1004	....	6.1299	6.2841	6.4143	6.5284	6.6314	6.7262	6.8149
	V	1.252	49.749	....	....	54.546	58.833	62.759	66.446	69.969	73.376
4000	U	1082.4	2601.3	....	....	2668.0	2726.7	2779.8	2829.3	2876.5	2922.2
(250.33)	H	1087.4	2800.3	....	....	2886.1	2962.0	3030.8	3095.1	3156.4	3215.7
	S	2.7965	6.0685	....	....	6.2288	6.3642	6.4817	6.5870	6.6834	6.7733

TABLE B.2. SUPERHEATED STEAM SI UNITS (Continued)

ABS PRESS KPA (SAT TEMP) DEG C		SAT WATER	SAT STEAM	425 (698.15)	450 (723.15)	475 (748.15)	500 (773.15)	525 (798.15)	550 (823.15)	600 (873.15)	650 (923.15)
				TEMPERATURE, DEG C (TEMPERATURE, K)							
2400 (221.78)	V	1.193	83.199	130.44	135.61	140.73	145.82	150.88	155.91	165.92	175.86
	U	949.066	2600.7	2984.5	3027.1	3069.9	3112.9	3156.1	3199.6	3287.7	3377.2
	H	951.929	2800.4	3297.5	3352.6	3407.7	3462.9	3518.2	3573.8	3685.9	3799.3
	S	2.5343	6.2690	7.1189	7.1964	7.2713	7.3439	7.4144	7.4830	7.6152	7.7414
2600 (226.04)	V	1.201	76.856	120.11	124.91	129.66	134.38	139.07	143.74	153.01	162.21
	U	968.597	2601.5	2982.3	3025.2	3068.1	3111.2	3154.6	3198.2	3286.5	3376.1
	H	971.720	2801.4	3294.6	3349.9	3405.3	3460.6	3516.2	3571.9	3684.3	3797.9
	S	2.5736	6.2387	7.0789	7.1568	7.2320	7.3048	7.3755	7.4443	7.5768	7.7033
2800 (230.05)	V	1.209	71.389	111.25	115.74	120.17	124.58	128.95	133.30	141.94	150.50
	U	987.100	2602.1	2980.2	3023.2	3066.3	3109.6	3153.1	3196.8	3285.2	3375.0
	H	990.485	2802.0	3291.7	3347.3	3402.8	3458.4	3514.1	3570.0	3682.6	3796.4
	S	2.6106	6.2104	7.0416	7.1199	7.1954	7.2685	7.3394	7.4084	7.5412	7.6679
3000 (233.84)	V	1.216	66.626	103.58	107.79	111.95	116.08	120.18	124.26	132.34	140.36
	U	1004.7	2602.4	2978.0	3021.3	3064.6	3107.9	3151.5	3195.4	3284.0	3373.9
	H	1008.4	2802.3	3288.7	3344.6	3400.4	3456.2	3512.1	3568.1	3681.0	3795.0
	S	2.6455	6.1837	7.0067	7.0854	7.1612	7.2345	7.3056	7.3748	7.5079	7.6349
3200 (237.45)	V	1.224	62.439	96.859	100.83	104.76	108.65	112.51	116.34	123.95	131.48
	U	1021.5	2602.5	2975.9	3019.3	3062.8	3106.3	3150.0	3193.9	3282.7	3372.8
	H	1025.4	2802.3	3285.8	3342.0	3398.0	3454.0	3510.0	3566.2	3679.3	3793.6
	S	2.6786	6.1585	6.9738	7.0528	7.1290	7.2026	7.2739	7.3433	7.4767	7.6039
3500 (242.54)	V	1.235	57.025	88.220	91.886	95.505	99.088	102.64	106.17	113.15	120.07
	U	1045.4	2602.4	2972.6	3016.4	3060.1	3103.8	3147.7	3191.8	3280.8	3371.2
	H	1049.8	2802.0	3281.3	3338.0	3394.3	3450.6	3506.9	3563.4	3676.9	3791.4
	S	2.7253	6.1228	6.9277	7.0074	7.0840	7.1580	7.2297	7.2993	7.4332	7.5607
3700 (245.75)	V	1.242	53.888	83.238	86.728	90.171	93.576	96.950	100.30	106.93	113.49
	U	1060.6	2602.1	2970.4	3014.4	3058.2	3102.1	3146.1	3190.4	3279.6	3370.1
	H	1065.2	2801.4	3278.4	3335.3	3391.9	3448.4	3504.9	3561.5	3675.2	3790.0
	S	2.7547	6.1004	6.8989	6.9790	7.0559	7.1302	7.2021	7.2719	7.4061	7.5339
4000 (250.33)	V	1.252	49.749	76.698	79.958	83.169	86.341	89.483	92.598	98.763	104.86
	U	1082.4	2601.3	2967.0	3011.4	3055.5	3099.6	3143.8	3188.2	3277.7	3368.4
	H	1087.4	2800.3	3273.8	3331.2	3388.2	3445.0	3501.7	3558.6	3672.8	3787.9
	S	2.7965	6.0685	6.8581	6.9388	7.0163	7.0909	7.1632	7.2333	7.3680	7.4961

TABLE B.2. SUPERHEATED STEAM SI UNITS (Continued)

ABS PRESS KPA (SAT TEMP) DEG C		SAT WATER	SAT STEAM	TEMPERATURE, DEG C (TEMPERATURE, K) 260 (533.15)	275 (548.15)	300 (573.15)	325 (598.15)	350 (623.15)	375 (648.15)	400 (673.15)	425 (698.15)
4200 (253.24)	V	1.259	47.307	48.654	51.438	55.625	59.435	62.998	66.392	69.667	72.856
	U	1096.3	2600.7	2620.4	2661.0	2721.4	2775.6	2825.8	2873.6	2919.7	2964.8
	H	1101.6	2799.4	2824.8	2877.1	2955.0	3025.2	3090.4	3152.4	3212.3	3270.8
	S	2.8231	6.0482	6.0962	6.1929	6.3320	6.4519	6.5587	6.6563	6.7469	6.8323
4400 (256.05)	V	1.266	45.079	45.853	48.601	52.702	56.409	59.861	63.139	66.295	69.363
	U	1109.8	2599.9	2611.8	2653.9	2716.0	2771.3	2822.3	2870.6	2917.1	2962.5
	H	1115.4	2798.3	2813.6	2867.8	2947.8	3019.5	3085.7	3148.4	3208.8	3267.7
	S	2.8487	6.0286	6.0575	6.1577	6.3006	6.4230	6.5315	6.6301	6.7216	6.8076
4600 (258.75)	V	1.272	43.038	43.278	46.000	50.027	53.643	56.994	60.167	63.215	66.172
	U	1122.9	2599.1	2602.9	2646.6	2710.4	2766.9	2818.7	2867.6	2914.5	2960.3
	H	1128.8	2797.0	2802.0	2858.2	2940.5	3013.7	3080.9	3144.4	3205.3	3264.7
	S	2.8735	6.0097	6.0190	6.1230	6.2700	6.3949	6.5050	6.6049	6.6972	6.7838
4800 (261.37)	V	1.279	41.161	....	43.604	47.569	51.103	54.364	57.441	60.390	63.247
	U	1135.6	2598.1	....	2639.1	2704.8	2762.5	2815.1	2864.6	2911.9	2958.0
	H	1141.8	2795.7	....	2848.4	2933.1	3007.8	3076.1	3140.3	3201.8	3261.6
	S	2.8974	5.9913	....	6.0887	6.2399	6.3675	6.4794	6.5805	6.6736	6.7608
5000 (263.91)	V	1.286	39.429	....	41.388	45.301	48.762	51.941	54.932	57.791	60.555
	U	1148.0	2597.0	....	2631.3	2699.0	2758.0	2811.5	2861.5	2909.3	2955.7
	H	1154.5	2794.2	....	2838.2	2925.5	3001.8	3071.2	3136.2	3198.3	3258.5
	S	2.9206	5.9735	....	6.0547	6.2105	6.3408	6.4545	6.5568	6.6508	6.7386
5200 (266.37)	V	1.292	37.824	....	39.330	43.201	46.599	49.703	52.614	55.390	58.070
	U	1160.1	2595.9	....	2623.3	2693.1	2753.4	2807.8	2858.4	2906.7	2953.4
	H	1166.8	2792.6	....	2827.8	2917.8	2995.7	3066.2	3132.0	3194.7	3255.4
	S	2.9431	5.9561	....	6.0210	6.1815	6.3147	6.4302	6.5338	6.6287	6.7172
5400 (268.76)	V	1.299	36.334	....	37.411	41.251	44.591	47.628	50.466	53.166	55.768
	U	1171.9	2594.6	....	2615.0	2687.1	2748.7	2804.0	2855.3	2904.0	2951.1
	H	1178.9	2790.8	....	2817.0	2909.8	2989.5	3061.2	3127.8	3191.1	3252.2
	S	2.9650	5.9392	....	5.9873	6.1530	6.2891	6.4066	6.5114	6.6072	6.6963
5600 (271.09)	V	1.306	34.946	....	35.617	39.434	42.724	45.700	48.470	51.100	53.630
	U	1183.5	2593.3	....	2606.5	2680.9	2744.0	2800.2	2852.1	2901.3	2948.7
	H	1190.8	2789.0	....	2805.9	2901.7	2983.2	3056.1	3123.6	3187.5	3249.0
	S	2.9863	5.9227	....	5.9537	6.1248	6.2640	6.3834	6.4896	6.5863	6.6761

TABLE B.2. SUPERHEATED STEAM SI UNITS (Continued)

ABS PRESS KPA (SAT TEMP) DEG C		SAT WATER	SAT STEAM	TEMPERATURE, DEG C (TEMPERATURE, K) 450 (723.15)	475 (748.15)	500 (773.15)	525 (798.15)	550 (823.15)	575 (848.15)	600 (873.15)	650 (923.15)
4200	V	1.259	47.307	75.981	79.056	82.092	85.097	88.075	91.030	93.966	99.787
(253.24)	U	1096.3	2600.7	3009.4	3053.7	3097.9	3142.3	3186.8	3231.5	3276.5	3367.3
	H	1101.6	2799.4	3328.5	3385.7	3442.7	3499.7	3556.7	3613.8	3671.1	3786.4
	S	2.8231	6.0482	6.9135	6.9913	7.0662	7.1387	7.2090	7.2774	7.3440	7.4724
4400	V	1.266	45.079	72.365	75.317	78.229	81.110	83.963	86.794	89.605	95.177
(256.05)	U	1109.8	2599.9	3007.4	3051.9	3096.3	3140.7	3185.3	3230.1	3275.2	3366.2
	H	1115.4	2798.3	3325.8	3383.3	3440.5	3497.6	3554.7	3612.0	3669.5	3785.0
	S	2.8487	6.0286	6.8892	6.9674	7.0426	7.1153	7.1858	7.2544	7.3211	7.4498
4600	V	1.272	43.038	69.063	71.903	74.702	77.469	80.209	82.926	85.623	90.967
(258.75)	U	1122.9	2599.1	3005.3	3050.0	3094.6	3139.2	3183.9	3228.8	3273.9	3365.1
	H	1128.8	2797.0	3323.0	3380.8	3438.2	3495.5	3552.8	3610.2	3667.8	3783.6
	S	2.8735	6.0097	6.8659	6.9444	7.0199	7.0928	7.1636	7.2323	7.2991	7.4281
4800	V	1.279	41.161	66.036	68.773	71.469	74.132	76.768	79.381	81.973	87.109
(261.37)	U	1135.6	2598.1	3003.3	3048.2	3092.9	3137.6	3182.4	3227.4	3272.7	3364.0
	H	1141.8	2795.7	3320.3	3378.3	3435.9	3493.4	3550.9	3608.5	3666.2	3782.1
	S	2.8974	5.9913	6.8434	6.9223	6.9981	7.0712	7.1422	7.2110	7.2781	7.4072
5000	V	1.286	39.429	63.250	65.893	68.494	71.061	73.602	76.119	78.616	83.559
(263.91)	U	1148.0	2597.0	3001.2	3046.3	3091.2	3136.0	3181.0	3226.1	3271.4	3362.9
	H	1154.5	2794.2	3317.5	3375.8	3433.7	3491.3	3549.0	3606.7	3664.5	3780.7
	S	2.9206	5.9735	6.8217	6.9009	6.9770	7.0504	7.1215	7.1906	7.2578	7.3872
5200	V	1.292	37.824	60.679	63.234	65.747	68.227	70.679	73.108	75.516	80.282
(266.37)	U	1160.1	2595.9	2999.2	3044.5	3089.5	3134.5	3179.5	3224.7	3270.2	3361.8
	H	1166.8	2792.6	3314.7	3373.3	3431.4	3489.3	3547.1	3604.9	3662.8	3779.3
	S	2.9431	5.9561	6.8007	6.8803	6.9567	7.0304	7.1017	7.1709	7.2382	7.3679
5400	V	1.299	36.334	58.297	60.772	63.204	65.603	67.973	70.320	72.646	77.248
(268.76)	U	1171.9	2594.6	2997.1	3042.6	3087.8	3132.9	3178.1	3223.4	3268.9	3360.7
	H	1178.9	2790.8	3311.9	3370.8	3429.1	3487.2	3545.1	3603.1	3661.2	3777.8
	S	2.9650	5.9392	6.7804	6.8604	6.9371	7.0110	7.0825	7.1519	7.2194	7.3493
5600	V	1.306	34.946	56.085	58.486	60.843	63.165	65.460	67.731	69.981	74.431
(271.09)	U	1183.5	2593.3	2995.0	3040.7	3086.1	3131.3	3176.6	3222.0	3267.6	3359.6
	H	1190.8	2789.0	3309.1	3368.2	3426.8	3485.1	3543.2	3601.3	3659.5	3776.4
	S	2.9863	5.9227	6.7607	6.8411	6.9181	6.9922	7.0639	7.1335	7.2011	7.3313

TABLE B.2. SUPERHEATED STEAM SI UNITS (Continued)

ABS PRESS KPA (SAT TEMP) DEG C		SAT WATER	SAT STEAM	TEMPERATURE, DEG C (TEMPERATURE, K) 280 (553.15)	290 (563.15)	300 (573.15)	325 (598.15)	350 (623.15)	375 (648.15)	400 (673.15)	425 (698.15)
6000 (275.55)	V	1.319	32.438	33.173	34.718	36.145	39.353	42.222	44.874	47.379	49.779
	U	1205.8	2590.4	2605.9	2638.4	2668.1	2734.2	2792.4	2845.7	2895.8	2944.0
	H	1213.7	2785.0	2804.9	2846.7	2885.0	2970.4	3045.8	3115.0	3180.1	3242.6
	S	3.0273	5.8908	5.9270	6.0017	6.0692	6.2151	6.3386	6.4475	6.5462	6.6374
6200 (277.70)	V	1.325	31.300	31.679	33.227	34.650	37.825	40.648	43.248	45.697	48.039
	U	1216.6	2588.8	2597.1	2630.8	2661.5	2729.2	2788.5	2842.4	2893.1	2941.6
	H	1224.8	2782.9	2793.5	2836.8	2876.3	2963.8	3040.5	3110.6	3176.4	3239.4
	S	3.0471	5.8753	5.8946	5.9721	6.0418	6.1911	6.3168	6.4272	6.5268	6.6188
6400 (279.79)	V	1.332	30.230	30.265	31.821	33.241	36.390	39.170	41.722	44.119	46.407
	U	1227.2	2587.2	2587.9	2623.0	2654.7	2724.2	2784.4	2839.1	2890.3	2939.2
	H	1235.7	2780.6	2781.6	2826.6	2867.5	2957.1	3035.1	3106.2	3172.7	3236.2
	S	3.0664	5.8601	5.8619	5.9425	6.0144	6.1675	6.2955	6.4072	6.5079	6.6006
6600 (281.84)	V	1.338	29.223	....	30.490	31.911	35.038	37.781	40.287	42.636	44.874
	U	1237.6	2585.5	....	2614.9	2647.7	2719.0	2780.4	2835.8	2887.5	2936.7
	H	1246.5	2778.3	....	2816.1	2858.4	2950.2	3029.7	3101.7	3168.9	3232.9
	S	3.0853	5.8452	....	5.9129	5.9872	6.1442	6.2744	6.3877	6.4894	6.5828
6800 (283.84)	V	1.345	28.272	....	29.226	30.652	33.762	36.470	38.935	41.239	43.430
	U	1247.9	2583.7	....	2606.6	2640.6	2713.7	2776.2	2832.4	2884.7	2934.3
	H	1257.0	2775.9	....	2805.3	2849.0	2943.3	3024.2	3097.2	3165.1	3229.6
	S	3.1038	5.8306	....	5.8830	5.9599	6.1211	6.2537	6.3686	6.4713	6.5655
7200 (287.70)	V	1.358	26.522	....	26.878	28.321	31.413	34.063	36.454	38.676	40.781
	U	1267.9	2579.9	....	2589.0	2625.6	2702.9	2767.8	2825.6	2878.9	2929.4
	H	1277.6	2770.9	....	2782.5	2829.5	2929.1	3013.1	3088.1	3157.4	3223.0
	S	3.1397	5.8020	....	5.8226	5.9054	6.0755	6.2132	6.3312	6.4362	6.5319
7600 (291.41)	V	1.371	24.949	....	....	26.204	29.297	31.901	34.229	36.380	38.409
	U	1287.2	2575.9	....	....	2609.7	2691.7	2759.2	2818.6	2873.1	2924.3
	H	1297.6	2765.5	....	....	2808.8	2914.3	3001.6	3078.7	3149.6	3216.3
	S	3.1742	5.7742	....	....	5.8503	6.0306	6.1737	6.2950	6.4022	6.4996
8000 (294.97)	V	1.384	23.525	....	....	24.264	27.378	29.948	32.222	34.310	36.273
	U	1306.0	2571.7	....	....	2592.7	2679.9	2750.3	2811.5	2867.1	2919.3
	H	1317.1	2759.9	....	....	2786.8	2899.0	2989.9	3069.2	3141.6	3209.5
	S	3.2076	5.7471	....	....	5.7942	5.9860	6.1349	6.2599	6.3694	6.4684

TABLE B.2. SUPERHEATED STEAM SI UNITS (Continued)

ABS PRESS KPA (SAT TEMP) DEG C		SAT WATER	SAT STEAM	TEMPERATURE, DEG C (TEMPERATURE, K) 450 (723.15)	475 (748.15)	500 (773.15)	525 (798.15)	550 (823.15)	575 (848.15)	600 (873.15)	650 (923.15)
6000 (275.55)	V	1.319	32.438	52.103	54.369	56.592	58.778	60.937	63.071	65.184	69.359
	U	1205.8	2590.4	2990.8	3036.9	3082.6	3128.2	3173.7	3219.3	3265.1	3357.4
	H	1213.7	2785.0	3303.5	3363.2	3422.2	3480.8	3539.3	3597.7	3656.2	3773.5
	S	3.0273	5.8908	6.7230	6.8041	6.8818	6.9564	7.0285	7.0985	7.1664	7.2971
6200 (277.70)	V	1.325	31.300	50.304	52.510	54.671	56.797	58.894	60.966	63.018	67.069
	U	1216.6	2588.8	2988.7	3035.0	3080.9	3126.6	3172.2	3218.0	3263.8	3356.3
	H	1224.8	2782.9	3300.6	3360.6	3419.9	3478.7	3537.4	3595.9	3654.5	3772.1
	S	3.0471	5.8753	6.7049	6.7864	6.8644	6.9393	7.0116	7.0817	7.1498	7.2808
6400 (279.79)	V	1.332	30.230	48.617	50.767	52.871	54.939	56.978	58.993	60.987	64.922
	U	1227.2	2587.2	2986.6	3033.1	3079.2	3125.0	3170.8	3216.6	3262.6	3355.2
	H	1235.7	2780.6	3297.7	3358.0	3417.6	3476.6	3535.4	3594.1	3652.9	3770.7
	S	3.0664	5.8601	6.6872	6.7692	6.8475	6.9226	6.9952	7.0655	7.1337	7.2649
6600 (281.84)	V	1.338	29.223	47.031	49.129	51.180	53.194	55.179	57.139	59.079	62.905
	U	1237.6	2585.5	2984.5	3031.2	3077.4	3123.4	3169.3	3215.2	3261.3	3354.1
	H	1246.5	2778.3	3294.9	3355.5	3415.2	3474.5	3533.5	3592.3	3651.2	3769.2
	S	3.0853	5.8452	6.6700	6.7524	6.8310	6.9064	6.9792	7.0497	7.1181	7.2495
6800 (283.84)	V	1.345	28.272	45.539	47.587	49.588	51.552	53.486	55.395	57.283	61.007
	U	1247.9	2583.7	2982.3	3029.3	3075.7	3121.8	3167.8	3213.9	3260.0	3353.0
	H	1257.0	2775.9	3292.0	3352.9	3412.9	3472.4	3531.5	3590.5	3649.6	3767.8
	S	3.1038	5.8306	6.6532	6.7361	6.8150	6.8907	6.9636	7.0343	7.1028	7.2345
7200 (287.70)	V	1.358	26.522	42.802	44.759	46.668	48.540	50.381	52.197	53.991	57.527
	U	1267.9	2579.9	2978.0	3025.4	3072.2	3118.6	3164.9	3211.1	3257.5	3350.7
	H	1277.6	2770.9	3286.1	3347.7	3408.2	3468.1	3527.6	3586.9	3646.2	3764.9
	S	3.1397	5.8020	6.6208	6.7044	6.7840	6.8602	6.9337	7.0047	7.0735	7.2058
7600 (291.41)	V	1.371	24.949	40.351	42.228	44.056	45.845	47.603	49.335	51.045	54.413
	U	1287.2	2575.9	2973.6	3021.5	3068.7	3115.4	3161.9	3208.4	3254.9	3348.5
	H	1297.6	2765.5	3280.3	3342.5	3403.5	3463.8	3523.7	3583.3	3642.9	3762.1
	S	3.1742	5.7742	6.5896	6.6742	6.7545	6.8312	6.9051	6.9765	7.0457	7.1784
8000 (294.97)	V	1.384	23.525	38.145	39.950	41.704	43.419	45.102	46.759	48.394	51.611
	U	1306.0	2571.7	2969.2	3017.6	3065.1	3112.2	3158.9	3205.6	3252.4	3346.3
	H	1317.1	2759.9	3274.3	3337.2	3398.8	3459.5	3519.7	3579.7	3639.5	3759.2
	S	3.2076	5.7471	6.5597	6.6452	6.7262	6.8035	6.8778	6.9496	7.0191	7.1523

TABLE B.2. SUPERHEATED STEAM SI UNITS (Continued)

ABS PRESS KPA (SAT TEMP) DEG C		SAT WATER	SAT STEAM	TEMPERATURE, DEG C (TEMPERATURE, K) 300 (573.15)	320 (593.15)	340 (613.15)	360 (633.15)	380 (653.15)	400 (673.15)	425 (698.15)	450 (723.15)
8400 (298.39)	V	1.398	22.231	22.469	25.058	27.203	29.094	30.821	32.435	34.337	36.[illegible]47
	U	1324.3	2567.2	2574.4	2651.1	2713.4	2767.3	2816.0	2861.1	2914.1	2964.7
	H	1336.1	2754.0	2763.1	2861.6	2941.9	3011.7	3074.8	3133.5	3202.6	3268.3
	S	3.2399	5.7207	5.7366	5.9056	6.0388	6.1509	6.2491	6.3376	6.4383	6.5309
8800 (301.70)	V	1.411	21.049	....	23.446	25.592	27.459	29.153	30.727	32.576	34.529
	U	1342.2	2562.6	....	2637.3	2702.8	2758.8	2808.8	2854.9	2908.9	2960.1
	H	1354.6	2747.8	....	2843.6	2928.0	3000.4	3065.3	3125.3	3195.6	3262.2
	S	3.2713	5.6948	....	5.8590	5.9990	6.1152	6.2162	6.3067	6.4092	6.5030
9200 (304.89)	V	1.425	19.964	....	21.952	24.110	25.961	27.625	29.165	30.966	32.668
	U	1359.7	2557.7	....	2622.7	2691.9	2750.0	2801.5	2848.7	2903.6	2955.5
	H	1372.8	2741.3	....	2824.7	2913.7	2988.9	3055.7	3117.0	3188.5	3256.1
	S	3.3018	5.6694	....	5.8118	5.9594	6.0801	6.1840	6.2765	6.3808	6.4760
9600 (307.97)	V	1.439	18.965	....	20.561	22.740	24.581	26.221	27.731	29.489	31.[illegible]45
	U	1376.7	2552.6	....	2607.3	2680.5	2741.0	2794.1	2842.3	2898.2	2950.9
	H	1390.6	2734.7	....	2804.7	2898.8	2977.0	3045.8	3108.5	3181.3	3249.9
	S	3.3315	5.6444	....	5.7637	5.9199	6.0454	6.1524	6.2470	6.3532	6.4498
10000 (310.96)	V	1.453	18.041	....	19.256	21.468	23.305	24.926	26.408	28.128	29.742
	U	1393.5	2547.3	....	2590.9	2668.7	2731.8	2786.4	2835.8	2892.8	2946.2
	H	1408.0	2727.7	....	2783.5	2883.4	2964.8	3035.7	3099.9	3174.1	3243.6
	S	3.3605	5.6198	....	5.7145	5.8803	6.0110	6.1213	6.2182	6.3264	6.4243
10400 (313.86)	V	1.467	17.184	....	18.024	20.282	22.121	23.726	25.185	26.870	28.446
	U	1410.0	2541.8	....	2573.4	2656.3	2722.2	2778.7	2829.3	2887.3	2941.5
	H	1425.2	2720.6	....	2760.8	2867.2	2952.3	3025.4	3091.2	3166.7	3237.3
	S	3.3889	5.5955	....	5.6638	5.8404	5.9769	6.0907	6.1899	6.3001	6.3994
10800 (316.67)	V	1.481	16.385	....	16.852	19.170	21.018	22.612	24.050	25.703	27.245
	U	1426.2	2536.2	....	2554.5	2643.4	2712.4	2770.7	2822.6	2881.7	2936.7
	H	1442.2	2713.1	....	2736.5	2850.4	2939.4	3014.9	3082.3	3159.3	3230.9
	S	3.4167	5.5715	....	5.6109	5.8000	5.9429	6.0604	6.1621	6.2744	6.3752
11200 (319.40)	V	1.496	15.639	....	15.726	18.124	19.987	21.573	22.993	24.619	26.128
	U	1442.1	2530.3	....	2533.8	2629.8	2702.2	2762.6	2815.8	2876.0	2931.8
	H	1458.9	2705.4	....	2710.0	2832.8	2926.1	3004.2	3073.3	3151.7	3224.5
	S	3.4440	5.5476	....	5.5553	5.7591	5.9090	6.0305	6.1347	6.2491	6.3515

TABLE B.2. SUPERHEATED STEAM SI UNITS (Continued)

ABS PRESS KPA (SAT TEMP) DEG C		SAT WATER	SAT STEAM	TEMPERATURE, DEG C (TEMPERATURE, K) 475 (748.15)	500 (773.15)	525 (798.15)	550 (823.15)	575 (848.15)	600 (873.15)	625 (898.15)	650 (923.15)
	V	1.398	22.231	37.887	39.576	41.224	42.839	44.429	45.996	47.544	49.076
8400	U	1324.3	2567.2	3013.6	3061.6	3108.9	3155.9	3202.9	3249.8	3296.9	3344.1
(298.39)	H	1336.1	2754.0	3331.9	3394.0	3455.2	3515.8	3576.1	3636.2	3696.2	3756.3
	S	3.2399	5.7207	6.6173	6.6990	6.7769	6.8516	6.9238	6.9936	7.0614	7.1274
	V	1.411	21.049	36.011	37.640	39.228	40.782	42.310	43.815	45.301	46.771
8800	U	1342.2	2562.6	3009.6	3058.0	3105.6	3152.9	3200.1	3247.2	3294.5	3341.8
(301.70)	H	1354.6	2747.8	3326.5	3389.2	3450.8	3511.8	3572.4	3632.8	3693.1	3753.4
	S	3.2713	5.6948	6.5904	6.6728	6.7513	6.8265	6.8990	6.9692	7.0373	7.1035
	V	1.425	19.964	34.298	35.872	37.405	38.904	40.375	41.824	43.254	44.667
9200	U	1359.7	2557.7	3005.6	3054.3	3102.3	3149.9	3197.3	3244.7	3292.1	3339.6
(304.89)	H	1372.8	2741.3	3321.1	3384.4	3446.5	3507.8	3568.8	3629.5	3690.0	3750.5
	S	3.3018	5.6694	6.5644	6.6475	6.7266	6.8023	6.8752	6.9457	7.0141	7.0806
	V	1.439	18.965	32.726	34.252	35.734	37.182	38.602	39.999	41.377	42.738
9600	U	1376.7	2552.6	3001.5	3050.7	3099.0	3146.9	3194.5	3242.1	3289.7	3337.4
(307.97)	H	1390.6	2734.7	3315.6	3379.5	3442.1	3503.9	3565.1	3626.1	3686.9	3747.6
	S	3.3315	5.6444	6.5392	6.6231	6.7028	6.7790	6.8523	6.9231	6.9918	7.0585
	V	1.453	18.041	31.280	32.760	34.196	35.597	36.970	38.320	39.650	40.963
10000	U	1393.5	2547.3	2997.4	3047.0	3095.7	3143.9	3191.7	3239.5	3287.3	3335.1
(310.96)	H	1408.0	2727.7	3310.1	3374.6	3437.7	3499.8	3561.4	3622.7	3683.8	3744.7
	S	3.3605	5.6198	6.5147	6.5994	6.6797	6.7564	6.8302	6.9013	6.9703	7.0373
	V	1.467	17.184	29.943	31.382	32.776	34.134	35.464	36.770	38.056	39.325
10400	U	1410.0	2541.8	2993.2	3043.3	3092.4	3140.8	3188.9	3236.9	3284.8	3332.9
(313.86)	H	1425.2	2720.6	3304.6	3369.7	3433.2	3495.8	3557.8	3619.3	3680.6	3741.8
	S	3.3889	5.5955	6.4909	6.5765	6.6574	6.7346	6.8087	6.8803	6.9495	7.0167
	V	1.481	16.385	28.706	30.106	31.461	32.779	34.069	35.335	36.580	37.808
10800	U	1426.2	2536.2	2989.0	3039.6	3089.0	3137.8	3186.1	3234.3	3282.4	3330.6
(316.67)	H	1442.2	2713.1	3299.0	3364.7	3428.8	3491.8	3554.1	3615.9	3677.5	3738.9
	S	3.4167	5.5715	6.4678	6.5542	6.6357	6.7134	6.7880	6.8599	6.9294	6.9969
	V	1.496	15.639	27.555	28.921	30.240	31.521	32.774	34.002	35.210	36.400
11200	U	1442.1	2530.3	2984.8	3035.8	3085.6	3134.7	3183.3	3231.7	3280.0	3328.4
(319.40)	H	1458.9	2705.4	3293.4	3359.7	3424.3	3487.7	3550.4	3612.5	3674.4	3736.0
	S	3.4440	5.5476	6.4452	6.5324	6.6147	6.6929	6.7679	6.8401	6.9099	6.9777

TABLE B.2. SUPERHEATED STEAM SI UNITS (Continued)

ABS PRESS KPA (SAT TEMP) DEG C		SAT WATER	SAT STEAM	TEMPERATURE, DEG C (TEMPERATURE, K) 340 (613.15)	360 (633.15)	380 (653.15)	400 (673.15)	420 (693.15)	440 (713.15)	460 (733.15)	480 (753.15)
11800 (323.36)	V	1.519	14.607	16.658	18.557	20.139	21.538	22.820	24.019	25.155	26.243
	U	1465.7	2521.0	2608.0	2686.3	2750.0	2805.3	2855.4	2902.1	2946.3	2983.8
	H	1483.6	2693.3	2804.6	2905.3	2987.6	3059.5	3124.7	3185.5	3243.1	3293.5
	S	3.4840	5.5121	5.6962	5.8579	5.9860	6.0943	6.1898	6.2763	6.3561	6.4305
12200 (325.91)	V	1.535	13.969	15.740	17.672	19.256	20.645	21.910	23.089	24.204	25.269
	U	1481.2	2514.5	2592.5	2675.2	2741.4	2798.2	2849.3	2896.8	2941.6	2984.6
	H	1499.9	2684.9	2784.5	2890.8	2976.3	3050.0	3116.6	3178.5	3236.9	3292.9
	S	3.5102	5.4884	5.6528	5.8236	5.9565	6.0678	6.1653	6.2532	6.3341	6.4094
12600 (328.40)	V	1.551	13.367	14.861	16.836	18.426	19.805	21.055	22.217	23.312	24.357
	U	1496.5	2507.7	2576.0	2663.7	2732.5	2790.9	2843.2	2891.4	2936.9	2980.4
	H	1516.0	2676.1	2763.2	2875.8	2964.6	3040.5	3108.5	3171.3	3230.6	3287.3
	S	3.5361	5.4646	5.6082	5.7889	5.9272	6.0416	6.1411	6.2305	6.3125	6.3888
13000 (330.83)	V	1.567	12.797	14.015	16.041	17.642	19.015	20.252	21.397	22.474	23.500
	U	1511.6	2500.6	2558.4	2651.6	2723.4	2783.5	2836.9	2886.0	2932.1	2976.1
	H	1532.0	2667.0	2740.6	2860.2	2952.7	3030.7	3100.2	3164.1	3224.2	3281.6
	S	3.5616	5.4408	5.5618	5.7539	5.8979	6.0155	6.1173	6.2082	6.2913	6.3685
13400 (333.19)	V	1.584	12.256	13.195	15.285	16.900	18.270	19.495	20.625	21.686	22.694
	U	1526.7	2493.2	2539.5	2639.1	2714.0	2776.0	2830.6	2880.5	2927.2	2971.7
	H	1547.9	2657.4	2716.3	2843.9	2940.5	3020.8	3091.8	3156.8	3217.8	3275.8
	S	3.5868	5.4168	5.5133	5.7183	5.8685	5.9897	6.0937	6.1862	6.2705	6.3486
13800 (335.51)	V	1.602	11.743	12.394	14.563	16.196	17.565	18.781	19.897	20.942	21.933
	U	1541.6	2485.5	2518.9	2625.9	2704.3	2768.3	2824.1	2874.9	2922.3	2967.4
	H	1563.7	2647.5	2689.9	2826.9	2927.9	3010.7	3083.3	3149.5	3211.3	3270.0
	S	3.6118	5.3925	5.4619	5.6820	5.8391	5.9641	6.0704	6.1645	6.2501	6.3291
14200 (337.76)	V	1.620	11.253	11.600	13.871	15.528	16.897	18.105	19.209	20.240	21.215
	U	1556.5	2477.3	2495.9	2612.1	2694.4	2760.4	2817.6	2869.3	2917.4	2962.9
	H	1579.5	2637.1	2660.6	2809.1	2914.9	3000.4	3074.6	3142.0	3204.8	3264.2
	S	3.6366	5.3679	5.4064	5.6449	5.8095	5.9385	6.0473	6.1431	6.2299	6.3099
14600 (339.97)	V	1.638	10.786	10.791	13.205	14.891	16.264	17.465	18.558	19.576	20.536
	U	1571.3	2468.8	2469.1	2597.6	2684.1	2752.4	2810.9	2863.6	2912.4	2958.5
	H	1595.3	2626.3	2626.6	2790.4	2901.5	2989.9	3065.9	3134.5	3198.2	3258.3
	S	3.6613	5.3431	5.3436	5.6069	5.7797	5.9130	6.0244	6.1220	6.2100	6.2910

TABLE B.2. SUPERHEATED STEAM SI UNITS (Continued)

ABS PRESS KPA (SAT TEMP) DEG C		SAT WATER	SAT STEAM	500 (773.15)	520 (793.15)	540 (813.15)	560 (833.15)	580 (853.15)	600 (873.15)	625 (898.15)	650 (923.15)
				TEMPERATURE, DEG C (TEMPERATURE, K)							
11800 (323.36)	V	1.519	14.607	27.293	28.312	29.305	30.278	31.232	32.172	33.328	34.467
	U	1465.7	2521.0	3030.1	3070.5	3110.3	3149.7	3188.8	3227.8	3276.4	3325.0
	H	1483.6	2693.3	3352.2	3404.6	3456.1	3507.0	3557.3	3607.4	3669.6	3731.7
	S	3.4840	5.5121	6.5009	6.5678	6.6320	6.6938	6.7535	6.8115	6.8818	6.9499
12200 (325.91)	V	1.535	13.969	26.296	27.291	28.260	29.208	30.138	31.052	32.177	33.283
	U	1481.2	2514.5	3026.3	3067.0	3107.1	3146.7	3186.0	3225.1	3273.9	3322.7
	H	1499.9	2684.9	3347.1	3400.0	3451.8	3503.0	3553.7	3604.0	3666.5	3728.8
	S	3.5102	5.4884	6.4805	6.5480	6.6126	6.6747	6.7348	6.7931	6.8637	6.9321
12600 (328.40)	V	1.551	13.367	25.362	26.334	27.281	28.206	29.112	30.003	31.098	32.175
	U	1496.5	2507.7	3022.4	3063.5	3103.8	3143.7	3183.2	3222.5	3271.5	3320.4
	H	1516.0	2676.1	3342.0	3395.3	3447.6	3499.1	3550.0	3600.5	3663.3	3725.8
	S	3.5361	5.4646	6.4605	6.5286	6.5936	6.6562	6.7166	6.7752	6.8461	6.9148
13000 (330.83)	V	1.567	12.797	24.485	25.437	26.362	27.265	28.150	29.019	30.086	31.135
	U	1511.6	2500.6	3018.5	3059.9	3100.6	3140.6	3180.4	3219.9	3269.0	3318.2
	H	1532.0	2667.0	3336.8	3390.6	3443.3	3495.1	3546.3	3597.1	3660.2	3722.9
	S	3.5616	5.4408	6.4409	6.5096	6.5752	6.6381	6.6989	6.7577	6.8289	6.8979
13400 (333.19)	V	1.584	12.256	23.659	24.592	25.497	26.380	27.245	28.093	29.134	30.157
	U	1526.7	2493.2	3014.6	3056.4	3097.3	3137.6	3177.5	3217.2	3266.6	3315.9
	H	1547.9	2657.4	3331.7	3385.9	3438.9	3491.1	3542.6	3593.6	3657.0	3720.0
	S	3.5868	5.4168	6.4218	6.4910	6.5571	6.6205	6.6816	6.7407	6.8122	6.8814
13800 (335.51)	V	1.602	11.743	22.882	23.796	24.683	25.547	26.392	27.221	28.238	29.236
	U	1541.6	2485.5	3010.7	3052.8	3094.0	3134.5	3174.7	3214.5	3264.1	3313.6
	H	1563.7	2647.5	3326.4	3381.2	3434.6	3487.1	3538.9	3590.2	3653.8	3717.1
	S	3.6118	5.3925	6.4030	6.4729	6.5394	6.6032	6.6646	6.7241	6.7959	6.8654
14200 (337.76)	V	1.620	11.253	22.147	23.045	23.914	24.760	25.587	26.398	27.392	28.367
	U	1556.5	2477.3	3006.7	3049.1	3090.6	3131.5	3171.8	3211.9	3261.7	3311.3
	H	1579.5	2637.1	3321.2	3376.4	3430.2	3483.1	3535.2	3586.7	3650.6	3714.1
	S	3.6366	5.3679	6.3846	6.4550	6.5221	6.5863	6.6481	6.7078	6.7800	6.8497
14600 (339.97)	V	1.638	10.786	21.453	22.335	23.187	24.017	24.827	25.620	26.592	27.545
	U	1571.3	2468.8	3002.7	3045.5	3087.3	3128.4	3169.0	3209.2	3259.2	3309.0
	H	1595.3	2626.3	3315.9	3371.6	3425.8	3479.0	3531.4	3583.3	3647.5	3711.2
	S	3.6613	5.3431	6.3665	6.4376	6.5051	6.5697	6.6319	6.6919	6.7644	6.8344

TABLE B.2. SUPERHEATED STEAM SI UNITS (Continued)

ABS PRESS KPA (SAT TEMP) DEG C		SAT WATER	SAT STEAM	TEMPERATURE, DEG C (TEMPERATURE, K) 360 (633.15)	370 (643.15)	380 (653.15)	400 (673.15)	420 (693.15)	440 (713.15)	460 (733.15)	480 (753.15)
15200 (343.19)	V	1.668	10.1246	12.248	13.181	13.988	15.371	16.565	17.643	18.643	19.584
	U	1593.5	2455.3	2574.4	2624.9	2668.0	2740.0	2800.7	2854.9	2904.8	2951.7
	H	1618.9	2609.2	2760.5	2825.3	2880.7	2973.7	3052.5	3123.0	3188.1	3249.4
	S	3.6981	5.3051	5.5476	5.6491	5.7345	5.8749	5.9903	6.0907	6.1807	6.2632
15600 (345.28)	V	1.689	9.7072	11.635	12.595	13.418	14.810	16.001	17.071	18.060	18.989
	U	1608.3	2445.8	2557.7	2611.5	2656.8	2731.6	2793.8	2849.0	2899.6	2947.2
	H	1634.7	2597.3	2739.2	2808.0	2866.2	2962.6	3043.4	3115.3	3181.4	3243.4
	S	3.7226	5.2793	5.5063	5.6142	5.7040	5.8495	5.9678	6.0701	6.1615	6.2450
16000 (347.33)	V	1.710	9.3075	11.036	12.030	12.871	14.275	15.464	16.527	17.506	18.423
	U	1623.2	2436.0	2539.9	2597.4	2645.2	2722.9	2786.8	2843.0	2894.5	2942.6
	H	1650.5	2584.9	2716.5	2789.9	2851.1	2951.3	3034.2	3107.5	3174.5	3237.4
	S	3.7471	5.2531	5.4634	5.5784	5.6729	5.8240	5.9455	6.0497	6.1425	6.2270
16400 (349.33)	V	1.733	8.9247	10.450	11.483	12.344	13.763	14.952	16.008	16.978	17.885
	U	1638.1	2425.7	2520.9	2582.6	2633.1	2714.0	2779.6	2837.0	2889.2	2938.0
	H	1666.5	2572.1	2692.3	2771.0	2835.5	2939.7	3024.8	3099.5	3167.7	3231.3
	S	3.7717	5.2267	5.4183	5.5417	5.6413	5.7984	5.9232	6.0294	6.1237	6.2093
16800 (351.30)	V	1.757	8.5535	9.8712	10.952	11.835	13.272	14.462	15.514	16.475	17.372
	U	1653.5	2414.9	2500.4	2567.0	2620.5	2704.8	2772.4	2830.9	2883.9	2933.3
	H	1683.0	2558.6	2666.2	2751.0	2819.4	2927.8	3015.3	3091.5	3160.7	3225.2
	S	3.7974	5.1994	5.3708	5.5037	5.6091	5.7728	5.9009	6.0093	6.1050	6.1918
17200 (353.22)	V	1.783	8.1912	9.2964	10.434	11.344	12.801	13.994	15.041	15.994	16.882
	U	1669.7	2403.5	2478.0	2550.6	2607.4	2695.4	2764.9	2824.7	2878.6	2928.6
	H	1700.4	2544.4	2637.9	2730.0	2802.5	2915.6	3005.6	3083.4	3153.7	3219.0
	S	3.8240	5.1713	5.3200	5.4644	5.5762	5.7469	5.8788	5.9894	6.0866	6.1745
17600 (355.11)	V	1.810	7.8394	8.7187	9.9272	10.868	12.348	13.545	14.588	15.535	16.414
	U	1685.8	2391.6	2453.2	2533.1	2593.7	2685.8	2757.4	2818.4	2873.2	2923.9
	H	1717.6	2529.5	2606.6	2707.8	2784.9	2903.1	2995.8	3075.2	3146.6	3212.8
	S	3.8504	5.1425	5.2649	5.4235	5.5425	5.7209	5.8566	5.9695	6.0684	6.1574
18000 (356.96)	V	1.840	7.4977	1.9465	9.4298	10.405	11.913	13.115	14.155	15.096	15.966
	U	1701.7	2378.9	1737.4	2514.5	2579.3	2675.9	2749.7	2812.1	2867.7	2919.1
	H	1734.8	2513.9	1772.4	2684.2	2766.6	2890.3	2985.8	3066.9	3139.4	3206.5
	S	3.8765	5.1128	3.9362	5.3808	5.5079	5.6947	5.8345	5.9498	6.0502	6.1405

TABLE B.2. SUPERHEATED STEAM SI UNITS (Continued)

ABS PRESS KPA (SAT TEMP) DEG C		SAT WATER	SAT STEAM	TEMPERATURE, DEG C (TEMPERATURE, K) 500 (773.15)	520 (793.15)	540 (813.15)	560 (833.15)	580 (853.15)	600 (873.15)	625 (898.15)	650 (923.15)
	V	1.668	10.1246	20.479	21.339	22.169	22.975	23.761	24.530	25.472	26.394
15200	U	1593.5	2455.3	2996.6	3040.0	3082.3	3123.7	3164.6	3205.2	3255.5	3305.6
(343.19)	H	1618.9	2609.2	3307.9	3364.4	3419.2	3473.0	3525.8	3578.0	3642.7	3706.8
	S	3.6981	5.3051	6.3399	6.4119	6.4803	6.5455	6.6082	6.6687	6.7417	6.8121
	V	1.689	9.7072	19.871	20.717	21.533	22.324	23.095	23.850	24.772	25.676
15600	U	1608.3	2445.8	2992.6	3036.3	3078.9	3120.6	3161.8	3202.5	3253.0	3303.3
(345.28)	H	1634.7	2597.3	3302.6	3359.5	3414.8	3468.9	3522.0	3574.5	3639.5	3703.8
	S	3.7226	5.2793	6.3225	6.3952	6.4641	6.5298	6.5928	6.6536	6.7270	6.7977
	V	1.710	9.3075	19.293	20.126	20.928	21.706	22.463	23.204	24.108	24.994
16000	U	1623.2	2436.0	2988.4	3032.6	3075.5	3117.5	3158.9	3199.8	3250.5	3301.0
(347.33)	H	1650.5	2584.9	3297.1	3354.6	3410.3	3464.8	3518.3	3571.0	3636.3	3700.9
	S	3.7471	5.2531	6.3054	6.3787	6.4481	6.5143	6.5777	6.6389	6.7125	6.7835
	V	1.733	8.9247	18.743	19.564	20.354	21.118	21.862	22.589	23.476	24.344
16400	U	1638.1	2425.7	2984.3	3028.8	3072.1	3114.3	3155.9	3197.1	3248.0	3298.7
(349.33)	H	1666.5	2572.1	3291.7	3349.7	3405.9	3460.7	3514.5	3567.5	3633.0	3697.9
	S	3.7717	5.2267	6.2885	6.3625	6.4325	6.4991	6.5629	6.6243	6.6983	6.7696
	V	1.757	8.5535	18.219	19.028	19.806	20.558	21.289	22.003	22.875	23.726
16800	U	1653.5	2414.9	2980.1	3025.1	3068.6	3111.2	3153.0	3194.3	3245.5	3296.4
(351.30)	H	1683.0	2558.6	3286.2	3344.7	3401.4	3456.6	3510.7	3564.0	3629.8	3695.0
	S	3.7974	5.1994	6.2718	6.3466	6.4171	6.4841	6.5483	6.6101	6.6844	6.7560
	V	1.783	8.1912	17.719	18.517	19.283	20.024	20.743	21.445	22.301	23.137
17200	U	1669.7	2403.5	2975.9	3021.3	3065.2	3108.0	3150.1	3191.6	3243.0	3294.1
(353.22)	H	1700.4	2544.4	3280.7	3339.8	3396.9	3452.4	3506.9	3560.5	3626.6	3692.0
	S	3.8240	5.1713	6.2554	6.3308	6.4019	6.4694	6.5340	6.5961	6.6708	6.7426
	V	1.810	7.8394	17.241	18.029	18.785	19.514	20.222	20.912	21.753	22.574
17600	U	1685.8	2391.6	2971.7	3017.5	3061.7	3104.8	3147.1	3188.9	3240.5	3291.7
(355.11)	H	1717.6	2529.5	3275.2	3334.8	3392.3	3448.3	3503.1	3556.9	3623.4	3689.0
	S	3.8504	5.1425	6.2392	6.3153	6.3870	6.4550	6.5199	6.5824	6.6574	6.7295
	V	1.840	7.4977	16.785	17.563	18.308	19.027	19.724	20.403	21.230	22.037
18000	U	1701.7	2378.9	2967.4	3013.6	3058.2	3101.6	3144.2	3186.1	3238.0	3289.4
(356.96)	H	1734.8	2513.9	3269.6	3329.8	3387.8	3444.1	3499.2	3553.4	3620.1	3686.1
	S	3.8765	5.1128	6.2232	6.3000	6.3722	6.4407	6.5061	6.5688	6.6442	6.7166

TABLE B.2. SUPERHEATED STEAM SI UNITS (Continued)

ABS PRESS KPA (SAT TEMP) DEG C		SAT WATER	SAT STEAM	TEMPERATURE, DEG C (TEMPERATURE, K) 380 (653.15)	390 (663.15)	400 (673.15)	410 (683.15)	420 (693.15)	440 (713.15)	460 (733.15)	480 (753.15)
18400 (358.77)	V	1.872	7.1647	9.9544	10.782	11.493	12.125	12.701	13.740	14.675	15.538
	U	1717.7	2365.6	2564.2	2619.5	2665.7	2705.9	2741.9	2805.6	2862.2	2914.3
	H	1752.1	2497.4	2747.4	2817.8	2877.2	2929.0	2975.6	3058.4	3132.2	3200.2
	S	3.9028	5.0820	5.4722	5.5794	5.6682	5.7446	5.8124	5.9302	6.0323	6.1237
18800 (360.55)	V	1.907	6.8386	9.5146	10.365	11.087	11.725	12.304	13.341	14.271	15.127
	U	1733.8	2351.3	2548.3	2606.8	2655.2	2696.8	2733.9	2799.1	2856.6	2909.4
	H	1769.7	2479.8	2727.2	2801.7	2863.6	2917.3	2965.2	3049.9	3124.9	3193.8
	S	3.9294	5.0498	5.4354	5.5486	5.6413	5.7205	5.7902	5.9106	6.0144	6.1071
19200 (362.30)	V	1.946	6.5173	9.0841	9.9593	10.695	11.340	11.922	12.959	13.884	14.733
	U	1750.4	2335.9	2531.6	2593.7	2644.3	2687.6	2725.8	2792.5	2851.0	2904.5
	H	1787.8	2461.1	2706.0	2784.9	2849.7	2905.3	2954.7	3041.3	3117.6	3187.4
	S	3.9566	5.0160	5.3973	5.5172	5.6141	5.6962	5.7679	5.8912	5.9967	6.0906
19600 (364.02)	V	1.989	6.1978	8.6617	9.5649	10.315	10.968	11.553	12.590	13.511	14.355
	U	1767.6	2319.2	2514.0	2580.0	2633.1	2678.0	2717.5	2785.8	2845.3	2899.5
	H	1806.5	2440.7	2683.7	2767.5	2835.3	2893.0	2943.9	3032.5	3110.2	3180.9
	S	3.9849	4.9801	5.3577	5.4850	5.5866	5.6717	5.7456	5.8717	5.9791	6.0743
20000 (365.70)	V	2.037	5.8766	8.2458	9.1805	9.9470	10.608	11.197	12.236	13.154	13.991
	U	1785.7	2300.8	2495.2	2565.7	2621.6	2668.3	2709.0	2778.9	2839.6	2894.5
	H	1826.5	2418.4	2660.2	2749.3	2820.5	2880.4	2932.9	3023.7	3102.7	3174.4
	S	4.0149	4.9412	5.3165	5.4520	5.5585	5.6470	5.7232	5.8523	5.9616	6.0581
20400 (367.36)	V	2.093	5.5484	7.8352	8.8052	9.5894	10.260	10.854	11.895	12.809	13.642
	U	1805.4	2280.1	2475.3	2550.8	2609.6	2658.3	2700.3	2772.0	2833.8	2889.5
	H	1848.1	2393.3	2635.2	2730.4	2805.2	2867.6	2921.7	3014.7	3095.1	3167.8
	S	4.0474	4.8984	5.2733	5.4181	5.5300	5.6220	5.7007	5.8330	5.9442	6.0421
20800 (368.98)	V	2.161	5.2050	7.4179	8.4380	9.2417	9.9225	10.522	11.566	12.478	13.305
	U	1827.6	2256.0	2453.4	2535.2	2597.2	2648.0	2691.4	2765.0	2827.9	2884.4
	H	1872.5	2364.2	2607.7	2710.7	2789.4	2854.3	2910.3	3005.6	3087.4	3161.2
	S	4.0841	4.8498	5.2265	5.3831	5.5010	5.5967	5.6781	5.8136	5.9269	6.0261
21200 (370.58)	V	2.249	4.8314	6.9956	8.0781	8.9032	9.5953	10.201	11.248	12.158	12.981
	U	1853.9	2226.5	2428.9	2518.8	2584.4	2637.4	2682.4	2757.9	2822.0	2879.3
	H	1901.5	2328.9	2577.2	2690.1	2773.1	2840.8	2898.6	2996.4	3079.7	3154.5
	S	4.1279	4.7917	5.1754	5.3471	5.4714	5.5712	5.6553	5.7943	5.9096	6.0103

TABLE B.2. SUPERHEATED STEAM SI UNITS (Continued)

ABS PRESS KPA (SAT TEMP) DEG C		SAT WATER	SAT STEAM	500 (773.15)	520 (793.15)	540 (813.15)	560 (833.15)	580 (853.15)	600 (873.15)	625 (898.15)	650 (923.15)
				TEMPERATURE, DEG C (TEMPERATURE, K)							
18400 (358.77)	V	1.872	7.1647	16.348	17.116	17.852	18.560	19.247	19.915	20.729	21.523
	U	1717.7	2365.6	2963.2	3009.8	3054.7	3098.4	3141.2	3183.4	3235.5	3287.1
	H	1752.1	2497.4	3263.9	3324.7	3383.2	3439.9	3495.4	3549.8	3616.9	3683.1
	S	3.9028	5.0820	6.2073	6.2849	6.3577	6.4267	6.4924	6.5555	6.6312	6.7040
18800 (360.55)	V	1.907	6.8386	15.929	16.689	17.415	18.114	18.791	19.449	20.250	21.030
	U	1733.8	2351.3	2958.8	3005.9	3051.2	3095.2	3138.3	3180.6	3232.9	3284.7
	H	1769.7	2479.8	3258.3	3319.6	3378.6	3435.7	3491.5	3546.3	3613.6	3680.1
	S	3.9294	5.0498	6.1916	6.2700	6.3434	6.4128	6.4790	6.5424	6.6185	6.6915
19200 (362.30)	V	1.946	6.5173	15.527	16.279	16.996	17.686	18.353	19.002	19.791	20.559
	U	1750.4	2335.9	2954.5	3002.0	3047.6	3091.9	3135.3	3177.9	3230.4	3282.4
	H	1787.8	2461.1	3252.6	3314.5	3374.0	3431.5	3487.6	3542.7	3610.4	3677.1
	S	3.9566	5.0160	6.1761	6.2552	6.3292	6.3992	6.4657	6.5295	6.6059	6.6793
19600 (364.02)	V	1.989	6.1978	15.142	15.885	16.594	17.275	17.934	18.573	19.350	20.106
	U	1767.6	2319.2	2950.1	2998.0	3044.1	3088.7	3132.3	3175.1	3227.8	3280.1
	H	1806.5	2440.7	3246.9	3309.4	3369.3	3427.3	3483.8	3539.1	3607.1	3674.1
	S	3.9849	4.9801	6.1608	6.2406	6.3153	6.3857	6.4527	6.5168	6.5936	6.6672
20000 (365.70)	V	2.037	5.8766	14.771	15.507	16.208	16.881	17.531	18.161	18.927	19.672
	U	1785.7	2300.8	2945.7	2994.1	3040.5	3085.4	3129.3	3172.3	3225.3	3277.7
	H	1826.5	2418.4	3241.1	3304.2	3364.7	3423.0	3479.9	3535.5	3603.8	3671.1
	S	4.0149	4.9412	6.1456	6.2262	6.3015	6.3724	6.4398	6.5043	6.5814	6.6554
20400 (367.36)	V	2.093	5.5484	14.415	15.144	15.838	16.502	17.144	17.766	18.521	19.254
	U	1805.4	2280.1	2941.2	2990.1	3036.9	3082.1	3126.2	3169.5	3222.7	3275.4
	H	1848.1	2393.3	3235.3	3299.0	3360.0	3418.8	3476.0	3531.9	3600.6	3668.1
	S	4.0474	4.8984	6.1305	6.2120	6.2878	6.3593	6.4271	6.4919	6.5694	6.6437
20800 (368.98)	V	2.161	5.2050	14.073	14.795	15.481	16.138	16.771	17.385	18.130	18.853
	U	1827.6	2256.0	2936.7	2986.1	3033.3	3078.8	3123.2	3166.7	3220.2	3273.0
	H	1872.5	2364.2	3229.5	3293.8	3355.3	3414.5	3472.1	3528.3	3597.3	3665.1
	S	4.0841	4.8498	6.1156	6.1978	6.2744	6.3463	6.4146	6.4798	6.5576	6.6322
21200 (370.58)	V	2.249	4.8314	13.743	14.459	15.138	15.787	16.413	17.019	17.754	18.467
	U	1853.9	2226.5	2932.2	2982.1	3029.6	3075.5	3120.2	3163.9	3217.6	3270.6
	H	1901.5	2328.9	3223.6	3288.6	3350.6	3410.2	3468.1	3524.7	3594.0	3662.1
	S	4.1279	4.7917	6.1008	6.1838	6.2610	6.3335	6.4022	6.4677	6.5460	6.6208

TABLE B.2. SUPERHEATED STEAM SI UNITS (Continued)

ABS PRESS KPA (SAT TEMP) DEG C		SAT WATER	SAT STEAM	TEMPERATURE, DEG C (TEMPERATURE, K) 380 (653.15)	390 (663.15)	400 (673.15)	410 (683.15)	420 (693.15)	440 (713.15)	460 (733.15)	480 (753.15)
21600 (372.15)	V	2.379	4.3918	6.5629	7.7248	8.5732	9.2777	9.8898	10.941	11.850	12.669
	U	1888.6	2186.7	2401.5	2501.7	2571.1	2626.5	2673.1	2750.7	2815.9	2874.1
	H	1940.0	2281.5	2543.2	2668.5	2756.2	2826.9	2886.7	2987.0	3071.9	3147.8
	S	4.1861	4.7154	5.1192	5.3098	5.4411	5.5453	5.6323	5.7750	5.8924	5.9945
22000 (373.69)	V	2.671	3.7278	6.1105	7.3771	8.2510	8.9689	9.5883	10.645	11.552	12.368
	U	1952.4	2113.6	2370.0	2483.6	2557.2	2615.2	2663.6	2743.3	2809.9	2868.9
	H	2011.1	2195.6	2504.4	2645.9	2738.8	2812.6	2874.6	2977.5	3064.0	3141.0
	S	4.2947	4.5799	5.0559	5.2711	5.4102	5.5190	5.6091	5.7556	5.8753	5.9789
22120 (374.15)	V	3.170	3.1700	5.9689	7.2738	8.1558	8.8779	9.4995	10.558	11.465	12.279
	U	2037.3	2037.3	2359.4	2478.0	2553.0	2611.8	2660.7	2741.1	2808.0	2867.3
	H	2107.4	2107.4	2491.5	2638.9	2733.4	2808.2	2870.9	2974.6	3061.6	3139.0
	S	4.4429	4.4429	5.0350	5.2593	5.4007	5.5110	5.6021	5.7498	5.8702	5.9742

ABS PRESS KPA (SAT TEMP) DEG C		SAT WATER	SAT STEAM	TEMPERATURE, DEG C (TEMPERATURE, K) 500 (773.15)	520 (793.15)	540 (813.15)	560 (833.15)	580 (853.15)	600 (873.15)	625 (898.15)	650 (923.15)
21600 (372.15)	V	2.379	4.3918	13.425	14.135	14.807	15.450	16.068	16.667	17.392	18.095
	U	1888.6	2186.7	2927.7	2978.0	3026.0	3072.2	3117.1	3161.1	3215.0	3268.3
	H	1940.0	2281.5	3217.7	3283.3	3345.8	3405.9	3464.2	3521.1	3590.7	3659.1
	S	4.1861	4.7154	6.0862	6.1700	6.2478	6.3208	6.3900	6.4559	6.5345	6.6096
22000 (373.69)	V	2.671	3.7278	13.119	13.822	14.488	15.124	15.736	16.327	17.043	17.737
	U	1952.4	2113.6	2923.1	2973.9	3022.3	3068.9	3114.0	3158.2	3212.4	3265.9
	H	2011.1	2195.6	3211.7	3278.0	3341.0	3401.6	3460.2	3517.4	3587.4	3656.1
	S	4.2947	4.5799	6.0716	6.1563	6.2347	6.3083	6.3779	6.4441	6.5231	6.5986
22120 (374.15)	V	3.170	3.1700	13.029	13.731	14.395	15.029	15.638	16.228	16.941	17.632
	U	2037.3	2037.3	2921.7	2972.7	3021.2	3067.8	3113.1	3157.4	3211.7	3265.2
	H	2107.4	2107.4	3209.9	3276.4	3339.6	3400.3	3459.0	3516.3	3586.4	3655.2
	S	4.4429	4.4429	6.0672	6.1522	6.2309	6.3046	6.3743	6.4406	6.5198	6.5953

APPENDIX

C

Physical Properties for Selected Compounds

Table C.1. Ideal Gas Heat Capacities (250 K - 1500 K)

$$C_P^* = a + bT + cT^2 + dT^3 \quad (T : \mathrm{K} \; ; \; C_P^* : \mathrm{J\,mol^{-1}\,K^{-1}})$$

Compound	a	b	c	d
Methane	2.328E+1	3.520E-2	3.270E-5	-1.836E-8
Ethane	8.582E+0	1.669E-1	-5.779E-5	4.851E-9
Propane	1.837E+0	2.816E-1	-1.319E-4	2.316E-8
n-Butane	1.034E+1	3.354E-1	-1.203E-4	1.394E-9
i-Butane	-5.677E+0	4.112E-1	-2.287E-4	5.102E-8
Ethylene	9.147E+0	1.346E-1	-5.717E-5	8.056E-9
Propylene	8.256E+0	2.174E-1	-9.692E-5	1.548E-8
Acetylene	2.121E+1	9.484E-2	-6.861E-5	1.946E-8
Benzene	-3.283E+1	4.711E-1	-3.005E-4	7.253E-8
Toluene	-2.742E+1	5.343E-1	-3.183E-4	7.205E-8
Methanol	2.114E+1	7.084E-2	2.586E-5	-2.850E-8
Ethanol	1.332E+1	1.971E-1	-6.454E-5	-5.224E-9
Acetone	1.352E+1	2.386E-1	-1.057E-4	1.606E-8
Ammonia	2.873E+1	1.798E-2	2.394E-5	-1.424E-8
H_2S	3.096E+1	7.683E-3	1.337E-5	-6.293E-9
O_2	2.866E+1	-2.380E-3	2.008E-5	-1.150E-8
N_2	3.081E+1	-1.255E-2	2.575E-5	-1.133E-8
Cl_2	2.638E+1	3.571E-2	-4.063E-5	1.609E-8
H_2	2.836E+1	4.943E-3	-9.201E-6	6.142E-9
HCl	3.010E+1	-6.692E-3	1.217E-5	-3.978E-9
NO	2.992E+1	-4.032E-3	1.394E-5	-5.868E-9
NO_2	2.532E+1	4.556E-2	-1.819E-5	-5.308E-10
SO_2	2.476E+1	6.380E-2	-4.643E-5	1.231E-8
SO_3	2.054E+1	1.295E-1	-1.038E-4	2.957E-8
CO	3.064E+1	-1.214E-2	2.713E-5	-1.247E-8
CO_2	1.973E+1	7.356E-2	-5.618E-5	1.722E-8
H_2O	3.224E+1	1.908E-3	1.057E-5	-3.602E-9

Table C.2. Solid Heat Capacities (Daubert and Danner)[1]

$$C_P^* = A + BT + CT^2 + DT^3 + ET^4 \quad (T\!: \mathrm{K}\,;\, C_P^*\!: \mathrm{J\,mol^{-1}\,K^{-1}})$$

Compound	A	B	C	D	E
n-Decane	-3.4900E+4	2.5280E+3	-1.2650E+1	2.8480E-2	
n-Dodecane	-4.0970E+4	2.8390E+3	-1.3232E+1	2.7972E-2	
Phenol	-5.9700E+3	1.0380E+3	-6.4670E+0	2.3040E-2	-2.6580E-5
NaCl	3.6710E+4	6.2770E+1	-6.6670E-2	2.8000E-5	
Water	-2.6249E+2	1.4052E+2			
Phosphoric Acid	-6.0200E+3	6.1300E+2	-1.4800E+0	2.3000E-3	

Table C.3. Liquid Heat Capacities[1] (Daubert and Danner)[2]

$$C_P^* = A + BT + CT^2 + DT^3 + ET^4 \quad (T\!: \mathrm{K}\,;\, C_P^*\!: \mathrm{J\,mol^{-1}\,K^{-1}})$$

Compound	A	B	C	D	E
n-Hexane	1.9820E+5	-3.8660E+2	1.2630E+0		
n-Heptane	2.1850E+5	-2.9680E+2	1.0610E+0		
n-Octane	2.4145E+5	-2.7780E+2	1.0620E+0		
Benzene	1.2944E+5	-1.6950E+2	6.4781E-1		
Toluene	1.4014E+5	-1.5230E+2	6.9500E-1		
Methanol	1.0580E+5	-3.6223E+2	9.3790E-1		
Ethanol	1.0264E+5	-1.3963E+2	-3.0341E-2	2.0386E-3	
Acetone	1.3560E+5	-1.7700E+2	2.8370E-1	6.8900E-4	
MEK	1.3230E+5	2.0087E+2	-9.5790E-1	1.9533E-3	
Ammonia	3.0094E+6	-4.3692E+4	2.4114E+2	-5.8560E-1	5.2953E-4
Water	2.7637E+5	2.0901E+3	8.1250E-0	-1.4116E-2	9.3701E-6

[1] At 101325 Pa (1 atm) when below the normal boiling point; at saturation pressure above it.

[2] Daubert, T.E., Danner, R.P. (eds.), *Physical and Thermodynamic Properties of Pure Chemicals, Data Compilation*, Hemisphere Publishing Corp., New York (extant 1992).

Table C.4. Critical Property and Omega Values

Compound	T_c (K)	P_c (bar)	z_c	ω
Methane	190.58	46.04	0.288	0.011
Ethane	305.33	48.71	0.284	0.099
Propane	369.82	42.49	0.280	0.152
n-Butane	425.18	37.97	0.274	0.193
i-Butane	408.10	36.48	0.283	0.176
n-Pentane	469.70	33.69	0.263	0.251
n-Hexane	507.30	30.12	0.262	0.299
n-Heptane	540.10	27.36	0.263	0.350
n-Octane	568.70	24.87	0.259	0.397
Benzene	562.16	48.98	0.271	0.211
Toluene	591.79	41.04	0.264	0.264
Methanol	512.64	80.92	0.224	0.564
Ethanol	513.92	61.37	0.240	0.635
i-Propanol	508.30	47.64	0.248	0.669
CO_2	304.19	73.81	0.274	0.228
SO_2	430.75	78.83	0.269	0.246
N_2	126.15	33.94	0.287	0.045
O_2	154.77	50.80	0.289	0.019
H_2S	373.55	90.07	0.282	0.100
NH_3	405.55	112.77	0.242	0.255
H_2O	647.14	220.64	0.230	0.344

Table C.5. Antoine Parameter Values

$$\ln P^s = A - \frac{B}{(C+T)}. \quad (P^s: \text{mmHg};\ T: \text{K})$$

Compound	A	B	C	T Range (K)
Ethane	15.6937	1520.15	-16.76	130 - 230
Propane	15.7863	1892.47	-24.33	180 - 280
n-Butane	15.7374	2151.63	-36.24	220 - 310
i-Butane	15.4481	1989.35	-36.31	210 - 310
n-Pentane	15.6663	2374.19	-46.24	270 - 350
n-Hexane	15.8366	2697.55	-48.78	245 - 370
n-Heptane	15.8737	2911.32	-56.51	270 - 400
n-Octane	16.1427	3220.75	-60.02	360 - 430
Propylene	15.7240	1821.01	-24.90	180 - 270
Benzene	15.8647	2769.42	-53.26	300 - 400
Toluene	15.9808	3076.65	-54.65	330 - 430
Methanol	18.3366	3477.90	-40.53	290 - 380
Ethanol	18.5645	3615.06	-48.60	300 - 380
n-Propanol	17.5026	3145.72	-80.95	320 - 410
i-Propanol	18.6867	3636.82	-53.66	300 - 400
Acetone	16.5145	2850.59	-40.82	290 - 370
Ammonia	17.1902	2227.37	-28.74	200 - 270
H_2O	18.3104	3826.36	-45.47	290 - 500
CCl_4	15.8572	2799.61	-46.37	290 - 380

APPENDIX

D

Lee - Kesler Tables

Table D.1a&b	**Compressibility Factor**
Table D.2a&b	**Residual Enthalpy**
Table D.3a&b	**Residual Entropy**
Table D.4a&b	**Fugacity Coefficient**

Table D.1a Compressibility Factor

Values of $z^{(0)}$

T_r	P_r 0.010	0.050	0.100	0.200	0.400	0.600	0.800
0.30	0.0029	0.0145	0.0290	0.0579	0.1158	0.1737	0.2315
0.35	0.0026	0.0130	0.0261	0.0522	0.1043	0.1564	0.2084
0.40	0.0024	0.0119	0.0239	0.0477	0.0953	0.1429	0.1904
0.45	0.0022	0.0110	0.0221	0.0442	0.0882	0.1322	0.1762
0.50	0.0021	0.0103	0.0207	0.0413	0.0825	0.1236	0.1647
0.55	0.9804	0.0098	0.0195	0.0390	0.0778	0.1166	0.1553
0.60	0.9849	0.0093	0.0186	0.0371	0.0741	0.1109	0.1476
0.65	0.9881	0.9377	0.0178	0.0356	0.0710	0.1063	0.1415
0.70	0.9904	0.9504	0.8958	0.0344	0.0687	0.1027	0.1366
0.75	0.9922	0.9598	0.9165	0.0336	0.0670	0.1001	0.1330
0.80	0.9935	0.9669	0.9319	0.8539	0.0661	0.0985	0.1307
0.85	0.9946	0.9725	0.9436	0.8810	0.0661	0.0983	0.1301
0.90	0.9954	0.9768	0.9528	0.9015	0.7800	0.1006	0.1321
0.93	0.9959	0.9790	0.9573	0.9115	0.8059	0.6635	0.1359
0.95	0.9961	0.9803	0.9600	0.9174	0.8206	0.6967	0.1410
0.97	0.9963	0.9815	0.9625	0.9227	0.8338	0.7240	0.5580
0.98	0.9965	0.9821	0.9637	0.9253	0.8398	0.7360	0.5887
0.99	0.9966	0.9826	0.9648	0.9277	0.8455	0.7471	0.6138
1.00	0.9967	0.9832	0.9659	0.9300	0.8509	0.7574	0.6353
1.01	0.9968	0.9837	0.9669	0.9322	0.8561	0.7671	0.6542
1.02	0.9969	0.9842	0.9679	0.9343	0.8610	0.7761	0.6710
1.05	0.9971	0.9855	0.9707	0.9401	0.8743	0.8002	0.7130
1.10	0.9975	0.9874	0.9747	0.9485	0.8930	0.8323	0.7649
1.15	0.9978	0.9891	0.9780	0.9554	0.9081	0.8576	0.8032
1.20	0.9981	0.9904	0.9808	0.9611	0.9205	0.8779	0.8330
1.30	0.9985	0.9926	0.9852	0.9702	0.9396	0.9083	0.8764
1.40	0.9988	0.9942	0.9884	0.9768	0.9534	0.9298	0.9062
1.50	0.9991	0.9954	0.9909	0.9818	0.9636	0.9456	0.9278
1.60	0.9993	0.9964	0.9928	0.9856	0.9714	0.9575	0.9439
1.70	0.9994	0.9971	0.9943	0.9886	0.9775	0.9667	0.9563
1.80	0.9995	0.9977	0.9955	0.9910	0.9823	0.9739	0.9659
1.90	0.9996	0.9982	0.9964	0.9929	0.9861	0.9796	0.9735
2.00	0.9997	0.9986	0.9972	0.9944	0.9892	0.9842	0.9796
2.20	0.9998	0.9992	0.9983	0.9967	0.9937	0.9910	0.9886
2.40	0.9999	0.9996	0.9991	0.9983	0.9969	0.9957	0.9948
2.60	1.0000	0.9998	0.9997	0.9994	0.9991	0.9990	0.9990
2.80	1.0000	1.0000	1.0001	1.0002	1.0007	1.0013	1.0021
3.00	1.0000	1.0002	1.0004	1.0008	1.0018	1.0030	1.0043
3.50	1.0001	1.0004	1.0008	1.0017	1.0035	1.0055	1.0075
4.00	1.0001	1.0005	1.0010	1.0021	1.0043	1.0066	1.0090

P_r							
1.000	1.200	1.500	2.000	3.000	5.000	7.000	10.000
0.2892	0.3470	0.4335	0.5775	0.8648	1.4366	2.0048	2.8507
0.2604	0.3123	0.3901	0.5195	0.7775	1.2902	1.7987	2.5539
0.2379	0.2853	0.3563	0.4744	0.7095	1.1758	1.6373	2.3211
0.2200	0.2638	0.3294	0.4384	0.6551	1.0841	1.5077	2.1338
0.2056	0.2465	0.3077	0.4092	0.6110	1.0094	1.4017	1.9801
0.1939	0.2323	0.2899	0.3853	0.5747	0.9475	1.3137	1.8520
0.1842	0.2207	0.2753	0.3657	0.5446	0.8959	1.2398	1.7440
0.1765	0.2113	0.2634	0.3495	0.5197	0.8526	1.1773	1.6519
0.1703	0.2038	0.2538	0.3364	0.4991	0.8161	1.1241	1.5729
0.1656	0.1981	0.2464	0.3260	0.4823	0.7854	1.0787	1.5047
0.1626	0.1942	0.2411	0.3182	0.4690	0.7598	1.0400	1.4456
0.1614	0.1924	0.2382	0.3132	0.4591	0.7388	1.0071	1.3943
0.1630	0.1935	0.2383	0.3114	0.4527	0.7220	0.9793	1.3496
0.1664	0.1963	0.2405	0.3122	0.4507	0.7138	0.9648	1.3257
0.1705	0.1998	0.2432	0.3138	0.4501	0.7092	0.9561	1.3108
0.1779	0.2055	0.2474	0.3164	0.4504	0.7052	0.9480	1.2968
0.1844	0.2097	0.2503	0.3182	0.4508	0.7035	0.9442	1.2901
0.1959	0.2154	0.2538	0.3204	0.4514	0.7018	0.9406	1.2835
0.2919	0.2237	0.2583	0.3229	0.4522	0.7004	0.9372	1.2772
0.4648	0.2370	0.2640	0.3260	0.4533	0.6991	0.9339	1.2710
0.5146	0.2629	0.2715	0.3297	0.4547	0.6980	0.9307	1.2650
0.6026	0.4437	0.3131	0.3452	0.4604	0.6956	0.9222	1.2481
0.6880	0.5984	0.4580	0.3953	0.4770	0.6950	0.9110	1.2232
0.7443	0.6803	0.5798	0.4760	0.5042	0.6987	0.9033	1.2021
0.7858	0.7363	0.6605	0.5605	0.5425	0.7069	0.8990	1.1844
0.8438	0.8111	0.7624	0.6908	0.6344	0.7358	0.8998	1.1580
0.8827	0.8595	0.8256	0.7753	0.7202	0.7761	0.9112	1.1419
0.9103	0.8933	0.8689	0.8328	0.7887	0.8200	0.9297	1.1339
0.9308	0.9180	0.9000	0.8738	0.8410	0.8617	0.9518	1.1320
0.9463	0.9367	0.9234	0.9043	0.8809	0.8984	0.9745	1.1343
0.9583	0.9511	0.9413	0.9275	0.9118	0.9297	0.9961	1.1391
0.9678	0.9624	0.9552	0.9456	0.9359	0.9557	1.0157	1.1452
0.9754	0.9715	0.9664	0.9599	0.9550	0.9772	1.0328	1.1516
0.9865	0.9847	0.9826	0.9806	0.9827	1.0094	1.0600	1.1635
0.9941	0.9936	0.9935	0.9945	1.0011	1.0313	1.0793	1.1728
0.9993	0.9990	1.0010	1.0040	1.0137	1.0463	1.0926	1.1792
1.0031	1.0042	1.0063	1.0106	1.0223	1.0565	1.1016	1.1830
1.0057	1.0074	1.0101	1.0153	1.0284	1.0635	1.1075	1.1848
1.0097	1.0120	1.0156	1.0221	1.0368	1.0723	1.1138	1.1834
1.0115	1.0140	1.0179	1.0249	1.0401	1.0747	1.1136	1.1773

Table D.1b Compressibility Factor

Values of $z^{(1)}$

T_r	P_r 0.010	0.050	0.100	0.200	0.400	0.600	0.800
0.30	-.0008	-.0040	-.0081	-.0161	-.0323	-.0484	-.0645
0.35	-.0009	-.0046	-.0093	-.0185	-.0370	-.0554	-.0738
0.40	-.0010	-.0048	-.0095	-.0190	-.0380	-.0570	-.0758
0.45	-.0009	-.0047	-.0094	-.0187	-.0374	-.0560	-.0745
0.50	-.0009	-.0045	-.0090	-.0181	-.0360	-.0539	-.0716
0.55	-.0314	-.0043	-.0086	-.0172	-.0343	-.0513	-.0682
0.60	-.0205	-.0041	-.0082	-.0164	-.0326	-.0487	-.0646
0.65	-.0137	-.0772	-.0078	-.0156	-.0309	-.0461	-.0611
0.70	-.0093	-.0507	-.1161	-.0148	-.0294	-.0438	-.0579
0.75	-.0064	-.0339	-.0744	-.0143	-.0282	-.0417	-.0550
0.80	-.0044	-.0228	-.0487	-.1160	-.0272	-.0401	-.0526
0.85	-.0029	-.0152	-.0319	-.0715	-.0268	-.0391	-.0509
0.90	-.0019	-.0099	-.0205	-.0442	-.1118	-.0396	-.0503
0.93	-.0015	-.0075	-.0154	-.0326	-.0763	-.1662	-.0514
0.95	-.0012	-.0062	-.0126	-.0262	-.0589	-.1110	-.0540
0.97	-.0010	-.0050	-.0101	-.0208	-.0450	-.0770	-.1647
0.98	-.0009	-.0044	-.0090	-.0184	-.0390	-.0641	-.1100
0.99	-.0008	-.0039	-.0079	-.0161	-.0335	-.0531	-.0796
1.00	-.0007	-.0034	-.0069	-.0140	-.0285	-.0435	-.0588
1.01	-.0006	-.0030	-.0060	-.0120	-.0240	-.0351	-.0429
1.02	-.0005	-.0026	-.0051	-.0102	-.0198	-.0277	-.0303
1.05	-.0003	-.0015	-.0029	-.0054	-.0092	-.0097	-.0032
1.10	-.0000	0.0000	0.0001	0.0007	0.0038	0.0106	0.0236
1.15	0.0002	0.0011	0.0023	0.0052	0.0127	0.0237	0.0396
1.20	0.0004	0.0019	0.0040	0.0084	0.0190	0.0326	0.0499
1.30	0.0006	0.0030	0.0061	0.0125	0.0267	0.0429	0.0612
1.40	0.0007	0.0036	0.0072	0.0147	0.0306	0.0477	0.0661
1.50	0.0008	0.0039	0.0078	0.0158	0.0323	0.0497	0.0677
1.60	0.0008	0.0040	0.0080	0.0162	0.0330	0.0501	0.0677
1.70	0.0008	0.0040	0.0081	0.0163	0.0329	0.0497	0.0667
1.80	0.0008	0.0040	0.0081	0.0162	0.0325	0.0488	0.0652
1.90	0.0008	0.0040	0.0079	0.0159	0.0318	0.0477	0.0635
2.00	0.0008	0.0039	0.0078	0.0155	0.0310	0.0464	0.0617
2.20	0.0007	0.0037	0.0074	0.0147	0.0293	0.0437	0.0580
2.40	0.0007	0.0035	0.0070	0.0139	0.0276	0.0411	0.0544
2.60	0.0007	0.0033	0.0066	0.0131	0.0260	0.0387	0.0512
2.80	0.0006	0.0031	0.0062	0.0124	0.0245	0.0365	0.0483
3.00	0.0006	0.0029	0.0059	0.0117	0.0232	0.0345	0.0456
3.50	0.0005	0.0026	0.0052	0.0103	0.0204	0.0303	0.0401
4.00	0.0005	0.0023	0.0046	0.0091	0.0182	0.0270	0.0357

P_r							
1.000	1.200	1.500	2.000	3.000	5.000	7.000	10.000
-.0806	-.0966	-.1207	-.1608	-.2407	-.3996	-.5572	-.7915
-.0921	-.1105	-.1379	-.1834	-.2738	-.4523	-.6279	-.8863
-.0946	-.1134	-.1414	-.1879	-.2799	-.4603	-.6365	-.8936
-.0929	-.1113	-.1387	-.1840	-.2734	-.4475	-.6162	-.8606
-.0893	-.1069	-.1330	-.1762	-.2611	-.4253	-.5831	-.8099
-.0849	-.1015	-.1263	-.1669	-.2465	-.3991	-.5446	-.7521
-.0803	-.0960	-.1192	-.1572	-.2312	-.3718	-.5047	-.6929
-.0759	-.0906	-.1123	-.1476	-.2160	-.3447	-.4653	-.6346
-.0718	-.0855	-.1057	-.1385	-.2013	-.3184	-.4270	-.5785
-.0681	-.0808	-.0996	-.1298	-.1872	-.2929	-.3901	-.5250
-.0648	-.0767	-.0940	-.1217	-.1736	-.2682	-.3545	-.4740
-.0622	-.0731	-.0888	-.1138	-.1602	-.2439	-.3201	-.4254
-.0604	-.0701	-.0840	-.1059	-.1463	-.2195	-.2862	-.3788
-.0602	-.0687	-.0810	-.1007	-.1374	-.2045	-.2661	-.3516
-.0607	-.0678	-.0788	-.0967	-.1310	-.1943	-.2526	-.3339
-.0623	-.0669	-.0759	-.0921	-.1240	-.1837	-.2391	-.3163
-.0641	-.0661	-.0740	-.0893	-.1202	-.1783	-.2322	-.3075
-.0680	-.0646	-.0715	-.0861	-.1162	-.1728	-.2254	-.2989
-.0792	-.0609	-.0678	-.0824	-.1118	-.1672	-.2185	-.2902
-.0223	-.0473	-.0621	-.0778	-.1072	-.1615	-.2116	-.2816
-.0062	0.0227	-.0524	-.0722	-.1021	-.1556	-.2047	-.2731
0.0220	0.1059	0.0451	-.0432	-.0838	-.1370	-.1835	-.2476
0.0476	0.0897	0.1630	0.0698	-.0373	-.1021	-.1469	-.2056
0.0625	0.0943	0.1548	0.1667	0.0332	-.0611	-.1084	-.1642
0.0719	0.0991	0.1477	0.1990	0.1095	-.0141	-.0678	-.1231
0.0819	0.1048	0.1420	0.1991	0.2079	0.0875	0.0176	-.0423
0.0857	0.1063	0.1383	0.1894	0.2397	0.1737	0.1008	0.0350
0.0864	0.1055	0.1345	0.1806	0.2433	0.2309	0.1717	0.1058
0.0855	0.1035	0.1303	0.1729	0.2381	0.2631	0.2255	0.1673
0.0838	0.1008	0.1259	0.1658	0.2305	0.2788	0.2628	0.2179
0.0816	0.0978	0.1216	0.1593	0.2224	0.2846	0.2871	0.2576
0.0792	0.0947	0.1173	0.1532	0.2144	0.2848	0.3017	0.2876
0.0767	0.0916	0.1133	0.1476	0.2069	0.2820	0.3097	0.3096
0.0719	0.0857	0.1057	0.1374	0.1932	0.2720	0.3135	0.3355
0.0675	0.0803	0.0989	0.1285	0.1812	0.2602	0.3089	0.3459
0.0634	0.0754	0.0929	0.1207	0.1706	0.2484	0.3009	0.3475
0.0598	0.0711	0.0876	0.1138	0.1613	0.2372	0.2915	0.3443
0.0565	0.0672	0.0828	0.1076	0.1529	0.2268	0.2817	0.3385
0.0497	0.0591	0.0728	0.0949	0.1356	0.2042	0.2584	0.3194
0.0443	0.0527	0.0651	0.0849	0.1219	0.1857	0.2378	0.2994

Table D.2a Residual Enthalpy

Values of $\left(\dfrac{H^\circ - H}{RT_c}\right)^{[0]}$

T_r	P_r 0.010	0.050	0.100	0.200	0.400	0.600	0.800
0.30	6.045	6.043	6.040	6.034	6.022	6.011	5.999
0.35	5.906	5.904	5.901	5.895	5.882	5.870	5.858
0.40	5.763	5.761	5.757	5.751	5.738	5.726	5.713
0.45	5.615	5.612	5.609	5.603	5.590	5.577	5.564
0.50	5.465	5.462	5.459	5.453	5.440	5.427	5.414
0.55	0.032	5.312	5.309	5.303	5.290	5.277	5.265
0.60	0.027	5.162	5.159	5.153	5.141	5.129	5.116
0.65	0.023	0.118	5.008	5.002	4.991	4.980	4.968
0.70	0.020	0.101	0.213	4.848	4.839	4.828	4.818
0.75	0.017	0.088	0.183	4.687	4.679	4.672	4.664
0.80	0.015	0.078	0.160	0.345	4.507	4.504	4.499
0.85	0.014	0.069	0.141	0.300	4.308	4.313	4.316
0.90	0.012	0.062	0.126	0.264	0.596	4.074	4.094
0.93	0.011	0.058	0.118	0.246	0.545	0.960	3.920
0.95	0.011	0.056	0.113	0.235	0.516	0.885	3.763
0.97	0.011	0.054	0.109	0.225	0.490	0.824	1.356
0.98	0.010	0.053	0.107	0.221	0.478	0.797	1.273
0.99	0.010	0.052	0.105	0.216	0.466	0.773	1.206
1.00	0.010	0.051	0.103	0.212	0.455	0.750	1.151
1.01	0.010	0.050	0.101	0.208	0.445	0.728	1.102
1.02	0.010	0.049	0.099	0.203	0.434	0.708	1.060
1.05	0.009	0.046	0.094	0.192	0.407	0.654	0.955
1.10	0.008	0.042	0.086	0.175	0.367	0.581	0.827
1.15	0.008	0.039	0.079	0.160	0.334	0.523	0.732
1.20	0.007	0.036	0.073	0.148	0.305	0.474	0.657
1.30	0.006	0.031	0.063	0.127	0.259	0.399	0.545
1.40	0.005	0.027	0.055	0.110	0.224	0.341	0.463
1.50	0.005	0.024	0.048	0.097	0.196	0.297	0.400
1.60	0.004	0.021	0.043	0.086	0.173	0.261	0.350
1.70	0.004	0.019	0.038	0.076	0.153	0.231	0.309
1.80	0.003	0.017	0.034	0.068	0.137	0.206	0.275
1.90	0.003	0.015	0.031	0.062	0.123	0.185	0.246
2.00	0.003	0.014	0.028	0.056	0.111	0.167	0.222
2.20	0.002	0.012	0.023	0.046	0.092	0.137	0.182
2.40	0.002	0.010	0.019	0.038	0.076	0.114	0.150
2.60	0.002	0.008	0.016	0.032	0.064	0.095	0.125
2.80	0.001	0.007	0.014	0.027	0.054	0.080	0.105
3.00	0.001	0.006	0.011	0.023	0.045	0.067	0.088
3.50	0.001	0.004	0.007	0.015	0.029	0.043	0.056
4.00	0.000	0.002	0.005	0.009	0.017	0.026	0.033

P_r							
1.000	1.200	1.500	2.000	3.000	5.000	7.000	10.000
5.987	5.975	5.957	5.927	5.868	5.748	5.628	5.446
5.845	5.833	5.814	5.783	5.721	5.595	5.469	5.278
5.700	5.687	5.668	5.636	5.572	5.442	5.311	5.113
5.551	5.538	5.519	5.486	5.420	5.288	5.154	4.950
5.401	5.388	5.369	5.336	5.270	5.135	4.999	4.791
5.252	5.239	5.220	5.187	5.121	4.986	4.849	4.638
5.104	5.091	5.073	5.041	4.976	4.842	4.704	4.492
4.956	4.945	4.927	4.896	4.833	4.702	4.565	4.353
4.808	4.797	4.781	4.752	4.693	4.566	4.432	4.221
4.655	4.646	4.632	4.607	4.554	4.434	4.303	4.095
4.494	4.488	4.478	4.459	4.413	4.303	4.178	3.974
4.316	4.316	4.312	4.302	4.269	4.173	4.056	3.857
4.108	4.118	4.127	4.132	4.119	4.043	3.935	3.744
3.953	3.976	4.000	4.020	4.024	3.963	3.863	3.678
3.825	3.865	3.904	3.939	3.958	3.910	3.815	3.634
3.658	3.732	3.796	3.853	3.890	3.856	3.767	3.591
3.544	3.652	3.736	3.806	3.854	3.829	3.743	3.569
3.376	3.558	3.670	3.758	3.818	3.801	3.719	3.548
2.573	3.441	3.598	3.706	3.782	3.774	3.695	3.526
1.796	3.283	3.516	3.652	3.744	3.746	3.671	3.505
1.627	3.039	3.422	3.595	3.705	3.718	3.647	3.484
1.359	2.034	3.030	3.398	3.583	3.632	3.575	3.420
1.120	1.487	2.203	2.965	3.353	3.484	3.453	3.315
0.968	1.239	1.719	2.479	3.091	3.329	3.329	3.211
0.857	1.076	1.443	2.079	2.807	3.166	3.202	3.107
0.698	0.860	1.116	1.560	2.274	2.825	2.942	2.899
0.588	0.716	0.915	1.253	1.857	2.486	2.679	2.692
0.505	0.611	0.774	1.046	1.549	2.175	2.421	2.486
0.440	0.531	0.667	0.894	1.318	1.904	2.177	2.285
0.387	0.466	0.583	0.777	1.139	1.672	1.953	2.091
0.344	0.413	0.515	0.683	0.996	1.476	1.751	1.908
0.307	0.368	0.458	0.606	0.880	1.309	1.571	1.736
0.276	0.330	0.411	0.541	0.782	1.167	1.411	1.577
0.226	0.269	0.334	0.437	0.629	0.937	1.143	1.295
0.187	0.222	0.275	0.359	0.513	0.761	0.929	1.058
0.155	0.185	0.228	0.297	0.422	0.621	0.756	0.858
0.130	0.154	0.190	0.246	0.348	0.508	0.614	0.689
0.109	0.129	0.159	0.205	0.288	0.415	0.495	0.545
0.069	0.081	0.099	0.127	0.174	0.239	0.270	0.264
0.041	0.048	0.058	0.072	0.095	0.116	0.110	0.061

Table D.2b Residual Enthalpy

Values of $\left(\dfrac{H^{\circ}-H}{RT_c}\right)^{[1]}$

T_r	P_r 0.010	0.050	0.100	0.200	0.400	0.600	0.800
0.30	11.101	11.100	11.098	11.095	11.088	11.081	11.074
0.35	10.652	10.651	10.651	10.649	10.646	10.643	10.640
0.40	10.120	10.120	10.120	10.120	10.120	10.120	10.120
0.45	9.513	9.514	9.514	9.515	9.518	9.520	9.522
0.50	8.867	8.868	8.869	8.871	8.875	8.879	8.883
0.55	0.080	8.213	8.214	8.216	8.222	8.227	8.232
0.60	0.059	7.568	7.570	7.573	7.579	7.585	7.592
0.65	0.045	0.247	6.949	6.952	6.959	6.966	6.973
0.70	0.034	0.185	0.415	6.360	6.366	6.373	6.381
0.75	0.027	0.142	0.306	5.796	5.803	5.809	5.816
0.80	0.021	0.110	0.234	0.542	5.266	5.271	5.277
0.85	0.017	0.087	0.182	0.401	4.753	4.754	4.758
0.90	0.014	0.070	0.144	0.308	0.751	4.254	4.248
0.93	0.012	0.061	0.126	0.265	0.612	1.236	3.941
0.95	0.011	0.056	0.115	0.241	0.542	0.994	3.737
0.97	0.010	0.052	0.105	0.219	0.483	0.837	1.616
0.98	0.010	0.050	0.101	0.209	0.457	0.776	1.324
0.99	0.009	0.048	0.097	0.200	0.433	0.722	1.154
1.00	0.009	0.046	0.093	0.191	0.410	0.675	1.034
1.01	0.009	0.044	0.089	0.183	0.389	0.632	0.940
1.02	0.008	0.042	0.085	0.175	0.370	0.594	0.863
1.05	0.007	0.037	0.075	0.153	0.318	0.498	0.691
1.10	0.006	0.030	0.061	0.123	0.251	0.381	0.507
1.15	0.005	0.025	0.050	0.099	0.199	0.296	0.385
1.20	0.004	0.020	0.040	0.080	0.158	0.232	0.297
1.30	0.003	0.013	0.026	0.052	0.100	0.142	0.177
1.40	0.002	0.008	0.016	0.032	0.060	0.083	0.100
1.50	0.001	0.005	0.009	0.018	0.032	0.042	0.048
1.60	0.000	0.002	0.004	0.007	0.012	0.013	0.011
1.70	0.000	0.000	0.000	-0.000	-0.003	-0.009	-0.017
1.80	-0.000	-0.001	-0.003	-0.006	-0.015	-0.025	-0.037
1.90	-0.001	-0.003	-0.005	-0.011	-0.023	-0.037	-0.053
2.00	-0.001	-0.004	-0.007	-0.015	-0.030	-0.047	-0.065
2.20	-0.001	-0.005	-0.010	-0.020	-0.040	-0.062	-0.083
2.40	-0.001	-0.006	-0.012	-0.023	-0.047	-0.071	-0.095
2.60	-0.001	-0.006	-0.013	-0.026	-0.052	-0.078	-0.104
2.80	-0.001	-0.007	-0.014	-0.027	-0.055	-0.082	-0.110
3.00	-0.001	-0.007	-0.014	-0.029	-0.058	-0.086	-0.114
3.50	-0.002	-0.008	-0.016	-0.031	-0.062	-0.092	-0.122
4.00	-0.002	-0.008	-0.016	-0.032	-0.064	-0.096	-0.127

P_r							
1.000	1.200	1.500	2.000	3.000	5.000	7.000	10.000
11.067	11.061	11.051	11.034	11.001	10.936	10.873	10.782
10.637	10.634	10.630	10.623	10.610	10.584	10.561	10.529
10.120	10.120	10.120	10.121	10.122	10.127	10.135	10.150
9.525	9.527	9.531	9.537	9.550	9.579	9.611	9.663
8.887	8.891	8.897	8.908	8.931	8.979	9.030	9.111
8.238	8.243	8.252	8.266	8.296	8.359	8.426	8.531
7.598	7.605	7.615	7.632	7.668	7.744	7.825	7.950
6.980	6.987	6.999	7.018	7.059	7.147	7.239	7.383
6.388	6.396	6.408	6.430	6.475	6.573	6.677	6.837
5.824	5.832	5.845	5.868	5.918	6.027	6.141	6.317
5.285	5.292	5.306	5.330	5.384	5.506	5.632	5.824
4.763	4.771	4.784	4.810	4.871	5.008	5.149	5.358
4.249	4.255	4.268	4.298	4.371	4.530	4.688	4.916
3.934	3.937	3.951	3.987	4.073	4.251	4.422	4.662
3.713	3.713	3.730	3.773	3.873	4.068	4.248	4.498
3.471	3.467	3.492	3.551	3.670	3.886	4.077	4.336
3.332	3.327	3.363	3.434	3.568	3.795	3.992	4.257
3.164	3.164	3.222	3.313	3.464	3.705	3.908	4.178
2.385	2.952	3.065	3.186	3.358	3.615	3.825	4.100
1.375	2.595	2.880	3.051	3.251	3.525	3.743	4.023
1.180	1.723	2.650	2.906	3.142	3.435	3.660	3.947
0.877	0.878	1.496	2.381	2.800	3.167	3.418	3.722
0.617	0.673	0.617	1.261	2.167	2.720	3.023	3.362
0.459	0.503	0.487	0.604	1.497	2.275	2.641	3.019
0.349	0.381	0.381	0.361	0.934	1.840	2.273	2.692
0.203	0.218	0.218	0.178	0.300	1.066	1.592	2.086
0.111	0.115	0.108	0.070	0.044	0.504	1.012	1.547
0.049	0.046	0.032	-0.008	-0.078	0.142	0.556	1.080
0.005	-0.004	-0.023	-0.065	-0.151	-0.082	0.217	0.689
-0.027	-0.040	-0.063	-0.109	-0.202	-0.223	-0.028	0.369
-0.051	-0.067	-0.094	-0.143	-0.241	-0.317	-0.203	0.112
-0.070	-0.088	-0.117	-0.169	-0.271	-0.381	-0.330	-0.092
-0.085	-0.105	-0.136	-0.190	-0.295	-0.428	-0.424	-0.255
-0.106	-0.128	-0.163	-0.221	-0.331	-0.493	-0.551	-0.489
-0.120	-0.144	-0.181	-0.242	-0.357	-0.535	-0.631	-0.645
-0.130	-0.156	-0.194	-0.257	-0.376	-0.567	-0.687	-0.754
-0.137	-0.164	-0.204	-0.269	-0.391	-0.591	-0.729	-0.836
-0.142	-0.170	-0.211	-0.278	-0.403	-0.611	-0.763	-0.899
-0.152	-0.181	-0.224	-0.294	-0.425	-0.650	-0.827	-1.015
-0.158	-0.188	-0.233	-0.306	-0.442	-0.680	-0.874	-1.097

Table D.3a Residual Entropy

Values of $\left(\frac{S^\circ - S}{R}\right)^{[0]}$

	P_r						
T_r	0.010	0.050	0.100	0.200	0.400	0.600	0.800
0.30	11.613	10.008	9.319	8.635	7.961	7.574	7.304
0.35	11.185	9.579	8.890	8.205	7.529	7.140	6.869
0.40	10.802	9.196	8.506	7.821	7.144	6.755	6.483
0.45	10.453	8.847	8.158	7.472	6.795	6.405	6.132
0.50	10.137	8.531	7.842	7.156	6.479	6.089	5.816
0.55	0.038	8.245	7.555	6.870	6.193	5.803	5.531
0.60	0.029	7.983	7.294	6.610	5.933	5.544	5.273
0.65	0.023	0.122	7.052	6.368	5.694	5.306	5.036
0.70	0.018	0.096	0.206	6.140	5.467	5.082	4.814
0.75	0.015	0.078	0.164	5.917	5.248	4.866	4.600
0.80	0.013	0.064	0.134	0.294	5.026	4.649	4.388
0.85	0.011	0.054	0.111	0.239	4.785	4.418	4.166
0.90	0.009	0.046	0.094	0.199	0.463	4.145	3.912
0.93	0.008	0.042	0.085	0.179	0.408	0.750	3.723
0.95	0.008	0.039	0.080	0.168	0.377	0.671	3.556
0.97	0.007	0.037	0.075	0.157	0.350	0.607	1.056
0.98	0.007	0.036	0.073	0.153	0.337	0.580	0.971
0.99	0.007	0.035	0.071	0.148	0.326	0.555	0.903
1.00	0.007	0.034	0.069	0.144	0.315	0.532	0.847
1.01	0.007	0.033	0.067	0.139	0.304	0.510	0.799
1.02	0.006	0.032	0.065	0.135	0.294	0.491	0.757
1.05	0.006	0.030	0.060	0.124	0.267	0.439	0.656
1.10	0.005	0.026	0.053	0.108	0.230	0.371	0.537
1.15	0.005	0.023	0.047	0.096	0.201	0.319	0.452
1.20	0.004	0.021	0.042	0.085	0.177	0.277	0.389
1.30	0.003	0.017	0.033	0.068	0.140	0.217	0.298
1.40	0.003	0.014	0.027	0.056	0.114	0.174	0.237
1.50	0.002	0.011	0.023	0.046	0.094	0.143	0.194
1.60	0.002	0.010	0.019	0.039	0.079	0.120	0.162
1.70	0.002	0.008	0.017	0.033	0.067	0.102	0.137
1.80	0.001	0.007	0.014	0.029	0.058	0.088	0.117
1.90	0.001	0.006	0.013	0.025	0.051	0.076	0.102
2.00	0.001	0.006	0.011	0.022	0.044	0.067	0.089
2.20	0.001	0.004	0.009	0.018	0.035	0.053	0.070
2.40	0.001	0.004	0.007	0.014	0.028	0.042	0.056
2.60	0.001	0.003	0.006	0.012	0.023	0.035	0.046
2.80	0.000	0.002	0.005	0.010	0.020	0.029	0.039
3.00	0.000	0.002	0.004	0.008	0.017	0.025	0.033
3.50	0.000	0.001	0.003	0.006	0.012	0.017	0.023
4.00	0.000	0.001	0.002	0.004	0.009	0.013	0.017

P_r							
1.000	1.200	1.500	2.000	3.000	5.000	7.000	10.000
7.099	6.935	6.740	6.497	6.182	5.847	5.683	5.578
6.663	6.497	6.299	6.052	5.728	5.376	5.194	5.060
6.275	6.109	5.909	5.660	5.330	4.967	4.772	4.619
5.924	5.757	5.557	5.306	4.974	4.603	4.401	4.234
5.608	5.441	5.240	4.989	4.656	4.282	4.074	3.899
5.324	5.157	4.956	4.706	4.373	3.998	3.788	3.607
5.066	4.900	4.700	4.451	4.120	3.747	3.537	3.353
4.830	4.665	4.467	4.220	3.892	3.523	3.315	3.131
4.610	4.446	4.250	4.007	3.684	3.322	3.117	2.935
4.399	4.238	4.046	3.807	3.491	3.138	2.939	2.761
4.191	4.034	3.846	3.615	3.310	2.970	2.777	2.605
3.976	3.825	3.646	3.425	3.135	2.812	2.629	2.463
3.738	3.599	3.434	3.231	2.964	2.663	2.491	2.334
3.569	3.444	3.295	3.108	2.860	2.577	2.412	2.262
3.433	3.326	3.193	3.023	2.790	2.520	2.361	2.215
3.259	3.188	3.081	2.932	2.719	2.463	2.312	2.170
3.142	3.106	3.019	2.884	2.682	2.436	2.287	2.148
2.972	3.010	2.953	2.835	2.646	2.408	2.263	2.126
2.167	2.893	2.879	2.784	2.609	2.380	2.239	2.105
1.391	2.736	2.798	2.730	2.571	2.352	2.215	2.083
1.225	2.495	2.706	2.673	2.533	2.325	2.191	2.062
0.965	1.523	2.328	2.483	2.415	2.242	2.121	2.001
0.742	1.012	1.557	2.081	2.202	2.104	2.007	1.903
0.607	0.790	1.126	1.649	1.968	1.966	1.897	1.810
0.512	0.651	0.890	1.308	1.727	1.827	1.789	1.722
0.385	0.478	0.628	0.891	1.299	1.554	1.581	1.556
0.303	0.372	0.478	0.663	0.990	1.303	1.386	1.402
0.246	0.299	0.381	0.520	0.777	1.088	1.208	1.260
0.204	0.247	0.312	0.421	0.628	0.913	1.050	1.130
0.172	0.208	0.261	0.350	0.519	0.773	0.915	1.013
0.147	0.177	0.222	0.296	0.438	0.661	0.799	0.908
0.127	0.153	0.191	0.255	0.375	0.570	0.702	0.815
0.111	0.134	0.167	0.221	0.325	0.497	0.620	0.733
0.087	0.105	0.130	0.172	0.251	0.388	0.492	0.599
0.070	0.084	0.104	0.138	0.201	0.311	0.399	0.496
0.058	0.069	0.086	0.113	0.164	0.255	0.329	0.416
0.048	0.058	0.072	0.094	0.137	0.213	0.277	0.353
0.041	0.049	0.061	0.080	0.116	0.181	0.236	0.303
0.029	0.034	0.042	0.056	0.081	0.126	0.166	0.216
0.021	0.025	0.031	0.041	0.059	0.093	0.123	0.162

Table D.3b Residual Entropy

Values of $\left(\frac{S^\circ - S}{R}\right)^{[1]}$

T_r	P_r 0.010	0.050	0.100	0.200	0.400	0.600	0.800
0.30	16.790	16.783	16.773	16.753	16.714	16.675	16.637
0.35	15.408	15.402	15.395	15.382	15.355	15.328	15.301
0.40	13.989	13.985	13.980	13.971	13.951	13.932	13.914
0.45	12.562	12.559	12.556	12.549	12.535	12.521	12.508
0.50	11.201	11.198	11.196	11.191	11.181	11.171	11.161
0.55	0.115	9.949	9.947	9.943	9.936	9.928	9.921
0.60	0.078	8.828	8.827	8.823	8.817	8.811	8.806
0.65	0.055	0.309	7.832	7.829	7.824	7.819	7.815
0.70	0.040	0.216	0.491	6.951	6.946	6.941	6.937
0.75	0.029	0.156	0.340	6.173	6.167	6.162	6.158
0.80	0.022	0.116	0.246	0.578	5.474	5.467	5.462
0.85	0.017	0.088	0.183	0.408	4.853	4.841	4.832
0.90	0.013	0.068	0.140	0.301	0.744	4.269	4.250
0.93	0.011	0.058	0.120	0.254	0.593	1.219	3.914
0.95	0.010	0.053	0.109	0.228	0.517	0.961	3.697
0.97	0.010	0.048	0.099	0.206	0.456	0.797	1.570
0.98	0.009	0.046	0.094	0.196	0.429	0.734	1.270
0.99	0.009	0.044	0.090	0.186	0.405	0.680	1.098
1.00	0.008	0.042	0.086	0.177	0.382	0.632	0.977
1.01	0.008	0.040	0.082	0.169	0.361	0.590	0.883
1.02	0.008	0.039	0.078	0.161	0.342	0.552	0.807
1.05	0.007	0.034	0.069	0.140	0.292	0.460	0.642
1.10	0.005	0.028	0.055	0.112	0.229	0.350	0.470
1.15	0.005	0.023	0.045	0.091	0.183	0.275	0.361
1.20	0.004	0.019	0.037	0.075	0.149	0.220	0.286
1.30	0.003	0.013	0.026	0.052	0.102	0.148	0.190
1.40	0.002	0.010	0.019	0.037	0.072	0.104	0.133
1.50	0.001	0.007	0.014	0.027	0.053	0.076	0.097
1.60	0.001	0.005	0.011	0.021	0.040	0.057	0.073
1.70	0.001	0.004	0.008	0.016	0.031	0.044	0.056
1.80	0.001	0.003	0.006	0.013	0.024	0.035	0.044
1.90	0.001	0.003	0.005	0.010	0.019	0.028	0.036
2.00	0.000	0.002	0.004	0.008	0.016	0.023	0.029
2.20	0.000	0.001	0.003	0.006	0.011	0.016	0.021
2.40	0.000	0.001	0.002	0.004	0.008	0.012	0.015
2.60	0.000	0.001	0.002	0.003	0.006	0.009	0.012
2.80	0.000	0.001	0.001	0.003	0.005	0.008	0.010
3.00	0.000	0.001	0.001	0.002	0.004	0.006	0.008
3.50	0.000	0.000	0.001	0.001	0.003	0.004	0.006
4.00	0.000	0.000	0.001	0.001	0.002	0.003	0.005

P_r							
1.000	1.200	1.500	2.000	3.000	5.000	7.000	10.000
16.598	16.559	16.501	16.405	16.214	15.838	15.469	14.927
15.274	15.248	15.208	15.142	15.012	14.757	14.511	14.154
13.895	13.876	13.848	13.803	13.713	13.540	13.376	13.144
12.494	12.481	12.462	12.429	12.367	12.251	12.144	11.998
11.151	11.142	11.128	11.105	11.063	10.986	10.920	10.836
9.914	9.907	9.897	9.881	9.853	9.806	9.770	9.732
8.800	8.795	8.788	8.777	8.759	8.735	8.723	8.720
7.810	7.807	7.801	7.794	7.784	7.778	7.785	7.811
6.933	6.930	6.926	6.922	6.919	6.928	6.952	7.002
6.155	6.152	6.149	6.146	6.149	6.174	6.213	6.285
5.458	5.455	5.452	5.452	5.461	5.501	5.555	5.648
4.826	4.822	4.820	4.822	4.839	4.898	4.969	5.083
4.238	4.232	4.230	4.236	4.267	4.351	4.442	4.578
3.893	3.885	3.883	3.896	3.941	4.046	4.151	4.300
3.658	3.647	3.648	3.669	3.728	3.851	3.966	4.125
3.406	3.391	3.401	3.437	3.517	3.661	3.788	3.957
3.264	3.247	3.268	3.318	3.412	3.569	3.701	3.875
3.093	3.082	3.126	3.195	3.306	3.477	3.616	3.795
2.313	2.868	2.967	3.067	3.200	3.387	3.532	3.717
1.306	2.513	2.784	2.933	3.094	3.297	3.450	3.640
1.113	1.655	2.557	2.790	2.986	3.209	3.369	3.565
0.820	0.831	1.443	2.283	2.655	2.949	3.134	3.348
0.577	0.640	0.618	1.241	2.067	2.534	2.767	3.013
0.437	0.489	0.502	0.654	1.471	2.138	2.428	2.708
0.343	0.385	0.412	0.447	0.991	1.767	2.115	2.430
0.226	0.254	0.282	0.300	0.481	1.147	1.569	1.944
0.158	0.178	0.200	0.220	0.290	0.730	1.138	1.544
0.115	0.130	0.147	0.166	0.206	0.479	0.823	1.222
0.086	0.098	0.112	0.129	0.159	0.334	0.604	0.969
0.067	0.076	0.087	0.102	0.127	0.248	0.456	0.775
0.053	0.060	0.070	0.083	0.105	0.195	0.355	0.628
0.043	0.049	0.057	0.069	0.089	0.160	0.286	0.518
0.035	0.040	0.048	0.058	0.077	0.136	0.238	0.434
0.025	0.029	0.035	0.043	0.060	0.105	0.178	0.322
0.019	0.022	0.027	0.034	0.048	0.086	0.142	0.254
0.015	0.018	0.021	0.028	0.041	0.074	0.120	0.210
0.012	0.014	0.018	0.023	0.035	0.065	0.104	0.180
0.010	0.012	0.015	0.020	0.031	0.058	0.093	0.158
0.007	0.009	0.011	0.015	0.024	0.046	0.073	0.122
0.006	0.007	0.009	0.012	0.020	0.038	0.060	0.100

Table D.4a Fugacity Coefficient

Values of $\left(\log \frac{f}{P}\right)^{[0]}$

T_r	P_r 0.010	0.050	0.100	0.200	0.400	0.600	0.800
0.30	-3.708	-4.402	-4.697	-4.985	-5.261	-5.412	-5.512
0.35	-2.472	-3.166	-3.461	-3.751	-4.029	-4.183	-4.285
0.40	-1.566	-2.261	-2.557	-2.847	-3.128	-3.283	-3.387
0.45	-0.879	-1.574	-1.871	-2.162	-2.444	-2.601	-2.707
0.50	-0.344	-1.040	-1.336	-1.628	-1.911	-2.070	-2.177
0.55	-0.008	-0.614	-0.911	-1.204	-1.488	-1.647	-1.755
0.60	-0.007	-0.269	-0.566	-0.859	-1.144	-1.304	-1.413
0.65	-0.005	-0.026	-0.283	-0.577	-0.862	-1.023	-1.132
0.70	-0.004	-0.021	-0.043	-0.341	-0.627	-0.789	-0.899
0.75	-0.003	-0.017	-0.035	-0.144	-0.430	-0.592	-0.703
0.80	-0.003	-0.014	-0.029	-0.059	-0.264	-0.426	-0.537
0.85	-0.002	-0.012	-0.024	-0.049	-0.123	-0.285	-0.396
0.90	-0.002	-0.010	-0.020	-0.041	-0.086	-0.166	-0.276
0.93	-0.002	-0.009	-0.018	-0.037	-0.077	-0.122	-0.214
0.95	-0.002	-0.008	-0.017	-0.035	-0.072	-0.113	-0.176
0.97	-0.002	-0.008	-0.016	-0.033	-0.067	-0.105	-0.148
0.98	-0.002	-0.008	-0.016	-0.032	-0.065	-0.101	-0.142
0.99	-0.001	-0.007	-0.015	-0.031	-0.063	-0.098	-0.137
1.00	-0.001	-0.007	-0.015	-0.030	-0.061	-0.095	-0.132
1.01	-0.001	-0.007	-0.014	-0.029	-0.059	-0.091	-0.127
1.02	-0.001	-0.007	-0.014	-0.028	-0.057	-0.088	-0.122
1.05	-0.001	-0.006	-0.013	-0.025	-0.052	-0.080	-0.110
1.10	-0.001	-0.005	-0.011	-0.022	-0.045	-0.069	-0.093
1.15	-0.001	-0.005	-0.009	-0.019	-0.039	-0.059	-0.080
1.20	-0.001	-0.004	-0.008	-0.017	-0.034	-0.051	-0.069
1.30	-0.001	-0.003	-0.006	-0.013	-0.026	-0.039	-0.052
1.40	-0.001	-0.003	-0.005	-0.010	-0.020	-0.030	-0.040
1.50	-0.000	-0.002	-0.004	-0.008	-0.016	-0.024	-0.032
1.60	-0.000	-0.002	-0.003	-0.006	-0.012	-0.019	-0.025
1.70	-0.000	-0.001	-0.002	-0.005	-0.010	-0.015	-0.020
1.80	-0.000	-0.001	-0.002	-0.004	-0.008	-0.012	-0.015
1.90	-0.000	-0.001	-0.002	-0.003	-0.006	-0.009	-0.012
2.00	-0.000	-0.001	-0.001	-0.002	-0.005	-0.007	-0.009
2.20	-0.000	-0.000	-0.001	-0.001	-0.003	-0.004	-0.005
2.40	-0.000	-0.000	-0.000	-0.001	-0.001	-0.002	-0.003
2.60	-0.000	-0.000	-0.000	-0.000	-0.000	-0.001	-0.001
2.80	0.000	0.000	0.000	0.000	0.000	0.000	0.001
3.00	0.000	0.000	0.000	0.000	0.001	0.001	0.002
3.50	0.000	0.000	0.000	0.001	0.001	0.002	0.003
4.00	0.000	0.000	0.000	0.001	0.002	0.003	0.004

P_r							
1.000	1.200	1.500	2.000	3.000	5.000	7.000	10.000
-5.584	-5.638	-5.697	-5.759	-5.810	-5.782	-5.679	-5.462
-4.359	-4.416	-4.479	-4.548	-4.611	-4.608	-4.531	-4.352
-3.464	-3.522	-3.588	-3.661	-3.735	-3.752	-3.694	-3.545
-2.784	-2.845	-2.913	-2.990	-3.071	-3.104	-3.062	-2.938
-2.256	-2.317	-2.387	-2.468	-2.555	-2.601	-2.572	-2.468
-1.835	-1.897	-1.969	-2.052	-2.145	-2.201	-2.183	-2.095
-1.494	-1.557	-1.630	-1.715	-1.812	-1.878	-1.869	-1.795
-1.214	-1.278	-1.352	-1.439	-1.539	-1.612	-1.611	-1.549
-0.981	-1.045	-1.120	-1.208	-1.312	-1.391	-1.396	-1.344
-0.785	-0.850	-0.925	-1.015	-1.121	-1.204	-1.215	-1.172
-0.619	-0.684	-0.760	-0.851	-0.958	-1.046	-1.062	-1.026
-0.479	-0.544	-0.620	-0.711	-0.820	-0.911	-0.930	-0.901
-0.359	-0.424	-0.500	-0.591	-0.700	-0.794	-0.817	-0.793
-0.296	-0.361	-0.437	-0.527	-0.637	-0.732	-0.756	-0.735
-0.258	-0.322	-0.398	-0.488	-0.598	-0.693	-0.719	-0.699
-0.223	-0.287	-0.362	-0.452	-0.561	-0.657	-0.683	-0.665
-0.206	-0.270	-0.344	-0.434	-0.543	-0.639	-0.666	-0.649
-0.191	-0.254	-0.328	-0.417	-0.526	-0.622	-0.649	-0.633
-0.176	-0.238	-0.312	-0.401	-0.509	-0.605	-0.633	-0.617
-0.168	-0.224	-0.297	-0.385	-0.493	-0.589	-0.617	-0.602
-0.161	-0.210	-0.282	-0.370	-0.477	-0.573	-0.601	-0.588
-0.143	-0.180	-0.242	-0.327	-0.433	-0.529	-0.557	-0.546
-0.120	-0.148	-0.193	-0.267	-0.368	-0.462	-0.491	-0.482
-0.102	-0.125	-0.160	-0.220	-0.312	-0.403	-0.433	-0.426
-0.088	-0.106	-0.135	-0.184	-0.266	-0.352	-0.382	-0.377
-0.066	-0.080	-0.100	-0.134	-0.195	-0.269	-0.296	-0.293
-0.051	-0.061	-0.076	-0.101	-0.146	-0.205	-0.229	-0.226
-0.039	-0.047	-0.059	-0.077	-0.111	-0.157	-0.176	-0.173
-0.031	-0.037	-0.046	-0.060	-0.085	-0.120	-0.135	-0.129
-0.024	-0.029	-0.036	-0.046	-0.065	-0.092	-0.102	-0.094
-0.019	-0.023	-0.028	-0.036	-0.050	-0.069	-0.075	-0.066
-0.015	-0.018	-0.022	-0.028	-0.038	-0.052	-0.054	-0.043
-0.012	-0.014	-0.017	-0.021	-0.029	-0.037	-0.037	-0.024
-0.007	-0.008	-0.009	-0.012	-0.015	-0.017	-0.012	0.004
-0.003	-0.004	-0.004	-0.005	-0.006	-0.003	0.005	0.024
-0.001	-0.001	-0.001	-0.001	0.001	0.007	0.017	0.037
0.001	0.001	0.002	0.003	0.005	0.014	0.025	0.046
0.002	0.003	0.003	0.005	0.009	0.018	0.031	0.053
0.004	0.005	0.006	0.008	0.013	0.025	0.038	0.061
0.005	0.006	0.007	0.010	0.016	0.028	0.041	0.064

Table D.4b Fugacity Coefficient

Values of $\left(\log \frac{f}{P}\right)^{[1]}$

T_r	P_r 0.010	0.050	0.100	0.200	0.400	0.600	0.800
0.30	-8.779	-8.780	-8.782	-8.785	-8.792	-8.799	-8.806
0.35	-6.526	-6.528	-6.530	-6.534	-6.542	-6.550	-6.558
0.40	-4.912	-4.914	-4.916	-4.920	-4.928	-4.936	-4.945
0.45	-3.726	-3.727	-3.729	-3.734	-3.742	-3.750	-3.758
0.50	-2.838	-2.839	-2.841	-2.845	-2.853	-2.861	-2.868
0.55	-0.013	-2.164	-2.166	-2.170	-2.177	-2.184	-2.192
0.60	-0.009	-1.644	-1.646	-1.650	-1.657	-1.664	-1.671
0.65	-0.006	-0.031	-1.241	-1.245	-1.252	-1.258	-1.265
0.70	-0.004	-0.021	-0.044	-0.927	-0.933	-0.940	-0.946
0.75	-0.003	-0.014	-0.030	-0.675	-0.682	-0.688	-0.694
0.80	-0.002	-0.010	-0.020	-0.043	-0.481	-0.487	-0.493
0.85	-0.001	-0.006	-0.013	-0.028	-0.321	-0.327	-0.332
0.90	-0.001	-0.004	-0.009	-0.018	-0.039	-0.199	-0.204
0.93	-0.001	-0.003	-0.007	-0.013	-0.029	-0.048	-0.141
0.95	-0.001	-0.003	-0.005	-0.011	-0.023	-0.037	-0.103
0.97	-0.000	-0.002	-0.004	-0.009	-0.018	-0.029	-0.042
0.98	-0.000	-0.002	-0.004	-0.008	-0.016	-0.025	-0.035
0.99	-0.000	-0.002	-0.003	-0.007	-0.014	-0.021	-0.030
1.00	-0.000	-0.001	-0.003	-0.006	-0.012	-0.018	-0.025
1.01	-0.000	-0.001	-0.003	-0.005	-0.010	-0.016	-0.021
1.02	-0.000	-0.001	-0.002	-0.004	-0.009	-0.013	-0.017
1.05	-0.000	-0.001	-0.001	-0.002	-0.005	-0.006	-0.007
1.10	-0.000	-0.000	0.000	0.000	0.001	0.002	0.004
1.15	0.000	0.000	0.001	0.002	0.005	0.008	0.011
1.20	0.000	0.001	0.002	0.003	0.007	0.012	0.017
1.30	0.000	0.001	0.003	0.005	0.011	0.017	0.023
1.40	0.000	0.002	0.003	0.006	0.013	0.020	0.027
1.50	0.000	0.002	0.003	0.007	0.014	0.021	0.028
1.60	0.000	0.002	0.003	0.007	0.014	0.021	0.029
1.70	0.000	0.002	0.004	0.007	0.014	0.021	0.029
1.80	0.000	0.002	0.003	0.007	0.014	0.021	0.028
1.90	0.000	0.002	0.003	0.007	0.014	0.021	0.028
2.00	0.000	0.002	0.003	0.007	0.013	0.020	0.027
2.20	0.000	0.002	0.003	0.006	0.013	0.019	0.025
2.40	0.000	0.002	0.003	0.006	0.012	0.018	0.024
2.60	0.000	0.001	0.003	0.006	0.011	0.017	0.023
2.80	0.000	0.001	0.003	0.005	0.011	0.016	0.021
3.00	0.000	0.001	0.003	0.005	0.010	0.015	0.020
3.50	0.000	0.001	0.002	0.004	0.009	0.013	0.018
4.00	0.000	0.001	0.002	0.004	0.008	0.012	0.016

P_r							
1.000	1.200	1.500	2.000	3.000	5.000	7.000	10.000
-8.813	-8.820	-8.831	-8.848	-8.883	-8.953	-9.022	-9.126
-6.566	-6.574	-6.586	-6.606	-6.645	-6.724	-6.802	-6.919
-4.953	-4.961	-4.973	-4.994	-5.034	-5.115	-5.194	-5.312
-3.766	-3.774	-3.786	-3.806	-3.846	-3.924	-4.001	-4.115
-2.876	-2.884	-2.896	-2.915	-2.953	-3.027	-3.101	-3.208
-2.199	-2.207	-2.218	-2.236	-2.272	-2.342	-2.411	-2.510
-1.678	-1.685	-1.695	-1.712	-1.746	-1.812	-1.875	-1.967
-1.272	-1.278	-1.288	-1.304	-1.336	-1.397	-1.456	-1.540
-0.952	-0.959	-0.968	-0.983	-1.013	-1.069	-1.123	-1.201
-0.700	-0.705	-0.714	-0.728	-0.756	-0.809	-0.858	-0.929
-0.498	-0.504	-0.512	-0.526	-0.551	-0.600	-0.645	-0.709
-0.338	-0.343	-0.351	-0.364	-0.388	-0.432	-0.473	-0.530
-0.210	-0.215	-0.222	-0.234	-0.256	-0.296	-0.333	-0.384
-0.146	-0.151	-0.158	-0.170	-0.191	-0.228	-0.262	-0.310
-0.108	-0.114	-0.121	-0.132	-0.151	-0.187	-0.219	-0.265
-0.075	-0.080	-0.087	-0.097	-0.116	-0.150	-0.180	-0.223
-0.059	-0.064	-0.071	-0.081	-0.099	-0.132	-0.162	-0.203
-0.044	-0.050	-0.056	-0.066	-0.084	-0.115	-0.144	-0.184
-0.031	-0.036	-0.042	-0.052	-0.069	-0.099	-0.127	-0.166
-0.024	-0.024	-0.030	-0.038	-0.054	-0.084	-0.111	-0.149
-0.019	-0.015	-0.018	-0.026	-0.041	-0.069	-0.095	-0.132
-0.007	-0.002	0.008	0.007	-0.005	-0.029	-0.052	-0.085
0.007	0.012	0.025	0.041	0.042	0.026	0.008	-0.019
0.016	0.022	0.034	0.056	0.074	0.069	0.057	0.036
0.023	0.029	0.041	0.064	0.093	0.102	0.096	0.081
0.030	0.038	0.049	0.071	0.109	0.142	0.150	0.148
0.034	0.041	0.053	0.074	0.112	0.161	0.181	0.191
0.036	0.043	0.055	0.074	0.112	0.167	0.197	0.218
0.036	0.043	0.055	0.074	0.110	0.167	0.204	0.234
0.036	0.043	0.054	0.072	0.107	0.165	0.205	0.242
0.035	0.042	0.053	0.070	0.104	0.161	0.203	0.246
0.034	0.041	0.052	0.068	0.101	0.157	0.200	0.246
0.034	0.040	0.050	0.066	0.097	0.152	0.196	0.244
0.032	0.038	0.047	0.062	0.091	0.143	0.186	0.236
0.030	0.036	0.044	0.058	0.086	0.134	0.176	0.227
0.028	0.034	0.042	0.055	0.080	0.127	0.167	0.217
0.027	0.032	0.039	0.052	0.076	0.120	0.158	0.208
0.025	0.030	0.037	0.049	0.072	0.114	0.151	0.199
0.022	0.026	0.033	0.043	0.063	0.101	0.134	0.179
0.020	0.023	0.029	0.038	0.057	0.090	0.121	0.163

APPENDIX

E

Computer Programs

The following computer programs are included in the supplied diskette:

PVT	Calculates the molar volume of pure compounds and mixtures using equations of state.
EOS	Calculates the vapor pressure and saturated volumes of pure compounds using cubic equations of state.
VLEEOS	Performs bubble point pressure calculations for binary mixtures using cubic equations of state.
EOSKIJ	Determines the k_{ij} value in cubic equations of state from binary vapor-liquid equilibrium data.
GAMMA	Evaluates activity coefficient values in binary systems from P-T-x-y data.
GAMPAR	Evaluates parameter values in the Wilson and UNIQUAC models from binary activity coefficient data.
VLEGPHI	Predicts ternary P and y values from T and x data using the Wilson, UNIQUAC and UNIFAC models.
KVALUE	Estimates the K-values of light hydrocarbons in multicomponent VLE mixtures using a 'Corresponding States Approach'.

Note: To use the programs see the information contained in the relevant to each program 'help' file, such as, for example, PVT.HLP.

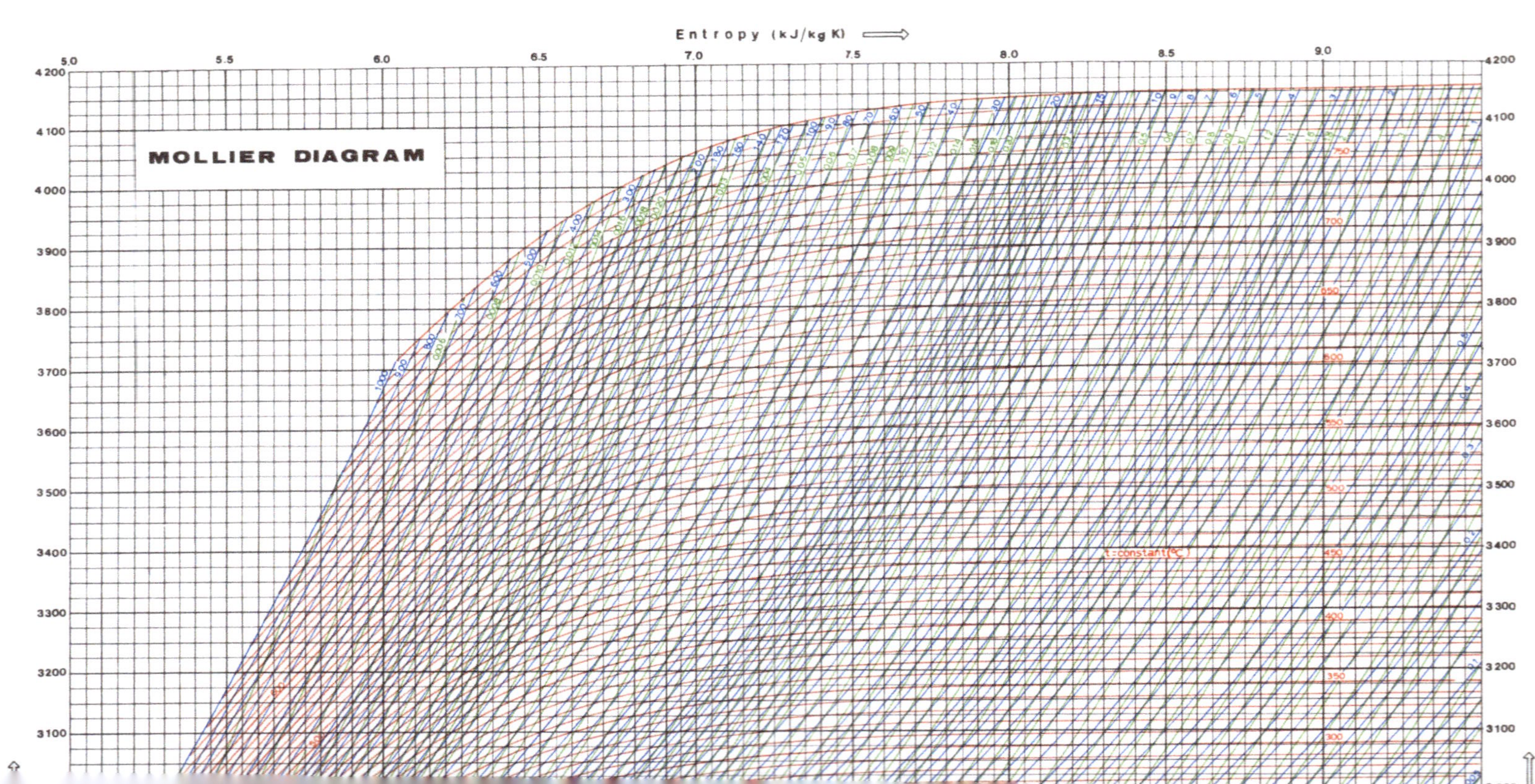
MOLLIER DIAGRAM
Entropy (kJ/kg K)
5.0
5.5
6.0
6.5
7.0
7.5
8.0
8.5
9.0
4200
4100
4000
3900
3800
3700
3600
3500
3400
3300
3200
3100
t=constant(°C)
750
700
650
600
550
500
450
400
350
300

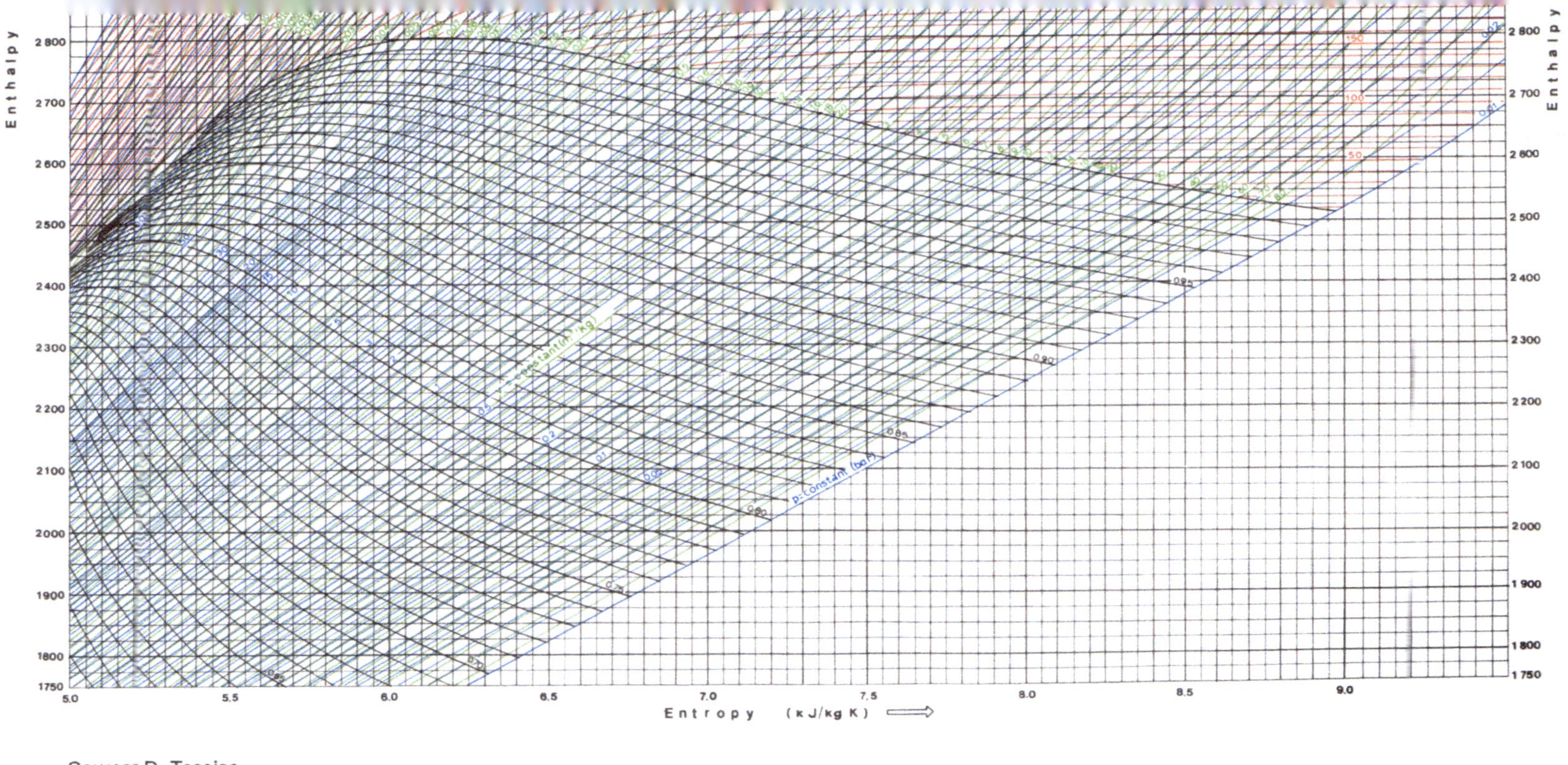

Source: D. Tassios

Index

Springer-Verlag and the Environment

We at Springer-Verlag firmly believe that an international science publisher has a special obligation to the environment, and our corporate policies consistently reflect this conviction.

We also expect our business partners – paper mills, printers, packaging manufacturers, etc. – to commit themselves to using environmentally friendly materials and production processes.

The paper in this book is made from low- or no-chlorine pulp and is acid free, in conformance with international standards for paper permanency.

MIX
Papier aus verantwortungsvollen Quellen
Paper from responsible sources
FSC® C105338

If you have any concerns about our products, you can contact us on
ProductSafety@springernature.com

In case Publisher is established outside the EU, the EU authorized representative is:
Springer Nature Customer Service Center GmbH
Europaplatz 3, 69115 Heidelberg, Germany

Printed by Libri Plureos GmbH
in Hamburg, Germany